The Physiology of
the Garden Pea

EXPERIMENTAL BOTANY

An International Series of Monographs

CONSULTING EDITOR

J. F. Sutcliffe

School of Biological Sciences, University of Sussex, England

FRONTISPIECE.   The garden pea—*Pisum sativum* L.

The Physiology of the Garden Pea

Edited by

J. F. SUTCLIFFE

School of Biological Sciences,
University of Sussex

and

J. S. PATE

Department of Botany,
University of Western Australia

1977

ACADEMIC PRESS
London   New York   San Francisco

A Subsidiary of Harcourt Brace Jovanovich, Publishers

ACADEMIC PRESS INC. (LONDON) LTD.
24/28 Oval Road,
London NW1

United States Edition published by
ACADEMIC PRESS INC.
111 Fifth Avenue
New York, New York 10003

Library of Congress Catalog Card Number: 76 24430
ISBN: 0 12 677550 8

PRINTED IN GREAT BRITAIN BY
COX & WYMAN LTD.
LONDON, FAKENHAM AND READING

Contributors

J. A. BRYANT, *Department of Botany, University College, Cardiff, P.O. Box 78, Cardiff CF1 1XL, Wales.*

A. M. FLINN, *School of Biological and Environmental Studies, The New University of Ulster, Coleraine BT52 1SA, N. Ireland.*

A. W. GALSTON, *School of Biological and Environmental Studies, Yale University, New Haven, Connecticut 06520, U.S.A.*

D. M. HARVEY, *John Innes Institute, Colney Lane, Norwich NR4 7UH, England.*

J. K. HEYES, *Department of Botany, Victoria University of Wellington, Wellington, New Zealand.*

P. H. LOVELL, *Department of Botany, The University of Auckland, Auckland, New Zealand.*

R. F. LYNDON, *Department of Botany, University of Edinburgh, The King's Buildings, Mayfield Road, Edinburgh EH9 3JH, Scotland.*

G. A. MARX, *New York State Agricultural Experiment Station, Department of Seed and Vegetable Sciences, Hedrick Hall, Geneva, New York 14456, U.S.A.*

S. MATTHEWS, *Department of Biology, University of Stirling, Stirling, Scotland.*

A. J. McCOMB, *Department of Botany, University of Western Australia, Nedlands, W.A. 6009, Australia.*

I. C. MURFET, *Botany Department, University of Tasmania, Hobart, Tasmania 7001.*

J. S. PATE, *Department of Botany, University of Western Australia, Nedlands, W.A. 6009, Australia.*

T. SACHS, *Botany Department, The Hebrew University, Jerusalem, Israel.*

J. F. SUTCLIFFE, *School of Biological Sciences, University of Sussex, Falmer, Brighton BN1 9QC, England.*

J. G. TORREY, *Cabot Foundation, Harvard University, Petersham, Massachusetts 01366, U.S.A.*

R. ZOBEL, *Department of Agronomy, Cornell University, Ithaca, New York, U.S.A.*

Preface

It is traditional in textbooks of plant physiology to discuss individual processes and the interactions between them against a background of information derived from a variety of plants. Although this approach has its merits, it has the disadvantage that important differences between plant species tend to become blurred or overlooked.

It occurred to us several years ago that it would be useful to have a comprehensive survey of the physiology of a particular plant, and we felt that the garden pea would be an excellent choice for the purpose. It is one of the most extensively and intensively studied of all flowering plants and research on peas has contributed greatly to knowledge in most fields of plant physiology. We were much encouraged by the enthusiastic response we received from others who have worked on peas, and everyone we asked readily agreed to contribute an account of research in his own field of interest. We asked contributors to confine themselves as far as possible to work that has been done on *Pisum sativum* L., referring to studies on other legumes or other plant species only when necessary for comparative purposes. In view of the uncertain taxonomic position of the field pea, *P. arvense* L., as indicated in Chapter 2, we have included some work on this plant, especially in Chapters 13 and 15, as if it were a cultivar of *P. sativum* L.

Thanks are due to our contributors for producing their manuscripts promptly and to the staff of Academic Press for their help in processing the material for the printer. Our secretaries, Miss N. Browning, Mrs B. Kilner and Mrs K. Holland have shouldered much of the burden of correspondence with contributors and the mechanics of putting the work together in its final form. We hope that our efforts will be useful not only to those who are researching on the garden pea, but to all students of plant physiology who seek an overall impression of the functioning of a particular plant.

January, 1977

J. F. SUTCLIFFE
J. S. PATE

Contents

Contents

1. History of the Use of the Pea in Plant Physiological Research

J. F. SUTCLIFFE

School of Biological Sciences, University of Sussex, England

I. IN PRAISE OF PEAS

There are certain organisms that are particularly suitable for biological research; *Escherichia coli*, *Drosophila melanogaster*, *Neurospora crassa* and *Amoeba proteus* spring readily to mind. Among flowering plants, *Pisum sativum* L. and *Zea mays* L. have often been chosen for study by botanists as representatives of dicotyledons and monocotyledons respectively.

The reasons for the popularity of the garden pea as research material are made very clear in later chapters of this monograph. Briefly, pea seeds of high viability are readily available throughout the year in most parts of the world. They are quiescent rather than dormant and germinate rapidly when placed under favourable conditions. There is no hard seed coat to be removed and no special conditions for germination, such as exposure to light or a fluctuating temperature, are required. The axis soon attains a sufficient size to provide material for biochemical study and yields preparations of high metabolic activity free of toxic substances (Chapter 3). Excised segments of root (Chapter 6) and shoot (Chapter 11) have considerable growth potential, and root tips can be cultured readily in defined media (Chapter 5). Because of the large food reserves stored in the cotyledons, pea seedlings can be grown for several weeks at low light intensity. They show marked etiolation effects and the dark-grown shoots are very responsive to light (Chapter 11).

The species contains a large amount of genetic variability (Chapter 2) which has been exploited in the study of growth (Chapters 5 and 9) and flowering (Chapter 14). Peas grow to a convenient, but not excessive, size and their life history is reasonably short. They do not undergo much lignification and are therefore amenable to experimental manipulation, including surgical treatments (Chapter 8). In common with most other legumes they form a symbiotic relationship with nitrogen-fixing bacteria and the pea root nodule system is a very convenient one for the study of nitrogen fixation (Chapter 13). Finally, as a bonus, the pea is an important crop plant and thus research on it may be of direct relevance to agriculture (Chapters 4 and 16).

II. SOME EARLY EXPERIMENTS ON PEAS

The pea seems to have commended itself to investigators from very early times. Stephen Hales reported in *Vegetable Staticks* (1727) that Mr (later Sir Robert) Boyle had discovered that pea seeds produced large quantities of gas ("air") when they were placed in closed containers. Hales himself carried out further experiments to confirm this and he studied the properties of the evolved gas, sometimes evacuating the container and at other times allowing an animal to respire, or a candle to burn, in it (Fig. 1.1: Hales' Fig. 35). Of another experiment (Fig. 1.1: Hales' Fig. 36) he said:

> I filled the strong *Hungary-water Bottle* (*b c*, Fig. 36) near half full of Pease and then full of water, pouring in first half an inch depth of Mercury; then I screwed at *b* into the bottle the long slender tube *a z* which reached down to the bottom of the bottle; the water was in two or three days all imbibed by the Pease and they thereby much dilated; the Mercury was also forced up the slender glass tube near 80 inches high; in which state the new generated air in the bottle was compressed with a force equal to more than two Atmospheres and an half; . . .

Hales concluded from such experiments that when a seed is sown in the ground the cotyledons imbibe moisture with great force and this squeezes water and nutrients into the radicle, causing it to grow. He distinguished between seeds, such as those of peas, in which the cotyledons remain below ground, and those in which they become foliage leaves.

Experiments on the production of carbon dioxide by germinating pea seeds, not very different from some of those carried out by Hales, were described by Sachs in his *Lectures on the Physiology of Plants* (1887) and by Pfeffer in *The Physiology of Plants* (1900). These experiments, together with the demonstration that germinating pea seeds swell, as Hales had shown (Fig. 1.1: Hales'

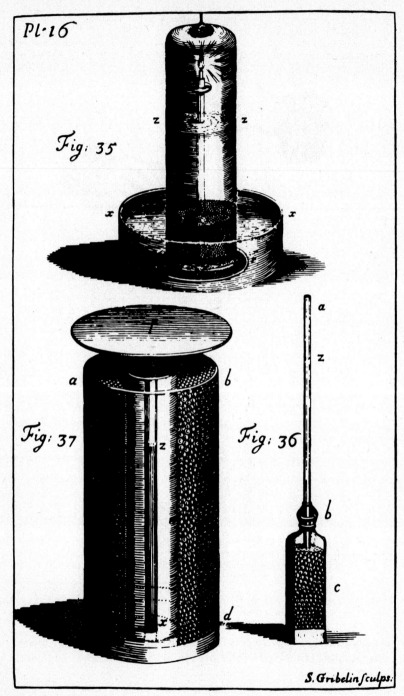

FIG. 1.1. Experiments on the production of gas by germinating pea seeds. (Reproduced from Hales, 1727)

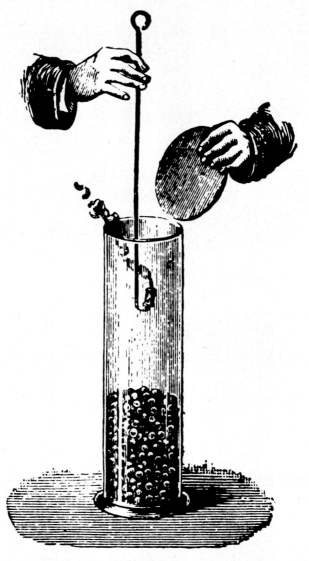

Fig. 1.2A. An experiment demonstrating that the gas produced by germinating peas will extinguish a glowing splint.

Fig. 37), became class experiments for students of plant physiology in schools and colleges in the latter part of the last century (Fig. 1.2) and even persist to the present day.

Sachs also described how he measured the production of heat by germinating pea seeds by enclosing them with a thermometer in a bell jar, and he noted

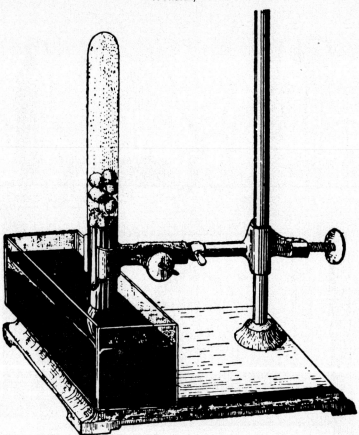

FIG. 1.2B. An experiment showing the production of carbon dioxide and ethanol by germinating pea seeds. (Both experiments are after Sachs; reproduced from Macdougal, 1902)

that 100–200 germinating seeds caused a temperature rise of $1 \cdot 5°C$. The method still commonly used for laboratory demonstrations of heat production by seeds involving the use of vacuum flasks was described by Ganong (1908), and by Peirce (1908). The performance of this experiment at school was my first introduction to practical plant physiology and to the use of peas in research.

III. USE OF PEAS IN STUDIES OF RESPIRATION

Sachs devised a method of measuring quantitatively the production of carbon dioxide in respiration by passing CO_2-free air over the respiring material and absorbing the CO_2 produced in lime water. Germinating peas were among the plant materials he investigated (Fig. 1.3).

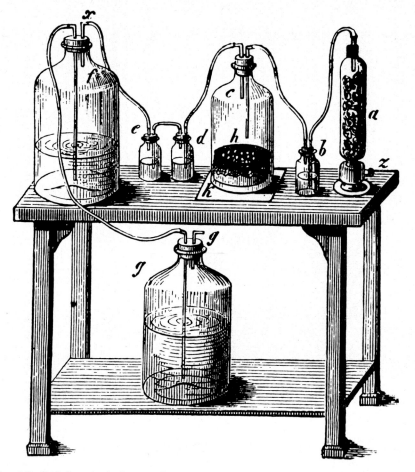

FIG. 1.3. Sachs' method of measuring the production of carbon dioxide by germinating
pea seeds. (Reproduced from Sachs, 1887)

Measurements of oxygen absorption and calculations of respiratory quotient (R.Q.) were made in the early years of this century on a large number of
different types of germinating seeds, including those of peas. The R.Q. of pea
seeds with the testa on was found to reach values as high as 4 in the early
stages of germination, and even with the testa off values in excess of unity
were recorded. It was concluded that the testa presents a barrier to diffusion
of oxygen and that peas may respire anaerobically in early germination with
production of ethanol (see Chapter 3)—a fact verified by several generations
of plant physiology students since that time.

Kuiper (1910) investigated the relationship between temperature and respiration in pea seedlings. He germinated pea seeds at 20°C and then exposed them to various temperatures between 0 and 50°C. Up to 30°C there was an increase in the rate of CO_2 evolution with increasing temperature, and at each temperature the rate remained fairly constant over a 6 h period (Fig. 1.4); the temperature coefficient (Q_{10}) was about 2. Above 30°C the rate of respiration was higher initially than at lower temperatures but it fell rapidly with time, especially above 40°C. A similar experiment was carried out later by Fernandes (1923) with the same result.

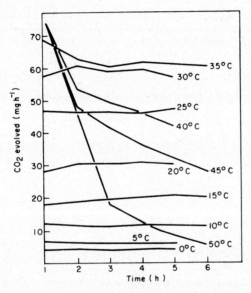

Fig. 1.4. Effect of temperature on the rate of respiration of pea seedlings. (After Kuiper, 1910)

Since those early days pea seedling tissues have been widely used in the study of the biochemistry of respiration (Beevers, 1961) and many of the enzymes involved in glycolysis and the Krebs cycle were obtained in a purified state first from peas (Stumpf, 1952). Studies on preparations from pea tissues played a prominent part in establishing the role of mitochondria and the Krebs cycle in intermediary metabolism (Price and Thimann, 1954; Davies, 1953). Brown and Broadbent (1950) and Sutcliffe and Sexton (1974) have made detailed studies of the changes in respiration rates of cells along the axis of pea roots (Fig. 1.5). Fowler and Ap Rees (1970) have examined the alternative pathways of carbohydrate oxidation during cell differentiation (see Chapter 6).

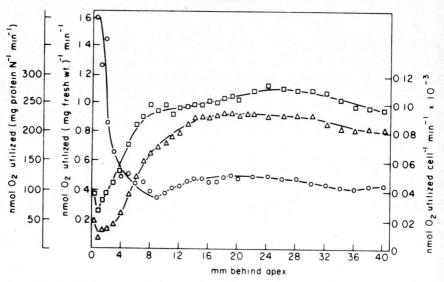

FIG. 1.5. Rates of oxygen uptake at various distances along the axis of young pea roots grown for about 5 days at 27°C. The results are expressed: per mg fresh wt. (○–○); per mg of protein N (□—□); and per cell (△—△). (Sutcliffe and Sexton, 1974)

IV. FLAVONOIDS IN PEAS

The role of flavonoids in plant physiology is uncertain. Some evidence that these substances play a part in disease resistance as phytoalexins comes from work on both peas and beans (Cruickshank and Perrin, 1964). It has been shown that pisatin (Fig. 1.6) is produced in *Pisum sativum* leaves in response to fungal infection and is not present in healthy tissues. Non-pathogenic fungi induce formation of toxic amounts of pisatin within 36 h after inoculation and are thus prevented from invading the plant. Pathogenic fungi are successful because they induce pisatin formation in amounts which are not toxic to them.

Another flavonoid, kaempferol-3-*p*-coumaroyl triglucoside (see Chapter 11, Fig. 11.8), which has been found in buds of etiolated pea seedlings, is a natural inhibitor of IAA oxidase (Mumford *et al.*, 1961). In contrast, quercetin glycosides (Fig. 11.8) which occur in pea leaves are cofactors to the same enzyme (Furuya *et al.*, 1962).

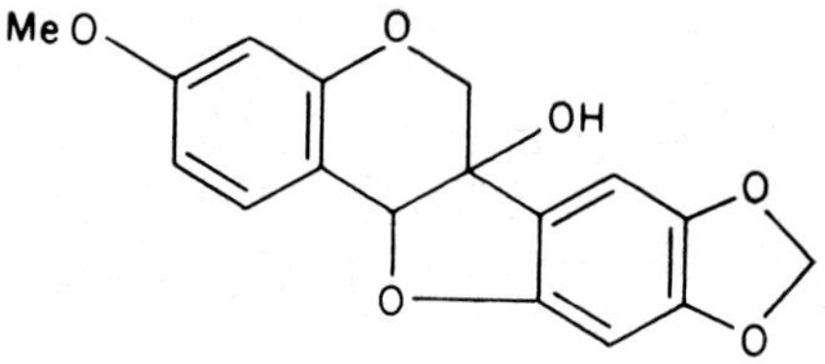

FIG. 1.6. Pisatin.

V. INVESTIGATIONS OF NITROGEN FIXATION AND NITROGEN METABOLISM

Towards the end of the nineteenth century the source of nitrogen for leguminous plants was still in doubt. The work of Boussingault early in the century and the field experiments started at Rothamsted by Lawes and Gilbert in 1843 had established that non-leguminous plants, such as cereals, required combined nitrogen and that nitrates and ammonium salts were both acceptable (Lawes *et al.*, 1861). However, leguminous plants were found to behave abnormally; it was shown that they can be grown without nitrogenous manure and yet accumulate large amounts of nitrogen and may actually enrich the soil in this element. Lachmann (1858) examined the structure of root nodules, which he guessed were involved in legume nutrition, and showed that they contained "*vibrionenartige*" organisms. In 1888, Hellriegel and Wilfarth published the results of their classical work on the nitrogen nutrition of plants, from which they concluded that peas and other legumes can fix atmospheric nitrogen only if they enter into a symbiotic relationship with nitrogen-fixing micro-organisms. Beyerinck (1888) was able to obtain pure cultures of a bacterium from root nodules and gave it the name *Bacillus radicicola*; Burk (1927) demonstrated that pea plants grown under sterile conditions do not fix atmospheric nitrogen. An account of more recent research of nitrogen fixation in peas is given in Chapter 13.

Virtanen *et al.* (1933) found that the roots of inoculated pea plants excreted various nitrogen compounds into the surrounding medium and showed that some 80% of the nitrogen released was in the amino and amide form. Since then pea tissues have been extensively used in the study of transamination and amino acid metabolism (Kritzman, 1939; Virtanen and Laine, 1941; Rautanen, 1948). They have also been employed in research on protein synthesis. Net synthesis of protein by ribosome preparations of plant tissues supplied with a range of amino acids was reported first for peas (Webster, 1955; Raacke, 1959), and pea extracts have been used to establish the existence of specific amino acid-activating enzymes (amino-acyl transfer RNA synthetases) in plants (Davis and Novelli, 1958; Webster, 1959). Some other aspects of research on protein synthesis in peas are discussed in Chapters 3, 13 and 15.

VI. GROWTH STUDIES

Young pea seedlings were used by Sachs and by Pfeffer to determine the

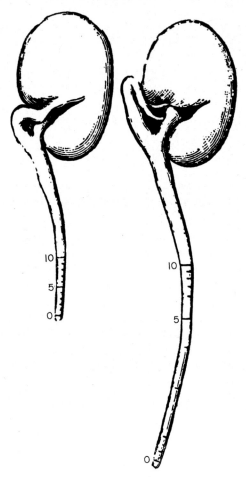

FIG. 1.7A. Determination of the region of maximum growth. Measurements are in mm.
From Sachs (1887)

region of maximum growth and of geotropic curvature in a root (Fig. 1.7).
The responses of shoot and main root to geotropic stimulation were demonstrated by Macdougal (1902); it is interesting to note that the lateral roots are
not apparently sensitive to gravity (Fig. 1.8).

Charles Darwin used pea seedlings in his studies of tropic responses in
plants. In *The Power of Movement in Plants* (1881) he described the effects of
attaching a small piece of card to the tips of young pea roots (Fig. 1.9) and
attributed the resulting curvatures to differential stimulation of growth by
contact with the solid object.

The circumnutation of shoot tips, internodes, leaf petioles and tendrils of

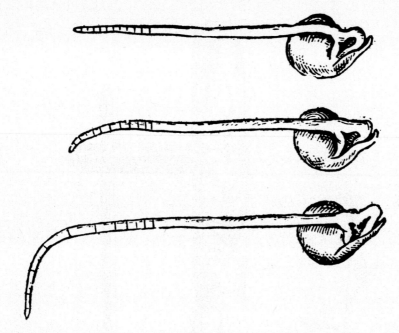

FIG. 1.7B. Determination of the region of geotropic curvature in the root. Measurements are in mm. After Pfeffer (1900)

pea was reported by Dutrochet (1843). Darwin (1888) studied this phenomenon in more detail and the results of one of his experiments are illustrated in Fig. 1.10. He showed that movement of young internodes and tendril tips in a series of ellipses enables tendrils to explore a large space in search of support. He denied that there was any circumnutation of the petioles such as Dutrochet has described. More recent research on tendril movement is discussed by Galston in Chapter 11.

Pea roots have played an important part in the development of the technique of root culture. Bonner and Addicott (1937) were able to grow 3–10 mm root tips in a defined medium for long periods of time, subculturing when necessary (see also Chapter 5). The processes of cell division and enlargement have been examined in detail in cultured root tips, in segments from the extension zone, and in intact roots of peas, by Brown and his collaborators (Brown and Rickless, 1949; Brown and Sutcliffe, 1950; Brown, 1951; Brown and Wightman, 1952; Wightman and Brown, 1953, see also Chapter 6).

Pieces of etiolated stem tissue are very responsive to auxins and Went (1934) made use of this in his split pea stem curvature assay. For this test,

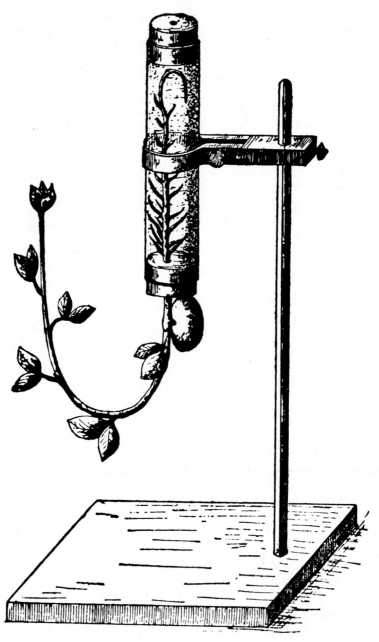

FIG. 1.8. Demonstration of the geotropic responses of pea roots and shoot. (Redrawn from Macdougal, 1902)

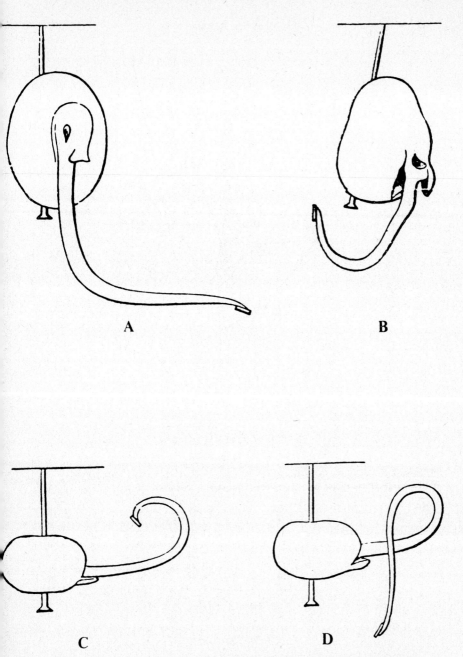

Fig. 1.9. Demonstration of pea-root curvatures induced by contact with a small piece of card fixed to one side of the apex. A and B deflections produced within 24 h in the growth of vertical radicles; C and D curvature of horizontal radicles caused by contact on the lower side of the tip; C after 21 h; D after 45 h. (From Darwin, 1881)

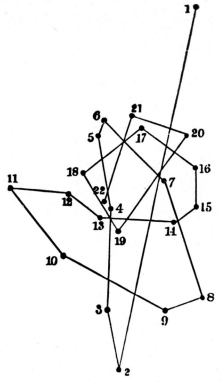

Side of room with window.

Fig. 6.

Diagram showing the movement of the upper internode of the common Pea, traced on a hemispherical glass, and transferred to paper; reduced one-half in size. (Aug. 1st.)

No.	H. M.		No.	H. M.		No.	H. M.	
1	8 46	A.M.	9	1 55	P.M.	16	5 25	P.M.
2	10 0	„	10	2 25	„	17	5 50	„
3	11 0	„	11	3 0	„	18	6 25	„
4	11 37	„	12	3 30	„	19	7 0	„
5	12 7	P.M.	13	3 48	„	20	7 45	„
6	12 30	„	14	4 40	„	21	8 30	„
7	1 0	„	15	5 5	„	22	9 15	„
8	1 30	„						

FIG. 1.10. Tracing of the circumnutation of an upper internode of a pea seedling. (From Darwin, 1888)

segments about 3 cm long are cut from the second or third internode of etiolated pea seedlings grown under controlled conditions. The segments are then slit longitudinally nearly to the base and when placed in water they curve outwards because of rapid uptake of water by the inner cells. If auxin is present in the solution growth of the epidermal cells is stimulated and as a result the outward curvature is reversed. By measuring the angle of curvature after a given time, it is possible to estimate the concentration of auxin in the solution after calibrating the assay using standard solutions. This method of assaying auxins has been extensively used by growth physiologists as an alternative to the *Avena* coleoptile section test (Audus, 1972).

The problem of apical dominance was investigated by the Snows in Oxford during the 1920s and 1930s using pea as their main experimental material. They showed that buds in the axils of lower leaves were inhibited by the presence of mature leaves higher up the stem and concluded (Snow, 1937) after a long series of ingenious experiments that although auxin is probably involved, it does not act directly on the inhibited bud as proposed by Thimann and Skoog (1934); rather it has an indirect effect by stimulating production in the stem of a secondary inhibiting influence which travels in a non-polar manner to the axillary bud and inhibits its growth. More recent research has lent further support to this view, as is discussed by Lovell in Chapter 10.

In the 1950s it was discovered that the gibberellins, which had been isolated earlier from a fungus, *Gibberella fujikuroi* (*Fusarium heterosporum*), had a dramatic effect on the growth of certain plants, notably dwarf peas. It soon became evident that gibberellins are natural constituents of flowering plants and important growth regulators, lack of which is responsible for the dwarf habit of some varieties. The pea cultivar 'Meteor' has been widely used as an assay system for gibberellins because stem growth can be correlated quantitatively with the amount of gibberellin applied.

There is increasing evidence for the importance of ethylene in the control of early growth of pea seedlings. Kang *et al.* (1967) found that the formation of the epicotyl hook in etiolated seedlings is a response to ethylene produced by the shoot apex. Red light, which is known to cause straightening of the epicotyl, also results in a sharp depression in the amount of ethylene produced. The effects of red light are reversed by far-red light which suggests that control is mediated by phytochrome (Goeschl *et al.*, 1967). Further discussion of the role of ethylene in the integration of growth in peas is to be found in Chapter 10, and its possible involvement in enzyme syntheses in cotyledons is mentioned in Chapter 3.

VII. FUTURE PROSPECTS

Despite the intensity with which the garden pea has been studied, large gaps in our knowledge of its physiology still remain. The water relations of *P. sativum* have received relatively little attention, although as early as the beginning of the nineteenth century de Saussure observed that pea plants became wilted if carbon dioxide was passed through the medium in which they were growing. Boczuk (1970) measured the transpiration rate of pea plants grown under controlled conditions in solution culture. She showed that transpiration on a per plant basis was reduced when sodium chloride was added to the medium and that this effect could not be accounted for entirely by the reduction in leaf surface area that occurred (Fig. 1.11). It was demonstrated that opening of the stomata was partially inhibited under saline conditions. There was no significant difference between the rates of transpiration on a unit surface area basis of plants grown in Hoagland solution

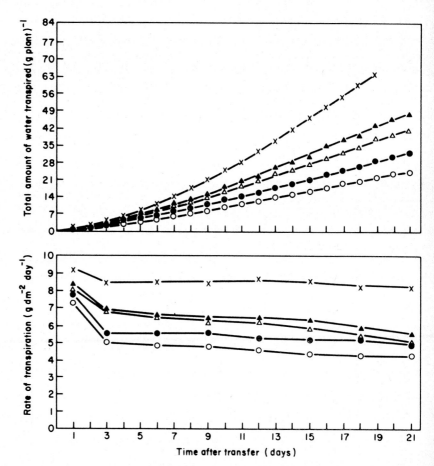

FIG. 1.11. Transpiration rates of pea plants during 3 weeks' growth in daily periods of light (20 klx; 19°C; 75% r.h.) and 8h darkness (18°C; 80% r.h.). Above, results expressed as g plant^{-1} day^{-1}; Below, results expressed as g (dm leaf area)$^{-2}$ day^{-1}. Key: x–x Hoagland solution; ▲–▲ 5× concentrated Hoagland solution; △–△ Hoagland solution + 63mM NaCl; ●–● 10× concentrated Hoagland solution; ○–○ Hoagland solution + 125mM NaCl.

+ 63 mM NaCl and Hoagland solution at 5× its normal concentration and, as these two solutions are approximately iso-osmotic, it was concluded that the reduction in transpiration in comparison with that of controls grown in normal Hoagland solution was caused by an effect of water stress. However, at a higher concentration of NaCl (125 mM), the effect of NaCl was greater than that of the corresponding Hoagland solution and it was concluded that there was a specific inhibitory effect of NaCl at the higher concentration on the opening of the stomata.

Nothing of particular interest has been discovered yet from the study of mineral nutrition in peas. Hylmö (1953) reported an exact quantitative relationship between transpiration and the uptake of ions by pea seedlings. Others have been unable to confirm this relationship, although an increase in uptake and in transport of ions to the shoot at high transpiration rates has sometimes been observed. The distribution of phosphate in intact pea plants in relation to growth of the various parts was examined by Linck (1955; see also Linck and Swanson, 1960). Some recent research on the distribution of inorganic elements in young pea seedlings and in developing fruits and seeds is described in Chapters 3 and 15 respectively. There is obviously scope for more research in these areas especially in relation to the genetic control of ion transport. As is pointed out in Chapter 12, the structure and physiology of the phloem in peas has not been studied extensively and further investigations of the relationships between translocation and growth are clearly desirable.

As subsequent chapters in this monograph and perusal of current literature will show, the garden pea continues to be a popular plant for research in plant physiology and many important questions still remain unanswered. It is hoped that the syntheses of recent research presented here will help to identify those areas in which further work is required, and to clarify the inter-relationships between the various physiological processes in this particular plant.

REFERENCES

AUDUS, L. J. (1972). "Plant Growth Substances". Leonard Hill, London.
BEEVERS, H. (1961). "Respiratory Metabolism in Plants". Row-Peterson and Co., Evanston, Illinois.
BEYERINCK, M. W. (1888). *Bot. Z.* **46**, 725–735; 741–750; 757–771; 781–790; 797–802.
BOCZUK, S. (1970). D. Phil. Thesis, University of Sussex, U.K.
BONNER, J. and ADDICOTT, F. J. (1937). *Bot. Gaz.* **99**, 144–170.
BROWN, R. (1951). *J. exp. Bot.* **2**, 196–210.
BROWN, R. and BROADBENT, D. (1950). *J. exp. Bot.* **1**, 249–263.
BROWN, R. and RICKLESS, P. (1949). *Proc. R. Soc.* B **136**, 110–125.
BROWN, R. and SUTCLIFFE, J. F. (1950). *J. exp. Bot.* **1**, 88–113.
BROWN, R. and WIGHTMAN, F. (1952). *J. exp. Bot.* **3**, 253–263.

18   J. F. Sutcliffe

BURK, D. (1927). *Pl. Physiol.* **2**, 83–89.
CRUICKSHANK, I. A. M. and PERRIN, D. R. (1964). *In* "Biochemistry of Phenolic Compounds". (J. B. Harborne, ed.), Academic Press, London and New York.
DARWIN, C. (1881). "The Power of Movement in Plants" (1966 Edition). Da Capo Press, New York.
DARWIN, C. (1888). "The Movements and Habits of Climbing Plants". John Murray, London.
DAVIES, D. D. (1953). *J. exp. Bot.* **4**, 173–183.
DAVIS, J. W. and NOVELLI, G. D. (1958). *Archs. Biochem. Biophys.* **75**, 299–308.
DUTROCHET, H. (1843). *C.r. hebd. Séanc Acad. Sci., Paris* **XVII**, 989.
FERNANDES, D. S. (1923). *Recl Trav. bot. néerl.* **20**, 107.
FURUYA, M., GALSTON, A. W. and STOWE, B. B. (1962). *Nature, Lond.* **193**, 456.
FOWLER, M. W. and AP REES, T. (1970). *Biochem. biophys. Acta* **20**, 33–44.
GANONG, W. F. (1908). *Bot. Gaz.* **46**, 193–202.
GOESCHL, J. D., PRATT, H. K. and BONNER, B. A. (1967). *Pl. Physiol.* **42**, 1077–1080.
HALES, S. (1727). "Vegetable Staticks" (1969 Edition). Macdonald and Co., London.
HELLRIEGEL, H. and WILFARTH, H. (1888). "Zeitscrift des Vereins f.d. Rubenzucker Industrie des Deutschen Reichs," pp 863–877. Berlin.
HYLMÖ, B. (1953). *Physiologia Pl.* **6**, 333–405.
KANG, B. G., YOCUM, C. S., BURG, S. P. and RAY, P. M. (1967). *Science, N.Y.* **156**, 958–959.
KRITZMAN, M. (1939). *Biokhimiya* **4**, 699.
KUIPER, K. J. (1910). *Recl Trav. bot. néerl.* **7**, 130.
LACHMANN, J. (1858). *Mitt. landw. Lehrenst., Poppelsdorf*, Vol. 1. Reprinted in *Biedermanns Zbl. Agrik Chem.* (1891), **20**, 837.
LAWES, J. B., GILBERT, J. H. and PUGH, E. (1861). *Phil. Trans. R. Soc. Ser. B* **151**, 431–577.
LINCK, A. J. (1955). Ph.D. Dissertation, Ohio State University, Columbus, Ohio.
LINCK, A. J. and SWANSON, C. A. (1960). *Pl. Soil* **12**, 57–68.
MACDOUGAL, D. T. (1902). "Elementary Plant Physiology". Longman, London and New York.
MUMFORD, F. E., SMITH, D. H. and CASTLE J. E. (1961). *Pl. Physiol.* **36**, 752–756.
PEIRCE, G. O. (1908). *Bot. Gaz.* **46**, 193–202.
PFEFFER, W. (1900). "The Physiology of Plants" (English Edition). Oxford University Press, Oxford.
PRICE, C. A. and THIMANN, K. V. (1954). *Pl. Physiol.* **29**, 495.
RAACKE, I. E. (1959). *Biochem. biophys. Acta* **34**, 1–9.
RAUTANEN, N. (1948). *Ann. Acad. sci. fen., Ser A* **33**, 1.
SACHS, J. VON (1887). "Lectures on the Physiology of Plants" (English Edition). Oxford University Press, Oxford.
SAUSSURE, T. DE (1804). "Recherches Chimiques sur la Végétation". Paris.
SNOW, R. (1937). *New Phytol.* **36**, 283–300.
STUMPF, P. K. (1952). *A. Rev. Pl. Physiol.* **3**, 17–34.
SUTCLIFFE, J. F. and SEXTON, R. (1974). *In* "Structure and Function of Primary Root Tissues". (J. Kolek, ed.). Veda, Bratislava.
THIMANN, K. V. and SKOOG, F. (1934). *Proc. R. Soc. B* **114**, 317–339.
VIRTANEN, A. I., HAUSEN, S. VON, and KARLSTROM, J. (1933). *Biochem. Z.* **258**, 106–117.
VIRTANEN, A. I., and LAINE, T. (1941). *Biochem. Z.* **308**, 213.

WEBSTER, G. C. (1955). *Pl. Physiol,* Suppl. **30,** 351.
WEBSTER, G. C. (1959). "Nitrogen Metabolism in Plants". Row Peterson, Evanston and New York.
WENT, F. W. (1934). *Proc. K. ned. Akad. Wet.* **37,** 547.
WIGHTMAN, F. and BROWN, R. (1953). *J. exp. Bot.* **4,** 184–196.

2. Classification, Genetics and Breeding

G. A. MARX

New York State Agricultural Experiment Station, Geneva, New York, U.S.A.

I. INTRODUCTION

The pea enjoys a unique place in the history of genetics as the plant that served Mendel with distinction during the course of his celebrated experiments. Since then genetic knowledge has accumulated to the extent that the pea now ranks among the genetically best known of all plants. But as the very presence of this monograph attests, the plant is also especially useful in physiological and biochemical studies (Chapter 1). Still, plant physiologists have generally confined their attention to a genetically narrow range of material. This chapter seeks in part to provide some background with respect to the history, phylogeny and classification of the pea; and in part to acquaint the non-geneticist with the rich array of heritable variation that the species offers, much of it being desirable material for investigation.

II. ORIGIN, HISTORY AND CLASSIFICATION

Although its ancient origin is obscure, the pea was surely among the earliest of cultivated plants. Indeed, archaeological evidence indicates that it was well under cultivation in the Near Eastern and Greek Neolithic settlements as early as 6000 B.C. (Zohary and Hopf, 1973). Evidently the pea, as well as other pulses, was a coeval of wheats and barley, the protein of the pulses complementing the starches of the cereals in the human diet (Hedrick *et al.*, 1928; Zohary and Hopf, 1973).

B

Unfortunately the archaeological remains of peas consist primarily of carbonized seeds. It is therefore more difficult to fix the time of domestication for peas than for cereals. The presence of a non-brittle rachis (non-shattering spikes) in the cereals provides clear-cut evidence of cultivation, for the non-shattering wheats and barley cannot survive in the wild (Zohary and Hopf, 1973). Among the few criteria available as evidence of domestication of the pea, Zohary and Hopf (1973) lay stress on the condition of the seed coat surface, a rough or granular testa being the primitive condition in contrast with the smooth surface which is characteristic of cultivated varieties. Thick and/or hard seed coats may also be taken as primitive characteristics, for they are associated with slow and erratic germination. Moreover, there is a character of peas that closely parallels the brittle *vs* non-brittle rachis of the cereals. Some primitive forms have dehiscent pods which upon maturity expel and scatter their seeds so that they cannot be effectively gathered.

Wild and primitive forms of peas are found in a vast region with its centre in the Near East but including areas of Central Asia, the Mediterranean and Abyssinia. These are areas of diversity that Vavilov (1951) interpreted as centres of origin, but this interpretation has lost much of its currency because genetic diversity alone does not satisfactorily mark the geographical place of origin (Harlan, 1970, 1971; Zohary, 1970). The dispersal of peas throughout much of the Old World cannot be dissociated from the distribution, migrations and activities of man.

According to most historical accounts (e.g. de Candolle, 1885), the garden pea has never been found in the wild. Presumably what is meant is that a pea very closely resembling the most advanced cultivars of the time has not been found. Much of the confusion and uncertainty concerning the ancestry of the pea may be ascribed to a faulty tack taken by the early botanists. They looked upon the garden pea (*Pisum sativum*) as an entity distinct from the field pea (*P. arvense*) and attempted to trace the ancestry of the former without much reference to the latter and without any knowledge of the genetic affinities between the two. They used classical taxonomic techniques which emphasized morphological criteria and they placed undue reliance on written historical accounts. The ancient writings did not clearly distinguish between garden peas and field peas, and in some instances did not even distinguish between peas and other pulses.

However, if we begin by emphasizing the similarities between *P. sativum* and *P. arvense* and not the differences, and if we recognize the close genetic affinity between the two, or in short, if we treat the two merely as variations of the same taxon and include the archaeological evidence, then the modern pea may simply be regarded as the product of a straightforward and limited evolutionary process. Doubtless man played an important part in this process. Certain characters such as white flowers and wrinkled seeds may be considered as advanced traits that are peculiar to esculents, and coloured flowers and round seeds as primitive characters peculiar to field peas, but there is, however, considerable overlapping of characters between the two types. The chief distinction between garden and field peas is that the former has a greater number of domestic attributes. It is therefore reasonable to suppose that the field pea is the immediate progenitor of the garden pea and that the field pea in turn was derived from one of the wild races. Ben-Ze'ev and Zohary (1973) argue that *P. humile* is the wild race of peas that best qualifies as the probable progenitor of the cultivated pea, but the evidence for this is not compelling.

Whatever the interpretation, history leads us back only as far as the oldest archaeological digs, and the seeds found there were borne by plants that seemingly were not very different from the most primitive forms found today. What is truly lost in antiquity is the record of evolutionary changes that occurred prior to the Neolithic Age.

Whether the crop was simply gathered or actually cultivated, peas offered a number of advantages to early man. The protein- and carbohydrate-rich seeds were easily stored and transported, and for most of recorded history peas have been grown primarily for the dry seed. But in contrast with modern peas, the seeds of the primitive forms often are hard and bitter, and may even contain toxic substances. Man evidently selected out and propagated mutations of agricultural value, including those that enhanced edible quality. Theophrastus, three centuries before the Christian era, gave an authoritative and detailed description of the pea and its culture, which suggests that prior to that time the crop was long known and commonly grown (Hedrick *et al.*, 1928).

The pea was called *pisos* by the Greeks and *pisum* by the Romans. When the plant was passed on to the English, it became "peason", then "pease" or "peasse", and finally "pea". It became a prominent crop in middle and northern Europe where the climate favoured its culture, and where in the Dark and Middle Ages it was grown almost as commonly as any of the cereals (Hedrick *et al.*, 1928). It was also one of the chief crops grown in England, and England is where, in modern times, significant improvements were made through breeding and selection. Thomas Andrew Knight, who hybridized peas as early as 1787, introduced a number of improved varieties with wrinkled seed. Most of the commercial varieties of today trace their origin to the varieties developed in England during the nineteenth century.

Peas belong to the order Fabales, family Leguminosae (Stebbins, 1974), *Vicia* and *Lathyrus* being the most closely allied genera. Predictably, the taxonomic treatment of peas has undergone change and modification since Linnaeus (1753) distinguished two species, *P. arvense* and *P. sativum*. Early systematists, observing the multiplicity of forms and emphasizing morphological differences rather than genetic compatibility, tended to recognize many species. Still, some such as Alefeld (1866) and Ascherson and Graebner

(1907) were inclined to accept only one or two species, and to reduce the variations to taxa of lower rank. Moreover, Wellensiek (1925), a geneticist, urged that all the mutually crossable cultivated forms be combined into one species. Indeed, Hedrick *et al.* (1928) were disposed to accept only two species: *P. sativum* and *P. fulvum*, a position taken later by Davis (1970). Recently Ben-Ze'ev and Zohary (1973) invoked genetic and cytogenetic data as well as other considerations in reaching the same conclusion. They cited greater reproductive isolation, differences in chromosome morphology, and the relative number of interchanges as reasons for retaining *P. fulvum* as a distinct species. *P. formosum*, once considered as a species of *Pisum*, is now recognized as belonging to a separate genus (Davis, 1970; Lamprecht, 1966). It may therefore appear that a consensus has been reached as regards classification: that there are two species, *P. sativum* and *P. fulvum*. However, Lamprecht (1966), who like Wellensiek was a geneticist, regarded *Pisum* as monospecific, a view shared by the present writer. The wild forms are regarded as races or ecotypes. Though *P. fulvum* shows comparatively greater genetic and cyto-genetic divergence than other races, the progenies from crosses with other peas are at least partially fertile and characters can be transferred without undue difficulty. Because the differences and the affinities are a matter of degree, the decision to recognize one or two species is a matter of personal interpretation. Absence of universal agreement is of little real consequence to the physiologist, for no matter how venturesome he may be in selecting material for investigation, the material is likely to fall within the range of variation generally accepted as *P. sativum*.

Even if *Pisum* is accepted as being monospecific, a problem remains concerning the species name. As Wellensiek (1925) and Lamprecht (1956, 1964, 1966) point out, Linnaeus recognized *P. arvense* and *P. sativum*. Since *P. arvense* has more primitive characteristics than *P. sativum* there can be little doubt that the former antedates the latter. On these grounds, it is more logical to accept *P. arvense* L. as the species name. This was formally proposed by Lamprecht (1966). However, because *P. sativum* is so solidly entrenched in the scientific literature, there is dubious merit in pressing for a change.

Often, as is the case in this volume, a modifying word such as "garden", "English" or "common" is used in conjunction with the word pea in order to identify it as a member of the genus *Pisum* and thereby to distinguish it from the "peas" of other genera such as *Cicer*, *Lathyrus* and *Vigna*. Moreover, peas are often described as either field peas or as garden peas, but the latter distinction is by no means clear-cut since some varieties are used in a dual capacity.

III. CYTOLOGY

Because the early and middle stages of the first meiotic prophase cannot be analysed in the pea, the plant has not become a favourite organism for classical cytological work. The chromosome number (2n = 14) was established early in this century (Cannon, 1903; Nemec, 1904; Strasburger, 1907), but little else of major significance was accomplished until the 1950s. It was then that Caroli and Blixt (1953) and Blixt (1958a,b) developed suitable techniques for pretreatment, fixation, staining, and chromosome measurement in mitotic chromosomes, which led to a generally acceptable description of the karyotype.

Chromosomal interchanges were found to be rather common. Lamprecht's Line 110 was shown to typify the common cultivated forms and was adopted as the standard karyotype (Blixt, 1958b). Some wild forms of peas carry one or more translocations but others have the standard arrangement (Lamprecht, 1966; Ben-Ze'ev and Zohary, 1973). A controversy (Lamprecht, 1948, 1958, 1960a; Lamm, 1949, 1951, 1956, 1960) concerning chromosome-linkage group relations was essentially reconciled by the work of Snoad (1966) and by the later findings of Lamm (1972). The cytological data were satisfactorily co-ordinated with the genetic (linkage) data (Blixt, 1959) and a translocation tester set was developed (Lamm and Miravalle, 1959; Lamm, 1974).

A current centre of activity for cytological study, especially the identification of mutants causing asynapsis, desynapsis, and other meiotic irregularities, is the laboratory of Gottschalk and his group at Bonn, West Germany (Gottschalk, 1971). For a general review of the cytology of *Pisum*, including the biochemical aspects, see Blixt (1972).

IV. GENETICS

A. General Considerations

Nearly 2000 mutants of *Pisum* have been reported (Blixt, 1974) but perhaps fewer than a fourth of these have been clearly established as being different, distinct and "workable" in Mendelian analyses. The first comprehensive linkage map, involving 37 genes distributed over all seven chromosomes, was presented by Lamprecht in 1948. The most recent map (Fig. 2.1), comprising 169 mutants, rivals in completeness the maps of such well studied plant species as maize, barley and tomatoes. The review by Yarnell (1962), the more recent comprehensive review by Blixt (1972), and the bibliographies appearing in the Pisum Newsletter (Vols 2 and 5) provide access to the majority of the genetical literature. Parenthetically, a statement persisting in the general

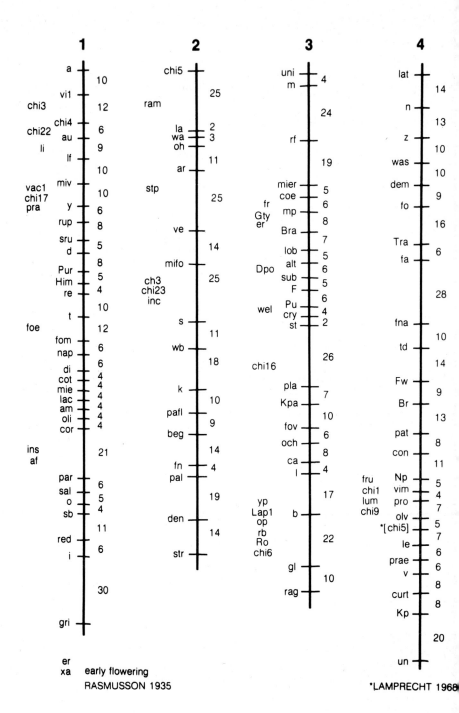

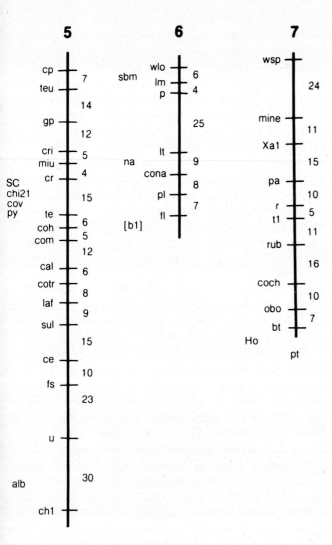

FIG. 2.1. A current gene map of *Pisum* (after Blixt, 1972). Genes set away from the lines have been located only as to chromosome.

literature is that the seven genes with which Mendel worked were located in seven different chromosomes. In fact, however, the seven genes reside in but four of the seven chromosomes (Lamprecht, 1968; Blixt, 1972, 1974).

Co-ordination of current genetical work is facilitated by the Pisum Genetics Association (PGA), a small, informal organization founded in 1969. The PGA encourages the exchange of genetic information and seeks to ensure the preservation of valuable genetic stocks. The Pisum Newsletter (PNL), an annual publication, serves as the medium of exchange.

The following discussion focuses exclusively on genetic differences that are controlled by major genes or the so-called oligogenes, i.e. those having clearly defined, visible, or easily detected effects. These, of course, constitute only a part, perhaps only a very small part, of the total heritable variation within the species. Much of the existing variation is based upon polygenic systems wherein differences depend not so much on individual gene action as on the cumulative effect of several or many genes, but this variation does not lend itself to mapping or simple genetic analysis.

The absence of discrete classes in segregating populations from a cross is frequently construed as presumptive evidence of polygenic control, but variation that is ostensibly quantitative may prove, in fact, to be qualitative. For example, heritable variation in flowering behaviour, often considered to be controlled largely by a polygenic system, has been shown by Murfet (see Chapter 14) to be controlled predominantly by an interacting system of major genes.

Since economic characters tend to be more closely identified with the polygenic system than with the oligogenic system, the former is of greater concern to plant breeders engaged in crop improvement. The two gene systems are not unrelated, however, for both share the same vehicles of inheritance, the chromosomes. Therefore, if major mutants occupy various known sites in all seven chromosomes, there must be, perforce, some relationship between these easily identified "landmarks" and the not-so-easily identified polygenes. Thus, the mapping of major genes serves not only as a reference in locating new or as yet unmapped major mutants, but also as an indirect means of locating and characterizing the polygenes that may be mutually associated with major genes. Thus, any mutant, regardless of its effect on form or function or its relationship with biological fitness, may be useful if it contributes to or facilitates the expansion of genetic knowledge.

Mutants may also be used to gain a fuller understanding of physiological processes. Physiological studies often involve the use of only one or a few cultivars. Findings from such studies sometimes are taken to represent peas in general rather than merely the varieties studied. In view of the great genetic diversity in the species, it is problematical just how representative a given line may be. Indeed, for some characteristics the differences within the species may be manifestly greater than those between species.

B. Some Specific Mutants

The mutants described in the following sections illustrate the diversity of *Pisum*. The treatment, however, is neither balanced nor comprehensive, for the mutants are principally those with which the author has some first-hand knowledge and experience, and those which, in his opinion, invite further investigation.

1. Foliage Mutants

The leaves of normal plants are composed of two rather large stipules, one or more pairs of leaflets, and several tendrils (Fig. 2.4). The mutant *st* sharply reduces the size of the stipules without directly affecting the other parts of the

FIG. 2.2. A plant showing modified foliar morphology resulting from the genotype of *af, st, Tl*

leaf (Wellensiek, 1925). Another mutant, *tl*, converts the tendrils into leaflets, giving rise to the phenotype known as "acacia" (White, 1917). A third recessive mutant, "afila" (*af*), acts to transform the leaflets into tendrils (Kujala, 1953). Neither *tl* nor *af* affects stipule size or shape, but profound morphological changes can result from the manipulation of just these three genes. For example, plants carrying both *af* and *st* are virtually leafless (Fig. 2.2), and yet the nearly denuded plants exhibit remarkable productivity in spite of the sharp reduction in leaf area. Evidently the stems, vestigial stipules, tendrils and pods are all capable of effective photosynthesis. Another bizarre phenotype is produced by adding *tl* to the above combination (Fig. 2.3). Here *af* interacts with *tl* to produce a cluster of small leaflets (Goldenberg, 1965). In all, eight distinct phenotypes are generated by the combinations of alleles at the three loci, creating striking differences in foliar configuration.

FIG. 2.3. Foliar configuration resulting from the gene combination *af, st, tl*.

The pea plant appears to be capable of adjusting to marked differences in foliage, and to some extent the leaf parts seem to be interchangeable. Near-isogenic lines have been developed to facilitate the measurement of the effects of *af*, *st* and *tl* singly and in combination (Marx, 1974). Isogenic lines mini-mize the confounding influence of background genotype, and so permit more valid comparisons among phenotypes.

As normal, non-mutant plants undergo the transition from the seedling to the adult stage, the number of leaflets per leaf increases from one pair to two or more. In contrast, plants carrying *up* (unipetiole) have a single pair of leaflets throughout their ontogeny (von Rosen, 1944), while the mutant "unifoliata" (*uni*) (Lamprecht, 1934) produces an even more dramatic change in leaf morphology (Fig. 2.4). In this case, the normally compound leaf is converted to a simple leaf, although a few bi- and tri-foliate leaves do appear when the mutant plants reach the flowering stage. Mutant plants invariably have sterile, pistilloid flowers.

Fig. 2.4. Plants with the mutant "unifoliata" (*uni*) have a single leaflet at each node (left), whereas leaflets on normal plants are paired (right).

2. Internode Length

Tall (*Le/-*) and dwarf (*le/le*) (White, 1917) peas are distinguished mainly by alternate alleles at the *Le* locus on chromosome 4. "Tall" and "dwarf" are familiar, if imprecise, terms, often used to describe the final height of the plant which is a secondary character compounded of internode length and internode number. Genetically controlled differences in internode length are best distinguished in seedling plants of the same age grown under uniform environmental conditions.

Genes other than *Le* influence internode length; two well studied loci are *La* and *Cry* (de Haan, 1927, 1930; Rasmusson, 1927; and Lamm, 1937, 1947). Dwarf (*le/le*) plants remain dwarf if they carry at least one dominant allele at either *La* or *Cry*. However, the combination *le, la, cry^c* yields a phenotype called "crypto-dwarf". In the very early stages of seedling development the height of crypto-dwarf plants is about equal to that of genetically tall (*Le*) plants, because the first one or two nodes of crypto-dwarfs are actually longer than those of *Le* plants. However, the length of succeeding internodes is greater in *Le* plants than in crypto-dwarfs, so the *Le* plants soon surpass the crypto-dwarfs in total height. Now, the *Cry* locus has a second recessive allele, *cry^s*. The combination *le, la, cry^s* produces a singular phenotype known as "slender". The internodes of slender plants are strikingly elongated, so much so that they are taller than *Le* plants, not only initially but well into the later stages of vegetative development. Slender plants exhibit a number of other characteristics: rapid germination; thin, weak and spindly stems; delayed flowering; malformed and abortive flowers; and poor seed-set. These characteristics are not unlike those seen in GA-treated normal dwarf peas. Table 2.1 summarizes the *Le, La, Cry* system.

TABLE 2.I. Internode length in peas as determined by three gene loci.

Genotype*			Phenotype
Le	*La*	*Cry*	
+	+	+	Tall
+	+	*cry^c* or *cry^s*	Tall
+	*la*	+	Tall
+	*la*	*cry^s*	Slender
le	+	+	Dwarf
le	+	*cry^c* or *cry^s*	Dwarf
le	*la*	+	Dwarf
le	*la*	*cry^c*	Crypto-dwarf
le	*la*	*cry^s*	Slender

* (+) indicates presence of at least one dominant allele at the specified locus.

3. Pigmentation

With few exceptions, the peas cultivated for human consumption are unpigmented. *A* is a dominant master gene or ground factor controlling antho-

cyanin production, pigmentation being inhibited by a/a (Tschermak, 1912). Although dominance at A is pre-requisite for pigment formation, it does not ensure pigmentation at a particular site. For example, axil coloration requires the presence not only of A but also of another dominant gene, D. Axils are unpigmented in plants with the constitution A, d, as well as in those with a, D and a, d.

A is also required for flower colour, but a plant may have white flowers despite the presence of A if the pigment inhibiting the recessive gene am is also present (de Hann, 1930). Wild-type flowers are purple-violet. Various modifications of flower colour are brought about by a series of recessive genes, some of the most thoroughly investigated (both genetically and chemically) of which are shown in Table 2.II. Other variations in flower

TABLE 2.II. The phenotypic expression of four well known flower colour genes of *Pisum* as expressed in an A background; i.e. a/a plants, regardless of the status of any other gene, have pure white flowers.

Status of flower colour locus*				Phenotype	(flower-colour)
B	*Ce*	*Cr*	*Am*		
+	+	+	+	Wild-type	(purple-violet)
b	+	+	+	Pink	
+	*ce*	+	+	Cerise	
+	+	*cr*	+	Crimson	
+	+	+	*am*	White†	

*(+) indicates presence of at least one dominant allele at the specified locus.
†A, am plants sometimes have flowers with a slight pink cast.

colour are known, and although they are genetically not yet well defined, they contribute to a broad range of phenotypic differences. In addition, some genes operate as modifiers of other flower colour genes, and still others affect the distribution of pigment in the different parts of the corolla.

Pod colour, too, is dependent in part on the presence of anthocyanin. The normal green pod (chlorophyll) becomes purple if two dominant genes, Pu and Pur, are simultaneously present together with A (Lamprecht, 1938, 1953). Again, a/a precludes pod coloration, so in a/a plants the pods remain green regardless of the status of Pu and Pur. The recessive gene, gp, inhibits chorophyll production in the pods, so they are yellow rather than green (White, 1917). Since the action of gp is not anthocyanin-dependent, the pods are yellow in a/a plants as well as in $A/-$ plants. However, gp interacts with Pu and Pur in $A/-$ plants to cause the otherwise purple pods to become red. The action and interaction of genes A, Pu, Pur and Gp are summarized in Table 2.III. A red pod colour somewhat similar to that brought about by the genotype A, Pu, Pur, gp is produced by the combination A, Pu, Pur, am. In the latter case, Gp (green) is present rather than gp (yellow), so the red pod

TABLE 2.III. Pod colour of peas as a function of genotype.

Genotype*				Phenotype (pod colour)	
Pu	*Pur*	*Gp*	*Am*	*A/–*	*a/a*
+	+	+	+	purple	green
+	*pur*	+	+	green	green
pu	+	+	+	green	green
+	+	*gp*	+	red	yellow
+	–	*gp*	+	yellow	yellow
pu	+	*gp*	+	yellow	yellow
+	+	+	*am*	red	green

*(+) in a column indicates the presence of at least one dominant allele at the specified locus.

results not so much from the physical interaction of purple (*Pu*, *Pur*) and yellow (*gp*) as it does from the dilution of the purple pigmentation through the action of *am*. The primary effect of *am* is to inhibit flower colour, transforming coloured flowers essentially to white; but *am* has a secondary (pleiotropic) effect on pod and axil coloration, converting the colour of both from purple to pinkish-red.

It should be apparent, then, that the presence or absence of anthocyanin, as determined by dominance or recessivity at the *A* locus, is critical to the expression of a number of genes and that *a/a* is therefore epistatic to these genes because it masks their effect.

Some seed coat patterns of *Pisum* are anthocyanin-dependent whereas other patterns are based on other pigments. These will not be considered here except to mention that the first evidence of pigment formation coincides approximately with physiological maturity of the seed. The seed is then still green in colour and not yet dry. At this stage, for example, a black pigment (chemically still unidentified) begins to form in the hilum of seeds if the plant carries the dominant gene *Pl*. The *Pl* gene therefore reliably marks a given stage in the maturation process of the seed as demonstrated by a correlation with physical and chemical measures of maturity (Marx and Duczmal, 1972). Pigment formation resulting from the action of other seed colour genes is also correlated with stage of maturity but each gene becomes expressed at a slightly different time and/or stage. Thus if the seed of a line carrying several different seed genes is observed throughout the maturation process, each gene will become expressed at a slightly different time and in a characteristic order.

4. Wax Mutants

Virtually the entire aerial portion of normal pea plants is coated with a thin layer of protective wax. A series of mutants, eight in number, affect the amount and distribution of wax, and they do so in a highly specific manner.

The action of *wlo*, for example, is to remove the wax exclusively from the upper surface of the leaflets, leaving the lower leaf surface, both surfaces of the stipules and the stem with a normal complement of wax (Nilsson, 1933). In contrast, *wsp* (Lamprecht, 1939) acts to prevent wax formation on all plant parts *except* the upper surface of the leaflets, producing a kind of "photographic negative" of *wlo*. These two mutants have additive effects, so a plant carrying both mutants is completely waxless. A completely waxless phenotype, ostensibly identical with that generated by the combination of *wlo* and *wsp*, is produced by a single mutant, *wel* (Marx, 1969). Conversely, *wex* results in a superabundance of wax on all aerial plant parts (Marx, 1969). Another mutant, *wp*, removes the wax coating from the pods only (Marx, 1971b).

5. Chlorophyll Mutants

In peas as in other plants, chlorophyll mutations are the most common class of spontaneous and induced mutations. Their number and diversity of expression call for orderly classification, and the schemes of Lamprecht (1952, 1960b) and Blixt (1961, 1972) were developed to meet this need. The few chlorophyll mutants selected for consideration here emphasize changes that occur during the course of plant development.

Seedlings that are homozygous recessive for the mutant *alt* have normal green foliage until they reach the 5–6 node stage; then, rather abruptly, they lose their capacity to produce chlorophyll, the terminal leaf tissue becomes bleached and the seedlings die (Lamprecht, 1955). Mutant seedlings recover and develop flowers, fruits and seeds if they are grafted to normal non-mutant plants. Partial recovery can also be achieved by applying extract from normal seeds (Acree and Marx, 1972). This suggests that a substance initially present in the seed, becomes exhausted in the mutant. Normal evidently contain a transmissible substance(s) capable of correcting deficiency.

Some mutants, such as those in the chlorina group, are pale green at time they emerge from the soil, but they die at about the 5–7 node stage without becoming chlorotic. Other mutants, with seemingly the same foliage colour, survive to maturity, and still others regain the normal green colour above the 5–7 node stage.

Normal, non-mutant plants lose chlorophyll gradually with the approach of maturity. Symptoms of senescence occur earlier and more abruptly in plants carrying the mutant *py* (Marx, 1971a). Another similar mutant, *o*, causes some paling of the tissues in the early stages of growth and development, but the paling is more extreme as the plants approach maturity (White, 1917). Then the pods and stems become yellow or yellowish-orange and the cotyledons of the seeds become pale yellow. This is not, however, the classic gene for cotyledon colour, which is so familiar and which has the symbol *I*

(White, 1916). The dry seeds of $I/$- plants are yellow, and those of $i\,i$ plants are green. $I/$- seeds cannot be distinguished from $i\,i$ seeds during most of the period of seed development; both contain chlorophyll and therefore both are green. It is only in the late stages of ovule development (just before the ovules begin to dry) that $I/$- seeds begin to lose chlorophyll *en route* to becoming yellow.

In this connection the principal difference between peas used for canning or for freezing is the relative intensity of the green in the edible, immature seed or "berry".* This distinction is brought about principally by the gene pair *Pa-pa* (Winge, 1936), *Pa* giving the pale green berries characteristic of varieties used for canning, and *pa* giving the darker green berry associated with varieties used for freezing. When varieties carrying *pa* are canned instead of frozen, the product is often considered objectionable (at least in the United States) because it has a characteristic bronzy cast associated with the degradation of chlorophyll to phaeophytin. Dark green (*pa*) *vs* pale green (*Pa*) berry colour is quite independent of green (*i*) *vs* yellow (*I*) cotyledons. The peas are usually processed well before the yellowing of the cotyledons associated with *I* takes place.

6. "Pathological" Mutants

Some mutants cause abnormalities that resemble pathological disorders. The presence of a pustular-like growth on the external surface of the pod is conditioned by the dominant gene *Np* (neoplastic pods, Fig. 2.5), but expression is greatly influenced by light, particularly by the quality of light (Nuttal and Lyall, 1964; Dodds and Matthews, 1966). According to Snoad (1969) the neoplasms are highly polyploid and growth apparently stems from induced mitotic activity of the subsidiary cells which surround the stomata. Subsidiary cells do not occur in the plant other than in the pods; neither do the neoplasms.

The effect of the recessive mutant *twp* (Marx, 1973) on early pod development is shown in Fig. 2.6. Tissue in the central region of the valve does not develop normally; it may be simply arrested or it may be necrotic. This, coupled with the normal development in the tissue surrounding the affected region, leads to differential growth characterized by a distortion and twisting of the immature pod. As the pods become mature, the symptoms diminish, due in part to the enlarging seeds and perhaps in part to compensatory growth of tissue surrounding the lesion. The histological nature of the phenomenon remains unknown.

Abnormal seed development is often associated with cytogenetic irregularity or environmental stress. However, a simply inherited seed disorder has been found in connection with some studies on the genetics of flower colour

---

*Berry is a horticultural appellation used to refer to the green, edible pea; the word is a convenient substitute for the more formal enlarged ovule or immature seed.

Fig. 2.5. The pod with pustular growths is from a plant carrying the mutant *Np*. The pod on the right is from a normal plant.

(Marx, unpublished). Although the symptoms vary somewhat from plant to plant and from seed to seed, the disorder is characterized by necrotic areas of varying size. Some seeds are affected so early in development and to such a degree that they never become fully developed.

7. *Miscellaneous Mutants*

At germination, the roots of mutant *age* (ageotropic) plants grow upward towards the soil surface (Blixt, 1970). After emerging from the soil the roots grow horizontally along the soil surface. Studies by Ekelund and Hemberg (1966) indicate that mutant behaviour is related to a supra-optimal auxin concentration.

A mutant recently isolated by L. G. Cruger (Del Monte Corporation) is distinctive in that the plants wilt dramatically when they encounter conditions

Fig. 2.6. The three pods on the left are from a plant possessing the mutant "twisted pod" (*twp*). The three pods on the right are from a normal plant.

of water stress. When favourable conditions are restored, the plants recover with little evidence of impaired growth and development. Whether abscisic acid is involved remains to be determined.

Several mutants cause drastic irregularities in flower structure accompanied by sterility. In most cases all the flowers on the plant are affected equally. One spontaneous mutant, yet unnamed and possibly identical with Mutant 251 which Gottschalk (1964) recovered after X-ray treatment, causes a malformation of the second flower on the first inflorescence. Other flowers are normal. Such specificity invites an explanation.

The action of certain mutants, including a number of chlorophyll mutants, suggest that the 5–7 node stage of seedling development has special physiological significance. Since 5–7 nodes are already preformed in the dry seed, this stage presumably marks the transition in growth and development from that which is largely dependent on stored reserves in the seed to that which is dependent on the assimilates manufactured by the plant itself. Even in normal, non-mutant plants there are observable signs of this transition, inasmuch as growth increases dramatically after the 5–7 node stage is reached.

Mutants that cause changes associated with this stage of development may be advantageous in certain physiological investigations because they provide an identifiable morphological or physiological landmark.

V. BREEDING

Although practical breeding programmes may proceed and indeed succeed with scarcely any reference at all to genetics, the potential for success is enhanced by a judicious blend of art and science.

Agronomically, dry peas and green peas are often considered as two separate crops, but many of the breeding objectives are similar. The accumulated genetic knowledge may apply regardless of agronomic use and the existing gene pool serves as a genus-wide resource. Genes for disease resistance, for example, may be used in the control of many of the 20 or more different diseases of pea. Also enlightened use of specific information such as linkage data may serve to improve breeding efficiency whatever the agronomic use of the crop may be.

Dry edible peas characteristically have round or smooth seeds R/R, Rb/Rb, whereas most green pea cultivars have wrinkled seeds $r\,Rb$, $R\,rb$, $r\,rb$. Seed shape, therefore, is one of the simply inherited characters that has pivotal importance in determining major alternative uses. Round seeds are imperative in dry edible peas because they contain relatively large amounts of amylopectin. Parchmented PV vs non-parchmented pV, Pv, pv pods provides another example of a simply inherited difference that effectively determines use—the edible podded (non-parchmented) peas are prized in certain, especially Oriental, dishes. Thus the association between genotype and use may be based on relatively small, often qualitative, genetic differences.

The relatively high protein content (c. 25%) of dry peas is an attractive attribute of the crop. Work is under way, world-wide, to increase the total protein content further through breeding and selection. In terms of protein quality, the sulphur-containing amino acids are first limiting, as they are also in other legumes. Regrettably, the genetic collections of peas have not been surveyed in search of mutants that are associated with higher levels of methionine or cystine.

Breeders of green peas for processing face some especially challenging problems. First, the relatively short life cycle of the plant limits its capacity to accumulate dry matter. Moreover, unlike most of the field crops which are harvested at the end of the maturation process, most of the vegetable crops are harvested *during* the maturation process. Dry peas fall into the former category and green peas into the latter. Quality in vegetables is of paramount importance, coequal, or nearly so, with yield. Quality is a function of maturity, which in turn is a function of time, and the vegetable breeder must therefore deal with time as a limiting factor. Peas remain within an optimal range of maturity for only a short period of time, literally only hours.

Until the peas reach a stage of prime maturity, yield and quality are positively correlated; thereafter they are negatively correlated. Moreover, prime quality usually anticipates maximal yield. A number of interacting variables contribute to the dynamics of the maturation process on a field scale. The total variation in size and maturity of the peas in the crop at any given time is attributable to inter-plant and to intra-plant variation, both being associated with genetic and non-genetic causes. Intra-plant variation is attributable in large measure to the normal indeterminate habit of the plant. Since the flowers and pods are borne at successive nodes over a period of time, the berries contained in the pods in the lower reproductive nodes are larger and more mature than those appearing at succeeding nodes. The presence of more pods and berries, each at a slightly different stage of growth and maturity, increases the variation and adds to the level of complexity.

Therefore, on a field scale, optimal quality is merely the most favourable average of the total variation. It occurs at a time when the greatest total number of peas falls within an acceptable range of quality. Economic considerations dictate that the crop grown for processing be harvested at one time. The peas that fall outside the limits of quality—either too mature or too immature—at the time of harvest do not contribute to an acceptable horticultural product. The balance of yield and quality therefore represents a "moving target". The time it takes for a crop to reach and remain in an acceptable stage of maturity depends not only on how the berries are distributed among and within plants, but also on the biochemistry of the individual berry. For example, the rate of maturation is greater for round-seeded than for wrinkled-seeded varieties, owing to a difference in the composition of starch, viz. the balance between amylose and amylopectin.

Several possible strategies to synchronize optimal quality and maximal yield more effectively suggest themselves. Since only the seeds that develop within a limited time frame contribute an acceptable horticultural yield, it seems advisable to breed and select plants with a more determinate habit, thereby minimizing wasteful distribution of assimilates to plant parts that will not add to the plant's economic yield. Means should then be sought to maximize productivity within this restricted number of reproductive nodes. Since both the number of pods per peduncle and the number of ovules per pod are subject to genetic control, an opportunity to effect increases in these components does exist. A practical limitation to this approach has been encountered, however. The physiological activities of the plant seem to be so closely co-ordinated that it is difficult to introduce major changes in morphological and physiological make-up without causing undesirable secondary effects. Thus, for example, attempts to increase the number of pods per node from one or two to three or more have met with limited success because the additional pods frequently abort under commercial competitive conditions. When the additional pods are retained, they often are poorly filled, so that the total of three pods may be no greater than a plant with a single pod at a node.

A promising approach towards increasing the intra-plant uniformity of maturity is through the use of a genetically determined property known as simultaneous flowering (Marx, 1972). The morphology of plants displaying this property is similar to normal plants inasmuch as the flowers and pods are separated spatially by being borne at successive nodes. The distinction arises from the fact that the flowers on the first three or four reproductive nodes tend to open at approximately the same time, and the pods and seeds tend therefore to be more uniform in maturity.

It has already been suggested that gene-controlled differences in the chemistry of the seed may affect rate of maturation. This has practical implication in breeding, not only in terms of improving edible quality but also in terms of improving the relationship between yield and quality. Hence, if acceptable edible quality can be retained until, or perhaps beyond, the time the seed reaches its maximal size and weight, this will result in a prolongation of the favourable positive correlation between yield and quality. This in turn may permit a delay in harvest in order to gain an increase in yield without a loss of quality.

Aside from the attempts to gain greater uniformity of maturity, plant breeders are seeking to enhance the biological efficiency of the plant. Most often, as a matter of practical necessity, breeding and selection are conducted in trellised nurseries where competition between plants is minimal. Yet, under commercial conditions, plants are exposed to a maximum of population stress. Under the latter conditions it would appear that many commercial cultivars have excessive foliage which may result in an unfavourable balance between vegetative and reproductive development; improving the cultural conditions merely leads to an increase in vegetative development at the expense of reproductive development. Some of the foliage mutants mentioned in a previous section, i.e. *af, st, tl,* are being intensively investigated in this connection.

REFERENCES

Acree, T. and Marx, G. A. (1972). *Experientia* **28**, 1505–1506.
Alefield, F. (1866). "Landwirtschaftliche Flora". Berlin.
Ascherson, P. and Graebner, P. (1907). *Synopsis mitteleur.* Flora 6 (II,) 1063–1067.
Ben-Ze'ev, N. and Zohary, D. (1973). *Israel J. Bot.* **22**, 73–91.
Blixt, S. (1958a). *Agri Hort. Genet.* **16**, 66–77.
Blixt, S. (1958b). *Agri Hort. Genet.* **16**, 221–237.

42   G. A. Marx

BLIXT, S. (1959). *Agri Hort. Genet.* **17**, 47–75.

BLIXT, S. (1961). *Agri Hort. Genet.* **19**, 402–477.

BLIXT, S. (1970). *Pisum Newsl.* **2**, 11–12.

BLIXT, S. (1972). *Agri Hort. Genet.* **30**, 1–293.

BLIXT, S. (1974). *In* "Handbook of Genetics" (R. C. King, ed.) Vol. II, pp 181–221. Plenum Press, New York and London.

CANNON, W. A. (1903). *Bull. Torrey bot. Club* **30**, 519–543.

CANDOLLE, ALPHONSE DE (1885). "Origin of Cultivated Plants". D. Appleton, New York.

CAROLI, G. and BLIXT, S. (1953). *Agri Hort. Genet.* **11**, 133–140.

DAVIS, P. H. (1970). *In* "Flora of Turkey", Vol. III, pp 370–373. Edinburgh University Press.

DODDS, K. S. and MATTHEWS, P. (1966). *J. Hered.* **57**, 83–85.

EKELUND, R. and HEMBERG, J. (1966). *Physiologia Pl.* **19**, 1120–1124.

GOLDENBERG, J. (1965). *Boln genet.* **1**, 27–31.

GOTTSCHALK, W. (1964). *Bot. Stud., Jena* **14**, 1–359.

GOTTSCHALK, W. (1971). "Die Bedeutung der Genmutationen fur die Evolution der Pflanzen; Fortschr. Evol.-Forschung". Vol. VI, pp 1–296. Fischer, Stuttgart.

HAAN, H. DE (1927). *Genetica* **9**, 481–498.

HAAN, H. DE (1930). *Genetica* **12**, 321–439.

HARLAN, J. R. (1970). *In* "Genetic Resources in Plants: Their Exploration and Conservation" (D. H. Frankel and E. Bennett, eds.), IBP Handbook No. 11 pp 19–32. F. A. Davis, Philadelphia.

HARLAN, J. R. (1971). *Science, N.Y.* **174**, 468–474.

HEDRICK, U. P., HALL, F. H., HAWTHORN, L. R. and BERGER, A. (1928). "Vegetables of New York. Vol. I: Peas of New York." Report of the N.Y. State Agricultural Experiment Station.

KUJALA, V. (1953). *Archivum Societatis Zoologicae Botanical Fennicae. "Vanamo"* **8**, 44–45.

LAMM, R. (1937). *Hereditas* **23**, 38–48.

LAMM, R. (1947). *Hereditas* **33**, 405–419.

LAMM, R. (1949). *Hereditas* **35**, 203–214.

LAMM, R. (1951). *Hereditas* **37**, 356–372.

LAMM, R. (1956). *Hereditas* **42**, 520–521.

LAMM, R. (1960). *Hereditas* **46**, 737–744.

LAMM, R. (1972). *Pisum Newsl.* **4**, 22.

LAMM, R. (1974). *Pisum Newsl.* **6**, 29.

LAMM, R. and MIRAVALLE, R. J. (1959). *Hereditas* **45**, 417–440.

LAMPRECHT, H. (1934). *Hereditas* **18**, 269–296.

LAMPRECHT, H. (1938). *Züchter* **10**, 150–157.

LAMPRECHT, H. (1939). *Hereditas* **25**, 459–471.

LAMPRECHT, H. (1948). *Agri Hort. Genet.* **6**, 10–48.

LAMPRECHT, H. (1952). *Agri Hort. Genet.* **10**, 1–18.

LAMPRECHT, H. (1953). *Agri Hort. Genet.* **11**, 40–54.

LAMPRECHT, H. (1955). *Agri Hort. Genet.* **13**, 103–114.

LAMPRECHT, H. (1956). *Agri Hort. Genet.* **14**, 1–4.

LAMPRECHT, H. (1958). *Agri Hort. Genet.* **16**, 38–48.

LAMPRECHT, H. (1960a). *Agri Hort. Genet.* **18**, 23–56.

LAMPRECHT, H. (1960b). *Agri Hort. Genet.* **18**, 135–168.

LAMPRECHT, H. (1964). *Agri Hort. Genet.* **22**, 243–255.

LAMPRECHT, H. (1966). "Die Entstehung der Arten und höheren Kategorien." Springer Verlag, Wien.
LAMPRECHT, H. (1968). "Die Grundlagen der Mendelschen Gesetze." Verlag Paul Parey, Berlin.
LINNAEUS, C. (1753). "Species Plantarum", Vol. II.
MARX, G. A. (1969). *Pisum Newsl.* **1**, 10–11.
MARX, G. A. (1971a). *Pisum Newsl.* **3**, 20.
MARX, G. A. (1971b). *Pisum Newsl.* **3**, 20–21.
MARX, G. A. (1972). *Pisum Newsl.* **4**, 28–29.
MARX, G. A. (1973). *Pisum Newsl.* **5**, 25–26.
MARX, G. A. (1974). *Pisum Newsl.* **6**, 60.
MARX, G. A. and DUCZMAL, K. (1972). *Pisum Newsl.* **4**, 32–33.
NEMEC, B. (1904). *Jb. wiss. Bot.* **39**, 645–730.
NILSSON, E. (1933). *Hereditas* **17**, 216–222.
NUTTAL, V. W. and LYALL, L. H. (1964). *J. Hered.* **55**, 184–186.
RASMUSSON, J. (1927). *Hereditas* **10**, 1–152.
ROSEN, G. VON. (1944). *Hereditas* **30**, 261–400.
SNOAD, B. (1966). *Genetica* **37**, 247–254.
SNOAD, B. (1969). *Pisum Newsl.* **1**, 19–20.
STEBBINS, G. L. (1974). "Flowering Plants: Evolution Above the Species Level." Belknap Press of Harvard University Press, Cambridge, Massachusetts.
STRASBURGER, E. (1907). *Jb. wiss. Bot.* **44**, 482–555.
TSCHERMAK, E. VON. (1912). *Z. Ind. Abst. Vererb.* **7**, 81–234.
VAVILOV, N.I. (1951). "The Origin, Variation, Immunity and Breeding of Cultivated Plants; Selected Writings." (Translated from the Russian by K. Starr Chester). *Chronica Botanica* **13**, 1–366.
WELLENSIEK, S. J. (1925). *Biblphia genet.* **2**, 343–476.
WHITE, O. E. (1916). *Am. Nat.* **50**, 530–547.
WHITE, O. E. (1917). *Proc. Am. phil. Soc.* **56**, 487–588.
WINGE, O. (1936). *C.r. Trav. Lab. Carlsberg Série physiologique* **21**, 271–293.
YARNELL, S. H. (1962). *Bot. Rev.* **28**, 465–537.
ZOHARY, D. (1970). *In* "Genetic Resources in Plants: Their Exploration and Conservation" (D. H. Frankel and E. Bennett, eds.), IBP Handbook No. 11, pp 33–42. F. A. Davis, Philadelphia.
ZOHARY, D. and HOPF, M. (1973). *Science, N.Y.* **182**, 887–894.

3.   Biochemistry of Germination and Seedling Growth

J. F. SUTCLIFFE
School of Biological Sciences, University of Sussex, England

J. A. BRYANT
Department of Botany, University College, Cardiff, Wales

I. INTRODUCTION

Peas, and especially *Pisum sativum*, have been widely used in studies of germination for a variety of reasons:

1. Peas are an important commercial crop and knowledge of the optimal conditions for germination and seedling establishment is essential for successful cultivation (see Chapter 4).

2. Mature seed of high viability is readily available in quantity at all times of the year in most parts of the world.

3. The ripe seed is quiescent rather than dormant and germinates rapidly and relatively uniformly when placed under favourable conditions. There are no special requirements such as exposure to light or fluctuating temperature and there is no hard seed coat to be removed.

4. The axis is sufficiently large for it to be removed relatively easily from the cotyledons at an early stage in germination and a small number of seeds contain sufficient material for biochemical investigation.

A minor disadvantage of pea seeds is the ease with which they become infected with fungi and bacteria because of leaching of organic materials through the testa (Chapter 4). This can be partially overcome by growing the seeds under carefully controlled moisture conditions, and in any case pea seeds are fairly easy to sterilize effectively (e.g. see Guardiola and Sutcliffe, 1971a; Robinson and Bryant, 1975a).

According to the broad definitions of Brown (1946) and Oota et al. (1953), germination is the heterotrophic stage of growth in which the seedling plant depends entirely on materials stored in the cotyledons or endosperm. This implies that if a seedling is grown in complete darkness it will never pass out of the germination phase. A different definition of germination variously expressed, for example, by Went (1961) and Mayer and Polyakoff-Mayber (1963), is that it is the process in which a quiescent seed becomes metabolically active and initiates the formation of a seedling from a dormant embryo. The first appearance of the radicle through the seed coat is chosen, in those seeds such as the pea in which the radicle emerges first, as the most convenient criterion of the transition from germination to early seedling growth. It is in this sense that the term germination is used in this chapter.

The events occurring during germination of lettuce seeds have been described by Evenari (1961) as a consecutive sequence of four separate phases. These are:

1. The imbibition phase
2. The activation phase
3. The mitotic phase
4. The protrusion phase

Similar stages have been distinguished in other seeds, e.g. by Went (1961) in mustard, tomato and oat, but it is difficult to separate stages 2–4 in peas. In what follows below only two phases are distinguished up to the time of radicle emergence.

The imbibition phase, characterized by a rapid and reversible uptake of water, occurs during the first few hours of soaking. As this phase declines the seed passes into the activation phase in which water uptake is low but there is increasing metabolic activity. During this stage there are dramatic ultrastructural changes (see below), RNA and protein synthesis probably begin and there is some increase in volume of the radicle leading to emergence. Growth of the axis, if defined as dry weight increase, does not occur until after emergence when there is a large increase in DNA and RNA contents of the axis. This is accompanied by rapid cell elongation leading to further radicle growth.

The post-germination stage of early seedling growth with which we shall also be concerned in this chapter involves continued growth of the axis, with concomitant development of photosynthetic activity, and increasing dependence on inorganic elements absorbed by the root system from the external medium. Mobilization of stored materials from cotyledons and their transport

o the axis also proceeds during this stage, which is taken to end with senescence and abscission of the cotyledons.

It should be mentioned that some investigators, working for the most part
n chemically orientated laboratories, have failed to separate the axis and the
cotyledons and have made analyses of the whole seed or seedling. It will
become apparent from the data discussed below that the usefulness of such
an approach is very limited because of the different physiological and biochemical states of the two parts at various stages of growth.

II. REHYDRATION OF THE DRY SEED

Mature, air-dry pea seeds have a water content of 10–15% which rises to
about 50% during the first 10 h of soaking. Kollöffel (1967) found that water
uptake in the imbibition phase was almost complete after about 20 h and
thereafter there was very little change in water content of the cotyledons up
to 70 h (cf. Fig. 3.5). During the initial phase of imbibition up to 6 h, McNair
(1966) found that dehydration of the axis and cotyledons had very little effect
on the subsequent ability of pea seeds (cv. 'Meteor') to germinate when water
was supplied again. However, between 6 h and 24 h after the start of imbibition the viability of the seed was progressively impaired by re-drying. The
response of excised axes to dehydration was found to be almost identical to
that of the axis attached to the intact seed. The transition from complete
desiccation resistance to maximum susceptibility in peas was found to correspond closely with the passage of the seed from quiescence to the initiation of
active growth. The development of susceptibility with rising moisture content in both intact seed and isolated axis is divisible into two apparently
separate phases. In the first phase, covering the initial stage of rapid hydration, the resistance of the seed to dehydration injury remains high. Subsequent
uptake of water to the hydration level at emergence causes a rapid decrease
in dehydration resistance, apparently in direct proportion to the increase in
hydration. During the early phase of imbibition most of the water taken up
is presumably used to rehydrate desiccated organic molecules, such as those
of nucleic acids and proteins. Later, free water begins to accumulate in cytoplasmic vacuoles and it is at this stage that the cells appear to become susceptible. The embryonic root and hypocotyl tissues are significantly more
susceptible to drying after 18 h imbibition than are the epicotyl and plumule,
perhaps because the cells of the former are at a more advanced stage of
development.

III. DEVELOPMENT OF RESPIRATORY ACTIVITY DURING IMBIBITION

The rate of gaseous exchange in air-dry pea seeds is extremely low (Smith and Gane, 1938). Oota (1957) and Czosnowski (1962) reported a significant increase in oxygen uptake 15 min after the start of imbibition. Köllöffel and Sluys (1970) showed that uptake of oxygen by pea cotyledons increased rapidly up to about 10 h and then more slowly until about 24 h, after which it increased rapidly again (Fig. 3.1). McNair (1966) found that during the

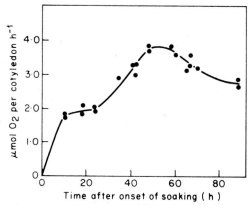

FIG. 3.1. Changes in oxygen uptake by pea cotyledons during germination and early seedling growth. (From Köllöffel and Sluys, 1970)

early stages of imbibition in cv. 'Meteor' there was a direct correlation between the level of hydration and respiratory exchange for all the seed tissues up to a moisture content of 52–53%. Mitochondrial fractions from both cotyledon and axis tissue showed increasing ability to oxidize succinate and malate during imbibition. In preparations from cotyledon tissue the increase continued only until approximately 12 h from the start of imbibition, but in particulate preparations from the axis the increase continued up to the time of radicle emergence. Since neither the total protein nitrogen content of the axis nor the protein nitrogen content of the mitochondrial fraction showed any net increase over this period it was deduced that the rise in response to added substrates represents a genuine increase in the specific activities of enzymes responsible for the oxidation of these substrates. Young et al. (1960) and Solomos et al. (1972) claimed that there was no significant increase in the amount of respiratory enzymes in peas during early germination. Köllöffel and Sluys (1970) found that during the first 24 h of germination the activity of succinate and malate oxidase increased whereas succinate and malate dehydrogenase activity hardly changed. There was, however, an increase in phosphorylation efficiency and in the degree of respiratory control, and it was concluded that the increase in respiratory capacity of mitochondria during the first day is attributable to an increase in the capacity of the electron transfer chain. Malholtra *et al.* (1973) concluded from feeding experiments with ^{3}H-thymidine that mitochondrial DNA is not synthesized in the early stage of germination and that development of mitochondrial structure is responsible for increased respiratory activity. Yoo (1970) observed dramatic ultrastructural changes in the later stages of imbibition including proliferation of endoplasmic reticulum and increased development of mitochondrial membranes. These changes precede any increase in mitochondrial numbers (Bain and Mercer, 1966; Nawa and Asahi, 1971) which contribute to the later increase in respiratory activity in both axis and cotyledon tissues. After emergence of the radicle, the rate of respiratory activity of the cotyledons increases for a time and then begins to decline, while that of the radicle continues to increase.

It is well known that before the testa splits, respiring pea seeds pass through a phase of partial anaerobiosis because of the low permeability of the seed-coat to oxygen (Fernandes, 1923; Spragg and Yemm, 1959). This period is characterized by a high respiratory quotient and by accumulation of ethanol (Cossins and Turner, 1962, 1963; Leblova *et al.*, 1969) and lactic acid (Cossins, 1964) in the tissues. Normally, the anaerobic phase is transitory and is terminated when the testa is ruptured by emergence of the radicle.

The enzyme systems involved in fermentation in pea cotyledons have been studied extensively, e.g. by Suzuki (1966), Cossins *et al.* (1968) and Ericksson (1968). Alcohol dehydrogenase (ADH) activity of cotyledon extracts normally begins to decline soon after emergence of the radicle (Davison, 1949; Goksoyr *et al.*, 1953; Cossins and Turner, 1959; Cossins *et al.*, 1968) but it remains high when seeds are allowed to germinate under anaerobic conditions (Kollöffel, 1968). Kollöffel found that extracts from air-dry seeds contain appreciable ADH activity and he concluded that this enzyme is synthesized during maturation of the seed and reactivated upon hydration. Homogenates of cotyledons from peas which have been growing for several days contain substances, possibly long-chain fatty acids, which inactivate the enzyme (Kollöffel, 1970b).

IV. METABOLISM OF NUCLEIC ACIDS AND PROTEINS IN THE EMBRYONIC AXIS

Examination of the growth and metabolism of the embryonic axis during germination and early seedling development reveals three distinct phases. The first phase is one of water uptake and initiation of metabolic activity and corresponds to the imbibition phase mentioned above. This leads to Phase II which is characterized by a greatly increased rate of metabolism.

During this phase, as already indicated, there are dramatic ultrastructural changes: the endoplasmic reticulum proliferates, the Golgi bodies and mitochondria show increased development of their membrane systems and it is also likely that there is some mitochondrial division (Yoo, 1970). The plastids, in contrast, show very little change during this period. There is very little change in the fresh weight of the axis or in nucleic acid content during Phase II (Fig. 3.2) but nevertheless the cells in the radicle elongate sufficiently to

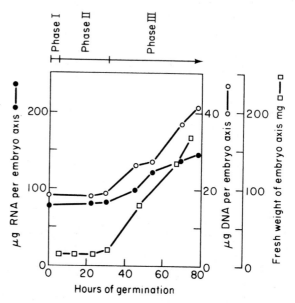

FIG. 3.2. Changes in the fresh weight and nucleic acid content of the embryonic axis during germination and early seedling growth. Data obtained for cv. 'Feltham First' grown at 22°C. (From Robinson and Bryant, 1975a)

cause it to emerge from the testa. This causes a further increase in respiration possibly because rupture of the testa allows a greater access of oxygen to cells of the axis. The transition from Phase II to Phase III after about 30 h from the onset of germination is marked by a dramatic increase in the fresh weight of the axis and by the start of massive net synthesis of both DNA and RNA (Fig. 3.2). The almost simultaneous onset of net synthesis of both nucleic acids at about the same stage in germination has been observed in other legumes including *Vicia faba* (Jakob and Bovey, 1969) and *P. arvense* (Chow, 1975).

The beginning of rapid net synthesis of nucleic acids is paralleled by a marked increase in the ability of the cells to incorporate radioactive precursors into nucleic acids. This is shown particularly well by the data for incorporation of ^{3}H-thymidine into DNA given in Table 3.I. The amount of precursor ncorporated into DNA during Phase II is low and may represent synthesis of mitochondrial DNA since mitochondrial development occurs during this phase (Yoo, 1970). It is also possible that the incorporation represents DNA turnover which has been shown to occur in roots of 5-day-old pea seedlings (Bryant *et al.*, 1974) and in roots of *V. faba* (MacLeod, 1972).

TABLE 3.I. Incorporation of radioactive precursors into nucleic acids of the embryonic axis during germination. Data were obtained with cv. 'Feltham First' germinated at 22°C. (From Robinson and Bryant, 1975a)

Time (h)	Incorporation axis^{-1} h^{-1} (d.p.m. $\times$ 10^{-2})	
	^{3}H-thymidine (DNA)	^{3}H-uridine (RNA)
4–8	16	189
21–25	44	144
29–33	199	299
45–49	1811	810

Analysis of labelled RNA by polyacrylamide gel-electrophoresis (Robinson and Bryant, 1975a) and by affinity chromatography indicate that rRNA, tRNA and poly-(A)-linked RNA (i.e. mRNA—see Grierson, 1976) all become labelled both in Phase II and Phase III. However, the relative proportion of the total radioactivity which is incorporated into poly-(A)-linked RNA is greater during Phase II than Phase III (Fig. 3.3). This suggests that during the period prior to the onset of massive net synthesis of RNA the synthesis of mRNA is relatively more rapid than that of other types of RNA, while in Phase III the situation is reversed.

The incorporation of precursors into RNA during Phase II may result from one or more of the following processes:

1. Syntheses of mitochondrial RNA: mitochondrial RNA species have very similar molecular weights to their cytoplasmic counterparts (Leaver and Harmey, 1973) and the two populations of RNA would not have been separated by the polyacrylamide gel-electrophoresis under the conditions employed. It is, however, extremely unlikely that only mitochondial RNA is synthesized in Phase II, as mitochondrial nucleic acids make up only 1% or less of the total cellular nucleic acids (Leaver and Harmey, 1973).

2. Turnover of RNA: the occurrence of RNA turnover in plants is a well established phenomenon (Trewavas, 1970; Grierson, 1976) and involves all types of cytoplasmic RNA.

3. It is also possible that the incorporation of precursors into RNA represents net RNA synthesis but at a rate which is too low to be detected by the methods used for estimation of total nucleic acid content. Net synthesis of mRNA, for example, can never be detected by measurement of total RNA content, and in view of the results obtained by affinity chromatography

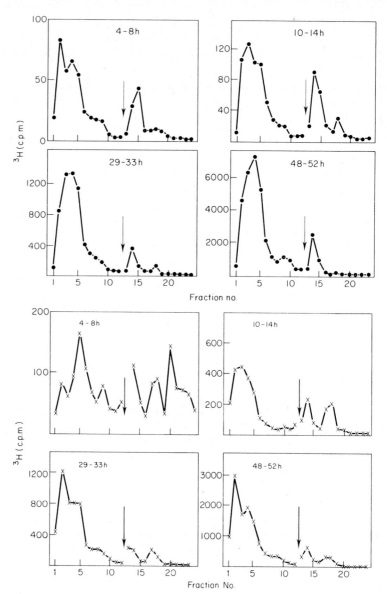

FIG. 3.3. Distribution of incorporated radioactivity in RNA, fractionated according to poly-(A) content. Seeds of cv. 'Feltham First' were germinated at 22°C and at the times indicated, samples of seeds were supplied with ^{3}H-adenine (above) or ^{3}H-uridine (below). RNA was extracted from the embryonic axis and applied to a column of poly-(U) sepharose. The column was washed with 700 mM NaCl (buffered) under conditions favouring formation of base pairs between complementary strands. Under these conditions RNA not containing poly-(A) was eluted and collected in fractions of 1·0 ml. The column was then washed with buffer containing no NaCl (arrow) causing elution of the poly-(A) containing RNA, which was collected in fractions of 1·0 ml. The radioactivity in each fraction was determined. (Unpublished results of S. C. Greenway and J. A. Bryant)

(mentioned above) it is likely that net synthesis of RNA does occur in this period.

A study of changes in the rate of incorporation of ^{3}H-leucine into protein in the embryonic axis of cv. 'Feltham First' has also indicated that there is a transition about 30 h after the onset of germination from a phase of low protein synthesis (Phase II) to one of high (Phase III; see Table 3.II). It is

TABLE 3.II. Incorporation of ^{3}H-leucine into protein in the embryonic axis during germination. Data were obtained with cv. 'Feltham First' germinated at 22°C. (S. C. Greenway and J. A. Bryant, unpublished)

Time (h)	Incorporation axis^{-1} h^{-1} (d.p.m. $\times$ 10^{-2})
4–8	14
10–14	15
29–33	232
31–35	266
48–52	666

likely that both RNA and protein synthesis begin at an early stage in germination, albeit at a low rate, but it is difficult at present to be sure which starts first. There is certainly no clear evidence for the existence of preformed mRNA which is able to initiate protein synthesis immediately after hydration, as has been reported for wheat embryos (Weeks and Marcus, 1970). On the other hand there is no definite indication that synthesis of RNA precedes protein synthesis. Similar results to those discussed here for *P. sativum* have been obtained in Bray's laboratory (Chow, 1975) in studies on *P. arvense*. Current work there, and also by one of us (J.A.B.), is aimed at examining more fully the inter-relationship between RNA and protein synthesis in the two species.

Although the rates of RNA and protein synthesis are low during Phase II it seems very likely that these processes are important for subsequent growth of the axis in Phase III. For example, the activities of both the chromatin-bound and soluble DNA polymerases increase markedly during Phase II (Fig. 3.4). Both types of enzyme have been extensively studied and characterized (Robinson and Bryant, 1975b; Stevens *et al.*, 1976; Stevens and Bryant, unpublished). On the basis of their properties and by comparison with mammalian DNA polymerases (Keir and Craig, 1973), it seems very likely that the soluble DNA polymerase is the enzyme involved in DNA replication. Support for this view comes from the finding that in roots of 5-day-old pea seedlings the activity of the soluble enzyme is higher in the meristematic cells, whilst the activity of the chromatin-bound enzyme is higher in the differentiating and differentated cells (Jenns and Bryant, unpublished). Thus it seems probable that the increase in activity of soluble DNA polymerase during Phase II is necessary for the onset of DNA replication. In wheat it has been shown that the increase in activity of soluble DNA poly-

c

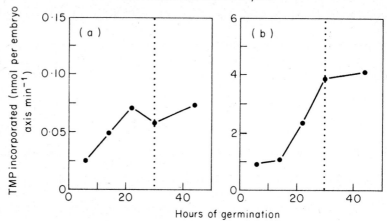

FIG. 3.4. Activity of (a) chromatin-bound DNA polymerase and (b) soluble DNA polymerase in the embryonic axis during germination and early seedling growth. The dotted vertical lines indicate the timing of the onset of net DNA synthesis. Data obtained for cv. 'Feltham First' grown at 22°C. (From Robinson and Bryant, 1975b)

merase is dependent upon protein synthesis, and further that inhibition of protein synthesis prevents the onset of DNA replication (Mory *et al.*, 1975). It would be interesting to investigate the situation in peas.

Whereas the DNA polymerases are synthesized during Phase II, certain other enzymes involved in nucleic acid metabolism increase in activity later. Thymidine kinase and uridine kinase, for example, increase in activity at the transition from Phase II to Phase III (Bryant, 1976). It is suggested that this argues against their being involved in the regulation of nucleic acid biosynthesis (cf. Schwarz and Fites, 1970). It is more likely that the increase in activity of the kinases is related to the export from the cotyledons to the axis of the nucleosides derived from breakdown of nucleic acids (see Section V below).

The rapid growth initiated in Phase II of germination leads to establishment of the root system and emergence of the shoot apex. Above ground, the plumular hook straightens and the leaves expand and become photosynthetic. The process of greening is biochemically complex and involves a close integration of gene expression in nuclear and chloroplast genomes. The plastids show marked changes in ultrastructure with the development of organized lamellae from the poorly defined membrane system of the proplastids. There is also a marked increase in size of the plastids as they develop into fully functional chloroplasts. These structural changes are paralleled by increases in the activities of enzymes of the Calvin cycle and of RNA polymerase, by an increased rate of synthesis of chloroplast RNA and DNA, and by a doubling in the number of chloroplast ribosomes (Ellis and Hartley, 1971).

Some of the components of the chloroplast are coded for by the chloroplast genome and synthesized within the chloroplast during greening. Such components include various chloroplast RNAs, the large sub-unit of Fraction I protein (i.e. ribulosediphosphate carboxylase), at least five membrane-bound proteins (probably structural proteins of the thylakoid membranes) and at least one protein of the chloroplast 70 s ribosome (Ellis and Hartley, 1971, 1974; Ellis, 1975). At present the mechanisms governing the integration of the nucleo-cytoplasmic system with the chloroplast during greening are not understood. Ellis (1975) has put forward a hypothesis for integration of the synthesis of the two sub-units of Fraction I protein. He suggests that the smaller sub-unit, after synthesis in the cytoplasm, is transported into the chloroplast where it acts as an initiation factor either for the transcription of the chloroplast gene coding for the large sub-unit, or for the translation of its mRNA on chloroplast ribosomes.

V. METABOLISM OF COTYLEDONS

Although the embryonic axis is the part destined to grow into the mature plant, much less is known about its metabolism during germination than is known about that of the cotyledons. This is probably because of the small size of the axis relative to the cotyledons which makes it somewhat more difficult to study. In the initial stages of germination the dry weight of the axis is only some 1·5% of that of the cotyledon pair in a seed. Because of the large bulk of the cotyledons relative to the axis, analyses of whole seeds after removal of the testa reflects closely those of the cotyledons. The cotyledons form the major site of storage of reserves in a pea seed. The major storage materials are starch (40–50%), proteins (3–5%) and phytin (mainly Ca and Mg salts of myo-inositol-hexaphosphate), together with small amounts of lipid. Soluble nucleotides and bases are also stored and in addition it seems reasonable to regard the RNA and DNA in cotyledons as food reserves since they are degraded to nucleotides, nucleosides and bases which are transported to the axis and utilized during germination. The elemental composition of a batch of ungerminated pea seeds (cv. 'Alaska') is given in Table 3.III.

Germination in peas is hypogeal and when depleted of nutrients the cotyledons shrivel and die. If they are exposed to light early in germination they undergo some greening and have a limited capacity for photosynthesis (see Chapter 12), but this is of little, if any, significance in the nutrition of the axis.

As with the embryonic axis, the metabolic changes occurring during germination and early seedling growth can be divided into three phases. However, the phases are not so clearly defined as in the case of the axis and transition from one to another does not coincide in time with change of

TABLE 3.III. The composition of ungerminated *P. sativum* seeds. 100 seeds (cv. 'Alaska') were soaked for 4 h in water distilled in glass and then separated into axis, cotyledons and testa. Results are expressed as mg per part or per whole seed. (From Guardiola, 1970)

Material	Whole Seed	Axis	Cotyledon Pair	Testa
Dry weight	174·6	1·97	158·8	13·8
Nitrogen	6·62	0·146	6·39	0·081
Phosphorus	0·817	0·023	0·786	0·008
Potassium	1·90	0·025	1·82	0·052
Sulphur	0·407	0·0079	0·392	0·006
Calcium	0·227	0·0022	0·134	0·091
Sodium	0·016	$8·7 \times 10^{-5}$	0·014	0·002

phase in the axis. During Phase I, which lasts for 20–24 h from the beginning of soaking there is rapid uptake of water and a dramatic increase in respiration (Kollöffel, 1967; Larson, 1968; Guardiola and Sutcliffe, 1971a; see Fig. 3.5). These increases appear to be the result of hydration of the cells and do not seem to depend on protein synthesis or any specific activation mechanism (Nawa and Asahi, 1971, 1973). However, there are significant changes in the ultrastructure of cotyledon cells during this period. Swift and Buttrose (1973) have shown by freeze-etching and thin-sectioning techniques that the protein bodies in air-dry tissues are coated with a convoluted structure which is apparently rich in lipid (see Fig. 15.4b). After brief periods of hydration the complex appears to expand and is transformed into a series of tubules; after 24 h imbibition it is no longer associated with the protein bodies. The surfaces of amyloplasts and plasmalemmae in the dry tissue were also found to be covered with lipid material which is dispersed in the early stages of germination. Only the nuclear envelopes appeared to be free of a lipid coat in the dry seed. Phase II begins about a day after the onset of soaking and during this period, which may last for several days, water content remains fairly constant and oxygen uptake continues to increase (Fig. 3.5). During Phase II the stored materials in the cotyledons begin to be degraded and as the levels of starch, protein N and phytic acid P fall, there is a concomitant increase in soluble carbohydrates, amino acids and amides and inorganic phosphate in the tissues. There is also a proliferation of cellular membrane systems, particularly of endoplasmic reticulum and Golgi bodies (Bain and Mercer, 1966). The mitochondria show increased internal organization and may also increase in number (Bain and Mercer, 1966; Nawa and Asahi, 1971), whereas by contrast, little structural change occurs in the proplastids. The cells are active in RNA synthesis and as in the embryonic axis, the synthesis involves all types of RNA (Chin *et al.*, 1972). The ribosome population becomes organized into polysomes although there is no apparent increase in the total number of ribosomes. Concomitant with the formation of polysomes there is a marked

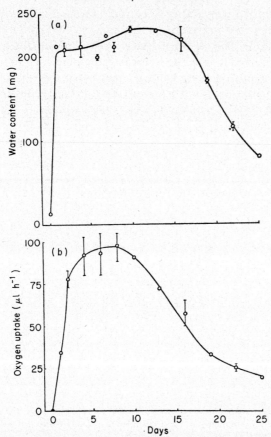

FIG. 3.5. Changes in (a) water content and (b) oxygen uptake in cotyledons of intact pea seedlings (cv. 'Alaska') grown for 16h in the light at 22°C and for 8h in the dark at 18°C in each daily cycle. The results are for three separate experiments and are expressed on a per seedling (2 cotyledons + testa) basis. (From Guardiola and Sutcliffe, 1971a)

increase in the rate of protein synthesis as measured either by the rate of incorporation *in vivo* of exogenously supplied labelled leucine or by the ability of ribosomal preparations to incorporate leucine into protein *in vitro* (Chin *et al.*, 1972). This increase in protein synthetic activity occurs against a background of net degradation of protein (presumably largely storage protein) (Chin *et al.*, 1972; Fig. 3.6). As in the case of the embryonic axis, it is difficult to decide whether or not the formation of polysomes and hence protein synthesis depends on the prior synthesis of mRNA. There is again no definitive evidence for the existence in pea cotyledons of latent mRNA which has been demonstrated in cotton seed cotyledons (Ihle and Dure, 1972). However, Gordon and Payne (1976) have detected latent mRNA in whole pea seeds, and it is likely that much of this is located in the cotyledons. During maturation of the seed prior to dehydration, polysome formation in the cotyledons is progressively inhibited, apparently by the non-availability of mRNA (Poulson and Beevers, 1973). Thus it seems likely that following rehydration, mRNA must be synthesized in order to instigate polysome formation and protein synthesis. It remains possible, however, that latent mRNA is present in the cotyledons of mature seeds but is unavailable for translation until rehydration occurs. Further work is required to clarify the situation.

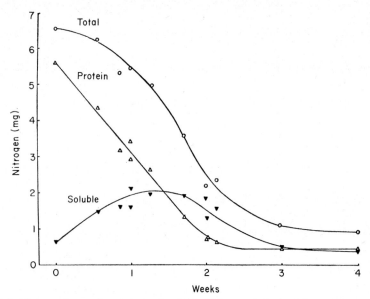

FIG. 3.6. Changes in total protein and soluble N in cotyledons of intact pea seedlings (cv. 'Alaska') grown as in Fig. 3.5. The results are expressed on a per seedling basis. (From Guardiola and Sutcliffe, 1971a)

In Phase III, catabolic processes predominate and during this stage the cotyledons become senescent. The transition from Phase II is difficult to define precisely but appears to occur about 5 days after soaking in cvs 'Victory Freezer' (Bain and Mercer, 1966), 'Alaska' (Fig. 3.5) and 'Feltham First' (Fig. 3.7), grown at 20–25°C. The phase is characterized by a decreasing rate of oxygen uptake (Fig. 3.5) and a progressive disorganization of the membrane systems, mitochondria and proplastids (Bain and Mercer, 1966). Extensive net degradation of protein, RNA (Fig. 3.7) and eventually of DNA is a characteristic feature of this phase. The nucleotides released from nucleic acids are, in part at least, degraded further to nucleosides and bases which are exported to the embryonic axis (Ross *et al.*, 1970; Silver and Gilmore, 1969). These degradative processes lead eventually to the death of the cotyledons some 20–35 days after the onset of germination.

Despite the extensive breakdown of macromolecules which occurs during

Phase III, there is still some synthesis of RNA and protein. Polysomes active in protein synthesis have been extracted from cotyledons of 10-day-old seedlings (Swain and Dekker, 1969) and RNA synthesis has been detected in cotyledons of 7-day-old seedlings (Ross *et al.*, 1970). Thus it seems likely that cessation of macromolecular syntheses occurs very late in cotyledon senescence. Support for this view comes from the finding that many enzymes show dramatic increases in activity late in Phase III (see below). These observations underline the conclusions made from findings on other senescent organs (e.g. detached leaves), namely that senescence, like other developmental processes, is a controlled sequence of reactions involving regulation of gene expression.

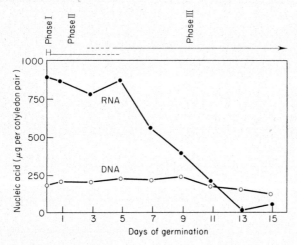

FIG. 3.7. Changes in the nucleic acid content of the cotyledons of cv. 'Feltham First' during germination at 22°C. (From Bryant *et al.*, 1976a)

One of the most widely investigated features of cotyledon metabolism is the behaviour of hydrolytic enzymes. A large number of hydrolases, including α- and β-amylase (Mayer and Shain, 1968; Yomo and Varner, 1973), amylopectin-1, 6-glucosidase (Mayer and Shain, 1968), acid phosphatase (Johnson *et al.*, 1973; Varner *et al.*, 1963), ATPase (Varner *et al.*, 1963), acid and alkaline ribonuclease and deoxyribonuclease (Bryant *et al.*, 1976a) and acid and alkaline protease (Beevers, 1968; Guardiola and Sutcliffe, 1971b) show dramatic increases in activity during Phase III (Fig. 3.8). In most cases there is a peak or plateau of activity followed by a decline in the later stages of cotyledon senescence. An exception to the pattern occurs in the case of acid ribonuclease (Fig. 3.8a) which has two peaks of activity, one at the transition from Phase II to Phase III and the other later on in Phase III. An almost identical pattern of development of acid ribonuclease activity has been found in cotyledons of *P. arvense* (Barker *et al.*, 1974).

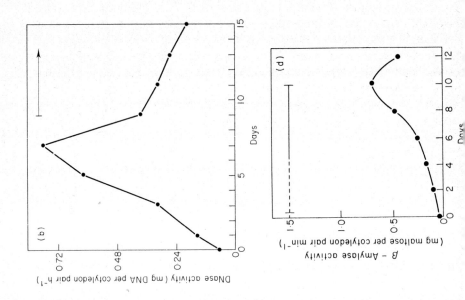

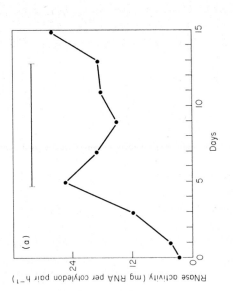

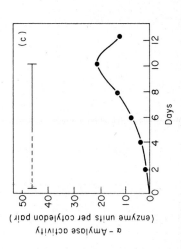

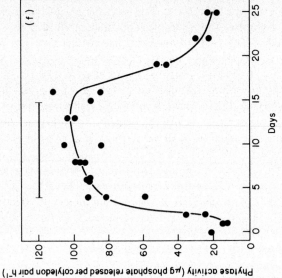

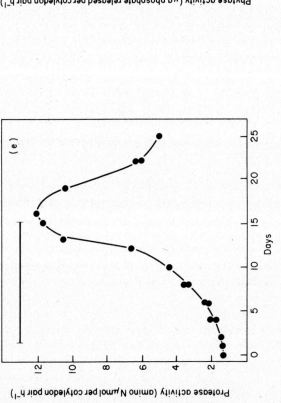

FIG. 3.8. Changes in the activities of hydrolytic enzymes in the cotyledons during germination. These include: (a) acid ribonuclease; (b) soluble deoxyribonuclease, both in cv. 'Feltham First' grown at 22°C (from Bryant *et al.*, 1976a); (c) α-amylase; (d) β-amylase, both in cv. 'Alaska' grown at 22°C (from Yomo and Varner, 1973); (e) protease and (f) phytase, both in cv. 'Alaska' grown at 20°C (from Guardiola and Sutcliffe, 1971 a, b). The horizontal bars indicate the periods of most rapid net degradation of substrate.

With the exception of phytase (Guardiola and Sutcliffe, 1971b), the peaks in enzyme activity show very little correlation with the periods of most rapid net degradation of substrate. For example, the most rapid decline in RNA content in the cotyledons of cvs 'Feltham First' and 'Meteor' occurs between days 5 and 13 while ribonuclease activity shows a peak at day 4 in cv. 'Meteor' and at day 5 in cv. 'Feltham First', with a second peak at day 15 or later in both cultivars. A difficulty in interpreting such observations is that measurements of total activity in tissue extracts may not reflect activity *in vivo*. There is obviously a very real possibility that much of the enzyme assayed *in vitro* is inactive *in vivo*, perhaps through being located in lysosomes.

A further difficulty arises through the existence of isoenzymes of a number of hydrolases. In germinating pea seeds, only the isoenzymes of acid phosphatase (Johnson *et al.*, 1973) and acid ribonuclease (Bryant *et al.*, 1976) have been investigated in any detail. The increase in activity of acid ribonuclease between day 2 and day 5 involves two isoenzymes, I and II. Both have molecular weights of around 17 000 but the two may be separated by chromatography on columns of carboxymethyl-cellulose (Fig. 3.9). During the decline in activity which occurs between day 5 and day 9, Isoenzyme II disappears. As activity increases again from day 9 onwards, two more isoenzymes appear. One has similar chromatographic properties to Isoenzyme II and is termed Isoenzyme IIa; the other is Isoenzyme III. The changes in total enzyme activity therefore reflect changes in a number of different isoenzymes. However, even when the activities of the different isoenzymes are measured it is still difficult to correlate them with the patterns of RNA metabolism.

The occurrence of protein synthesis throughout the period in which increases in enzyme activity occur is of course consistent with the view that at least some of the increases in activity are mediated by *de novo* synthesis. However, this has not been unequivocally demonstrated for any of the enzymes in question. The increase in activity of a number of hydrolases is known to be inhibited by cycloheximide. These include acid phosphatase (Johnson *et al.*, 1973) and Isoenzyme II of acid ribonuclease (Bryant *et al.*, 1976). Both α- and β-amylases have been shown to become labelled when seedlings are supplied with ^{35}S-sulphate (which enters protein as ^{35}S-methionine) (Mayer and Shain, 1968) and acid phosphatase incorporates deuterium from exogenously supplied deuterium oxide (Johnson *et al.*, 1973). These data are a clear indication that the increases in activity are dependent on protein synthesis. However, in the absence of any data on enzyme turnover rates it cannot be stated definitely that *de novo* enzyme synthesis is involved in the observed increases in enzyme activity (see Trewavas, 1976, for detailed discussion of this point).

On the other hand there is clear evidence that the increases in activity of certain hydrolases are not mediated by *de novo* synthesis. The increase in

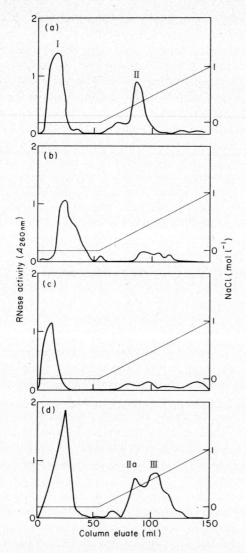

Fig. 3.9. Isoenzymes of acid ribonuclease in cotyledons of cv. 'Feltham First' grown at 22°C. Acid ribonuclease was partially purified by differential precipitation with $(NH_4)_2SO_4$ and then applied to columns of carboxymethyl-cellulose. The columns were washed with acetate buffer (pH 5·4) and then with a linear gradient of 0 to 1·0 M NaCl in acetate buffer. Fractions of the column eluate were assayed for ribonuclease: (a) after 5 days of germination, (b) after 5 days germination in the presence of cycloheximide, (c) after 9 days of germination and (d) after 15 days of germination (cf. Fig. 5a). (From Bryant *et al.*, 1976b)

activity of amylopectin-1, 6-glucosidase, for example, is not accompanied by incorporation of exogenously supplied ^{35}S-sulphate and is caused by a specific proteolytic cleavage of an inactive enzyme precursor (Mayer and Shain, 1968). The increase in activity of Isoenzyme I of acid ribonuclease in pea cotyledons between days 2 and 5 is not inhibited by cycloheximide at concentrations which cause an 80% inhibition of the incorporation of leucine into cotyledon protein and which completely prevent growth of the embryonic axis (Fig. 3.9). Similarly in *P. arvense*, acid ribonuclease does not incorporate deuterium from deuterium oxide, although under the same conditions other proteins become density labelled (Barker *et al.*, 1974). Thus far, there is no indication of the nature of the mechanism regulating ribonuclease activity in either *P. sativum* or *P. arvense*. However, it seems likely that the increases in the activities of different hydrolases are brought about in different ways, some perhaps by *de novo* synthesis and others by the operation of post-translational mechanisms. The pea cotyledon is an excellent system for study of the control of gene expression at different levels.

VI. CONTROL OF COTYLEDON METABOLISM BY THE EMBRYONIC AXIS

In the intact seedling, the sequence of changes in cotyledon metabolism is correlated with metabolic changes in the axis but the relationship between the timing of the various phases in the two parts appears to be quite complex. Phase I (the "hydration" phase) is of similar duration in both axis and cotyledons, but Phase II is much more prolonged in the cotyledons and does not end until Phase III (the phase of rapid growth) is well under way in the axis. The main significance of Phase III in the cotyledons seems to lie in the release of stored materials in an orderly fashion for use in growth of the axis.

Treatments such as the feeding of additional nutrients to the roots, etiolation, or the application of gibberellic acid (all of which enhance axis growth), lead to a speeding-up of the changes in cotyledon metabolism, with more rapid depletion of nutrients and earlier senescence (Guardiola, 1970; Guardiola and Sutcliffe, 1972). Removal of the axis during germination prolongs Phase II in the cotyledons and delays senescence (Varner *et al.*, 1963; Guardiola and Sutcliffe, 1971a). The degradation of starch (Sprent, 1968) and protein (Chin *et al.*, 1972; Yomo and Varner, 1973) is greatly reduced (Fig. 3.10).

Varner *et al.* (1963) concluded that changes in metabolic activity in the cotyledons are programmed by stimuli arriving from the growing axis and they suggested that a gibberellin might be involved as in germinating barley seeds (Paleg, 1960). However, it has not yet been possible to replace the influence of the axis by supplying known growth-regulating substances (Guardiola and Sutcliffe, 1972) or by adding extracts of pea axis tissue to the incuba-

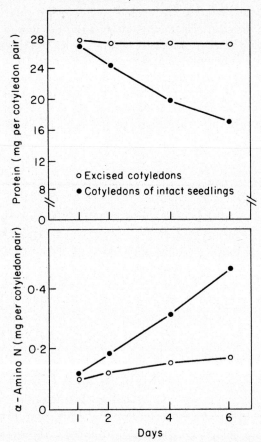

FIG. 3.10. Effect of removal of the axis (after 3 h) on degradation of protein in pea cotyledons, Data obtained for cv. 'Burpeeana' germinated at 20°C. (From Chin *et al.* 1972)

tion medium (Yomo and Varner, 1973). Yomo and Varner (loc. cit.) showed that when pea seedlings were germinated in vermiculite soaked in 1% casein hydrolysate, after day 8 there was only about half as much protease activity in the treated cotyledons as there was in the cotyledons of seedlings fed with water. Correlated with this decrease in activity, they found that only about half as much nitrogen had left the treated cotyledons during this period compared with the control. From their observations they concluded that, although a hormonal control mechanism is not totally ruled out, regulation of protease activity is most likely to be by feed-back inhibition of enzyme synthesis (or activation?) by accumulation of soluble nitrogen compounds resulting from protease activity. In support of this conclusion they reported that when one-half of each cotyledon was removed from a seedling growing

under normal conditions, the free amino acid content of the attached pieces
(i.e. attached to the axis) decreased to about a third of that in control cotyledons and protease activity increased by 1·5–2·0×. Guardiola and Sutcliffe
(1971a) reached a different conclusion from a detailed study of changes in
total protein- and soluble-N levels and of protease activities in excised cotyledons of cv. 'Alaska'. Cotyledons excised at day 4 showed a rapid reduction in
proteolytic activity but the soluble-N content did not rise significantly above
that of the control for a further 8 days (Fig. 3.11). This observation, coupled

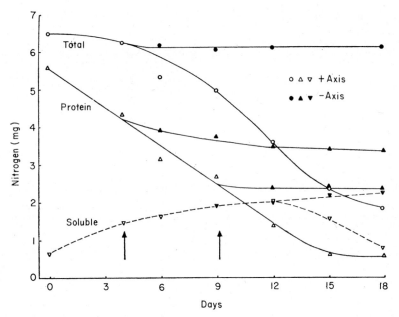

FIG. 3.11. Changes in the N fractions in cotyledons of light-grown pea seedlings
(cv. 'Alaska') excised on day 4 (closed signs) compared with those in cotyledons of
intact seedlings. The values for protein-N content of cotyledons excised on day 9 have
also been plotted. The results are expressed on a per seedling basis. (From Guardiola
and Sutcliffe, 1971a)

with the fact that the soluble-N content is increasing in whole cotyledons of
intact seedlings at the time protease activity is rising, suggests that soluble-N
content does not regulate protease synthesis in the manner suggested by
Yomo and Varner. Rather, Guardiola and Sutcliffe favoured control by
a growth regulator mechanism, despite their inability to demonstrate it
directly.

There is similar uncertainty about the control of phytase activity in pea
cotyledons. The possibility that accumulation of inorganic phosphate inhibits
phytase activity in wheat endosperm has been put forward by Sartirana and
Bianchetti (1967). However, the fact that phytase activity increases rapidly

in pea cotyledons up to day 5 of germination while the concentration of inorganic phosphate is increasing (Fig. 3.12), but does not begin to fall again until the level of inorganic phosphate has decreased appreciably suggests that feed-back control is not the operative mechanism.

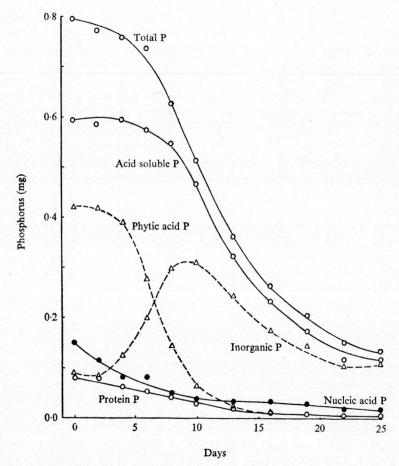

FIG. 3.12. Changes in the levels of various P fractions in cotyledons of intact pea seedlings (cv. 'Alaska') grown as in Fig. 3.5. The results are expressed on a per seedling basis. (From Guardiola and Sutcliffe, 1971b)

It is undoubtedly dangerous to transfer the hormonal-control concept derived from work on cereal grains to dicotyledonous seeds without substantial supporting evidence, because of differences in structure and physiology between the scutellum/aleurone/endosperm system and the cotyledon. Cotyledon cells are capable of appreciable growth in terms both of increased volume and protein synthesis when the axis is removed, even though their ability to do so declines more rapidly than in the intact seedling (see below). It seems likely that they are able to synthesize for themselves, if only to a limited extent, a variety of growth regulators. The developing seeds in cultures of isolated fruits have been shown to synthesize both gibberellins (Balder *et al.*, 1965) and cytokinins (Hahn *et al.*, 1974). Application of gibberellic acid has been found to stimulate axis growth for a short time in decotylized seedlings and so cotyledons may be a source of gibberellin for the young seedling (Shininger, 1972). As the development of hydrolytic activity is mainly associated with the senescence phase it seems most likely that a hormone which controls senescence will also regulate the activity of these enzymes. It is of interest in this connection that Ihle and Dure (1972) found that the increase in protease activity observed to occur in germinating excised embryos of immature cotton seeds was inhibited by abscisic acid (ABA). Yomo and Varner (1973) reported that ABA at concentrations as low as 5×10^{-4} mM inhibits the development of amylase activity in cotyledons excised from 6-day-old pea plants. ABA at a concentration of 10^{-2} mM which inhibited amylase synthesis by 90%, did not inhibit O_2 uptake or ^{14}C-leucine incorporation into protein by the cotyledons, and so may have a quite specific role in controlling amylase activity, at least in the early stages of cotyledon senescence. There was no inhibition by ABA of amylase formation in cotyledons excised after day 6.

The effect of axis removal on the development of enzyme activities in excised pea cotyledons depends very much on the conditions under which the cotyledons are incubated. Varner *et al.* (1963) reported that excised cotyledons incubated on moist filter paper in Petri-dishes failed to show any increase in α-amylase, acid phosphatase or ATPase activity. However, when cultured in Erlenmeyer flasks which have a much greater internal air space than Petri-dishes, both α- and β-amylases showed greater increases than in the intact seedling, and of the enzymes investigated only the development of protease activity was prevented under these conditions. Similarly, Bryant and Haczycki (1976) have found that excised pea cotyledons incubated in Petri-dishes fail to show the normal development of acid phosphatase and acid ribonuclease activities, whereas excised cotyledons maintained in large culture vessels show greater increases in both these enzymes than do cotyledons of intact seedlings (Fig. 3.13). Since it has also been shown that incubation of seeds in Petri-dishes inhibits the onset of net synthesis of RNA and DNA in the axis and delays growth (Robinson, 1975), results obtained with cotyledons incubated in Petri-dishes must be interpreted with caution. The apparently abnormal behaviour of cotyledons incubated in a confined space suggests that a metabolically active volatile substance produced by the cotyledon cells may be involved. The role of ethylene as a regulator of growth processes, including senescence, is becoming increasingly evident (Pratt and Goeschl, 1969).

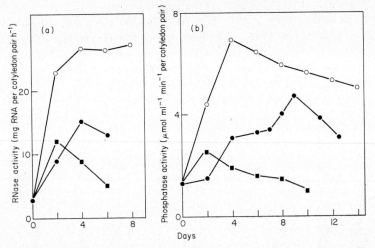

FIG. 3.13. Effect of culture conditions on the development of (a) acid ribonuclease and (b) acid phosphatase in excised cotyledons. Pea seeds (cv. 'Meteor') were imbibed at 1°C. Axes were then removed and the excised cotyledons were incubated either in Petri dishes (■—■—■—■) or in large glass culture vessels (○—○—○—○) at 22°C (a) or 25°C (b). Ribonuclease and phosphatase were assayed at intervals in the excised cotyledons and in the cotyledons of intact seedlings, imbibed at 1°C and then grown at 22° or 25°C (●—●—●—●) (From Bryant and Haczycki, 1976)

From experiments in which cotyledons were incubated in large vessels it appears that removal of the axis has different effects on the development of different enzyme activities. With some enzymes, notably proteases, removal of the axis during the first day of germination completely prevents the increase in activity (Yomo and Varner, 1973), and even when axis removal is delayed until protease activity has begun to appear, further increase in activity of the enzyme is reduced (Guardiola and Sutcliffe, 1971a). In contrast to this, many other enzymes, including α- and β-amylases (Yomo and Varner, 1973), acid phosphatase and acid ribonuclease (Fig. 3.13) show a greater rise in activity in excised cotyledons than in the cotyledons of intact seedlings. Further, the activity of acid ribonuclease in excised cotyledons fails to show the decline between days 4–5 and 9 that is a characteristic of normal cotyledon development. Thus, with regard to the activities of this group of enzymes, the axis has a moderating rather than a stimulating influence. It is interesting that this group of enzymes includes some, e.g. α- and β-amylase and acid phosphatase, which may be synthesized *de novo* in pea cotyledons and one (acid ribonuclease) which certainly is not. This suggests that the restraining influence of the axis on enzyme activity may be exerted at the levels both of protein synthesis and activation of preformed protein.

The moderating influence of the axis may continue even after enzyme activity has reached its maximum. In seedlings of cv. 'Alaska', grown at

20°C, phytase activity exhibits a broad plateau extending from about day 4 to day 16, after which activity falls back to the level observed in the early stages of germination (Fig. 3.14). Removal of the axis at day 4 prevents this later decrease. The simplest explanation of this is that the decrease in phytase activity (and perhaps the activities of other hydrolases) is brought about by proteolytic degradation, and that the reduction of protease activity caused by removal of the axis inhibits enzyme breakdown.

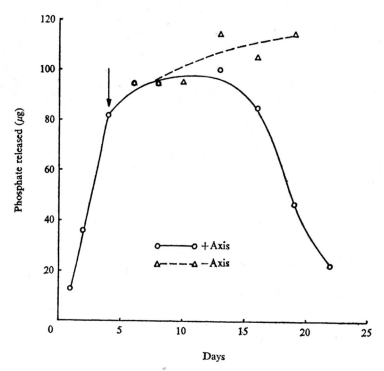

FIG. 3.14. Effect of axis removal at 4 days (arrow) on the subsequent decline of phytase activity in cotyledons of cv. 'Alaska' maintained at 20°C. (From Guardiola and Sutcliffe, 1971b)

Study of nucleic acid and protein metabolism in excised cotyledons indicates that the onset of RNA synthesis, polysome formation and protein synthesis is not affected by axis removal (Chin et al., 1972). In excised cotyledons the number of polysomes and rates of RNA and protein synthesis decline much more rapidly than in cotyledons of intact seedlings (Fig. 3.15), but it must be emphasized that, even in excised cotyledons, protein and RNA synthesis can be detected at least as late as 6 days after the onset of hydration. Such observations indicate that the processes of RNA and protein synthesis in cotyledons cannot be wholly dependent on the embryonic axis.

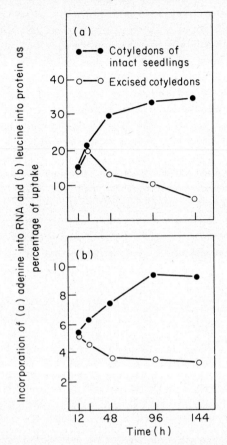

FIG. 3.15. The effect of axis removal (at 3 h) on (a) RNA and (b) protein synthesis in cotyledons of cv. 'Burpeeana' germinated at 20°C (From Chin *et al.* 1972)

The relationship between the reduced rate of RNA and protein synthesis, and the greater than normal increase in activities of certain enzymes in excised cotyledons, is a matter of speculation, but it may be significant that protease activity is reduced. It is clear that integration of cotyledonary metabolism with growth and metabolic activity of the axis is complex and there is considerable scope for further investigation.

VII. CONTROL OF THE TRANSPORT OF MATERIALS FROM THE COTYLEDONS

As has been mentioned above, there is no clear correlation between the activity of various hydrolases and the movement of soluble products into the

axis. It has been shown (Figs 3.11 and 3.12) that the levels of soluble N and
inorganic P increase in pea cotyledons up to about day 9 and this is clearly
due to the more rapid hydrolysis of stored protein and phytin than of transport. After day 9, when growth of the axis is still proceeding at an increasing
rate and the rate of hydrolysis of stored materials is beginning to decline,
the levels of soluble N and inorganic P in the cotyledons start to fall. It is
evidently not the amount of *in vitro* enzyme activity (whether protease or
phytase) or even the level of *in vivo* hydrolysis that regulates the rate at which
materials are transported into the axis: control must lie, at least partly, in the
axis itself.

It has been shown (Sutcliffe, 1962) that movement of potassium ions from
the cotyledons into the developing shoot and root of pea seedlings grown in
10^{-1} mM $CaCl_2$ solution is closely related to the growth of each organ (Fig.
3.16). Relatively more K moved into the root than into the shoot during the
first week after germination when the root was growing rapidly, but when
growth of the shoot got under way in the second week, the situation was

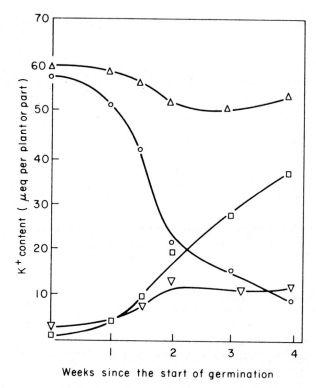

FIG. 3.16. K content of cotyledons ($\bigcirc$), shoots ($\square$), roots ($\triangledown$) and whole plants ($\triangle$)
of *P. sativum* during growth for 4 weeks at 20°C in the dark in 10^{-1} mM $CaCl_2$ solution.
(From Sutcliffe, 1962)

reversed. Removal of the shoot after a few days of growth caused an immediate reduction in the rate of depletion of the cotyledons, but the amount of K transported to the roots was unaffected. This was taken to indicate that in the early stages of growth there is little or no competition between shoot and root sinks for the available K. This conclusion was supported by the observation of Guardiola (1973) that removal of the shoot from 7-day-old seedlings did not affect movement of K into the root during a subsequent 3 week period (Table 3.IV). Fresh weight and dry matter content of the roots

TABLE 3.IV. The influence of the shoot on growth and accumulation of mineral elements in the roots of pea plants (cv. 'Alaska') grown without an external source of nutrient. Plants were de-shooted when 1 week old. Growth parameters of the root are expressed as a percentage of those from intact plants. Values marked with an asterisk(*) differ significantly (1% level) from the untreated control plants. (From Guardiola, 1973)

| | Weeks after excision of the shoot | | |
	1	2	3
Fresh weight	77*	79*	77*
Dry weight	85*	76*	71*
Protein nitrogen	96	96	94
Total nitrogen	93	124*	126*
Phosphorus	97	133*	130*
Sulphur	101	112*	139*
Potassium	94	100	100

of de-shooted plants was lower than in intact plants, but protein-N content was not affected, suggesting that protein synthesis rather than volume or dry weight increment might be the factor regulating the rate of K movement from the cotyledons to the root. It is assumed that the bulk, if not all, of the material moving out of cotyledons does so in the phloem as movement in xylem sap is likely to be slow and in the opposite direction. When seedlings are grown in a $CaCl_2$ solution, the amount of K in the xylem sap is likely to be low and this coupled with the low rate of transpiration of cotyledons means that the amount of K transported into the cotyledons is probably negligible (Collins and Sutcliffe, 1977). Therefore the net change in K content can be taken to be equivalent to the total amount exported in the phloem.

The contents of total N, P and S in the roots were significantly higher 2–3 weeks after excision of the shoot than in roots of intact plants kept under the same conditions (Table 3.IV). These increases are an indication either of increased movement of these elements out of the cotyledons into the roots of de-shooted plants, perhaps because of reduced competition, or of secondary redistribution from root to shoot via xylem or phloem in the intact plant.

Guardiola (1970) found that treatments, such as the supply of additional nutrients to the roots, which stimulated growth of the axis, also increased the movement of materials from the cotyledons. When plants were kept in the dark to stimulate shoot growth relative to that of the roots, a larger proportion of the total material transported moved to the shoot (Table 3.V), but the amount moving into the roots was unaffected. Differential stimulation of shoot growth by gibberellic acid in a dwarf variety (cv. 'Progress') had the same effects (Garcia-Luis and Guardiola, 1975).

When seedlings of cv. 'Alaska' are grown in a $CaCl_2$ solution at 20–25°C

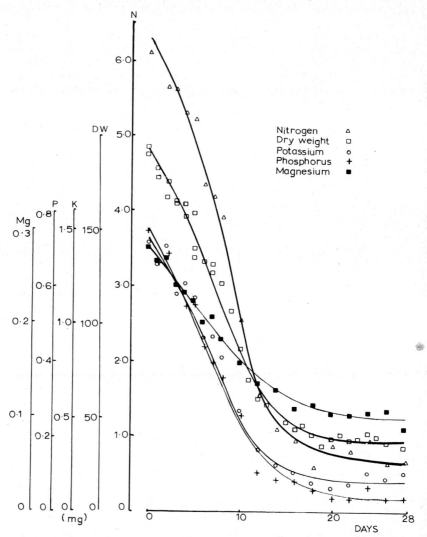

FIG. 3.17. Changes in the amounts of various elements and dry matter in the cotyledons of young pea seedlings (cv. 'Alaska') during growth for 28 days in 16 h of light at 22°C and 8 h of darkness at 18°C in each 24 h cycle. (From Collins and Sutcliffe, 1977.)

TABLE 3.V. Transport of mineral elements from pea cotyledons (cv. 'Alaska') into the shoot during 4 weeks of growth. Data expressed as percentages of the total amount transported from the cotyledons recovered in the shoot. (From Guardiola, 1973)

Element	Age of the seedlings (weeks)			
	1	2	3	4
Dark-grown Seedlings				
Nitrogen	63·3	72·6	83·3	88·8
Phosphorus	62·2	71·9	84·1	88·6
Potassium	56·1	69·5	77·7	81·4
Sulphur	65·3	84·3	91·0	93·3
Light-grown Seedlings				
Nitrogen	43·4	57·8	73·9	73·1
Phosphorus	44·2	54·6	73·6	78·4
Potassium	31·3	45·8	62·3	62·6
Sulphur	40·0	66·0	78·9	79·6

the cotyledons become almost depleted in dry matter and various mineral elements after 3–4 weeks (Fig. 3.17). The small amount of dry matter that remains is mainly cellulose and much of this and the residual amounts of mineral elements are located in the testa. When the amount of a particular element remaining in the cotyledons at a given time is plotted against dry matter content, a linear relationship is observed for each element (Fig. 3.18). This relationship is not affected by changes in growth rate of the axis caused by experimental treatments, even when the quantities of materials exported from the cotyledons vary greatly. Guardiola and Sutcliffe (1972) thought that there might be some deviation from linearity during the first week of growth but this has been disproved by the more recent data presented in Fig. 3.18 which clearly shows that the relationship is linear for each of the elements studied throughout the whole period of seedling growth.

When the amounts of each element exported in a given time are expressed as percentages of the total exported in 4 weeks and plotted against the amounts of dry matter exported in the same time, again expressed as a percentage of the total exported, all the points fall on the same straight line (Fig. 3.19). It is evident that the various elements are being exported during seedling growth in constant proportions relative to one another and to total dry matter, and when 50% of the available dry matter has been exported the supply of each of the elements examined has been depleted to the same extent. Thus although the quantity of materials in the cotyledons gradually decreases during growth and there are profound changes in composition as a result of enzyme activity, the ratio of N:P:K:Mg:Na:dry matter does not alter appreciably from the time of germination until the cotyledons are depleted. It is concluded that in the pea, food materials are stored in the cotyledons

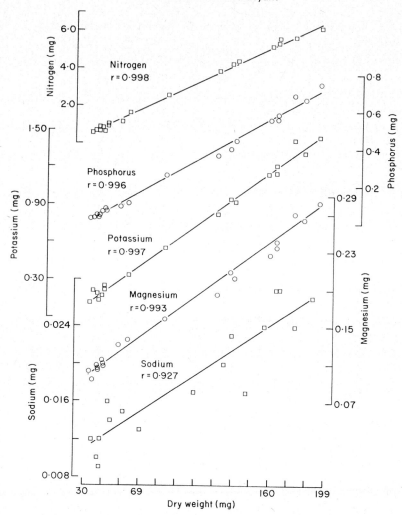

FIG. 3.18. Amounts of various elements present in the cotyledons after various times plotted against cotyledon dry weight. (From Collins and Sutcliffe, 1977)

during seed development in approximately the proportions in which they are required in subsequent growth of the seedling, and they are transported from the cotyledons in these proportions throughout early seedling growth.

This is in marked contrast to the situation in the *Avena* grain in which Baset and Sutcliffe (1975) showed that there is a non-linear relationship between export of various elements and dry matter from the endosperm to the growing axis over a period of 7 days. During this period the reserves become almost depleted, and the relationship with dry matter is found to differ for each element. Sutcliffe (1976a,b) has interpreted the results of Baset, Guard-

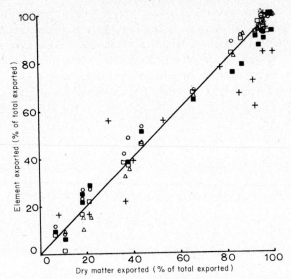

Phosphorus (o), Potassium (□), Magnesium (■), Sodium (+),
Nitrogen (△).

FIG. 3.19. Amounts of various elements exported from the cotyledons after various times plotted against the amounts of dry matter exported in the same time. (Data from Fig. 3.17; amounts exported are expressed as percentages of the total exported in 4 weeks)

iola and others in terms of what has been called the solute potential gradient hypothesis. The hypothesis is based on the assumption that solutes move from sources to sinks through sieve tubes in response to gradients of electro-chemical potential ("solute potential") between the solutions bathing the free space at the two ends of the system. Transport occurs from a place where the concentration of a solute is high to one where it is less high and each solute moves independently in response to the existing gradient (cf. Mason *et al.*, 1936). The amount of an individual solute transported in unit time will be determined by the solute potential difference and the resistance presented by the connecting sieve tubes.

Collins and Sutcliffe (1977) have suggested that the same basic mechanism operates in peas but that solute potentials at the source (i.e. in the cotyledons) are controlled not by hydrolase activity or substrate concentration as in oat endosperm, but by the rate of release of soluble products from the storage cells to the free space. Individual materials, whether sugars, soluble-N compounds, phosphate, K or Mg are evidently released through the plasma membranes in constant proportions, although at a changing rate during early axis growth, and this is coupled with a constancy of the proportions in which these same materials are removed from the free space at the sink. Support for the view that movement of each solute is regulated independently even though the ratios remain constant comes from experiments in which the concentration of one element in the free space is altered artificially and that for the others remains constant. Collins has found that when KCl is injected into the free space of pea cotyledons, movement of K from the cotyledons is stimulated and the linear relationship between K content and dry matter content is disturbed (Fig. 3.20). The experiment was carried out with

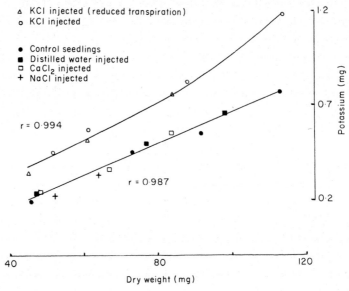

FIG. 3.20. Effect of injecting various solutions into pea cotyledons on the relationship between the export of K and dry matter to the axis. (O. D. G. Collins, unpublished)

plants under normal conditions, and also with plants in an atmosphere of high humidity in which transpiration was reduced, with the same result. This makes it unlikely that a significant amount of the K was being exported in the xylem. Injection of solutions of other salts or of distilled water into the cotyledons did not disturb the linear relationship between K and dry matter export, although, as might be expected, the injection of water into the free space lowered the rate of export of all elements temporarily, presumably by diluting the free space solution.

If the hypothesis is correct it should be possible to reduce the transport of a particular element into a leaf on the growing axis by increasing the concentration of this element in the free space at the sink. Preliminary experiments (Collins, 1976, unpublished) have indicated that when KCl solutions are injected into the free space of a leaf at a stage when it is importing K through the phloem, the transport of K is slowed down and may even be reversed.

The transport of Ca from pea cotyledons is a matter of some interest because of the widely held view that Ca is immobile in the phloem (e.g. see

Canny, 1973). Guardiola and Sutcliffe (1972) found that 26% of the total Ca content was exported from pea cotyledons during a 4-week growth period compared with 72% of dry matter and 82% of K. It is not likely that there is any transport of Ca from cotyledons in the xylem because, as already indicated, the flow of xylem sap is presumably in the other direction. Experiments in which shoots of 9-day-old pea seedlings were steam-girdled showed that movement of Ca as well as of other elements from the cotyledons was severely inhibited and it was concluded that Ca is transported in the phloem.

Attempts to increase the amount of Ca mobilized from pea cotyledons have been unsuccessful. Ferguson and Bollard (1976) found that [45]Ca injected into the cotyledons was not transported to the axis and concluded that only Ca associated with phytin is available for transport. It appears that Ca released from phytin is protected in some way from being bound at cytoplasmic or cell wall sites and is thus released into the free space where it is taken up by the sieve tubes. It seems unlikely that Ca is transported in ionic form; Cosgrove (1966) suggested that Ca may be released from phytin as soluble undissociated oligophosphates or it is possibly transported in the phloem in a chelated form.

When pea seedlings are grown in a $CaCl_2$ solution, Ca taken up by the roots is transported, presumably in the xylem, into the cotyledons. This makes it difficult to study the relationship between the transport of available Ca and dry matter from cotyledons in the way that has been done for other elements. Collins (1975, unpublished) has shown that although pea plants do not grow as well in distilled water as in a $CaCl_2$ solution they will survive for about 2 weeks and during this time the relationship between Ca and dry matter export is linear as is shown in Fig. 3.21. In contrast to Ferguson and Bollard (1976),

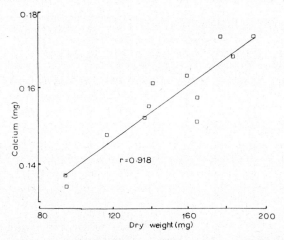

FIG. 3.21. Amounts of Ca present in pea cotyledons at various times during growth of seedlings for 2 weeks in distilled water, plotted against the dry weight of the cotyledons. (O. D. G. Collins, unpublished)

Collins has found that injection of $CaCl_2$ into cotyledons can lead to increased Ca export if sufficient Ca is supplied to saturate binding sites in the cotyledon tissue. This preliminary evidence suggests that Ca transport from cotyledons and perhaps also from leaves is limited by the non-availability of a fraction of the Ca through binding to sites in the cell walls and elsewhere rather than by immobility in the phloem.

REFERENCES

BAIN, J. M. and MERCER, F. V. (1966). *Aust. J. biol. Sci.* **19**, 69–84.
BALDER, B., LANG, A. and AGATEP, A. O. (1965). *Science, N.Y.* **147**, 155–157.
BARKER, G. R., BRAY, C. M. and WALTER, T. J. (1974). *Biochem. J.* **142**, 211–219.
BASET, Q. A. and SUTCLIFFE, J. F. (1975). *Ann. Bot.* **39**, 31–41.
BEEVERS, L. (1968). *Phytochemistry* **7**, 1837–1844.
BROWN, R. (1946). *Ann. Bot.* **10**, 73–84.
BRYANT, J. A. (1976). *Ann. Bot.* **40**, 1317–1320.
BRYANT, J. A. and HACZYKI, S. J. (1976). *New Phytol.* **77**, 757–760.
BRYANT, J. A., GREENWAY, S. C. and WEST, G. A. (1976a). *Planta* **130**, 137–140.
BRYANT, J. A., GREENWAY, S. C. and WEST, G. A. (1976b). *Planta* **130**, 141–144.
BRYANT, J. A., WILDON, D. C. and WONG, D. (1974). *Planta,* **118**, 17–24.
CANNY, M. J. (1973). "Phloem Translocation". Cambridge University Press, Cambridge, U.K.
CHIN, T. Y., POULSON, R. and BEEVERS, L. (1972). *Pl. Physiol.* **49**, 482–489.
CHOW, T. Y. (1975). Ph.D. Thesis, University of Manchester, Manchester, U.K.
COLLINS, O. D. G. and SUTCLIFFE, J. F. (1977). *Ann. Bot.* **41**, 163–171.
COSGROVE, D. J. (1966). *Rev. pure appl. Chem.* **16**, 209.
COSSINS, E. A. (1964). *Nature, Lond.* **203**, 989–990.
COSSINS, E. A., KOPALA, L. C., BLAWACKY and SPRANK, O. M. (1968). *Phytochemistry,* **7**, 1125–1134.
COSSINS, E. A. and TURNER, E. R. (1959). *Nature, Lond.* **183**, 1599–1600.
COSSINS, E. A. and TURNER, E. A. (1962). *Ann. Bot.* **26**, 591–597.
COSSINS, E. A. and TURNER, E. R. (1963). *J. exp. Bot.* **14**, 290–298.
CZOSNOWSKI, J. (1962). *Acta Soc. bot. pol.* **31**, 135–146.
DAVISON, D. C. (1949). *Proc. Linn. Soc. N.S.W.* **74**, 28–36.
ELLIS, R. J. (1975). *Phytochemistry* **14**, 89–93.
ELLIS, R. J. and HARTLEY, M. R. (1971). *Nature, New Biology* **233**, 193–196.
ELLIS, R. J. and HARTLEY, M. R. (1974). *In* "Nucleic Acids" (K. Burton, ed.), Medical and Technical Publishing Co., Lancaster, U.K.
ERIKSSON, C. E. (1968). *J. Fd. Sci.* **33**, 525–532.
EVENARI, M. (1961). *Proc. int. Seed Test. Ass.* **26**, 597.
FERGUSON, I. M. and BOLLARD, E. G. (1976). *Ann. Bot.* **40**, 1047–1055.
FERNANDES, D. S. (1923). *Recl Trav. bot. néerl.* **20**, 107–156.
GARCIA-LUIS, A. and GUARDIOLA, J. L. (1975). *Ann. Bot.* **39**, 325–330.
GOKSOYR, J., BOERI, E. and BONNICHSEN, R. K. (1953). *Acta chem. scand.* **7**, 657–662.
GORDON, M. E. and PAYNE, P. I. (1976). *Planta* **130**, 209–273.
GRIERSON, D. (1976). *In* "Molecular Aspects of Gene Expression in Plants" (J. A. Bryant, ed.). Academic Press, London and New York.
GUARDIOLA, J. L. (1970). D.Phil. Thesis, University of Sussex, Brighton, U.K.
GUARDIOLA, J. L. (1973). *Acta bot. neerl.* **22**, 55–68.

GUARDIOLA, J. L. and SUTCLIFFE, J. F. (1971a). *Ann. Bot.* **35**, 791–807.
GUARDIOLA, J. L. and SUTCLIFFE, J. F. (1971b). *Ann. Bot.* **35**, 809–823.
GUARDIOLA, J. L. and SUTCLIFFE, J. F. (1972). *J. exp. Bot.* **23**, 322–337.
HAHN, H., DE ZACKS, R. and KENDE, H. (1974). *Naturwissenschaften* **61**, 170.
IHLE, J. N. and DURE, L. S. (1972). *J. biol. Chem.* **247**, 5048–5055.
JAKOB, K. M. and BOVEY, F. (1969). *Expl Cell Res.* **54**, 118–126.
JOHNSON, C. B., HOLLOWAY, B. R., SMITH, H. and GRIERSON, D. (1973). *Planta* **115**, 1–10.
KEIR, H. M. and CRAIG, R. K. (1973). *Biochem. Soc. Trans.* **1**, 1073–1077.
KOLLÖFFEL, C. (1967). *Acta bot. neerl.* **16**, 111–122.
KOLLÖFFEL, C. (1968). *Acta bot. neerl.* **17**, 70–77.
KOLLÖFFEL, C. (1970a). *Planta,* **91**, 321–328.
KOLLÖFFEL, C. (1970b). *Acta bot. neerl.* **19**, 539–545.
KOLLÖFFEL, C. and SLUYS, J. V. (1970). *Acta bot. neerl.* **19**, 503–508.
LARSON, L. A. (1968). *Pl. Physiol.* **43**, 255–259.
LEAVER, C. J. and HARMEY, M. A. (1973). *Biochem. Soc. Symp.* **38**, 175–194.
LEBLOVA, S., ZIMAKOVA, I., SOFRAVA, D. and BARTHOVA, J. (1969). *Biologia Pl.* **11**, 417–423.
MACLEOD, R. D. (1972). *Z. Pflanzenphysiol.* **68**, 379–381.
MAHOLTRA, S. S., SOLOMOS, T. and SPENCER, M. (1973). *Planta,* **114**, 169–184.
MASON, T. G., MASKELL, E. J. and PHILLIS, E. (1936). *Ann. Bot.* **50**, 23–58.
MAYER, A. M. and POLYAKOFF-MAYBER, A. (1963). "The Germination of Seeds". Pergamon Press, Oxford.
MAYER, A. M. and SHAIN, Y. (1968). *Science, N.Y.* **162**, 1283–1284.
McNAIR, D. J. (1966). Ph.D. Thesis, University of Edinburgh, Edinburgh, U.K.
MORY, Y. Y., CHEN, D. and SARID, S. (1975). *Pl. Physiol.* **55**, 437–442.
NAWA, Y. and ASAHI, T. (1971). *Pl. Physiol.* **48**, 671–674.
NAWA, Y. and ASAHI, Y. (1973). *Pl. Cell Physiol, Tokyo* **14**, 607–610.
OOTA, Y. (1957). *Physiologia Pl.* **10**, 910.
OOTA, Y., FUJII, R. and OSAWA, S. (1953). *J. Biochem., Tokyo* **40**, 649.
PALEG, L. G. (1960). *Pl. Physiol.* **35**, 293–299.
POULSON, R. and BEEVERS, L. (1973). *Biochim. biophys. Acta* **308**, 381–389.
PRATT, H. D. and GOESCHL, J. D. (1969). *A. Rev. Pl. Physiol.* **20**, 541–584.
ROBINSON, N. E. and BRYANT, J. A. (1975a). *Planta* **127**, 63–68.
ROBINSON, N. E. and BRYANT, J. A. (1975b). *Planta* **127**, 69–75.
ROSS, C., CODDINGTON, R. L., MURRAY, M. G. and BLEDSOE, C. S. (1970). *Pl. Physiol.* **47**, 71–75.
SARTIRANA, M. L. and BIANCHETTI, R. (1967). *Physiologia Pl.* **20**, 1066–1075.
SCHWARZ, O. J. and FITES, R. C. (1970). *Phytochemistry* **9**, 1899–1905.
SHININGER, T. L. (1972). *Pl. Physiol.* **99**, 341–344.
SILVER, A. V. and GILMORE, V. (1969). *Phytochemistry* **8**, 2295–2299.
SMITH, A. J. M. and GANE, R. (1938). *Rep. Fd Invest. Bd* (for 1937), 231.
SOLOMOS, T., MALHOTRA, S. S., PRASAD, S., MALHOTRA, S. K. and SPENCER, M. (1972). *Can. J. Biochem. Physiol.* **50**, 725–732.
SPRAGG, S. P. and YEMM, E. W. (1959). *J. exp. Bot.* **10**, 409–425.
SPRENT, J. I. (1968). *Planta* **80**, 299–301.
STEVENS, C., JENNS, S. M. and BRYANT, J. A. (1975). *Biochem. Soc. Trans.* **3**, 1126–1128.
SUTCLIFFE, J. F. (1962). "Mineral Salts Absorption in Plants". Pergamon Press, Oxford.

Sutcliffe, J. F. (1976a). *In* "Perspectives in Experimental Biology" (N. Sunderland, ed.), Vol. II, pp 433–444. Pergamon Press, Oxford.
Sutcliffe, J. F. (1976b). In "Encyclopaedia of Plant Physiology" (U. Luttge and M. G. Pitman, eds.), Vol. II B, pp. 394–417. Springer-Verlag, Heidelberg.
Suzuki, Y. (1966). *Phytochemistry* **5**, 761–765.
Swain, R. R. and Dekker, E. E. (1969). *Pl. Physiol.* **44**, 319–325.
Swift, J. G. and Buttrose, M. S. (1973). *Planta* **109**, 61–72.
Trewavas, A. J. (1970). *Pl. Physiol.* **45**, 742–751.
Trewavas, A. J. (1976). *In* "Molecular Aspects of Gene Expression in Plants" (J. A. Bryant, ed.). Academic Press, London, U.K.
Varner, J. E., Balce, L. V. and Huang, R. C. (1963). *Pl. Physiol.* **38**, 89–92.
Weeks, D. P. and Marcus, A. (1970). *Biochim. biophys. Acta* **232**, 671–684.
Went, F. W. (1961). *Proc. int. Seed Test. Ass.* **26**, 674.
Yomo, M. and Varner, J. E. (1973). *Pl. Physiol.* **51**, 708–713.
Yoo, B. Y. (1970). *J. Cell Biol.* **45**, 158–171.
Young, J. L., Huang, R. C., Vanecko, S., Marks, J. D. and Varner, J. E. (1960). *Pl. Physiol.* **35**, 288–299.

4. Field Emergence and Seedling Establishment

S. MATTHEWS

Department of Biology, University of Stirling, Scotland

I. INTRODUCTION

The field emergence and seedling establishment of any crop is the result of successful germination and early seedling growth in conditions that are variable, fluctuate and are seldom ideal for growth. Emergence in peas is usually considered to have taken place when the plumule appears above ground. Seedling establishment has never been clearly defined as a stage of growth and will be loosely interpreted to include any phase in which the term seedling is used.

For convenience, the development of the pea from a sown seed to an established seedling will be divided into four stages: imbibition, germination (using the physiologist's definition—the appearance of the radicle; see Chapter 3), pre-emergence seedling growth and post-emergence seedling growth. Pea seeds could fail to produce established seedlings through the unsuccessful completion of any one of these stages. The considerable literature on the field emergence and establishment of pea seeds will be examined with these stages in mind.

Most crop seeds that are sown in Europe (including the Soviet Union), North America and Australasia have been subjected to some form of seed testing. Seed testing has been used on a large scale in the developed countries since the 1920s, having begun in Saxony in 1869 (Justice, 1965). The major objective in seed testing is the determination of the suitability of a seed lot for sowing (McKay, 1972), a seed lot being the term used in seed testing and the seed trade to denote a uniform batch of seeds within one cultivar from which representative samples are drawn for testing. Seed lots may differ in their place of production and their subsequent treatment. There is often an upper limit to the size of seed lots stipulated by law, in the case of peas in the U.K. and the European Economic Community, 20 tonnes is the maximum (Anon., 1973).

The laboratory germination test, which determines the percentage of seeds which can produce seedlings capable of continued development into mature plants when germinated under standardized conditions of substrate, moisture supply and temperature (Wellington, 1966), is the most important laboratory test. As far as the seed trade is concerned the germination test is the touchstone of seed quality, its main feature being reproducibility. There is a minimum germination expected of most crop seeds in the seed trade, for example, in the current regulations for the U.K. which are in line with those governing the European Economic Community, the minimum standard for peas is 80% (Anon., 1973). Before 1973 the seed regulations operating in the U.K. (Anon., 1961) discouraged the use of seed lots of peas with laboratory germination of less than 80% without actually prohibiting their sale.

Most of the work done by agronomists and seed testers on the emergence and establishment of peas has been done on seed that has been subjected to a laboratory germination test. Thus, unless otherwise stated, in any work on pea seed emergence it is reasonable to assume that the germination of the seed material under optimal conditions is at least 80%. An actual laboratory germination figure for each seed lot under investigation is often given and these are usually in excess of 80%. Therefore, as far as the substance of this chapter is concerned any problems associated with dormancy or the inability to germinate under any circumstances are not under consideration, even though they would obviously influence emergence. This chapter deals with the problems of emergence and establishment of viable pea seeds in the field.

II. REPORTS ON EMERGENCE AND SEEDLING ESTABLISHMENT

Reports on the emergence and establishment of peas have for the most part been on emergence and have arisen from the results of two types of inquiry: the examination by seed testers of the relationship between laboratory germination and field emergence, and investigations into the effect of chemical seed dressings on emergence and associated micro-organisms by agronomists and plant pathologists. Both types of inquiry began over 50 years ago and provide a valuable background to the more physiological studies of the last 10 years.

A. The Relation between Laboratory Germination and Field Emergence

Although seed testers have been and still are devotees of the laboratory germination test their aim has always been to provide information about the probable emergence of seeds in the field. For most crop species the correlation between laboratory germination and field emergence appears good, despite highly variable sowing conditions (Essenburg and Schoorel, 1962). However, in peas there can be considerable discrepancies between the laboratory germination of a seed lot and its emergence in the field. Seed testers have generated considerable information about the field emergence of peas in evaluations of both normal laboratory germination tests and other tests as indicators of field emergence.

As early as 1903, Hiltner (1903) drew attention to complaints about the poor emergence of peas, particularly in wet conditions. Eastham (1925) referred to the relation between the laboratory germination of peas and their field emergence as an important problem in England. In Norway, peas were commonly found to exhibit poor emergence (Rasmusson, 1926), and in Denmark, Stahl (1936) found that field emergence varied greatly between seed lots, soils and sowing occasions. Gadd (1936) working in Sweden examined the relation between field emergence and laboratory tests, including a germination test, over a period of 5 years beginning in 1931. Typical findings were those of 1933 when among 15 lots of which laboratory germinations were in excess of 80% the field emergence ranged from 28–86% (Gadd, 1936). In North America, Munn (1926) had also reported large discrepancies between laboratory germination and field emergence. More detailed comparisons

D

were made later in the U.S.A. by Clark and Little (1954, 1955) who found that when samples from 114 seed lots with laboratory germinations in excess of 80% were sown in the field in 1954, the range in emergence was from 31–85%. This trial included seven cultivars all of which showed some discrepancy between laboratory germination and emergence. Many field experiments in the U.K. since 1964 have confirmed these discrepancies between the laboratory and the field (Perry, 1967; Matthews and Bradnock, 1967; Matthews and Whitbread, 1968; Perry, 1970; Bradnock and Matthews, 1970).

B. The Role of Micro-organisms

In some of the early reports on discrepancies between laboratory germination and emergence, reference is made to soil infection as a cause of emergence failure (Hiltner, 1903; Gadd, 1936). This observation accorded with the emergence improvement that was almost invariably achieved by dressing seed with fungicidal chemicals starting with copper compounds (Horsfall et al., 1934) and organomercurials (Clayton, 1927; Jones, 1931; Baylis et al., 1943), and more recently with the purely organic compound, captan (N-(trichloromethylthio)-cyclohex-4-ene-1, 2-dicarboximide) (Jacks, 1959) which is currently the most extensively used dressing for pea seeds.

Although seed-borne fungi such as Ascochyta pisi, A. pinodella and Mycosphaerella pinoides reduce emergence of peas and are controlled by fungicidal dressings (Jones, 1927; Wallen and Skoko, 1950), most investigations have pointed to the soil as the source of fungal infection. Pythium spp., a facultative parasite ubiquitous in soils, have been isolated from peas that have failed to emerge in the field in many countries including the U.S.A. (Haenseler, 1925), the U.K. (Baylis, 1941; Matthews and Whitbread, 1968), Australia (Kerling, 1947; Flentje, 1964), Mexico (Ewing, 1959), New Zealand (Jacks, 1963) and Canada (Harper, 1964). Kerling (1947) isolated P. debaryanum, P. polymorpha and an isolate thought to be either P. debaryanum or P. ultimum. Baylis (1941) isolated both P. ultimum and P. debaryanum from infected peas and found that both could cause pre-emergence mortality when introduced into sterilized soil. Flentje (1964) found P. ultimum to be the main associated species as did Ewing (1959) and Matthews and Whitbread (1968). Fusarium spp. have been isolated in a few cases (Flentje, 1964; Harper, 1964) but for the most part the important species linked with pre-emergence mortality in peas have been P. ultimum and P. debaryanum, particularly the former.

Baylis (1941) was the first to examine the aetiology of pre-emergence mortality in peas in any detail. He isolated Pythium spp. between 1 and 3 weeks after sowing from embryo axes excised from seeds that had failed to germinate; some isolations were also made from germinated but unemerged seedlings. Flentje (1964) observed that infection took place between 48 and

6 h after sowing and that most of the infected seed had produced a radicle, initial infection having occurred at the point of attachment of the axis to the cotyledons from where it spread into the shoot and cotyledons. Matthews Table 4.I) found that both ungerminated seeds and seeds with just radicles were infected after 7 days in the field. In a series of isolations from seeds held in wet unsterilized field soil at 15°C in the laboratory, cotyledons were found to become infected before embryo axes, infection occurring in cotyledons as early as 24 h after sowing (Matthews, 1971). The seed lot with the least ability to emerge in the field showed high levels of cotyledon infection Matthews, 1971), an observation that has been made many times since Matthews, unpublished data). A major cause of pre-emergence mortality in peas appears to be infection by *Pythium* spp. which infects and kills seeds before germination or in the early part of pre-emergence seedling growth.

TABLE 4.I. Stage of germination of 120 pea seeds and seedlings recovered from the field 14 days after sowing and the incidence of fungal infection in unemerged seeds. (Matthews, unpublished)

Stage of germination	Numbers in each stage	Number from which *Pythium* spp. has been isolated
Ungerminated	44	31
Germinated (some radicle growth)	40	31
Emerged (both radicle and plumule)	36	–

Fungicidal chemicals nearly always improve emergence but some discrepancy between laboratory germination and emergence has often remained despite fungicidal protection (Clayton, 1927; Jacks, 1963; Perry, 1967; Matthews and Bradnock, 1967). This observation has prompted the suggestion (Perry, 1967) that seed death from physiological causes contributes to the failure in emergence. However, other explanations are possible, for instance, the protection may not be complete, especially in the case of a seed lot highly disposed to fungal infection, a suggestion made by Matthews and Bradnock (1967) on the evidence of their data.

In work in which fungicides have been used on pea seeds sown in the field, not only did the chemicals improve emergence, but they also, according to several investigators, enhanced the "vigour" or size of the emerging seedlings (Clayton, 1927; Ogilvie and Mulligan, 1932; McNew, 1943a,b; Flentje, 1946; Jacks, 1963). This suggests that infection can not only prevent emergence but may also reduce the growth rate of those seedlings that do emerge. This may subsequently affect the survival of the seedlings especially when environmental conditions are adverse.

Stahl (1934) found that seed lots in which a high proportion of bacteria-infected seeds developed on filter paper, following a 24 h soaking in water, tended to emerge poorly in the field. Seed-borne bacteria have also been reported to reduce the laboratory germination of pea seeds that have broken seed coats (Virgin, 1940). Wellington (1962) noted a marked tendency for the cotyledons of seed lots showing low emergence in unsterilized soil to rot in compost. Although he did no isolations from cotyledons, bacteria from either the seed or the compost may have been involved in this rotting. Perry (1969a) commented that the cotyledons of seed from lots with poor field emergence potential were first colonized by bacteria and then fungi when sown in field soil and stated that the bacteria were pathogenic to seeds sown in sterile conditions. Bacteria, either seed- or soil-borne, may not be a major cause of emergence failure but their association with some seeds may be a sign of a defective condition.

C. Post-Emergence Seedling Loss

Little information has been published on post-emergence seedling loss and this may indicate that it is a minor problem compared with pre-emergence mortality. In his appraisal of the importance of seed-borne fungi, Jones (1927) noted that both *Ascochyta pinodella* and *Mycosphaerella pinoides* could cause death of seedlings after emergence. Field experiments in the U.S.A. (Wade 1941) revealed considerable differences between breeding lines in their susceptibility to frost damage. Observations were made on a spring sowing which encountered air temperatures below 0°C on 3 days, and on an autumn sowing, which was subjected to 8 days on which air frost occurred. The lowest temperature recorded was −7°C. The three commercial cultivars included in the trials were very sensitive to frosts, the damage rating given to them indicated that many seedlings were completely lost.

In an attempt to achieve a more even supply of harvested peas for process-ing factories, many more earlier sowings are made in the U.K. than used to be the case (McKay, 1966; Perry, 1967). This is also likely to have occurred in other countries. As a result large numbers of seedlings may encounter frosts, and post-emergence seedling loss from frost damage may be more common than is generally realized. Dry soil conditions during the time of seedling growth could also be a major hazard in some parts of the world.

III. SEED GENOTYPE AND CONDITION IN RELATION TO EMERGENCE

Many of the reports of emergence problems in peas have contained observa-tions pertinent to the more recent physiological investigations. Two types of observation can be distinguished: those relating to the seed and those describ-ing the influence of soil conditions. Differences in emergence have been reported both between and within cultivars, and seemingly, both seed geno-type and seed condition influence emergence performance.

A. Seed Genotype

Poor emergence has been reported as being more a feature of wrinkle-seeded cultivars than smooth-seeded ones (Jones, 1931; Hull, 1937; Walker, 1940; Baylis, 1941; McNew, 1943; Anon., 1950; Natti and Schroeder, 1955; Flentje, 1964). This difference is also remarked upon by growers. The wrinkle-seeded cultivars referred to in the work cited above were of the so-called 'classic" wrinkle-seeded cultivars (Marshall, 1966) such as 'Kelvedon Wonder' and 'Lincoln' which have round, radially split starch grains (Gregory, 1903). Kooistra (1962) reported a further two types of wrinkle-seeded peas. He recognized in the two cultivars 'Cennia' and 'Alsweet' a "new" type of wrinkled pea in which the starch grains were oval and unsplit, like those of smooth-seeded cultivars. The progeny of the crosses between the classic and the new wrinkle-seeded cultivars contained the third type, the so-called "double" wrinkled peas, which were phenotypically similar to the classic wrinkled peas but genetically different.

The chemical composition of the four types of pea also differs. Bonney and Fischbach (1945) found that the dried seeds of 13 "classic" wrinkle-seeded cultivars contained about twice as much sucrose as the smooth-seeded cultivar 'Alaska'. The water-absorptive capacity of seeds, measured after 24 h soaking, was found to be much higher in wrinkled seeds than smooth ones and tended to be high in seed types containing low levels of starch and high levels of sugar (Kooistra, 1962). Smooth peas contain more starch than do wrinkled seeds and there are also differences between all four types of seed in the amylose content of the starch (Kooistra, 1962).

Flentje and Saksena (1964) found that a wrinkle-seeded cultivar released about 50 times more sucrose into the soaking water than did a smooth-seeded cultivar. On the basis of this and other evidence they suggested that the considerable pre-emergence mortality seen for wrinkle-seeded peas was attributable to high levels of sucrose being exuded into soil in the first 24 h of germination which enhanced the growth of *Pythium* in the soil and thereby increased the chance of infection. In support of this suggestion Singh (1965) found that the presence of pea seeds in unsterilized soil greatly increased the number of propagules of *P. ultimum* in the soil in the vicinity of the seed. Flentje and Saksena (1964) emphasized that the differences in the levels of sucrose exuded (often referred to as leached by physiologists) into water between the two seed types were far greater than would have been expected from the reported differences in the sucrose contents of the dried seeds, and concluded that the total amount of sucrose present in the seed was not the major reason for the differences in exudate which they found. The readiness with which sugars and other solutes are released into water from dried seeds of all four types of pea needs clarification. There may be fundamental differences in the way in which solutes are retained within the cells of the embryos.

B. Seed Condition

Differences in emergence ability have also been noted between seed lots within wrinkle-seeded cultivars (Clark and Little, 1955; Perry, 1967; Matthews and Bradnock, 1967; Bradnock and Matthews, 1970). Perry (1969a) grew seed from several cultivars, mostly wrinkle-seeded types, in the glasshouse and in all but one cultivar ('Gregory's Surprise') the field emergence of this carefully grown seed was uniformly high, despite extremely adverse sowing conditions. Thus, since differences in emergence ability exist within cultivars and since seed with high emergence potential can be produced from several cultivars in which variability is commonly found, the source of the variability in emergence must arise from differences in the history of the seed lots before sowing.

The wholeness of the testa or seed coat is one aspect of seed condition that has been examined by several investigators in relation to the variability in emergence found among seed lots of peas. Hulbert and Whitney (1934) reported that seed coat injuries caused in various ways—threshing, drilling and improper irrigation during hot weather (which causes so-called sun scald) —reduced germination and increased the proportion of weak seeds. Hynes and Wilson (1939) suggested a link between the incidence of cracks in the seed coat and poor field emergence and Ledingham (1946) thought that seed with defective coats showed the greatest improvements in emergence when dressed with organomercurials. Flentje and Saksena (1964) increased the incidence of cracks in the testa both by machine harvesting and by deliberately cracking with a needle and found that both types of injury increased the exudation of solutes and reduced field emergence, but the treatment inducing the higher solute exudation did not produce the greater mortality (Table 4.II).

TABLE 4.II. Emergence in soil and exudation into water of samples of a wrinkle-seeded cultivar of peas after different harvesting and post-harvesting treatments affecting the wholeness of the testa. (From Flentje and Saksena, 1964)

	Seed with cracked testa (%)	Total soluble solids exuded per 100 g seeds (mg)	% emergence in soil at 18% moisture content
Sample treatment	2	714	86·6
+ crack near micropyle	100	1470	43·3
+ crack opposite micropyle	100	2102	48·3
Machine harvested	64	2064	24·8

Wellington (1962) found no association between the percentage of peas with cracked testas and emergence ability in unsterilized soil. In three out of four sowings in 1965 Perry (1967) found that the incidence of seed coat injuries was not directly related to field emergence. However, in a comparison of two contrasted seed lots he did link poor emergence with microscopic cracks in the testa and increased *Pythium* infection by removing sections from the testa (Perry, 1973). Removing the testa completely (Matthews, 1971) did not significantly increase mortality in soil of two seed lots but the testa was not removed until after 17 h soaking in water which may have eliminated the significance of the testa as a barrier to infection in seed lots that emerge well. The wholeness of the testa may be a factor determining emergence under some circumstances but the evidence does not convincingly suggest that it is the major cause of the variability in the emergence found between seed lots within cultivars.

The data from some investigations suggest that seed lots whose laboratory germinations are only just above 80% tend to be more prone to emergence failure than those with higher laboratory germination percentages (Clark and Little, 1955; Perry, 1967). The comment has often been made *en passant* that seed lots showing so-called low "vitality" or "vigour" were the ones emerging poorly in the field (Jones, 1931; Baylis *et al.*, 1943; Hutton, 1944; Ledingham, 1946; Clark and Little, 1955). The implication seems to have been that emergence potential was reflected in the so-called "vigour" of the seed lot, suggesting a link between emergence ability and the physiological condition of the seed.

IV. SOIL CONDITIONS AND EMERGENCE

Put in simple terms, in order to successfully germinate and emerge in soil, a seed needs an adequate supply of water and oxygen, and a temperature suitable for growth. The situation in peas is complicated by the involvement of soil-borne fungi and by the existence of differential emergence ability between genotypes and between seed lots within genotypes. Soil conditions could therefore reduce emergence directly by not fulfilling the physiological requirements for germination and pre-emergence seedling growth, and indirectly through their effect on seed infection. A review of reports on the influence of soil conditions on pea seed emergence is used to reveal the conditions that have been seen to reduce emergence and then the nature of the influence of these conditions is discussed.

A. Reports on the Influence of Soil Conditions

Although not dealing specifically with soil conditions, several studies on emergence have included comments on the influence of soil temperature and moisture on emergence, either in the field or in containers of unsterilized soil. Early work on the field emergence of peas (Haenseer, 1925; Horsfall, 1938; Ogilvie and Mulligan, 1933) contained evidence of the adverse effects of high soil moisture. Later, emergence was found to be reduced in soils either prepared in containers at high moisture contents (McNew, 1943a,b) or watered after sowing both in the field (Angell, 1952) and in containers (Jones, 1931). In Australia Norris and Hutton (1943) noted that the emergence in the wetter half of their field experiment was 18 % lower than the emergence in the rest of the plots. Also in Australia, Ledingham (1946) and Hutton (1944) found little association between rainfall and emergence but as the sowings were made progressively later in the season, when the air and soil temperatures were high, they noted a decline in emergence, possibly because the soil was not only warmer but also drier.

Other work has dealt in more detail with the influence of soil conditions on emergence using two or more cultivars (Hull, 1937; Baylis 1941; Baylis *et al.*, 1943; Flentje, 1964; Flentje and Saksena, 1964) or a range of seed lots (Perry, 1967; Heydecker and Kohistani, 1969; Perry, 1970; Carver, 1972). These experiments include both field sowings and experiments done in containers.

In a series of sowings over 3 years Hull (1937) found that, although low soil temperatures prolonged the time needed for emergence and tended to lower emergence, high soil moisture was a much more important adverse influence. Reduced emergence was largely attributable to soil-borne infection and was more a feature of the wrinkle-seeded cultivar that he used than the smooth-seeded one (Hull, 1937). Baylis (1941) tested the pathogenicity of *Pythium* isolates in soil containers at a range of moisture contents and found the mortality induced in the soil increased as the moisture level increased. The highest moisture, which was in excess of any he had encountered in the field, prevented the emergence of all seeds in all treatments, including those with no pathogen. This failure of emergence at high moistures he interpreted as death from physiological causes, presumably lack of oxygen.

In later work on the effects of fungicides Baylis *et al.* (1943) confirmed that high soil moisture in the field was particularly conducive to severe pathogenic attack, especially of wrinkle-seeded cultivars, and implied that some of the differences observed between wrinkle-seeded cultivars were attributable to differences in the condition of the seed. In a container experiment, they found that high soil moisture, achieved by watering after sowing, had a deleterious effect only if the water was added soon after sowing during the early stages of germination. Watering had little effect when done 3 days after sowing. Flentje (1964) and Flentje and Saksena (1964) working in Australia found, similarly, that increasing moisture increased infection especially in a wrinkle-seeded cultivar. Flentje (1964) also observed that emergence was much reduced in a late sowing when the soil moisture was below permanent wilting point (8 % for this particular soil). In these conditions many of the seeds that failed to emerge had imbibed but had either not germinated or had failed to develop much after germination. At the same time many of these seeds showed no signs of infection.

In more recent work in the U.K. attention has been given to the response to soil conditions of different seed lots within cultivars. In five sowings done over 2 years (1964 and 1965) in Scotland, Perry (1967) found that the average emergence of all seed lots being tested was highest in the later sowings in both seasons, and that the emergence improvement produced by a fungicidal dressing was greatest in the early plantings. He also noted that some seed lots emerged well at all sowings and did not improve with dressing whereas others showed marked improvements with both later sowing and seed dressing. In extensive sowings throughout the U.K. of five seed lots of each of two wrinkle-seeded cultivars, Perry (1970) found that although the general level of emergence of the seed lots at the different sowings varied greatly the rank order of the lots remained the same.

Carver (1972) compared the emergence of eight seed lots of a wrinkle-seeded cultivar on nine sowing dates at three adjacent sites in Scotland (Fig. 4.1). As soil temperatures increased the time taken to reach 50% of ultimate emergence fell and the level of ultimate emergence increased for all lots. The emergence response to sowing date differed between the lots: two emerged well at all sowings, including the early ones when emergence was slow because of low temperatures; three improved as sowings were made progressively later, but never emerged well; the rest were low in the early sowings and improved considerably, achieving the emergence level of the better lots in the later sowings. Reduced emergence was associated with high rainfall at the first and last sowings, especially for the three poorer lots. At these times soil moisture was around field capacity. At one of the later sowings, when the emergence of all the seed lots was low compared with the previous and subsequent sowings, reduced emergence seemed to be caused by the lack of sufficient water, the sowing having been made in the middle of a long dry spell which affected all the seed lots.

The same grouping of the seed lots that was made on the basis of their response to sowing date was also seen in their response to soil moisture and temperature in containers of unsterilized soil (Carver, 1972). All seeds failed to emerge in conditions where the highest moisture content (40%) was combined with the lowest temperature (5°C), but the better lots were able to emerge at this moisture level, which was in excess of field capacity (33%), at higher temperatures. The death of some of the seeds under these extremely adverse conditions of low temperature and high moisture may be attributable to causes other than fungal infection.

B. Possible Effects of Soil Conditions

The evidence presented suggests that three soil conditions reduce emergence: low temperatures and both high and low moistures. The first two conditions, low temperature and high moisture, will be considered at this stage only in

D*

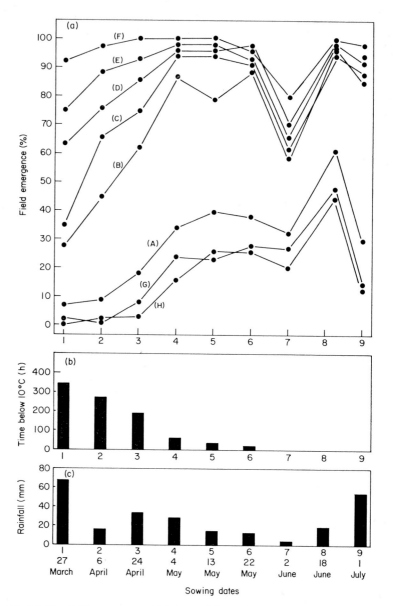

FIG. 4.1. (a) The field emergence of eight seed lots of peas planted on nine occasions during 1970, expressed as a percentage of laboratory germination; (b) the total hours below 10°C for the 2 weeks following each of the nine planting dates; (c) the total rainfall (mm) in the week following each planting date.

general terms. A more detailed analysis of the response of the seed and the pathogen to environmental conditions will follow in section VI, after a description in section V of the physiological nature of the variability found among seed lots. The aim of this sub-section is to clarify the way in which these soil conditions influence the environment of the seed. Although temperature and moisture interact with respect to their effects on emergence (Carver, 1972) the three conditions are more conveniently discussed separately.

1. Low Temperature

Low temperatures could adversely influence emergence through their direct effect on germination and pre-emergence seedling growth, resulting in their prevention and retardation. Such temperatures may also act indirectly through their effects on fungal infection by influencing either the host or the pathogen or both. All of these possibilities will be discussed in section VI.

2. High Soil Moisture

Soil has been described as a three-phase system consisting of solids, liquids and gases in varying proportions (Currie, 1973). The solid phase consists of minerals and organic matter which in a good seed-bed would be aggregated into crumbs. The pores between the crumbs (macropores) and the much smaller pores within them (micropores) are filled with water (the liquid phase) and air (the gas phase) in varying proportions. After heavy rain when a soil is saturated, most of the macropores will be filled with water, and as the soil drains, and also water evaporates from it, the water will be replaced in these macropores by air from the soil surface. At field capacity, at the end of drainage, the macropores tend to be filled with air whilst the crumbs remain saturated—any further water loss is from within the crumbs. Seeds are usually sown into the macropores but they may get incorporated into the matrix of the crumbs with time and compaction (Currie, 1973). The seed is therefore dependent upon the macropores and, to a lesser extent, the micropores for its supply of oxygen and water. High moistures could influence emergence adversely through the effect that water has on the availability of oxygen or because of the direct effect of water (Perry and Harrison, 1970).

Provided the seed is in a macropore, which is likely in the case of large seeds such as the pea, oxygen diffusion will be through air at all moisture contents up to approximately field capacity. Below field capacity oxygen would only be limiting if its content in the soil air fell greatly, which does not seem to happen near the surface in arable soils (Russell and Appleyard, 1915). Above this moisture level, as the macropores become progressively filled with water, the seed will come to rely more on diffusion through water for its oxygen supply. The rate of diffusion of oxygen in water is over 300 000 times slower than its diffusion in air because both the solubility and the

diffusion coefficients of oxygen in water are low (Goddard and Bonner, 1960).
Thus, only in soils at or above field capacity, when the macropores start to
fill with water, would soil moisture begin to reduce the availability of oxygen
to the seed.

Measurements of oxygen diffusion in soil with oxygen electrodes have
shown that the rate of diffusion begins to decline with increases in moisture
at water potentials of more than -0.08 bar (Stolzy and Letey, 1964). Field
capacity has been variously characterized in water potential terms as the
equivalent of -0.05, -0.10 (Marshall, 1959) or even -0.33 bar (Kohnke,
1968). This variation underlines the empirical nature of the term field capacity.
However, these estimates indicate that the moisture conditions which lead
to reductions in measured rates of oxygen diffusion are around and above
field capacity.

The frequency and duration of periods of time when seed-bed conditions
are at such moisture levels depend on rainfall (amount and distribution over
time) and soil structure, including bulk density which affects pore volume.
Over the winter months (October to March) which include the sowing months
of February and March, there is on average an estimated total excess of rainfall over evaporation in all parts of the U.K. Also data for particular sites
(Anon., 1967; Winter, 1974) show this to be true for the months of February
and March when calculated separately. Thus, soil moistures in excess of field
capacity are possible for pea seeds at sowing but the level and duration
involved would depend on day to day rainfall and soil drainage.

There is also evidence from the literature on field experiments that it is soil
moisture conditions near or above field capacity which cause the greatest pre-
emergence mortality (Flentje, 1964). Hull (1937) found that when soil moisture
was at its highest (25%), emergence was at its lowest (50%), and although not
stated this moisture level was probably above field capacity, being the highest
recorded during that winter in a series of weekly samplings. Furthermore,
the adverse effect that watering or rainfall just after sowing has on emergence
(Baylis et al., 1943; Angell, 1952) indicates that moisture conditions at field
capacity and above during germination and early seedling growth are particularly detrimental. In support of the effect of rain on soil oxygen availability Erickson and Van Doren (1960) found that oxygen diffusion rates
measured in field soils after very heavy rain were at their lowest (30×10^{-8}
g^{-1} cm^{-2} min^{-1}) which was a level that had been seen to reduce emergence in
sugar beet. For some reason maintained moistures in unsterilized soil in
containers reduce emergence in peas even at levels below field capacity (Flentje,
1964; Carver, 1972).

3. Low Soil Moisture

In peas, as in most seeds, the biochemical processes leading to germination
do not begin until some cellular hydration has taken place (Kollöffel, 1967;

see also Chapter 3). Uptake of water into pea seeds, which can take place over the whole surface of the seed coat but more readily through the chalaza (Spurny, 1973) and the micropyle (Brown, 1965), is by the physical process of imbibition. Water is taken up by hydrophilic colloids which have a very low matric potential and in consequence an almost equally low water potential. Imbibition can take place at low temperatures (below 4°C), in the absence of oxygen and even in seeds that are completely dead.

Seed imbibition is possible even when water is presented at extremely low water potentials; wheat seeds have been seen to imbibe and germinate when supplied with water in the vapour phase where the water potential was −32 bar (Owen, 1952). In peas Manohar and Heydecker (1964) used solutions of polyethylene glycol to show that imbibition and germination could take place where the water was presented at water potentials of −5 bar but not at −10 bar. Recently, Hadas and Russo (1974) found water had been taken up into pea seeds from water potentials as low as −30 bar. Currie (1973) apparently illustrated that the uptake of water into pea seeds could occur by vapour diffusion and suggested that in many soil situations the seed has little direct contact with free water and may have to rely on the mechanism of vapour diffusion. However, in the field, fluctuating temperatures are the norm and the condensation of water vapour out of the soil atmosphere on to the surface of the seed could be important.

Observations in dry soils suggest that imbibition and germination are not prevented in soils which have been referred to as being at permanent wilting point with a water potential of around −15 bar. Doneen and McGillivray (1943) and Flentje (1964) found that in very dry soils most pea seeds had imbibed and many had germinated even if they had not emerged. Thus, the important adverse consequence of lack of soil water in the field seems to be an inadequate supply during the pre-emergence seedling growth of peas. Because of the very low water potentials of dry seeds drought is probably seldom a problem during imbibition and germination. Much more of a problem is lack of water in subsequent growth when uptake is by osmosis in which higher, and therefore less effective, water potentials are involved.

V. LABORATORY ASSESSMENT OF EMERGENCE

Reference has already been made to the term "vigour" which is commonly used to describe the differences in emergence ability found between seed lots and even genotypes (Perry, 1972). Normally, vigour is used to describe physiologically based differences between seeds and not, for instance, differences in the condition of the seed coat or contamination with micro-organisms (Perry, 1972). Much has been written on seed vigour (Woodstock, 1973; Heydecker, 1972) including that of peas and this forms a useful background against which to review the laboratory measures that have been used to assess field emergence in peas.

In a consideration of the sequence of events that occur in various types of seeds which have been placed in hot and/or humid atmospheres, leading to loss of viability after short storage periods of several weeks or even days, Delouche (1969) has pointed out (and others have quoted him since, e.g. Heydecker, 1972; Roberts, 1973) that many changes occur in seeds before the ability to germinate is lost. The properties resulting from these changes (in so-called "aged" seeds) only reveal themselves in particular circumstances and are similar to many of the properties that are characteristic of seed lots with poor field emergence ability.

In order of appearance, the changes that have been observed by Delouche and his colleagues are: a reduction in the ability of seeds to retain solutes when placed in water; a decline in respiration and biosynthesis; slower germination and growth, including greater variability; increased sensitivity to environmental stresses; and increased susceptibility to infection by micro-organisms. Many of these properties found in "aged" but viable seeds would clearly influence the ability of the seeds to emerge in the field under conditions of stress in the presence of pathogens. In peas, attempts have been made to relate seed characteristics, performance in growth and stress tests, and levels of solute leaching and respiration, to field emergence among commercially available seed lots of wrinkle-seeded cultivars where the major differences in physiological condition exist.

A. Seed Characteristics

Emergence in the field has been related in some work to an abnormal condition of pea cotyledons which is characterized by a concave area on the adaxial surface of the cotyledons of imbibed seeds. The condition was first observed by Myers (1948) in Australia who called the condition "hollow-heart". Since then hollow-heart has been reported in other countries among commercially available seed lots of peas (Noble, 1960; Moore, 1964; Perry and Howell, 1965; Gane and Biddle, 1973).

Noble and Howell (1962) suggested that the disorder caused poor field emergence. Perry (1967) found that, although seed lots with a high incidence of hollow-heart emerged poorly in the field, low emergence was not always associated with the occurrence of the disorder. Gane and Biddle (1973) found that dry seeds with a triangular shape in cross-section tended to have hollow-heart and emerged less well than more squarely shaped seeds. A more common observation has been the slow growth of peas with hollow-heart (Myers, 1948; Noble, 1960; Moore, 1964; Harrison and Perry, 1973). Harrison and Perry (1973) suggested that the reduced seedling growth was associated with the presence of a growth inhibitor in the affected cells, which were dead, in the concave area of the cotyledons. They also showed (Perry and Harrison, 1973) that symptoms developed during imbibition, especially rapid imbibition, and that a cause of the disorder was high ambient temperatures during seed maturation on the plant.

"Marsh spot" is another physiological disorder of pea seeds in which brown spots develop on the adaxial surfaces of the cotyledons (Reynolds, 1955). The cause is manganese deficiency in the parent plant. Although seed samples with a high incidence (20%) of Marsh spot have shown reduced germination (Bruijn, 1933) and seedlings from affected seeds may be small and abnormal (Pethybridge, 1934), it is not considered to be associated with poor field emergence (Lacey, 1934; Wellington, 1962).

B. Growth and Stress Tests

Many of the growth tests used to indicate emergence ability in peas have been done under stress conditions of one sort or another. One of the earliest, in which seeds were soaked for 24 h before being set to germinate on filter paper where counts of bacteria-infected peas were made, was possibly a test of the tolerance of seeds to anaerobic conditions (Stahl, 1934, 1936). In the U.S.A. Clark and Baldauf (1958) recommended a cold test in which peas are incubated at 4–6°C for 1 week with unsterilized soil and then subsequently germinated at 16–18°C. Apparently, this was a test of resistance to infection under cold conditions. Caldwell (1960) showed some significant correlations between field emergence and emergence in very wet sterilized sand at 30°C. Possibly, this was a test of tolerance of not only high temperatures but also anaerobic conditions. Perry (1969b) successfully used a seedling evaluation test in which counts were made of well-grown seedlings after germination at 20°C in sterilized calcined aluminium silicate particles. Heydecker (1969) summarized a series of tests done in various seed testing centres on three seed lots of peas which differed in emergence ability. The tests included: germination in solutions of low water potential; various soil tests; categorizations of seeds by vital staining using tetrazolium chloride; and evaluations of seedling sizes after various lengths of time in germination tests. The tests were highly variable in predicting the relative emergence performances of the three seed lots.

C. Leaching of Solutes into Water

1. Relation between leaching and field emergence

Some of the earliest measurements of solutes leached from seeds into water (Fick and Hibbard, 1925; Hibbard and Miller, 1928) included work on the loss of electrolytes from pea seeds of different ages which showed a range of viabilities (97 down to 63%). Leachates from these samples also showed a range in conductivity readings from 62–181 μmho indicating a greater loss of electrolytes from some samples (those of low viability) than others. Although not commented upon by Hibbard and Miller (1928) those samples with laboratory germinations in excess of 80% exhibited a wide range in electrolyte leaching (62–103 μmho) which did not entirely correspond with their germination percentages. This was one of the first indications that differences in the ability to retain solutes existed among seed samples containing a high proportion of viable seeds. Since then, differences in solute retention have been reported between highly viable samples of maize (Tatum, 1954) and castor beans (Thomas, 1960). In both crops, leaching was inversely related to emergence in soil.

Flentje and Saksena (1964) found that the emergence ability of different genotypes of peas could be related to leaching, which they called exudation. A few years later, it was discovered that the level of leaching into water from seed lots of the same cultivar could be used to detect seed of low field emergence potential (Matthews and Bradnock, 1967; Matthews and Whitbread, 1968). The inverse relation between field emergence ability and the leaching of electrolytes among highly viable seed lots of peas has been confirmed for several cultivars (Perry, 1970; Bradnock and Matthews, 1970; Gane and Biddle, 1973) and is now used in seed testing on a routine basis. Bradnock and Matthews (1970) found that the inclusion of seed weight improved correlations between electrical conductivity and emergence within some cultivars. As a result the electrical conductivity which is determined after a 24 h soaking of pea seeds in de-ionized water, is usually expressed per unit weight of air-dry seeds.

Seed lots that differ in the levels of solute leaching which they exhibit do so with respect to electrolytes, sugars (Matthews and Whitbread, 1968), particularly sucrose (Matthews and Carver, 1971), amino acids (Matthews and Carver, 1971) and potassium (Matthews and Rogerson, 1976). This suggests that differences in the ability of seeds to retain solutes in general are involved.

2. The origin of differential solute leakage

The testa is undoubtedly a barrier to the release of solutes from pea seeds into water. Larson (1968) and Simon (1974) reported greater loss of solutes from seeds of round-seeded cultivars when the testa was removed, and incisions in the testa were also found to increase leaching (Flentje and Saksena, 1964; Table 4.II). Simon (1974) proposed that the testa acts as more than a barrier to the solutes that have been released from the embryo. He found that, when seed coats were dissected off under water after a period of soaking, the concentration of electrolytes released into the water did not rise to the level that would have been expected if the embryo had been uncovered from the start. He suggested that in intact seeds, an accumulation of solutes trapped between the testa and the embryo slows down the rate of release.

It is therefore possible that differences in leaching between seed lots are associated with the effectiveness of the testa as a barrier to leakage. However, other evidence suggests that this is not the case (Matthews 1971; Matthews and Rogerson, 1976). When the testas were removed from two seed lots after 17 h soaking and the seeds were subsequently soaked for a further 24 h, the differences between the lots in the readiness with which they released electrolytes were maintained, even though the removal of seed coats from the poorer lot significantly increased leaching. Recently, the significance of the seed coat has been examined again in comparisons of the leaching of solutes from eight seed lots of each of two cultivars. Removal of the seed coat after a 6 h soaking increased leaching over the next 24 h but the distinctions between the seed lots were always maintained (Table 4.III). Furthermore, differences in leaching between lots remained after the careful removal of the seed coat from air-dry seed. It appears that the condition of the embryo is crucial in determining the level of leaching from the seeds of different lots.

TABLE 4.III. The effect of testa removal on the loss of electrolytes into water (mean conductivities are expressed as μmho cm^{-1} (g seed)$^{-1}$ $\pm$ standard error) for pea seed (cv. 'Jade'). Figures in parentheses are the number of seeds included in each mean. (From Matthews and Rogerson, 1976)

Seed lot	Seeds intact	Testa removed
A	524 ± 43 (24)	747 ± 62 (25)
B	478 ± 31 (23)	641 ± 35 (24)
C	403 ± 48 (25)	544 ± 33 (25)
D	367 ± 32 (22)	461 ± 40 (25)
E	344 ± 37 (23)	360 ± 27 (23)
F	322 ± 37 (22)	397 ± 40 (21)
G	217 ± 24 (23)	299 ± 20 (24)
H	191 ± 29 (22)	333 ± 61 (22)

When dead areas of tissue were produced on the abaxial surfaces of pea cotyledons by sub-zero temperatures, leaching was increased (Matthews, 1971). This suggests that a cause of differences in leaching between lots might be the extent of dead tissue on the cotyledons since seeds with food reserves in cotyledons can tolerate considerable amounts of dead tissue in their cotyledons without losing their ability to germinate (Moore, 1973). This possibility has been examined in more detail recently (Matthews and Rogerson, 1976). When the leaching of electrolytes from individual seeds of eight lots was related to the staining of seeds with the vital stain tetrazolium chloride, which produces a red formazan in the presence of active dehydrogenases (Roberts, 1951), seeds with dead areas on the abaxial surfaces of their cotyledons were seen to release the most electrolytes (Table 4.IV). But, significant differences in the retention of solutes were also seen between the seeds from the eight lots that had stained completely. Moreover, the mean conductivity readings of the seeds from each lot possessing no areas of dead tissue were positively and significantly correlated with the means obtained when all seeds in each lot were included. Thus, although areas of dead cotyledon tissue will contribute to a high mean level of leaching in a seed lot, the major difference between seed lots lies in the readiness with which living cotyledon cells release solutes to water, assuming that staining with tetrazolium chloride is a valid indication of living cells.

TABLE 4.IV. The mean electrical conductivity of seed soak water (μmho cm^{-1} g seed^{-1} $\pm$ standard error) for three categories of seed based on the staining of the abaxial cotyledon surfaces with tetrazolium chloride for eight seed lots of peas (cv. 'Jade'). Figures in parentheses are the numbers of seed in each category. (From Matthews and Rogerson, 1976)

Seed lots	Completely stained	More than half stained	Less than half stained
A	460 $\pm$ 32 (20)	841 $\pm$ 110 (4)	(0)
B	387 $\pm$ 31 (9)	533 $\pm$ 44 (13)	592 $\pm$ 12 (2)
C	210 $\pm$ 25 (11)	398 $\pm$ 54 (6)	671 $\pm$ 67 (8)
D	254 $\pm$ 39 (10)	440 $\pm$ 28 (11)	536 $\pm$ 109 (2)
E	291 $\pm$ 74 (20)	696 $\pm$ 65 (3)	(0)
F	220 $\pm$ 27 (13)	456 $\pm$ 58 (7)	515 $\pm$ 130 (2)
G	153 $\pm$ 9 (16)	344 $\pm$ 41 (8)	(0)
H	166 $\pm$ 28 (18)	303 $\pm$ 88 (4)	(0)

Simon and Raja Harun (1972) found that, when dry embryos of a smooth-seeded cultivar of peas were immersed without their seed coats for 1 min in each of a series of aliquots of water in turn, there was a dramatic fall in the quantity of solutes (electrolytes, sugars and potassium) released into the water over the first few minutes. This and other evidence led them to propose that the plasmalemma and tonoplast of cells in dry seeds were leaky. They suggested that lack of hydration so affected the molecular architecture of the membranes that it made them ineffective, and that upon hydration, membrane integrity was restored and the membranes became effective barriers to the passage of solutes out of cells.

Comparisons of the leaching of solutes from the dry embryos of seed lots, which differ in the loss of electrolytes from intact seeds over 24 h, have been made (Matthews and Rogerson, 1976) in the manner of Simon and Raja Harun (1972). Seed coats were removed either by carefully chipping them from dry seeds or by dissection from seeds after soaking for 6 h, followed by drying over calcium chloride which restores the rapid early release phenomenon (Simon and Raja Harun, 1972). All seed lots tested yielded similar curves to those found previously two of which are illustrated in Fig. 4.2. Thus, there was no failure in the restoration of effective membranes upon hydration. However, differences between seed lots in the release of solutes were seen from the first minute of immersion (Fig. 4.2). Since, according to the hypothesis of Simon and Raja Harun, it is in the first few minutes of immersion that

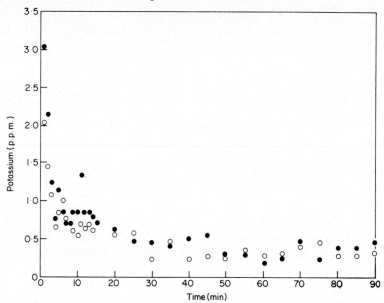

FIG. 4.2. Time course of leakage of potassium from the air-dry embryos of two seed lots of peas (cv. 'Kelvedon Wonder') which show differences in the levels of leaching from intact seeds over 24 h (high ● ; low ○).

the plasmalemma is ineffective as a barrier to leakage, the cells of the seed lots must differ in the amount of solutes that can be leached from within the cytoplasm. No correlation was seen between the amount of extractable solutes (potassium, sugars and electrolytes in general) and leaching over 24 h for eight seed lots, indicating that the leaching differences were attributable to the ability of the cytoplasm to retain solutes and not to the quantity of solutes present (Matthews and Rogerson, 1976). The retention in the cytoplasm could be associated with the form of the solutes but more likely, since both sugars and potassium are retained either equally well or badly, it is the effectiveness with which membranes within the cytoplasm retain solutes that is important. This hypothesis presumes that membranes within the cytoplasm of dried seeds are to some extent effective.

The sub-cellular observations on the membranes within the cotyledon cells of ripening and germinating pea seeds are of interest in this context (Bain and Mercer, 1966a,b). During the development of the ripening pea seeds most membranes become difficult to distinguish (Bain and Mercer, 1966a). The tonoplast becomes indistinct on day 30 after flowering and at about day 40 the endoplasmic reticulum fragments leaving only a few scattered vesicles. Using electron micrographs of cells in ripening pea seeds that were drying out (45 days after flowering) to give an approximation of the cotyledon cells before imbibition, Bain and Mercer (1966b) suggested that the cytoplasm in the air-dried seeds was undifferentiated and vesiculated. Thus, the only membranes that are at all likely to persist in the dry seeds in an unmodified form are those bounding these scattered vesicles. Consequently, if membranes within the cytoplasm are involved in the retention of solutes by seeds then the solutes might well be held within these vesicles and the distinctions between seeds which result in differences in leaching could be found in either the condition of the membranes bounding these vesicles or in the number and sizes of the vesicles themselves. Some evidence for the existence of vesicles associated with protein bodies in air-dry cotyledon tissue of peas has been provided by Swift and Buttrose (1973) (see Fig. 15.4, pp. 444–445).

D. Respiration

Respiration measurements have been related to the physiological condition (or vigour) of samples of several crop seeds (Woodstock, 1973). Almost always, the vigour of the seed had been impaired by an inflicted treatment such as gamma irradiation, mechanical damage, low temperatures or prolonged storage. In peas, for instance, Woodstock and Combs (1967) found that mechanical damage reduced seedling growth and respiration which were measured 18 h into germination. They also noted that the respiration rate 6 h after imbibition of three lots of pea seeds was correlated with their emergence in soil, the lot showing the best emergence having the highest oxygen uptake.

Only recently have the respiration rates of commercially available seed lots of peas been related to field emergence (Carver and Matthews, 1975). A positive correlation (0.97, $P < 0.001$) was found between field emergence (a mean derived from nine sowings) and oxygen uptake after 24 h on moist filter paper for eight seed lots of the cultivar 'Dark Skinned Perfection' (Fig. 4.3). Even though the two seed lots emerging least well showed the highest respiratory quotients (R.Q.) after 24 h, R.Q. was not significantly correlated with emergence. Respiration was also inversely related to leaching, low respiration being a feature of the seed lots that were readily leached.

E. The Significance of Physiological Condition

The variability in physiological condition which is revealed in measurements of leaching and respiration may be particularly significant when seeds are sown in the field, where soil conditions are seldom ideal. Deficiencies in the provision of the physiological requirements of the seeds may lead to failure to emerge in seeds that are readily leached and respire slowly in laboratory tests because these seeds lack resilience. Such a cause of failure to emerge which does not involve soil pathogens will be discussed in the next section.

Alternatively, the physiological condition may be important in determining the ability of the seeds to avoid infection by *Pythium* spp. Seed lots from which

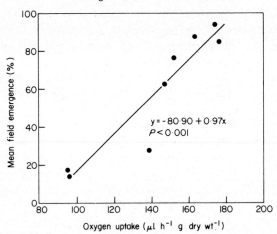

Fig. 4.3. Relationship between oxygen uptake 24 h from the start of germination and mean percentage field emergence for eight seed lots of peas (cv. 'Dark Skinned Perfection'). Oxygen uptake was determined for six replicates of four seeds and field emergence is the mean of 200 seeds sown at each of the nine sowing dates.

sugars and amino acids are readily leached into water may show similar characteristics in soil leading to a promotion of pathogen growth in the vicinity of the seed which, in turn, increases the chance of infection occurring (Matthews and Whitbread, 1968; Perry, 1973). This is an extension of the hypothesis of Flentje and Saksena (1964) which was put forward to explain differences between cultivars. In support of this hypothesis Perry (1973) found that, when seeds were coated with agar containing sugar or leachates and then placed in sand inoculated with *P. ultimum*, the level of disease in these coated seeds was significantly increased.

Another suggestion involving the fungal pathogen was prompted by observations on the disease in soil (Matthews, 1971). Cotyledons are always infected in soil before embryo axes and seed lots showing differences in emergence ability also show corresponding differences in the extent of cotyledon infection. It was suggested, therefore, that the major difference between seed lots lay in the susceptibility of their cotyledons to infection. Once cotyledons became infected the fungal hyphae would have a food base (Garrett, 1960), which would enhance the pathogen's ability to infect the embryo axis.

The evidence in favour of this hypothesis is circumstantial. Dead areas produced on the abaxial surfaces of cotyledons not only increase leaching but also increase cotyledon infection and reduce emergence drastically. However, high levels of leaching are not only found in seeds with areas of dead tissue. In many seed lots, cotyledons containing a large proportion of cells that are leaky but not dead, may be in a condition which renders them unable to resist infection. Small disks of cotyledon material placed in unsterilized soil adjacent to seeds increased mortality dramatically from 14% (control) to over 80% (Matthews, 1971). Measurements of leaching from these disks into water indicated that the leachates from the disks would only supplement the seed leachates to a very small extent. Furthermore, no effect on mortality was seen when disks were removed from alongside the seeds after 23 h, and it is in the first 24 h when most of the electrolytes are leached from the disks into water—less than 20% of the total electrolytes lost in 48 h comes out during the second day of soaking (Matthews, 1971). The disks, which were all colonized by *Pythium* spp. after 24 h in soil, could act as sites for initial infection, simulating a susceptible cotyledon, from which hyphae grow and infect the embryo axis. This hypothesis depends upon the existence of differences in cotyledon susceptibility between seed lots exhibiting a range of emergence abilities and Perry (1973) has questioned this. He found no differences in cotyledon susceptibility between seeds of two contrasted lots. However, he did not use soil as a source of inoculum but inoculated with *P. ultimum* and such a high inoculum might be able to overcome the resistance of most cotyledons.

Suspensions of cotyledon fragments of peas are able to produce a fungal growth inhibitor called pisatin (see Fig. 1.6, p. 8) in the presence of *Phytophthora megaspermae* (Keen and Sims, 1973). Pisatin is the phytoalexin produced by peas in response to infection which has been implicated as a major factor in the disease resistance of peas to fungal infection generally (Cruickshank, 1963). Pisatin has been found to be produced by whole cotyledons when the cotyledons were inoculated with *P. ultimum* and preliminary work has shown that seed lots showing little cotyledon infection in soil tend to produce the highest levels of pisatin (Powell, 1974). Although the adoption of a hypothesis concerned with cotyledon resistance would entail dismissing leachates as a major factor, the release of sugars and amino acids may still be important through their effects in soil and at sites of infection.

VI. PHYSIOLOGICAL RESPONSES TO GERMINATION CONDITIONS

The soil conditions of low temperature and high moisture which lead to low emergence through the effects of temperature, superabundance of water or limited oxygen availability (section IVB) may act directly on the viability of the seed or indirectly through their effect on fungal infection. Infection could be influenced by effects on the host rate of growth, leaching of solutes, respiration or the ability to resist infection. On the other hand, the major factor could be the influence of conditions on the activity of the pathogen. The effects of environmental conditions during germination on viability, leaching, respiration and ability to resist infection are examined and the significance of environmental conditions in the soil where both the seed and the fungal pathogen may be affected, is then discussed.

A. Effect on Viability and Rate of Emergence

The effect of temperature on the germination of pea seeds in quasi-sterile conditions has been investigated on several occasions. Kotowski (1926) found that peas emerged well (80%), albeit slowly, at 4°C. The optimum temperature was 18°C, when all seeds emerged, and the worst temperature was the highest used, 30°C, when 70% of the seeds germinated. This deleterious effect of high temperature, also noted more recently in soil emergence tests at controlled temperatures (Singh and Dhaliwal, 1972), may have been responsible for the low emergences in hot conditions reported by Ledingham (1946) and Hutton (1944). Gadd (1936) showed that germinating seed lots at 10°C rather than at 20°C did not affect their germination capacity. In tests at low temperatures (Torfarson and Nonnecke, 1959) emergence in vermiculite was significantly reduced in each of 2 years by a temperature just below 2°C but the greatest difference recorded between two temperatures (2° and 10°C) was only 18%. In comparisons of emergence in molochite at 4° and 20°C on several seed lots, Perry (1969a) found that the emergence of the poorest lot was depressed most by the low temperature (germination was lower by 19%) and the lot with the best field emergence ability was not at all affected. All the unemerged seeds at low temperatures were, in fact, dead and not just slow to emerge. Alternating temperatures between 10° and 14°C have been found to reduce germination compared with constant temperatures of either 10° or 14°C (Gulliver and Heydecker, 1973) which may be indicative of a more marked effect of temperature in the field where it normally fluctuates.

Leach (1947) suggested that the pre-emergence mortality of seeds caused by fungi such as *P. ultimum* varied in severity at different temperatures because the outcome of the interaction between pathogen and host depended upon the ratio of their rates of growth, the effect of temperature being different for the two rates. Those conditions relatively more suitable for the seed than the fungus led to emergence and vice versa. On this basis he successfully predicted that in his artificially inoculated substrate the lowest emergence would be achieved between 12° and 20°C; 4°C gave the highest emergence counts. If, however, the vulnerable stages of the pea seed are, as have been shown, prior to germination and just after, then a protraction of this phase of growth would lead to high levels of infection even if conditions did not favour the pathogen either. Some support for this is found in the work of Harper *et al.* (1955), who found that maize cultivars, which are able to germinate and grow relatively rapidly at low temperatures in early sowings in the U.K., showed the least pre-emergence mortality, presumably because they spent less time at a stage when the seeds and seedlings were vulnerable. Furthermore, the evidence from the field is that it is temperatures below 10°C that are most commonly associated with low levels of emergence in peas (Perry, 1967; Carver, 1972).

Perry and Harrison (1970) have suggested that, in wet, cold soils, pea seeds that are low in vigour may be further weakened and killed by the rapid inrush of water. This suggestion was based on the observation that seed survival was significantly less after a 24 h soaking in water at 1°C (59%) compared with survival after 24 h either between moist papers (80%) or in polyethylene glycol solutions at the same temperature. Excessively rapid imbibition appeared to be the detrimental effect of soaking; lack of oxygen was not considered important at such a low temperature. Larson (1968) had previously postulated that the increased leaching and reduced seedling growth which he found when pea seeds were soaked without their testas, was caused by the broaching of cell membranes through the rapid inflow of water. Simon (1974) has questioned this explanation because, firstly, Simon and Raja Harun (1972) found that after soaking dry pea embryos in 1 M sucrose and then placing them in water, there was no rapid leakage although imbibition was very fast, and, secondly, the effect of rapid imbibition on leakage was not long-lasting which would have been expected, unless the membranes were re-assembled.

Low oxygen diffusion rates in soils and atmospheres containing low levels of oxygen can affect emergence in crop seeds. Unger and Danielson (1965) noted that germination, i.e. appearance of the radicle, took place even in atmospheres containing no oxygen, but the subsequent growth of the radicle was reduced by atmospheres containing less oxygen than air. Higher than normal concentrations of carbon dioxide did not reduce germination or adversely affect growth. In peas, Spragg and Yemm (1959) have shown that before the testa is ruptured the R.Q. of germinating seeds can become as high as 3 while oxygen uptake remains low. This suggests that partially anaerobic conditions exist within the testa until germination, and therefore low oxygen availability might not be expected to affect germination, since it normally takes place under anaerobic conditions, even in air. Erickson and Van Doren (1960) have noted reduced emergence in peas in soils in which oxygen diffusion is low, after germination has occurred. Very wet soils, in excess of field capacity at a water potential of about −0·02 bar (in which the oxygen diffusion rate is very low), have been found to reduce the emergence of maize without affecting germination (Wengel, 1966). If the observations on maize also apply to peas (and this seems likely from the work of Erickson and Van Doren) then the first direct effect of reduced oxygen availability caused by high soil moisture would be on pre-emergence seedling growth and not on germination. A direct adverse effect on viability after 24 h in very wet sand at 20°C has been reported for a seed lot of peas with low emergence potential, involving a reduction from 100 to 67% (Carver, 1972; Table 4.V). Apparently, very high levels of moisture have to be reached before oxygen availability is so limited that emergence is reduced by the direct effect of the lack of oxygen.

TABLE 4.V. The effect on three seed lots of peas of 24 h in sand at four moisture contents (4, 11, 17 and 22%) at 20°C. Observations were made on the subsequent germination in sand, the mortality in soil (each a mean of five replicates of ten seeds), and *Pythium* spp. infection of cotyledon pieces in soil (% of 80 pieces). (From Carver, 1972)

Seed lot	% germination				% mortality				% infection			
% moisture content	*4*	*11*	*17*	*22*	*4*	*11*	*17*	*22*	*4*	*11*	*17*	*22*
A	88	100	88	67	78	90	77	62	12	12	16	50
C	100	100	100	83	20	10	5	26	0	12	30	36
F	93	93	93	93	3	2	9	20	0	0	0	6

B. Effect on Leaching

Leaching has been found to be greater at low temperatures than at high ones in mung beans (Kraft and Erwin, 1967) and cotton (Hayman, 1969), but in peas no significant effect of temperature was seen by Schroth *et al.* (1966). Perry and Harrison (1970) found temperature had no effect on leaching from living pea seeds, but Carver (1972) noted a consistently greater level of leaching into sand of several moisture levels at 5°C compared with 20°C for six seed lots of peas in the second day of germination.

The leaching of sugars from peas into soils increased with increasing moisture according to Kerr (1964) who measured the leaching of sugars indirectly by determining the loss in weight of the seeds. He suggested that this effect was due to an increase in the rate of diffusion of sugars in soils at high moisture contents. Carver (1972) found some increase in leaching from seeds with increasing sand moisture but no great increases were involved except in the difference between the driest sand and the other moisture levels. In an examination of the reasons for high pre-emergence mortality in soybeans in wet soils, Brown and Kennedy (1966) found some evidence of an increase in the leaching of amino acids and glucose from seeds held in atmospheres containing low concentrations of oxygen (4 and 8%) compared with normal air. No similar work has been done on peas or any other seeds.

C. Effect on Respiration and Resistance

The respiration rates of six seed lots of peas were measured following removal from moist sand 2 days after the onset of germination, and were found to decline as sand moisture content increased (Carver and Matthews, 1975). The moisture level at which the uptake began to decline differed between lots (Table 4.VI). Seed lots that had emerged well in the field maintained their oxygen uptake until the very wettest treatments whereas the uptake of the other lots started to decline at lower moisture levels. The conditions that produced a fall in the oxygen uptake of a seed lot also produced a rise in R.Q. indicating that a major effect of an increase in moisture was a reduction in oxygen availability. Until the wettest conditions the seed lots with the highest R.Q. values tended to be the ones with the least emergence ability but under the most anaerobic conditions the better lots reverted to the most vigorous anaerobic respiration, as indicated by R.Q. (Carver and Matthews, 1975) and ethanol determinations following the sand treatments (Carver, 1972). Thus, seed lots that emerge well in the field and are better able to retain solutes when soaked in water, respire more actively aerobically and exhibit a higher level of anaerobic respiration under anaerobic conditions than do seed lots with poor field emergence potential.

TABLE 4.VI. Mean oxygen uptakes (μl h^{-1} g dry wt.$^{-1}$) determined for six replicates of four seeds for each of six seed lots of peas (cv. 'Dark Skinned Perfection') after 24 h in sand at each of four moisture contents (4, 11, 17 and 22%) and at two temperatures (5° and 20°C). (From Carver and Matthews, 1975)

Temperature	% moisture content	Seed lots*					
		A	B	C	D	E	F
5°C	4	31$_a$	42$_a$	35$_a$	55$_a$	54$_a$	104$_a$
	11	28$_a$	52$_a$	38$_a$	78$_b$	77$_b$	103$_a$
	17	31$_a$	69$_a$	36$_a$	90$_b$	82$_b$	95$_a$
	22	22$_a$	65$_a$	36$_a$	80$_b$	84$_b$	104$_a$
20°C	4	301$_a$	209$_a$	263$_a$	282$_a$	493$_a$	270$_a$
	11	196$_b$	257$_a$	240$_a$	270$_a$	494$_a$	291$_a$
	17	166$_b$	183$_b$	180$_b$	167$_b$	450$_a$	255$_a$
	22	142$_b$	149$_b$	133$_b$	111$_b$	240$_b$	130$_b$

*Within each temperature and seed lot, means with subscripts in common are not significantly different ($P > 0.05$) as determined by Duncan's multiple range test following analysis of variance.

In a small experiment, Carver (1972) showed that 24 h in wet sand at either 5° or 20°C led to an increase in the *Pythium* infection of cotyledons in soil for seed lots with poor field emergence ability (Table 4.V). In some cases an increase in mortality in soil was also seen after wet pre-treatments. It appears that only 24 h in anaerobic conditions in wet sand can increase the susceptibility of the cotyledons to *Pythium* infection in seed lots that are in a relatively poor physiological condition. A link between respiratory metabolism and cotyledon resistance to infection has been tentatively suggested (Carver and Matthews, 1975).

D. Significance to Emergence in Soil

The reduction of germination and emergence by the direct effects of low temperatures, freely available water and low oxygen availability, is possible in the field, particularly in the case of seed lots that are in a poor physiological condition. However, the extent of the reductions that have been observed does not seem sufficiently great to explain the considerable discrepancies that have been found between laboratory germination and field emergence. Nevertheless, some soil conditions, for instance water-logging at low temperatures, could drastically reduce emergence without the involvement of micro-organisms. It is more likely that the environmental conditions that have been seen to reduce emergence do so by further weakening already poor seeds so that they become more susceptible to soil pathogens, notably *Pythium* spp.

At the same time such conditions may directly affect the pathogen and its ability to infect living tissues. Low temperatures reduce the growth of *P. ultimum* (Leach, 1947) but under these circumstances mortality is increased. Kerr (1964) found that high soil moistures did not increase the vegetative growth of *P. ultimum*. He also pointed out that *P. ultimum*, the most likely causal fungus of pre-emergence mortality in peas, did not produce zoospores and could not, therefore, take advantage of free water to spread more quickly through soils. Low oxygen atmospheres do not reduce the growth of *P. ultimum* until below 4% but neither do they enhance the growth of the fungus (Brown and Kennedy, 1966). Griffin (1963) examined the growth of *P. ultimum* at different water potentials in artificial soils and found no increase in fungal growth as potentials increased from -0.4 to -0.001 bar (calculated from his data). This range covers all the various estimates of field capacity and above. Thus, there does not seem to be evidence of any increase in the growth of *P. ultimum* as soil moisture increases.

Low temperatures increase the time spent at a stage when seeds and seedlings are vulnerable to infection, and they could also increase leaching. High moisture could increase the incidence of infection by also increasing leaching from seeds or by increasing the susceptibility of the cotyledons to infection.

VII. PRODUCTION FACTORS INFLUENCING PHYSIOLOGICAL CONDITION

Perry (1969a) suggested two possible causes of poor physiological condition in viable pea seeds: the harvesting and drying of immature seeds; and deterioration in unfavourable storage conditions.

A. Seed Age at Harvest

In the U.S.A., Fields and King (1963) found that pea seeds could be harvested while immature at moisture contents of 45%, and artificially dried at temperatures up to 45°C, with little adverse effects. The seed quality was superior to seeds taken from plants that had been field dried after being cut at the same seed moisture contents. Perry (1969a) reported an experiment in which glasshouse-grown plants were harvested at various stages (from before the time the pods had begun to wrinkle until the plants had dried out) and dried the plants in a hot air-stream at $30°$–$35°C$. Seed from the first harvest had a low emergence capacity in a quasi-sterile germination media, while the rest of the harvests yielded seeds which gave emergence counts in excess of 99 %. However, a count of vigorous seedlings, which had previously been found to relate to field emergence in commercial seed lots, increased as the harvests were made progressively later, which led Perry (1969a) to suggest that immature harvesting was a cause of poor physiological condition and low emergence in pea seeds.

In a series of harvests of pea seeds from field plots, Matthews (1973) found that only when the seed moisture content had fallen to less than 70% of fresh weight could seeds be harvested and dried in a heated glasshouse and still remain viable. Moisture content was influenced both by rainfall before harvest and by seed age. Comparisons of changes in dry weight accumulation and moisture content with time suggested that it was only after some slow dehydration had taken place on the plant that seeds could withstand enhanced rates of drying off the plant. The mortality in soil continued to decline with increasing age at harvest beyond the age when viability had reached its optimum which prompted the suggestion that early harvesting and drying could produce viable seeds with low soil emergence potential. In later experiments (Matthews, 1974) in which glasshouse-grown seeds were dried at different ages in vacuum desiccators, the need for some slow moisture loss whilst the seeds were still on the plant was again indicated as necessary for the seeds to develop the ability to withstand severe desiccation. The cessation in moisture increase in the seeds was followed by a slowing-down in physiological activity as measured by oxygen uptake. This is similar to the findings of Kollöffel (1970) who noted that the loss of water in developing pea seeds was accompanied by a fall in respiration which began to decline when moisture increase had ceased (see Chapter 3). On the basis of observations in lima beans (Klein and Pollock, 1968) and castor beans (Marré, 1967), it has been suggested (Thomas, 1972) that a gradual increase in water stress in the seed is partially responsible for the fall in respiration observed in developing seeds. However, this suggestion needs further examination since respiration appears to start to fall before the decline in the moisture content of the whole seed begins (Kollöffel, 1970; Matthews, 1974).

The nutrition of the parent plant can influence the growth of the seedlings of the subsequent generation under some circumstances. Austin and Longden (1965) found that if pea plants were grown in low levels of phosphate, then the phosphorus content of the seeds was reduced. Low phosphorus seeds produced significantly smaller seedlings compared with those produced from high phosphorus seeds when grown in culture media low in phosphate. Where parent plants are grown in soils low in phosphate and seed from them are sown in similar soils, reduced seedling growth may result and this may adversely influence the ability of seedlings to emerge and establish.

B. Storage

In the 1 or 2 years over which pea seeds are normally stored before sowing it is generally considered that there is little loss in viability (Perry, 1969a). In storage experiments peas tend to be one of the better crop species at maintaining viability even when stored under relatively unfavourable conditions, but longevity varies between cultivars and harvest years (James et al., 1967). Peas do, however, show losses of viability in a relatively short time (24–45 days) at seed moistures in excess of 12% when stored at 35° and 45°C (Abdalla and Roberts, 1969). Such storage treatments are often regarded as simulations of storage under more normal conditions for much longer periods (years). Abdalla and Roberts (1969) found that the seeds which did germinate after such storage treatments (moist seed kept at high temperatures) showed reduced growth rates in the seedling stages. Thus, reduced seedling vigour in viable pea seeds can result from deterioration in storage treatments and may arise before any significant loss in viability has occurred. Harman and Granett (1972) found that storing peas in an atmosphere with 93% r.h. at 25°C produced only a slight fall in viability over 14 weeks but increased both the leaching of solutes from the seed and the susceptibility of the seed to storage fungi. In recent work (Powell and Matthews, 1977) in which similar storage conditions were used, a significant increase in leaching was found after 6 weeks. After storage the seeds were viable and abaxial surfaces of the cotyledons stained completely but intensely with tetrazolium chloride which has been found to be a characteristic of some seeds that have poor field emergence potential (Matthews, 1971). Furthermore, when tested for the rapid early-release-of-solutes phenomenon (Simon and Raja Harun, 1972) the deteriorated seeds showed higher levels of leaching from the first minute than did control seeds. This greater leaching was not associated with higher levels of extractable solutes in the seeds and therefore an impairment in the effectiveness of membranes within the cytoplasm was indicated. Perhaps some of the membrane lipid changes found by Koostra and Harrington (1969) in cucumber seeds during moist storage also occur in peas. They found that there was a marked fall in the phosphorus content of the membrane lipids after only 1 week of moist storage. Many of the changes seen in artificially aged pea seeds indicate a deterioration of membranes and these changes result in properties—intense tetrazolium staining and ready leaching of solutes —that are similar to those found in commercially available seed lots with low field emergence potential. Prolonged storage under unfavourable conditions could be a cause of poor physiological condition, especially if the condition of the seed is already impaired before the start of a period of storage.

VIII. AGRONOMIC SIGNIFICANCE

The increasing interest in the use of specific plant populations and arrangements to achieve maximum yields per unit area (Bleasdale, 1964) has made the prediction of field emergence in peas a practical necessity. At the same time, there has been a tendency in recent years, certainly in the U.K., to begin sowing earlier so as to spread the harvesting and processing of the crop later in the year. This has placed more seeds in unfavourable soil conditions than used to be the case. Consequently, any techniques of seed evaluation which can select out material with especially good field emergence potential is of real practical benefit. Physiological comparisons of seed lots of peas have produced a seed-leaching test and, potentially at least, a test based on respiratory differences between seed lots. Both these tests give a valuable guide to the emergence potential of a seed lot which the normal laboratory germination test alone cannot give. Such tests may also be helpful in evaluating the potential of seeds to be stored for long periods without any significant deterioration in physiological condition.

It has been suggested that differences in vigour and emergence potential in highly viable seed lots, which are revealed in various laboratory tests, also result in yield differences because of a reduced yield potential in the plants which successfully emerge (Perry, 1972; Heydecker, 1972). Abdalla and Roberts (1969) found no evidence for this in their work when viable seeds (those remaining after an accelerated ageing treatment in moist storage) were grown at the same plant density in the field as seeds that had not been aged, even though the moist storage had reduced the growth rate of the seedlings. Perry (1968) did find a yield advantage in seed lots that had emerged well compared with less vigorous lots in plots that had been thinned to the same plant density. Yield differences may arise under some circumstances but further experimentation is needed.

IX. SUMMARY AND CONCLUSIONS

The major problem in pea seed emergence and establishment is the poor emergence of some commercially available seed lots which are in a poor physiological condition, especially in cold, wet soils. This seed condition reveals itself in the poor retention of solutes (when seeds are placed in water) and in low respiration. Poor solute retention appears to be attributable to defective membranes within the cells.

Although death from the direct effect of the inadequate provision of the physiological requirements of the emerging seed, such as oxygen and water, might occur under some extreme circumstances, the more important cause of failure to emerge appears to be infection by the soil-borne fungus *Pythium ultimum* before or just after germination. The suggestion is made that low temperatures and high soil moistures combine to both prolong the time when the seed is vulnerable to infection and increase the susceptibility of the seed to the pathogen. Low resistance to infection in the cotyledons of seeds that are in poor physiological condition is considered more important than the leaching of nutrients into the soil which might stimulate fungal growth. It is suggested that seeds which are viable but in poor physiological condition are produced by the harvesting and drying of immature seeds and by prolonged storage in unfavourable conditions.

REFERENCES

ABDALLA, F. H. and ROBERTS, E. H. (1969). *Ann. Bot.* **33**, 169–184.
ANGELL, H. R. (1952). *J. Aust. Inst. agric. Sci.* **18**, 99–101.
ANON (1950). *Rep. Waite agric. Res. Inst.* (1943–1947).
ANON. (1961). "The Seeds Regulations, 1961". H.M.S.O., London.
ANON. (1967). "Potential Transpiration for Use in Irrigation and Hydrology in the U.K. and Republic of Ireland." H.M.S.O., London.
ANON. (1973). "The Vegetable Seeds Regulations, 1973". H.M.S.O., London.
AUSTIN, R. B. and LONGDEN, P. C. (1965). *Nature, Lond.* **205**, 819.
BAIN, J. M. and MERCER, F. V. (1966a). *Aust. J. biol. Sci.* **19**, 49–67.
BAIN, J. M. and MERCER, F. V. (1966b). *Aust. J. biol. Sci.* **19**, 69–84.
BAYLIS, G. T. S. (1941). *Ann. appl. Biol.* **28**, 210–218.
BAYLIS, G. T. S., DESHPANDE, R. S. and STOREY, I. F. (1943). *Ann. appl. Biol.* **30**, 19–26.
BLEASDALE, J. K. A. (1964). *Rep. natn. Veg. Res. Stn* 1964, 36.
BONNEY, V. G. and FISCHBACH, H. (1945). *J.Ass. off. agric. Chem.* **28**, 409–417.
BRADNOCK, W. T. and MATTHEWS, S. (1970). *Hort. Res.* **10**, 50–58.
BROWN, R. (1965). *In* "Encyclopedia of Plant Physiology" (W. Ruhland, ed.). Springer-Verlag, Heidelberg.
BROWN, G. E. and KENNEDY, B. W. (1966). *Phytopathology* **56**, 407–411.
BRUIJN, H. L. G. DE (1933). *Tijdschr. Pl Ziekt.* **39**, 281–318.
CALDWELL, W. P. (1960). *Proc. Ass. off. Seed Analysts N.Am.* **50**, 130–136.
CARVER, M. F. F. (1972). Ph.D. Thesis, University of Stirling, Scotland.
CARVER, M. F. F. and MATTHEWS, S. (1975). *Seed Sci. Technol.* **3**, 871–879.
CLARK, B. E. and BALDAUF, D. (1958). *Proc. Ass. off. Seed Analysts N.Am.* **48**, 133–135.
CLARK, B. E. and LITTLE, H. B. (1954). *Bull. N.Y. St. agric. Exp. Stn* **760**.
CLARK, B. E. and LITTLE, H. B. (1955). *Bull. N.Y. St. agric. Exp. Stn* **770**.
CLAYTON, E. E. (1927). *Bull. N.Y. St. agric. Exp. Stn* **554**.
CRUICKSHANK, I. A. M. (1963). *A. Rev. Phytopathol.* **1**, 351–374.
CURRIE, J. A. (1973). *In* "Seed Ecology" (W. Heydecker, ed.), p. 463. Butterworths, London.

ELOUCHE, J. C. (1969). *Bull. Miss. agric. Exp. Stn* **1721**, 16–18.

DONEEN, J. D. and McGILLIVRAY, J. A. (1943). *Pl. Physiol.* **18**, 524–529.

EASTHAM, A. (1925). *In* "Report of the Fourth International Seed Testing Congress", p. 12. H.M.S.O., London.

ERICKSON, A. E. and VAN DOREN, D. M. (1960). *Int. Congr. Soil Sci.* **3**, 428–434.

ESSENBURG, J. F. W. and SCHOOREL, A. F. (1962). *Lit Overz. Cent. LandbPubl. LandbDocum., Wageningen,* **26**.

EWING, E. E. (1959). *Diss. Abstr.* **20**, 1518.

FIELDS, R. W. and KING, T. H. (1963). *Proc. Minn. Acad Sci.* **30**, 128–130.

FICK, G. L. and HIBBARD, R. P. (1925). *Mich. Acad. sci. Arts Letters* **5**, 95.

FLENTJE, N. T. (1946). *J. Dep. Agric. S. Aust.* **1**, 246–249.

FLENTJE, N. T. (1964). *Aust. J. biol. Sci.* **17**, 651–654.

FLENTJE, N. T. and SAKSENA, H. K. (1964). *Aust. J. biol. Sci.* **17**, 665–675.

GADD, I. (1936). *Proc. int. Seed Test. Ass.* **8**, 159–210.

GANE, A. J. and BIDDLE, A. J. (1973). *Ann. appl. Biol.* **74**, 239–247.

GARRETT, S. D. (1960). *In* "Biology of Root-infecting Fungi". Cambridge University Press, U.K.

GODDARD, D. R. and BONNER, W. D. (1960). *In* "Plant Physiology" (F. C. Steward, ed.), Vol. 1A, p. 209. Academic Press, New York and London.

GREGORY, P. P. (1903). *New. Phytol.* **2**, 226–228.

GRIFFIN, D. M. (1963). *Biol. Rev.* **38**, 141–166.

GULLIVER, R. L. and HEYDECKER, W. (1973). *In* "Seed Ecology" (W. Heydecker, ed.), pp 433–462. Butterworths, London.

HADAS, A. and RUSSO, D. (1974). *Agron. J.* **66**, 643–647.

HAENSELER, C. M. (1925). *Rep. New Jers. St. agric. Exp. Stn* **45**, 403–414.

HARMAN, G. E. and GRANETT, A. L. (1972). *Physiol. Pl. Pathol.* **2**, 271–278.

HARPER, F. R. (1964). *Can. J. Pl. Sci.* **44**, 531–537.

HARPER, J. L., LANDRAGIN, P. A. and LUDWIG, J. W. (1955). *New Phytol.* **54**, 107–118.

HARRISON, J. G. and PERRY, P. A. (1973). *Ann. appl. Biol.* **73**, 103–109.

HAYMAN, D. S. (1969). *Can. J. Bot.* **47**, 1521–1523.

HEYDECKER, W. (1969). *Proc. int. Seed Test. Ass.* **34**, 751–774.

HEYDECKER, W. (1972). *In* "Viability of Seeds" (E. H. Roberts, ed.), pp 209–252. Chapman and Hall, London.

HEYDECKER, W. and KOHISTANI, M. R. (1969). *Ann. appl. Biol.* **64**, 153–160.

HIBBARD, R. P. and MILLER, E. V. (1928). *Pl. Physiol.* **3**, 335–352.

HILTNER, L. (1903). *Arb. biol. BundAnst. Land-u. Forstw.* **3**, 1–102.

HORSFALL, J. G. (1938). *Bull. N.Y. St. agric. Exp. Stn* **683**.

HORSFALL, J. G., NEWHALL, A. G. and GUTTERMAN, C. E. F. (1934). *Bull. N.Y. St. agric. Exp. Stn* **643**.

HULL, R. (1937). *Ann. appl. Biol.* **24**, 681–689.

HULBERT, H. W. and WHITNEY, G. M. (1934). *J. Am. Soc. Agron.* **26**, 876–884.

HUTTON, E. M. (1944). *J. Coun. scient. ind. Res. Aust.* **18**, 71–74.

HYNES, H. J. and WILSON, R. P. (1939). *Agric. Gaz. N.S.W.* **50**, 657–659.

JACKS, H. (1959). *N.Z. Jl agric. Res.* **2**, 306.

JACKS, H. (1963). *N.Z. Jl agric. Res.* **6**, 115–117.

JAMES, E., BASS, L. N. and CLARK, D. C. (1967). *Proc. Am. Soc. hort. Sci.* **91**, 521–528.

JONES, L. K. (1927). *Bull. N.Y. St. Agric. Exp. Stn* **547**.

JONES, L. K. (1931). *J. agric. Res.* **42**, 25–32.

JUSTICE, O. L. (1965). *Proc. int. Seed Test. Ass.* **30**, 3–13.

KEEN, N. T. and SIMS, J. J. (1973). *Proc. 2nd int. Pl. Path. Congr.*
KERLING, L. C. P. (1947). *Trans. R. Soc. S. Aust.* **71**, 253–258.
KERR, A. (1964). *Aust. J. biol. Sci.* **17**, 676–685.
KLEIN, S. and POLLOCK, B. M. (1968). *Am. J. Bot.* **55**, 658–672.
KOHNKE, H. (1968). "Soil Physics". McGraw Hill, New York.
KOLÖFFEL, C. (1967). *Acta bot. neerl.* **16**, 111–122.
KOLÖFFEL, C. (1970). *Planta* **91**, 321–328.
KOOISTRA, E. (1962). *Euphytica* **11**, 357–373.
KOOSTRA, P. T. and HARRINGTON, J. F. (1969). *Proc. int. Seed Test. Ass.* **34**, 329–340.
KOTOWSKI, F. (1926). *Proc. Am. Soc. hort. Sci.* **23**, 176–184.
KRAFT, J. M. and ERWIN, D. C. (1967). *Phytopathology* **57**, 866–868.
LACEY, M. S. (1934). *Ann. appl. Biol.* **21**, 621.
LARSON, L. A. (1968). *Pl. Physiol.* **43**, 255–259.
LEACH, L. D. (1947). *J. agric. Res.* **75**, 161–179.
LEDINGHAM, R. J. (1946). *Scient. Agric.* **26**, 248–257.
MANOHAR, M. S. and HEYDECKER, W. (1964). *Rep. Univ. Nottingham Sch. Agric.* 1963: 65–72.
MARRÉ, E. (1967). *In* "Current Topics in Developmental Biology" (A. Monroy and A. A. Moscana, eds.), Vol. 2, pp 75–105. Academic Press, London and New York.
MARSHALL, T. J. (1959). *Tech. Commun. Commonw. Bur. Agric.* **50**, pp.
MARSHALL, H. H. (1966). *Can. J. Pl. Sci.* **46**, 545–551.
MATTHEWS, S. (1971). *Ann. appl. Biol.* **68**, 177–183.
MATTHEWS, S. (1973). *Ann. appl. Biol.* **73**, 211–219.
MATTHEWS, S. (1974). *Ann. appl. Biol.* **75**, 93–105.
MATTHEWS, S. and BRADNOCK, W. T. (1967). *Proc. int. Seed Test. Ass.* **32**, 553–563.
MATTHEWS, S. and CARVER, M. F. F. (1971). *Proc. int. Seed Test. Ass.* **36**, 307–312.
MATTHEWS, S. and ROGERSON, N. (1976). *J. exp. Bot.* **47**, 119–125.
MATTHEWS, S. and WHITBREAD, R. (1968). *Pl. Path.* **17**, 11–17.
McKAY, D. B. (1966). *J. nat. Inst. agric. Bot.* **10**, 42–46.
McKAY, D. B. (1972). *In* "Viability of Seeds" (E. H. Roberts, ed.), pp 114–149. Chapman and Hall, London.
McNEW, G. L. (1943a). *Phytopathology* **33**, 9.
McNEW, G. L. (1943b). *Canner,* **96**, 14, 16, 30, 32, 35.
MOORE, R. P. (1964). *Newsl. Ass. off. Seed Analysts N.Am.* **38**, 12.
MOORE, R. P. (1973). *In* "Seed Ecology" (W. Heydecker, ed.), p. 347. Butterworths, London.
MUNN, M. T. (1926). *Proc. Ass. off. Seed Analysts N.Am.* **55**.
MYERS, A. (1948). *Proc. int. Seed Test. Ass.* **14**, 35–37.
NATTI, J. J. and SCHROEDER, W. T. (1955). *Bull. N.Y. State Agric. Exp. Stn* **771**.
NOBLE, M. (1960). *Proc. int. Seed Test. Ass.* **25**, 536.
NOBLE, M. and HOWELL, P. (1962). *Seed Trade Rev.* **14**, 171.
NORRIS, D. O. and HUTTON, E. M. (1943). *Aust. J. of Council for Sci. Ind. Res.* **16**, 149–154.
OGILVIE, L. and MULLIGAN, B. O. (1933). *Rep. agric. hort. Res. Stn Univ. Bristol* **132**, 103–120.
OWEN, P. C. (1952). *J. exp. Bot.* **3**, 188–204.
PERRY, D. A. (1967). *Proc. int. Seed Test. Ass.* **32**, 3–12.

E

PERRY, D. A. (1969a). *Proc. int. Seed Test. Ass.* **34**, 221–232.
PERRY, D. A. (1969b). *Proc. int. Seed Test. Ass.* **34**, 265–272.
PERRY, D. A. (1970). *J. agric. Sci., Camb.* **74**, 343–348.
PERRY, D. A. (1972). *Hort. Abstr.* **42**, 334–342.
PERRY, D. A. (1973). *Trans. Br. mycol. Soc.* **61**, 135–144.
PERRY, D. A. and HARRISON, J. G. (1970). *J. exp. Bot.* **21**, 504–512.
PERRY, D. A. and HARRISON, J. G. (1973). *Ann. appl. Biol.* **73**, 95–101.
PERRY, D. A. and HOWELL, P. J. (1965). *Pl. Path.* **14**, 111–116.
PETHYBRIDGE, G. H. (1934). *J. Minist. Agric. Fish.* **41**, 833.
POWELL, A. A. (1974). B.A. thesis, University of Stirling, Scotland.
POWELL, A. A. and MATTHEWS, S. (1977). *J. exp. Bot.* (in press).
RASMUSSON, J. (1926). *Nord. JordbrForsk.* **8**, 724–735.
REYNOLDS, J. D. (1955). *J. Sci. Fd. Agric.* **6**, 725–734.
ROBERTS, E. H. (1973). *Seed Sci. Technol.* **1**, 529–545.
ROBERTS, L. W. (1951). *Science, N.Y.* **113**, 692–693.
RUSSELL, E. J. and APPLEYARD, A. (1915). *J. agric. Sci., Camb.* **7**, 1–48.
SCHROTH, M. N., WEINHOLD, A. R. and HAYMAN, D. S. (1966). *Can. J. Bot.* **44**, 1429–1432.
SIMON, E. W. (1974). *New Phytol.* **73**, 377–420.
SIMON, E. W. and RAJA HARUN, R. M. (1972). *J. exp. Bot.* **23**, 1076–1085.
SINGH, R. S. (1965). *Mycopath. Mycol. appl.* **27**, 155–160.
SINGH, N. T. and DHALIWAL, G. S. (1972). *Pl. Soil* **37**, 441–444.
SPRAGG, S. P. and YEMM, E. W. (1959). *J. exp. Bot.* **10**, 409–425.
SPURNY, M. (1973). *In* "Seed Ecology" (W. Heydecker, ed.), p. 367. Butterworth, London.
STAHL, C. (1934). *Tiddskr. Pl Avl*, **39**, 673–680.
STAHL, C. (1936). *Tiddskr. Pl Avl* **41**, 139–148.
STOLZY, L. H. and LETEY, J. (1964). *Adv. Agron.* **16**, 249–279.
SWIFT, J. G. and BUTTROSE, M. S. (1973). *Planta* **109**, 61–72.
TATUM, L. A. (1954). *Agron. J.* **46**, 8–10.
THOMAS, C. A. (1960). *Science, N.Y.* **131**, 1045–1046.
THOMAS, H. (1972). *In* "Viability of Seeds" (E. H. Roberts, ed.), pp 360–396. Chapman and Hall, London.
TORFARSON, W. E. and NONNECKE, I. L. (1959). *Can. J. Pl. Sci.* **39**, 119–124.
UNGER, P. W. and DANIELSON, R. E. (1965). *Agron. J.* **57**, 56–58.
VIRGIN, W. J. (1940). *Phytopathology* **30**, 790–791.
WADE, B. L. (1941). *Proc. Am. Soc. hort. Sci.* **38**, 530–534.
WALKER, J. C., DELWICHE, E. J. and SMITH, P. G. (1940). *J. agric. Res.* **61**, 143–147.
WALLEN, V. R. and SKOKO, A. J. (1950). *Can. J. Res. Ser. C* **28**, 623–636.
WELLINGTON, P. S. (1962). *J. natn. Inst. agric. Bot.* **9**, 160–169.
WELLINGTON, P. S. (1966). *Jl R. agric. Soc.* **127**, 164–186.
WENGEL, R. W. (1966). *Agron. J.* **58**, 69–72.
WINTER, E. J. (1974). *In* "Water, Soil and the Plant". Macmillan, London.
WOODSTOCK, L. W. (1973). *Seed Sci. Technol.* **1**, 127–157.
WOODSTOCK, L. W. and COMBS, M. F. (1967). *Proc. Ass. off. Seed Analysts N. Am.* **57**, 144–148.

5. Root Growth and Morphogenesis

J. G. TORREY

Cabot Foundation, Harvard University, Massachusetts, U.S.A.

R. ZOBEL

Department of Agronomy, Cornell University, New York, U.S.A.

I. ROOT MORPHOLOGY

Within about 48 h after the seed begins to imbibe water at 25°C, the pea seed radicle breaks the seed coat and elongates downwards into the substrate. The first visible lateral roots are formed on the seedling radicle towards the base of the root within 1–2 cm of the cotyledon attachments. Germination is

hypogeal, i.e. the cotyledons remain buried in the substrate and the radicle
extends downwards while the epicotyl or young, elongating shoot emerges
above ground. Laterals are initiated at a nearly constant distance from the
root tip as the root elongates, occurring in acropetal sequence and forming
three rows or ranks along the longitudinal axis. Thus, the external array of
lateral roots reflects the precision of their origin in the pericycle in a radial
direction opposite the three internal primary xylem poles of the triarch
vascular cylinder.

Figure 5.1 is a shadowgraph of a root system from a typical pea plant
which has not nodulated. The mature pea root system is made up of a
primary or tap root, first laterals or secondary roots and second order

FIG. 5.1. Shadowgraph of the root system of a 4-week-old pea plant, cv. 'Little Marvel'
grown with the roots in nutrient mist. Note that the secondary roots are the same length
as the primary root and that all the roots are highly branched. × 1/3.

laterals or tertiary roots. There may be a further sub-division of the roots
into third and fourth order laterals, but these are only rarely observed in plants
cultured for observation of the root system. These lateral roots serve primarily
as the absorbing region of the root. The physiological mechanisms which
control the initiation and development of lateral roots of all orders are,
however, identical.

In field peas, according to Hayward (1938), the dominant primary root
axis may reach a depth of up to 4 ft (1·2 m) or more in the soil. The radicle

etains a dominant role in root development, with the formation of a strong
tap root system. In a field situation, root nodules develop either on the main
root axis or on the major lateral roots beginning about 10 or 12 days after
germination, and are formed in large numbers, depending upon the genetics
of the infective *Rhizobium* or the variety of the host plant (see Chapter 13)
At the base of the root near its attachment to the shoot, limited thickening
of the root axis may occur due to activity of a vascular cambium. A detailed
description of the morphology of the pea plant is given by Hayward (1938).

II. THE FORMATION AND ORGANIZATION OF THE ROOT APEX

The radicle is initiated in early embryogeny so that the mature seed possesses
a well organized root apex at the radicle end of the embryo. Early stages in
the embryology of *Pisum sativum*, including the development of distinctive
paired multinucleated suspensor cells and the formation of the early globular
embryo were described by Cooper (1938). According to Reeve (1948), the
embryo radicle begins to show organization at the time the embryonic cotyledons are formed. Then, as the embryonic axis elongates, a well defined root
cap with tiered columella becomes evident and thereafter clear delineation of
the future central cylinder and of the future cortex is apparent. In the fully
mature radicle apex in longitudinal section, approximate limits of the root
cap, the central cylinder and the future cortex–epidermis can be discerned.

The organization of the root apical meristem in *Pisum* during seedling
growth has been a matter of considerable controversy. The root apex in pea
is described as typical of the "transverse meristem" or open type (Fig. 5.2A),
an observation going back to Janczewski (1874), who first described the
zonation of the apical meristem as a cambium-like transversal generative
layer in the sub-apical region of the apex, giving rise to the root cap in the
distal direction and the central cylinder, cortex and epidermis in the proximal
direction. A review of the anatomical and embryological literature on the
root apex of *Pisum* was made by Popham (1955a) and a detailed cellular
analysis of the terminal 0·5 mm of the root apex was described by Torrey
(1955) in relation to his surgical experiments on the apex.

With the discovery by Clowes (1956) of the existence of the quiescent
centre, a population of cells in the root apex located proximal to the root cap
junction which are relatively inactive metabolically and mitotically, a re-
assessment of earlier ideas about the apical organization was necessary. The
transverse meristem type was especially interesting in this context and has
been studied by Clowes (1961, 1967) in *Pisum* and in *Vicia*. Although Abbott
(1972) stated that he was unable to demonstrate a quiescent centre in cultured
pea roots, Webster and Langenauer (1973) reported a quiescent centre after
autoradiography of excised roots of pea in culture medium.

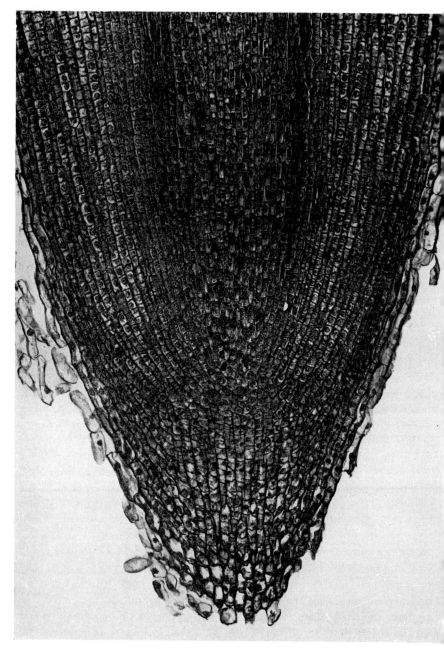

FIG. 5.2A. Median longitudinal section of seedling root tip of *Pisum sativum*, cv. 'Alaska', excised from seed 48 h after germination. × 175. (From Torrey, 1955)

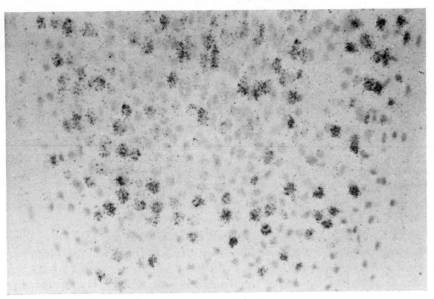

FIG. 5.2B. Autoradiograph of median longitudinal section of seedling root tip of *Pisum sativum*, cv. 'Alaska', excised and cultured in White's medium with 2 % sucrose for 72 h and provided with ^{3}H-thymidine from 48 to 72 h. Note clearly delimited quiescent centre and well-labelled nuclei of root cap initials. × 350. (Photo by D. Stetka, courtesy of P. Webster)

From anatomical studies and autoradiographic analyses (especially of median longitudinal sections of seedling pea roots), the following picture emerges. The seedling root cap is about 250–300 μm long from the extreme tip to the root cap initials; these initials form a disk-shaped cellular layer giving rise to the aligned cells of the central columella which occupies the central two-thirds of the root tip. From autoradiographs made of longitudinal sections of pea roots fed ^{3}H-thymidine for 24 h (Clowes, 1961), one can discern a flattened quiescent centre in the shape of a concave–convex lens proximal to the root cap initials (Fig. 5.2B). The distal surface of the centre is separated from the root cap initials by a distinct boundary defined by thickened trans-verse cell walls especially apparent at the base of the columella. The proximal surface of the disk-shaped quiescent centre is defined by labelled cells ar-ranged in a shallow arc which give rise to the cells forming the central cylinder, the cortex and the epidermis. No continuity of cell lineages exists across the root-cap junction although the separation of meristematic layers which is so apparent in roots such as *Zea* and *Convolvulus* is not evident in *Pisum*. Thus arises the general comment that the histogenic boundaries in the *Pisum* root apex are indistinct. As in other roots studied, the relative height and breadth of the quiescent centre may change during ontogeny so that the zonation of meristematic layers may be more or less apparent. These variations are evident in the published literature and are part of the basis for the controversy about root organization in *Pisum*. According to Clowes (1972), the rate of mitosis in the quiescent centre of seedling pea roots is only about one-tenth that of the surrounding cells.

III. CELL CYCLE ACTIVITY IN THE ROOT APEX

Analyses of mitotic behaviour in the root apex of *Pisum* have been favoured by virtue of the relatively large size of the apex, its ease of access from seeds and the large size and excellent staining properties of the chromosomes. Wilson chose *Pisum* as a cytological test system (Wilson *et al.*, 1959) and reported the influence of a number of drugs on mitotic activity in pea root tips.

An elaborate analysis of the cell cycle using pea root apices has been made by Van't Hof and his associates [see Van't Hof and Kovacs (1972) for references]. This approach based on cytological squashes, mitotic counts and average responses of whole populations at the molecular and cell level, ignores the tissue relationships and produces a picture of the average cell unrelated to its place or individual activity in the apex. Valuable information about cellular activity comes from such an analysis which must ultimately be brought to bear on the organization of the apex as a tissue system.

Using the cell maceration technique of Brown and Rickless (1949) which allowed an accurate assessment by direct count of cell numbers, Brown (1951) determined the overall length of the cell cycle and the duration of each phase of cell division in the pea root apex at different temperatures. At 15°C he reported an overall cycle time of 25·5 h; at 25°C, 15·9 h; and at 30°C, 14·4 h.

Van't Hof and his associates used quite different techniques with pea root apices. A method devised by Wilson *et al.* (1959) for culturing excised pea root tips in synthetic nutrient medium for use in mitotic studies was applied by Van't Hof (1966), first to study DNA synthesis in the cells of the pea root apex, and thereafter (Van't Hof and Kovacs, 1972) for the regulation of the cell cycle. They found that in excised root tips in culture deprived of a carbo-hydrate substrate (usually sucrose), most cell division ceased within 24 h. After 72 h all DNA synthesis and cell division had stopped. Return to a carbohydrate-containing medium restored mitotic activity, the starvation period having served to synchronize cell cycle events in a portion of the apical cell population. Based on mitotic indices derived from the determination of the percentage of labelled interphase cells when provided with ^{3}H-thymidine and labelled mitotic figures, Van't Hof and Kovacs (1972) proposed that in the pea root apex, the mitotic cycle is governed by two principal control points, one regulating the transition from G_1 into the DNA-synthesis phase, and the other controlling the transition from G_2 into mitosis. These control points involve regulation of metabolic events, which depend upon available carbohydrate and aerobic processes to provide energy for, among other steps, new protein synthesis. Cells deprived of carbohydrate may arrest in either G_1 or in G_2; when allowed to renew the mitotic cycle upon the addition of carbohydrate, at least two sub-populations of cells in the apex can be distinguished. To date, no satisfactory evidence has been provided that allows one to relate the populations inferred from statistical data by Van't Hof and his colleagues to anatomically defined sub-populations such as those of the apical initials and of the quiescent centre.

For a comprehensive study of the ultrastructure of the root apex of *P. sativum*, the reader is referred to Sitte (1958).

IV. PRIMARY TISSUE DIFFERENTIATION

Depending on the cultural conditions and age of the primary root, the root cap may vary in length from 250 to 450 μm from the extreme tip to the root-cap initials. As the most clearly designated region of the root apex, the root-cap junction or initials make a convenient base line for determining levels of tissue differentiation (Popham 1955a,b; Torrey 1955). Proximal to the root-cap junction, the first clear primary tissue pattern is the complete procambial cylinder within the central region of the root axis which becomes apparent from 40 μm in cultured root tips to 80–220 μm in seedling roots. The first evident vascular tissue pattern discerned in transverse sections of the primary seedling root is the radial enlargement of late metaxylem elements which are blocked out as close as 175–330 μm proximal to the root-cap junction, making the triarch pattern characteristic of the primary radicle axis. The first proto-phloem elements mature at alternate radii to the future primary xylem at 300–500 μm proximal to the root-cap junction. The first mature xylem, characterized by secondary wall formation and lignification of the protoxylem elements at the periphery of the central cylinder, may mature as close as 5 mm proximal to the root-cap junction or as distant as 17·5 mm, depending on the rate of root elongation.

Root cell elongation occurs at a maximum rate from about 1·5–3·0 mm proximal to the root apex (Brown and Sutcliffe, 1950). At this level mature primary phloem has formed but no mature xylem is present.

The triarch pattern of primary vascular tissue differentiation (Fig. 5.3) appears to be very stable in the seedling radicle and the primary tap root of the pea plant. However, the primary vascular tissue arrangement in the lateral roots of the primary root is very variable as is shown in Fig. 5.4 (Torrey and Wallace, 1975). Arrangements of diarch, triarch, tetrarch and pentarch vascular patterns were observed in lateral roots, and within a given lateral root, transitions from diarch through triarch to tetrarch were observed as

E*

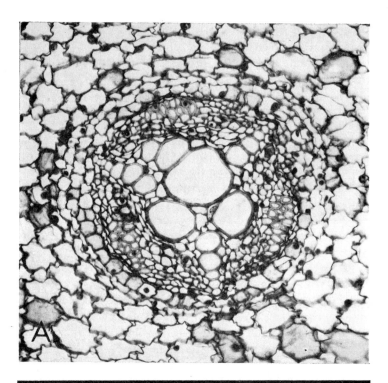

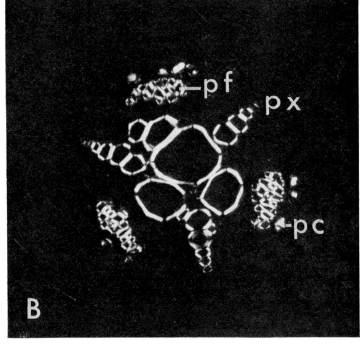

well as reductions from tetrarch to triarch or triarch to diarch. The factors
controlling the very stable pattern in the main root axis are clearly variable
in lateral axes.

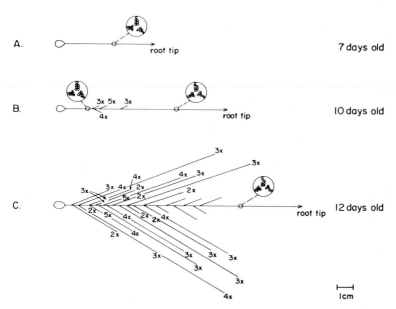

FIG. 5.4. Diagrammatic representation of primary vascular pattern (diarch, 2x; triarch, 3x; tetrarch, 4x; pentarch, 5x) seen in section in the primary root and in lateral
roots in seedlings of *P. sativum* of different ages. (From Torrey and Wallace, 1975)

At the level of mature primary vascular tissues in the seedling pea root at
1 cm or more proximal to the root apex, the three radiating strands of
primary xylem may be joined at the centre by fully differentiated and mature
metaxylem vessels forming a continuous xylem tissue. Under some conditions, the innermost cells of the procambial cylinder may remain as un-
differentiated parenchyma. At alternate radii are primary phloem bundles
characterized most strikingly by phloem fibres with thickened walls at the
outer periphery adjacent to the single-celled pericycle. Groups of sieve tube
elements and companion cells occur in the primary phloem area centripetal
to the fibrous bundle. Outside the pericycle is a single-layered endodermis
with not-always-easily-distinguished Casparian strips. The endodermis is the

FIG. 5.3. A. Transverse section of the primary root of *P. sativum*, cv. 'Alaska', showing
the mature primary tissues of the central cylinder. The triarch pattern of primary xylem
with large central late metaxylem elements is apparent. × 250. B. Photograph of section
shown in A, taken with polarizing lenses.
    Key: px—primary xylem; pf—primary phloem fibres; pc—calcium oxalate crystal in
pericyclic cell external to the primary phloem. × 250.

innermost layer of the wide cortex which is comprised of large vacuolated parenchyma cells with an extensive intercellular-space network. The outer surface layer is compact and well defined and shows root hairs at a level proximal to the region of maturing xylem about equivalent to the appearance of Casparian thickening.

Levels of tissue differentiation are dependent on the environmental conditions in which the roots develop, upon the genetic constitution of the plant, and upon the age of the plant. Aerated roots grow faster and show tissue differentiation occurring farther from the root apex than do more slowly growing non-aerated roots.

V. EXPERIMENTAL MANIPULATION OF VASCULAR TISSUE PATTERNS

The longitudinal course of tissue differentiation and maturation is affected by the rate of root elongation and probably other parameters as noted above. Any manipulation which affects root elongation may disturb or change the pattern of longitudinal tissue development. Thus Torrey (1953) showed that different metabolic inhibitors which reduced root elongation had different effects on tissue maturation. Thus indol 3 yl acetic acid (IAA) at 5×10^{-6}M at pH 5·0 inhibited root elongation by 90% but accelerated xylem tissue maturation. In contrast, 2,4-dinitrophenol at 10^{-4}M at pH 5·0, which inhibited root elongation by 90%, completely inhibited the maturation of both primary phloem and primary xylem, presumably blocking differentiation by uncoupling aerobic respiration from phosphorylation and energy transfer systems within the root.

Radial patterns of primary vascular tissue formation have also been studied in roots of pea. One experimental approach involved the cultivation in vitro of small excised root tips or excised pieces from the root. Reinhard (1954) showed that excised root tips of 0·7 mm length containing the root cap initial region, the proximal initials and the quiescent centre (not then identified as such) could be cultured successfully with the development of normal roots. Any other excised pieces below this length were not successfully cultured. Torrey (1954) was successful in culturing 0·5-mm root tips excised from cultured roots of pea. He found that such tiny excised tips required enriched levels of the vitamins thiamine and nicotinic acid, increased sucrose concentrations and the addition of micronutrients not required for cultivation of 5-mm root tips.

In most of the roots grown from 0·5-mm tips, the normal triarch vascular pattern was observed. In 20% of the roots studied a reduced pattern, either diarch or even monarch, was observed. During elongation a transition from monarch to diarch or to triarch was observed. From this observation it was concluded that the control of pattern formation resides in the apical meristem itself and that, as its activities changed, so also could the pattern change.

In related studies Torrey (1957) followed the regeneration of new root apices in cultured pea roots grown after 0·5-mm root tip removal and cultured in control medium in the presence of IAA. Regeneration occurred within 3 days, establishing a new elongating root with no striking external discontinuity. Internally, vascular tissue formation depended upon the hormonal constitution of the medium. In control medium, many roots showed the normal restitution of the triarch pattern with complete or partial continuity of the internal vascular tissues. Root tips regenerated in 10^{-2}mM IAA showed a hexarch vascular pattern with twice the number of xylem and phloem strands. Elongation in the high auxin concentration was markedly reduced. If such hexarch roots were transferred to media containing lower IAA concentrations, i.e. 10^{-3}mM, the vascular pattern of the elongating roots showed a transition back through pentarch and tetrarch towards the normal triarch condition.

From measurements of root diameter at different levels of tissue maturation in roots of different vascular patterns it was concluded that the auxin had influenced pattern formation by increasing the diameter of the procambial cyclinder of the root apex at the level where pattern formation was determined. A detailed discussion of primary vascular tissue pattern determination has been presented by Torrey (1965).

VI. SECONDARY TISSUE FORMATION

According to Hayward (1938), the seedling root of pea shows limited secondary thickening. A vascular cambium arises first in the procambial tissues lying between the primary phloem bundles and the primary xylem, extends around the poles of the protoxylem involving divisions of the pericycle, and ultimately forms a concentric cylindrical layer. At this stage there is usually a complete central region of primary xylem. Secondary xylem in the seedling root may be limited to a relatively small number of large xylem vessels.

In experimental manipulation of the pea root, one can induce a well developed vascular cambium with relatively extensive secondary xylem formation (Fig. 5.5). Torrey and Shigemura (1957) showed that excised root tips of pea taken from germinating seeds, when grown in culture medium containing 10^{-2}mM IAA, produced considerable secondary xylem and phloem. Tips from excised pea roots which had been grown in culture for one week then sub-cultured as first-transfer tips and treated with 10^{-2}mM IAA showed no cambial formation. It was believed that the lack of response in the sub-cultured tips resulted from the loss by dilution of factors coming from the cotyledons.

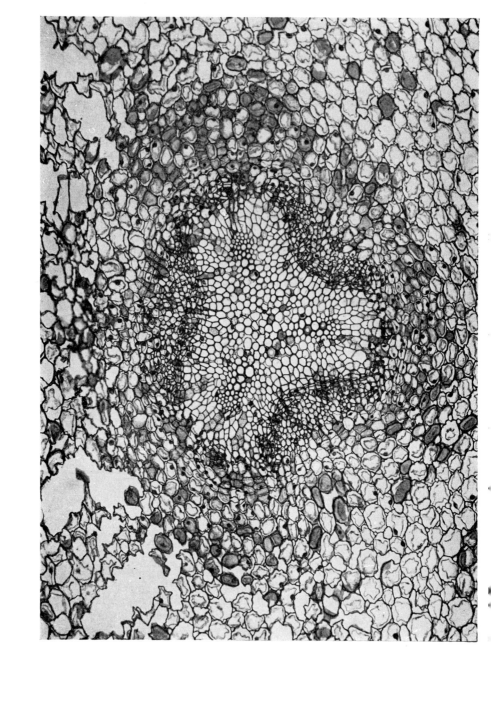

Fries (1954) found that "decotylized" pea seedlings formed roots with only primary vascular tissues, while pea roots excised from the seedling and grown in isolation showed secondary thickening at the basal region of the root and were seen to possess a much more extensive proportion of xylem tissue when studied in cross-section. Supplementing the medium of the decotylized plants with arginine at 0·3 mM increased the amount of xylem formation. Fries interpreted the results in terms of a depletion of the root by the attached shoot of factors needed for its normal development.

In further studies of the excised pea root system, Torrey (1963) reported that excised and sub-cultured first-transfer tips could be induced to form a vascular cambium and produce secondary tissues if the excised root was provided with nutrients and hormones from the proximal end by basal feeding. Cultured first-transfer tips formed secondary tissues when provided with the usual pea-root medium supplemented with 8% sucrose and IAA at 10^{-2}mM; the response could be augmented by adding growth factors including adenine sulphate at 10^{-1}mM and a mixture of organic nitrogen compounds comprising arginine, glutamic acid, glycine, asparagine, aspartic acid and urea, all at 1 mM. Cytokinins were not tested in this system. It was presumed that these factors or some of them at least might be provided from the cotyledons in the germinating seed. Sustained basal feeding of materials to the roots allowed the necessary stimuli for vascular cambium activity to progress acropetally along the root axis so that as much as 60% of the length of the root showed secondary thickening and at the root base there was more radial thickening than is normally found in the intact plant. It seems clear from these experiments that secondary vascular tissue formation in the root depends on nutrients and hormones which originate in the cotyledons and shoot, and that the expression of cell division and differentiation is dependent on the inter-related demands of the shoot and root.

The complexity of the inter-relationship between secondary and primary vascular tissue formation was shown in studies by Torrey (1951) on cambial formation of pea roots after decapitation. Here new lateral roots initiated proximal to the decapitation were associated with the formation of extensive new secondary vascular tissues under stimuli presumably arising in part from the lateral root meristems. Secondary thickening in this case progressed with time in a basipetal direction, supporting the idea that the stimuli for cambial formation arose in the lateral root meristems and moved in a proximal direction, i.e. from root apex towards the base.

VII. HORMONES: EVIDENCE FOR THEIR OCCURRENCE AND FORMATION IN THE ROOT

The root is a non-photosynthetic tissue system, typically underground and, as a heterotrophic organ, is dependent upon the green, photosynthetic shoot for a continuous supply of carbohydrates (probably in the form of sucrose) for its carbon and energy source, and also perhaps for other products of photosynthesis or of other syntheses in the shoot. The dependence of the root on the shoot for substrates and metabolites is demonstrated with excised root culture. Pea roots require a carbohydrate (preferably sucrose) and in addition to inorganic ions, require the organic compounds, thiamine and niacin, which are both B vitamins (Bonner and Addicott, 1937; Addicott and Devirian, 1939). In the absence of thiamine, HCl at 0·1 mg l^{-1} in the medium, excised pea roots can be grown for the first week, but first-transfer tips stop dividing and die. Thus thiamine can be characterized as a hormone which is normally provided from the shoot by photosynthetic activity and transported to the root apex, where it is essential for cell division activity (Addicott 1939). The root itself is unable to make thiamine; Bonner and Buchman (1938) were able to demonstrate that excised pea roots in culture were able to utilize the thiazole half of thiamine, presumably by synthesizing the pyrimidine portion of the molecule. Nicotinic acid is also an essential factor for the growth of excised pea roots grown in culture, since they appear to be unable to synthesize this molecule when grown in isolation in nutrient culture.

It is a striking fact that excised roots of pea grown in culture show no other organic nutrient requirement (although they are not easily propagated indefinitely in culture as will be discussed later). The question arises as to whether the root itself is capable of synthesizing other plant hormones which are usually presumed to play a role in cell growth and organ development.

A. Auxins

Considerable effort has gone into the question as to whether roots form auxins, and further still, whether auxins are required for root elongation or root development. Even today these questions remain unanswered or only partially answered. Indirect evidence suggests that auxin, probably IAA, plays a role in lateral root initiation in pea roots (cf. pp. 140–144; functions in the activation of the vascular cambium as discussed above; and plays a role in vascular tissue pattern formation (See also Chapter 8).

The presence of auxin in the pea root tip was claimed shortly after the discovery of auxin. Nagao (1936) placed excised seedling root tips on 10% glucose-agar blocks and showed that diffusible auxin demonstrable by the *Avena* coleoptile curvature was also present in the pea root apex. Thimann (1936) inferred from experiments with auxin-treatment of seedling roots of pea that auxin is not made in appreciable amounts by the root tip but that it is transported to the tip. In a series of experiments in which he extracted auxin from cultured pea root tips and bases, van Overbeek (1939) concluded that pea roots do synthesize auxin in culture, with the highest concentrations occurring in the tip. This evidence, based on careful ether extraction and *Avena* coleoptile curvature tests, may still be the best evidence of the presence and probable synthesis of auxin in pea roots. Even today, there seem to be only two bona fide demonstrations of the presence of IAA in root tips of any species, as demonstrated by rigorous chemical extractions, purification and characterization by gas chromatography and mass spectrometry. These reports are for *Zea mays* L. roots (cf. Greenwood *et al.*, 1973).

Exogenously applied auxin (IAA) suffers a variety of fates when provided to pea roots. In the early work by Tang and Bonner (1948) pea roots were shown to cause a rapid disappearance of IAA by an enzyme system they characterized as IAA oxidase. Van der Mast (1970) later found two IAA oxidase fractions in pea roots, one with peroxidase activity and one without this activity. IAA can be metabolized by pea roots to indol-3yl-acetaldehyde (Morris *et al.*, 1969) and perhaps also to oxindoles (Tuli and Moyed, 1967) and other oxidation products which have not been shown to play a definitive role in root development.

In a series of papers Andreae and his co-workers elucidated the action and fate of auxin in pea roots and by extension in pea plants as a whole (Andreae and Good, 1955, 1957; Andreae and van Ysselstein, 1960; Andreae *et al.*, 1961; Andreae, 1967). When auxin is applied to roots or root explants, it is taken up by the roots and metabolized or transported. The primary product of auxin metabolism is indol-acetyl-aspartate, a non-mobile conjugation product, and a lesser product results from the decarboxylation of IAA. According to these workers, α-naphthalene acetic acid (NAA) when applied to pea roots is not decarboxylated but does undergo conjugation, and 2,4-dichlorophenoxyacetic acid (2,4-D) is not affected by either process. These observations would explain many of the differences in their mode of action. Andreae and co-workers showed further that the internal concentration of auxin does not control the response to auxin; as long as the external supply of auxin is present, the response, whether it be inhibition or stimulation, will persist. The response to applied auxins is uniform enough to allow the use of pea root elongation inhibition by auxin as a bioassay for auxins (Audus and Thresh, 1953).

The auxin in roots may be transported from the cotyledons in seedlings or from the shoot in mature plants (cf. McDavid *et al.*, 1972). The evidence for polar transport of auxin in the root from the basal part of the root towards the root apex is now quite strong (cf. Batra *et al.*, 1975) and one must presume this situation applies to the pea root as well as to other systems where the matter has been more thoroughly studied. Morris *et al.* (1969) and Hillman and Phillips (1970) have studied the transport of IAA in the root of pea and confirm this view.

B. Ethylene

One fascinating aspect of the auxin role in root growth is the dilemma over whether the effect is one of direct auxin action or the result of auxin-stimulated ethylene production. Burg and his collaborators (Burg and Burg, 1966; Chadwick and Burg, 1967, 1970) ascribed almost all of the auxin effects on root growth and morphogenesis to ethylene. Scott and Norris (1970) studied the effect of exogenously applied auxins and ethylene to determine if they did indeed have identical effects on the pea root system. Their results demonstrated that all auxin effects except lateral root induction were mimicked by ethylene. Many researchers have noticed increases in adventitious root numbers after ethylene application; the lack of effect of ethylene on lateral root initiation indicates, therefore, a basic physiological difference between lateral and adventitious root initiation.

Andreae *et al.* (1968) questioned the role of ethylene in the auxin inhibition of root growth. They pointed out that ethylene effects are dependent on culture conditions to a much greater extent that IAA effects, and that growth inhibition by ethylene is not reversible after 16 h of growth in contrast to that by IAA. The results reported by Chadwick and Burg (1970) showing ethylene production after a lag period following treatment with IAA, are not compatible with the induction of inhibition immediately after IAA application. It can be concluded that, although auxin and ethylene are intimately inter-related, the inhibition of root elongation is not due to auxin-induced ethylene production, but is due to auxin action alone. No clear picture of the normal physiological role of either ethylene or auxin came from this work, although an involvement of ethylene in auxin-controlled geotropic responses in pea roots was suggested. Konings (1969) showed that IAA applied to pea seedling roots was rapidly transported acropetally, influenced geotropic response and could be modified by 2,3,5-tri-iodobenzoic acid (TIBA) treatment. No effort to examine ethylene effects was reported in this work. The possibility cannot be excluded that ethylene, which serves as a potent growth regulator when applied exogenously to pea roots, may serve a role at exceedingly low concentrations as a natural plant hormone required for normal plant growth and development. However, by exposing pea seedling roots to ethylene-depleted and ethylene-enriched air, Barlow (1976) concluded that it was unlikely that ethylene was a natural regulator of either DNA synthesis or cell division in the root apex, although he did not rule out the possibility that ethylene might play a role in cell growth (cf. p. 250).

The preoccupation during the 1930s with auxin as the main hormone in plants was changed (if slowly) by the "caline" theory proposed by Went (1938) from extensive studies of seedling development in the pea plant. Went suggested that auxin interacted with other hormone-like factors in the plant which had their origin in different organs of the plant and were transported to other parts of the plant where they had their effect. Roots were suggested as sites for auxin formation but also as sources of "caulo-caline", a factor made in the roots which moves to the shoot and together with auxin controls bud development, stem elongation, etc. This prophetic view of hormonal interaction remains poorly documented today, but a number of specific hormones have been identified since Went's day and some insight into their role has been gained. Evidence seems compelling, if still quite inadequate, that roots produce gibberellins, cytokinins, probably abscisic acid and perhaps other compounds which serve hormonal functions (cf. Torrey, 1976).

C. Gibberellins

Evidence for occurrence and formation of gibberellins in roots is sparse. Although there is a general belief that roots contain and actually synthesize gibberellins, this being based on rather indirect evidence obtained especially in studies of sunflower (cf. Crozier and Reid, 1971), good evidence for gibberellin synthesis in pea roots is lacking. Carr and Reid (1968) reported that "cycocel" (CCC), a compound believed to interfere with gibberellin synthesis, when applied to the root system of de-topped pea plants, substantially reduced the gibberellin activity of the exuded xylem sap. The presumption was that gibberellin was synthesized in the root in the absence of CCC. This root-originating gibberellin was believed to influence shoot growth in the seedling pea.

Radley (1961) reported that root nodules induced by the bacterium *Rhizobium* on roots of *Pisum* were rich sources of gibberellins, especially when compared with the root tissues of the same plants. The significance of this finding, which needs confirmation, is unclear, especially since treatment of roots (in *Phaseolus*) by basal feeding of excised roots with gibberellic acid resulted in 85% inhibition of nodule initiation (Cartwright, 1967).

D. Cytokinins

Cytokinin occurrence in pea seedling roots and in pea root callus was demonstrated by direct extraction and bioassay (Short and Torrey, 1972a,b). From evidence based on co-chromatography and mobility in different solvents, it was concluded that zeatin was the most active cytokinin in pea roots, occurring as the free base, as the riboside and ribotide, and in much lower levels as a component of transfer RNA extractable from the roots. The terminal $1\cdot0$ mm of the tip contained over 40 times more cytokinin than the next 5-mm-long root segments.

Babcock and Morris (1970) reported the isolation and characterization of cytokinins from hydrolysates of tRNA extracted from pea roots. The cytokinins were identified as *cis*-zeatin riboside and isopentenyladenosine (IPA). Transport of cytokinins in the xylem sap has been reported in many species including *Pisum arvense* L. (Carr and Burrows, 1966). McDavid *et al.* (1973) showed that root tip removal accelerated senescence of older leaves and reduced their rate of CO_2 fixation, an effect reversed by treatment of the shoot with synthetic cytokinins, suggesting a regulatory role exerted on the shoot by hormones from the root. Carr and Burrows (1966) postulated that cytokinins were normally produced in the root apex and transported to the shoot where they modify shoot growth and development. This view coupled to evidence for the directional transport of auxin, has stimulated the proposal by some researchers that a double concentration gradient involving auxin moving down from the shoot and cytokinins moving up from the root may be responsible for the ultimate control over localization of tissue differentiation and sites of root and nodule formation. Torrey's (1962) study of auxin-purine interactions on the initiation of lateral roots provides support for this hypothesis. However, with the existence of abscisic acid and gibberellic acid, not to mention ethylene, one would expect that the situation is really much more complex than this simplistic hypothesis seems to indicate.

E. Abscisic Acid

Tietz (1971) reported the presence of abscisic acid (ABA) at low concentrations in roots of 20-day-old pea seedlings, much lower than that which occurred in the shoots. Application of exogenous ABA to tap roots of peas produced inhibition of root elongation, induction of lateral roots and root swelling (Tietz, 1973). ABA is considered by many to be identical with the β-inhibitor complex and possibly the inhibitors found by Torrey (1952) in the root apex of pea. With this compound, as with the other growth regulators, not enough is known of the site of synthesis vs transport to other sites. Phillips (1971a) reported that ABA applied to the roots of pea at 2×10^{-3}mM inhibited nodulation by 60% without affecting root or shoot growth. This treatment had no effect on root hair infection nor on rhizobial growth in culture but markedly reduced cytokinin-stimulated polyploid mitotic activity in cultured pea root cortical explants. He suggested (Phillips, 1971b) that ABA might serve as the naturally occurring cotyledonary inhibitor which influences the time and place of nodule formation in pea seedlings. He also demonstrated that by precise timing, the same compound could be used to increase nodulation. He attributed this to ABA having two distinct effects, being able: (a) to increase the number of infection threads; and (b) to decrease cell proliferation after infection. By very short-term treatments he was able to increase the number of infection threads and then reduce the amount of ABA in order to allow development of an increased number of nodules.

VIII. ISOLATED ROOT CULTURE

Much of the available information about root growth and morphogenesis comes from isolated (excised) root cultures and the many manipulations possible. Since the first attempts of Haberlandt (1902) to culture excised plant parts, most of the media requirements for culturing excised pea roots over extended periods of time have been elucidated. Much of the initial work utilized Knop's solution with sugars and poorly defined extracts added for improved growth. Some of the early observations were quite phenomenal; for instance, Scheitterer (1931) demonstrated that 0·5 to 1·0-mm pea root apices placed in sterile Knop's solution and 2% glucose produced a six-fold increase in length in 6 days. However, in all the studies on culture of excised roots up to the middle 1930s, culturing root tips for more than several weeks proved impossible. During repeated sub-culture of the main apex on Knop's medium and glucose the growth rate became very low. Sub-cultures of lateral apices produced during culture gave little or no growth. It was therefore impossible to establish pure clones of pea root material, or to maintain roots in culture for extended periods.

Bonner and Addicott (1937) succeeded in developing a defined medium in which relatively long initial root tips, 3–10 mm in length, could be cultured and sub-cultured for extended periods. They found that a medium containing 4% sucrose, inorganic salts and vitamin B_1 (thiamine) would support growth during repeated sub-culturing for 3 months. The roots eventually became thin and ceased producing lateral roots. The addition of a mixture of amino acids improved the growth of the roots over extended periods. They were, however, unable to successfully culture excised lateral roots which were produced in culture. Addicott and Bonner (1938) demonstrated that the addition of nicotinic acid further improved the growth of pea roots. Modifications of the medium developed by Bonner and Devirian (1939), itself a modification of the original Bonner and Addicott (1937) medium (cf. Table 5.I), are in use in many of the laboratories currently working with excised

TABLE 5.I. Culture media devised for growing excised pea roots.

	Bonner–Addicott (1937) Medium (mg l^{-1})	Bonner–Devirian (1939) Medium (mg l^{-1})	Torrey (1954) Medium (for 0·5-mm tips) (mg l^{-1})
$Ca(NO_3)_2 \cdot 4H_2O$	236	242	242
$MgSO_4 \cdot 7H_2O$	36	42	42
KNO_3	81	85	85
KCl	65	61	61
KH_2PO_4	12	20	20
$Fe_2(SO_4)_3$	2		
Ferric tartrate		1·5	1·5($FeCl_3$)
Sucrose	40 g l^{-1}	40 g l^{-1}	60 g l^{-1}
Thiamine·HCl	0·1 mg l^{-1}	0·1 mg l^{-1}	1·0
Nicotinic acid		0·5 mg l^{-1}	5·0
$MnSO_4 \cdot H_2O$			4·5
$ZnSO_4 \cdot 7H_2O$			1·5
H_3BO_3			1·5
$CuSO_4 \cdot 5H_2O$			0·04
$Na_2MoO_4 \cdot 2H_2O$			0·25

pea root cultures (Fig. 5.6). A key factor in the development of a medium which allows maximum growth of excised root tissues was the establishment of a pH optimum. The initial pH of the Bonner–Devirian medium (BD) was 4·8 after autoclaving. An experiment conducted by Torrey (1954) demonstrated a pH optimum around 5·0 with a sharp drop-off in growth rate at values above 5·2 or below 4·8. Naylor and Rappaport (1950) found that roots of different pea varieties behaved very differently in the same nutrient medium, and they were unable to culture pea roots through many sub-transfers. No success in establishing potentially unlimited culture of pea roots has yet been reported.

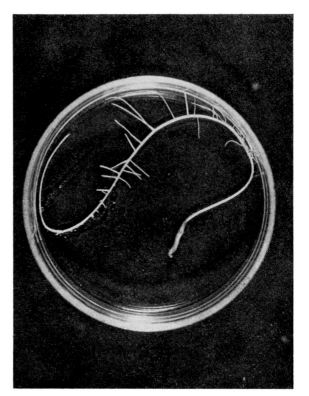

FIG. 5.6. An excised root of *P. sativum* grown from a 5-mm root tip for 3 weeks in a sterile nutrient medium in a 10-cm Petri-dish. Note typical acropetal pattern of lateral root formation.

A. Culture of Small Apices

Several researchers (Bonner and Addicott, 1937; Addicott and Bonner, 1938; Addicott, 1941; Wightman and Brown, 1953) demonstrated that the vitamins, nicotinic acid and thiamine, are needed for normal cellular divisions in the meristem of excised pea roots. The addition of these vitamins to a mixture of minerals and sugar produced a medium which allowed much better growth— a 2-mm root tip could grow to nine times its length in one week. Despite this improved growth, root tips shorter than 0·5 mm frequently developed into non-differentiated cellular proliferations (callus). This led these researchers to postulate the need for mature tissues for normal growth of immature, un-differentiated tissues. Root tips shorter than 3 mm have little mature vascular tissue, and those shorter than 0·5 mm consist only of the root cap and meri-stem, including the quiescent centre. This view found its origin with Haber-landt's work and was supported by that of Bonner and Addicott (1937) and others in the field. Scheitterer (1931), however, had achieved quite respectable growth with 0·5–1·0 mm root tips on Knop's solution plus sugar.

In an attempt to clarify the situation Torrey (1954) confirmed Scheitterer's work by growing 0·5-mm root tips on BD medium plus yeast extract. With this single additive (1 g l^{-1}) he found that 80% of the root tips would grow to 6 mm in 1 week and then grow normally thereafter. Upon further investiga-tion, he found that by increasing the concentrations of sucrose, nicotinic acid and thiamine and adding small quantities of several minor elements he could get normal, reproducible growth of 0·5-mm root tips on a totally defined medium (Table 5.I).

Several conclusions can be inferred from the above data:

1. The shoot normally produces nicotinic acid and thiamine and transports them along with sucrose or other sugars to the root, allowing normal growth and development.

2. Micronutrients are absorbed by the roots and retained in older root tissues.

3. These nutrients are required for maintaining normal growth of the apical meristem of the root.

4. It is possible to grow pea roots in excised (axenic) root culture and to determine from their responses to differing media what the role of the shoot is in the control of normal root growth and morphogenesis.

IX. LATERAL ROOT FORMATION

There are four categories of roots which develop on intact plants:

1. The primary root which is the extension of the embryonic radicle

2. Lateral roots which arise from meristematic tissues in existing roots (usually pericycle) to form secondary and tertiary branching patterns

3. Adventitious roots which arise from non-root tissues or tissues in mature roots which have undergone secondary thickening and have lost the capacity to develop lateral roots

4. Basal roots which arise from the basal region or transition zone of the plant (not yet demonstrated in pea, cf. Zobel, 1975).

The present discussion is limited to the growth and development of the primary root and physiological controls over initiation of lateral and adventitious roots. Experience indicates that, although there are differences in their patterns of initiation, the different types of root behave in an essentially identical fashion after initiation.

Lateral roots are initiated as the result of tangential division of cells in the pericycle. The site of initial division is opposite a primary xylem pole in peas. At the same time several of the neighbouring endodermal cells undergo division as if in response to the division of the pericyclic cells. Because of their endogenous origin, it has not been determined precisely how many pericyclic cells contribute to the final meristem of the lateral root. The divisions in the pericycle, together with those of the endodermis, are joined by sub-divided cortical cells (Fig. 5.7) which are incorporated into the primordium forming the root cap, quiescent centre and procambial tissues. The root primordium pushes its way through the cortex, and can then be identified as a lateral root. There are several lines of evidence which indicate that root initiation is a two-phase process in both lateral and adventitious roots (Pecket, 1957a; Eriksen, 1974; Mohammed and Eriksen, 1974). The first phase is the initiation of the root primordium and the second is the stimulation of the primordium to begin elongation as an intact root.

A. Lateral Formation in Intact Plants

Intact plants form laterals beginning at the base of the root and progressing down the root as the root grows, maintaining a distance of several centimetres between the apex and the most recent lateral. In young seedlings, removal of the cotyledons severely suppresses the initiation of lateral roots on the primary root (Rippel, 1937; Torrey, 1949). When laterals do form after cotyledon removal, they begin first at the root apex and then develop progressively farther back towards the base. Decapitation of the intact root on a normal seedling causes the initiation of laterals beginning at the decapitated apical end and progressing towards the base. Earlier studies by Thimann (1936) and Thimann and Went (1934) demonstrated that when auxin was substituted for an excised shoot the resultant reduction in root growth was alleviated and a slight increase above the controls was noted in the number of lateral roots formed. Auxin, however, did not substitute for excised cotyledons in the induction of lateral roots, and excision of the shoot without removal of the cotyledons had little effect on the initiation of lateral roots (Torrey, 1949).

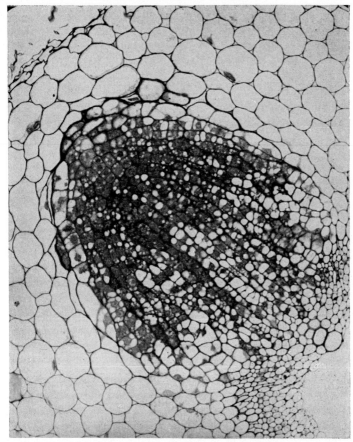

FIG. 5.7. Transverse section of a pea seedling root, showing a young lateral-root primordium cut in median longitudinal section. Notice the position of origin opposite the primary xylem strand. The lateral root is within a few layers of the outside, growing through cells of the root cortex. × 200. (Courtesy of M. McCully)

In a series of tests on intact pea seedlings, Torrey (1949) examined the effect of removal of the shoot, one or both cotyledons, or portions of cotyledons on the initiation of lateral roots. His data clearly showed a 50% reduction in the number of lateral roots with the removal of one cotyledon or half of each cotyledon, while total removal of cotyledons reduced the number of laterals to zero at 8 days (5 days after removal). When at least one-half of each cotyledon was present, the primary root had essentially normal growth rates, but upon total removal of the cotyledons, the growth rate was reduced to half, and additional removal of the shoot reduced the growth rate even further.

All of this evidence strongly supports the idea that a substance, either produced or stored in the cotyledons, is transported to the roots and there is responsible for the initiation of lateral roots. Auxin application to intact shootless and cotyledonless plants indicates that this hormone is not the lateral-root-initiation substance but that it can interact with it to induce greater numbers of roots. Further, the decapitation experiment suggests an inhibitory influence from the root apex. Is a substance which is inhibitory to lateral-root initiation produced in the root apex, or does the lateral-root-promoting substance moving from the cotyledons accumulate at the cut surface?

B. Lateral Formation in Excised Root Culture

1. Auxin effects

By using excised root culture Torrey (1949, 1950, 1956, 1959, 1962) succeeded in answering some of the above questions but has left several still unanswered. When initial 3- to 4-mm root tips are placed into culture, they respond to IAA treatment by producing lateral roots in a typical acropetal sequence. Root tips 10 mm long, excised from roots grown for 1 week in excised root culture do not respond, however, to auxin treatment, and will not form laterals until they have been cultured for at least 6 weeks in the absence of IAA. After setting up and analysing the results of a series of experiments based on the above phenomena, Torrey (1950) concluded that although auxin was required for the initiation of lateral roots, another substance, initially provided by the cotyledons and later by the root apex, was necessary. Because roots which had never been given exogenous auxin were able after an extended period of growth to initiate lateral roots, it was assumed that pea roots do produce some auxin in the apex. Pecket (1957a) and Nagao (1942) both came to the same conclusions as Torrey, and Pecket concluded further that two factors were involved in the responses described by Torrey. By analysing the initiation of lateral roots and separating the phenomena into two separate steps, namely primordial initiation and lateral-root emergence from the parent root, Pecket (1957a) concluded that these steps are controlled by separate factors or complexes of factors, both of which move in an apical direction. Later, Pecket (1957b) analysed these two phenomena in relation to auxin stimulation of lateral-root development and concluded that auxin probably stimulates an increase in the primordial initiation factor(s) while interacting with it in the initiation of lateral primordia. On the other hand, no evidence was presented on the nature of the emergence factor.

2. Other factors

Not only are auxin and one or more unidentified substances involved in the control of lateral-root initiation but other compounds can become limiting in excised root culture, and presumably these are normally provided by the shoot, or in some cases, the root apex. Torrey (1959) identified a need for adequate supplies of thiamine, nicotinic acid, adenine and several micronutrients for optimum lateral-root initiation. Presumably, the nicotinic acid and thiamine are normally provided by the shoot or mature-root tissues when adequate substrates are present, and the adenine is provided by the root apex. The role of a cytokinin as an apex-produced compound which acts in harmony with auxin in the initiation of lateral roots was suggested by Skoog and Miller (1957), and was considered as a potential mediator in the auxin–adenine synergistic control of lateral-root initiation by Torrey (1962). Extraction of compounds from the root tip yields one or more inhibitor(s) (Libbert, 1957; Torrey, 1956, 1959) which may play an important role in the mediation of lateral-root initiation. These inhibitors, still not unequivocally identified, presumably include compounds which affect auxin metabolism, as well as potential growth regulators such as cytokinins and ABA.

3. Light

One other aspect of the control of lateral-root initiation should be mentioned, because there appears to be a phytochrome-mediated system involved in the production of lateral roots and controlling root elongation. White light inactivates substances moving from the cotyledons to the roots, and also effectively stops lateral-root initiation as well as reducing elongation in excised roots. It can also inhibit effectively either auxin treatment or root decapitation. The most effective wave length in this inhibition is red light (Torrey, 1952). Furuya and Torrey (1964) demonstrated that the red-light inhibition is reversed by far-red light treatment, and concluded that a phytochrome system is involved. In their study the system producing lateral roots after root-tip excision was far more sensitive to the far-red, red light manipulations than the cotyledonary system. This may be due to the presence of additional pigments in the older tissues.

X. ADVENTITIOUS ROOT FORMATION

Adventitious roots, though not of regular occurrence in many varieties of pea, do form upon treatment of the intact plant or stem cuttings and pieces with auxin. Adventitious roots, in contrast to lateral roots, do not form at a specific site or in any specific tissue. They are found to originate from cell divisions in parenchyma cells near the vascular cambium of the stem, leaf, or mature-root tissue from which they are produced. These parenchyma cells may be descendants of an embryological pericycle, or part of the phloem, or a vascular ray. The site of origin can be at any position around the stem in the vicinity of the parental vascular tissues. The sequence of development follows that of lateral roots with the root cap and vascular connections completed prior to the emergence of the root from the parent tissues. After formation, adventitious roots grow out through the parent tissues, occasionally bending to by-pass especially tough sclerenchyma or other tissue regions, and push the surrounding cells aside. Once formed, there is no evident anatomical difference between lateral and adventitious roots. Adventitious roots in turn may produce lateral roots opposite the primary xylem pole.

Adventitious roots, like their lateral-root counterparts, require auxin for their initiation, and follow a two-phase developmental pattern. As with lateral roots, the first phase, primordium initiation, requires auxin and an unidentified additional substance (or substances). The second phase, root emergence and initial elongation, appears to be independent of auxin (Mohammed and Eriksen, 1974). The data of Eriksen (1974) indicate that cytokinins from the roots inhibit adventitious-root initiation but support root emergence. This hypothesis is supported by data of Kaminek (1967, 1969). Other compounds, such as gibberellins, may play a role in the control of auxin-stimulated adventitious-root initiation. Brian et al. (1960) demonstrated that exogenous GA_3 (gibberellic acid) prevented or reduced adventitious-root initiation, presumably by inhibiting the mitotic divisions giving rise to the primordia. Blahova (1969) demonstrated a reduction in gibberellin content of cuttings which could be correlated with the initiation of adventitious primordia. The physiological control of adventitious-root initiation in peas appears to be similar to that of lateral root initiation, and both involve complex hormonal and metabolic balances.

The initiation of roots from explants and callus tissues may be a special case of adventitious-root initiation. Torrey and Shigemura (1957), for example, described the production of callus tissues from vascular cambia of pea root tips. These calli then developed root primordia and roots spontaneously when placed on auxin-free media. Torrey (1967) reported the progressive loss in the capacity of pea callus tissues to initiate organized roots over long periods of culture. The loss was correlated with the development of aneuploidy and high polyploidy in the tissues. Because of the special origin of the callus tissues, these roots may be classified as laterals rather than adventitious roots, although there is no physiological or anatomical basis for any distinction in classification.

XI. FORMATION OF ROOT NODULES

A shadowgraph of a pea root system the same age as that of Fig. 5.1 is presented in Fig. 5.8; this pea plant, however, has been inoculated with the nodule-producing bacteria *Rhizobium leguminosarum*. The reader's attention is directed to the reduced number of lateral roots in the region of nodulation and the relatively normal numbers of laterals below the region of nodule development. Nodulation in the pea involves a rather complex series of events, beginning with the baterium penetrating the tip of a root hair. Details of the infection process in peas have been described by Libbenga and Harkes (1973). The infection is accomplished through the formation of an infection thread which then penetrates into the cells of the root cortex. The inner cortical tissues respond to the advance of the infection thread with the initiation of cellular divisions in polyploid or endomitotic cortical parenchyma cells far in advance of the thread itself. These cells are not pericyclic derivatives and are predominantly polyploid.

FIG. 5.8. Shadowgraph of the root system of a pea plant grown in the same manner as that in Fig. 5. 1 but inoculated with an effective strain of *Rhizobium leguminosarum*. Note the relative absence of lateral roots in the region of nodulation. × 1/3.

After it has induced a meristematic tissue made up of polyploid cells, the infection thread branches, penetrating all the cells of the meristematic region. Bacteria are released into the invaded polyploid cells and are eventually transformed to bacteroids. At the same time the cortical cells on the periphery of this meristematic region and towards the outside of the root continue to divide and a meristematic region is formed which gradually progresses outwards until it emerges from the root, forming a typical rounded nodule. Just behind this meristematic region, the infection thread continues to infect the cellular derivatives and release bacteria. This process continues until a mature nodule has formed (Fig. 5.9) and is terminated only when various developmental, physiological or genetic factors become limiting.

In some cases, elongate and branched nodules are formed, apparently by the splitting of the meristem into two or more separate meristems. The pre dominant form of young nodules is a small round sphere often slightl elongated [cf. Bond (1948) for a more complete description of nodule forma tion]. Anatomically, the nodule consists of a central region filled with bacter oids, surrounded by a cortical layer containing vascular tissue, and finally a outer layer of epidermal-like cells. At the apex is a meristematic tissue whic adds to the length of the nodule as it develops. The base of the nodule ha vascular connections to the vascular system of the root. A comprehensive study of the development of rhizobial nodules in pea using light and electro microscopy, and analyses of the related hormonal stimuli involved, have bee published recently (Newcomb, 1976; Syōno *et al.*, 1976; Syōno and Torrey 1976).

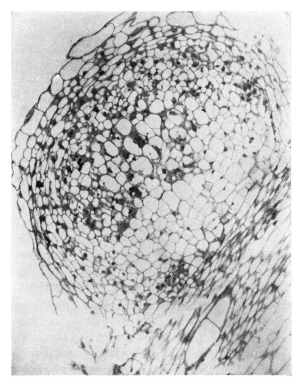

Fɪɢ. 5.9. Longitudinal section of a young rhizobial root nodule on the seedling root of *Pisum sativum* cv. 'Little Marvel' showing the central bacteroid-containing tissue and the surrounding uninfected cortical tissue. ×100

A. Nodule *vs* Lateral Root

For the most part nodules arise in the cortical tissues opposite a xylem pole, but may be found originating as well in the cortical tissues opposite the phloem pole. As is demonstrated in Fig. 5.8, there appears to be little lateral root initiation in the region where nodulation takes place. This observation has led, in the past, to the hypothesis that nodules are transformed lateral root apices. As was discussed by Libbenga *et al.* (1973), this hypothesis is not supported by the existing facts concerning nodule and lateral root development. Lateral roots are always formed opposite the xylem poles while nodules are not so precisely restricted. It has been found (Zobel, unpublished) that genetic modification of the pea plant by mutagenesis will yield a wide range of differing nodulation patterns, including root systems with normal numbers of lateral roots and nodules in the region of nodulation, or conversely, with neither nodules nor laterals in the region of expected nodulation. The lack of lateral roots in the region of nodulation in a normal plant may be associated with the metabolic and hormonal changes induced in the root by the infection of *Rhizobium*, resulting in a suppression of lateral-root initiation.

B. Physiology of Nodule Initiation

Torrey (1961), Phillips and Torrey (1970, 1972), Phillips (1971a,b) and Libbenga and Torrey (1973) demonstrated a rather complex system of hormonal gradients and sites of hormone production as the primary determining factors for the initiation and localization of nodulation. Torrey (1961) and Libbenga and Torrey (1973) developed a specialized tissue-culture explant system by isolating the cortex and epidermis of pea roots and culturing small explants on defined media. In the cortical tissue they demonstrated a need for auxin and cytokinin for the production of cortical proliferations. Phillips and Torrey (1970, 1972) showed that *Rhizobium* growing in synthetic nutrient medium produced and released a cytokinin into the medium. They proposed that proliferation in the seedling root cortex during nodule initiation was stimulated by rhizobial-cytokinin release. Phillips (1971a,b) later demonstrated that ABA inhibited polyploid mitosis in pea root segments, and also prevented cellular proliferation in nodule development. He further demonstrated the presence of a cotyledonary inhibitor of nodulation which he thought might be identical with ABA. Using this information and the results of some extraction studies of substances in the vascular cylinder of the root, Libbenga *et al.* (1973) developed a hypothesis which stated that an unknown cell-division factor (isolated only in crude extracts) exists in a transverse gradient within the root cortex and that this gradient, in conjunction with auxins and cytokinins released by the *Rhizobium* and present in the root, is responsible for the localization and ultimate induction of nodulation.

C. Genetic Variability

Many differing nodule types have been observed in peas, most of these du
to differing strains of *Rhizobium*. In a controlled environment, different strain
of *Rhizobium* may be used to inoculate a series of pea plants and produc
nodules which can be either the normal round unbranched type, or larg
and branched, or small and ineffective in nitrogen fixation, or of any of
number of differing types. In the experimental mutagenesis study mentione
before, pea varients have been observed which mimic every type of nodul
variation produced by the *Rhizobium* variants, including normal-sized nodule
without leghaemoglobin, and large branched nodules.

For further discussion of nitrogen fixation by root nodules, see Chapter 13

XII. GENETICS OF PEA ROOTS

The genetic control of root growth and development has not been investigate
to any great extent. All of the characteristics discussed in this chapter may b
considered to be under some form of genetic control but the exact nature o
this control has yet to be determined. It was recently reported (Contin an
Marx, 1974a) that a series of inbred pea lines demonstrated resistance t
being uprooted. This character appeared to be correlated with either tap-roo
thickness or number of lateral roots on the tap root. After a series of tests
the authors concluded that tap-root thickness was highly heritable and tha
tap-root branching was much less so. In a companion article, Contin an
Marx (1974b) described an inbred line which showed an increase in th
pattern of the vascular system of primary roots from 3 to 4 vascular strands
The genetic modification involved here may be useful in the study of th
control of vascular pattern.

Based on morphological or visual observations of progeny from mutagen
esis studies, Zobel (1975) reported that 30 % of the genes in plants are respon
sible for root characteristics. Further to this, he concluded that 10 % of th
plant genome is responsible for determining the characters of the root system
alone, 20 % for coding both root and shoot characters, and 70 % for strictl
shoot characters including chlorophyll and anthocyanin characters. In a study
of pea roots and nodulation in peas, Zobel (1975) determined that any
modification of nodule characteristics resulting from the use of mutant strain
of *Rhizobium* could be mimicked with mutation in the host plant. In his study
pea lines were found which had reduced numbers of secondary roots, and
there were several which had increased numbers of both secondary and
tertiary roots.

One aspect of pea roots little investigated and related to the genetics o
roots is the development of overall root-form and branching pattern. Casua

bservation of different pea varieties demonstrates the variation in branching
atterns which can be achieved with different varieties. The work of Contin
nd Marx illustrates this matter and gives some idea of the genetic nature
nd complexities involved. Yorke and Sagar (1970) and Klasova *et al.* (1972)
escribed the complex process which enters into the control of overall pattern
1 pea roots. The latter authors stated that the time up until initiation of any
ateral on the primary root after seed imbibition is genetically fixed and
haracteristic of each variety. Yorke and Sagar (1970) described a sequence of
oot development in which the onset of lateral roots caused a reduction in
longation of the root from which they arose, resulting in a cyclic type of
rowth pattern and in a clustering of lateral roots. It would appear from these
nd other results that the study of the genetic control of root growth should
ield considerable amounts of information, leading to the production of
nutant lines. It is possible that these lines could then shed some light on the
omplex physiological basis for the control of root growth and morphogenesis.

REFERENCES

ABBOTT, A. J. (1972). *J. exp. Bot.* **23**, 667–674.
ADDICOTT, F. T. (1939). *Bot. Gaz.* **100**, 836–841.
ADDICOTT, F. T. (1941). *Bot. Gaz.* **102**, 576–581.
ADDICOTT, F. and BONNER, J. (1938). *Science, N.Y.* **88**, 577–588.
ADDICOTT, F. T. and DEVIRIAN, P. S. (1939). *Am. J. Bot.* **26**, 667–671.
ANDREAE, W. A. (1967). *Can. J. Bot.* **45**, 736–753.
ANDREAE, W. A. and GOOD, N. E. (1955). *Pl. Physiol.* **30**, 380–382.
ANDREAE, W. A. and GOOD, N. E. (1957). *Pl. Physiol.* **32**, 566–572.
ANDREAE, W. A. and VAN YSSELSTEIN, M. W. H. (1960). *Pl. Physiol.* **35**, 225–232.
ANDREAE, W. A., ROBINSON, J. R. and VAN YSSELSTEIN, M. W. H. (1961). *Pl.
    Physiol.* **36**, 783–787.
ANDREAE, W. A., VENIS, M. A., JURSIC, F. and DUMAS, T. (1968). *Pl. Physiol.*
    43, 1375–1379.
AUDUS, L. J. and THRESH, R. (1953). *Physiologia Pl.* **6**, 451–465.
BABCOCK, D. F. and MORRIS, R. O. (1970). *Biochemistry, N.Y.* **9**, 3701–3705.
BARLOW, P. W. (1976). *Planta* **131**, 235–243.
BATRA, M. W., EDWARDS, K. L. and SCOTT, T. K. (1975). *In* "The Development
    and Function of Roots" (J. G. Torrey and D. Clarkson, eds), pp 299–325.
    Academic Press, London and New York.
BLAHOVA, M. (1969). *Flora* **160**, 493–499.
BOND, L. (1948). *Bot. Gaz.* **109**, 411–434.
BONNER, J. and ADDICOTT, F. T. (1937). *Bot. Gaz.* **99**, 144–170.
BONNER, J. and BUCHMAN, E. R. (1938). *Proc. natn. Acad. Sci. U.S.A.* **27**, 431–438.
BONNER, J. and DEVIRIAN, P. S. (1939). *Am. J. Bot.* **26**, 661–665.
BRIAN, P. W., HEMMING, H. G. and LOWE, D. (1960). *Ann. Bot.* **24**, 407–419.
BROWN, R. (1951). *J. exp. Bot.* **2**, 96–110.
BROWN, R. and RICKLESS, P. (1949). *Proc. R. Soc.* B **136**, 110–125.
BROWN, R. and SUTCLIFFE, J. F. (1950). *J. exp. Bot.* **1**, 88–113.

F

BURG, S. P. and BURG, E. A. (1966). *Proc. natn. Acad. Sci. U.S.A.* **55**, 262-26

CARR, D. J. and BURROWS, W. J. (1966). *Life Sci.* **5**, 2061-2077.

CARR, D. J. and REID, D. M. (1968). *In* "Biochemistry and Physiology of Pla Growth Substances" (F. Wightman and G. Setterfield, eds), pp 1169-118 Runge Press, Ottawa, Canada.

CARTWRIGHT, P. M. (1967). *Wiss. Z. Univ. Rostock* **16**, 537-538.

CHADWICK, A. V. and BURG, S. P. (1967). *Pl. Physiol.* **42**, 415-420.

CHADWICK, A. V. and BURG, S. P. (1970). *Pl. Physiol.* **45**, 192-200.

CLOWES, F. A. L. (1950). *New Phytol.* **49**, 248-268.

CLOWES, F. A. L. (1956). *J. exp. Bot.* **7**, 307-312.

CLOWES, F. A. L. (1961). *In* "Recent Advances in Botany", pp 791-794. Un versity of Toronto Press, Canada.

CLOWES, F. A. L. (1967). *Phytomorpholology.* **17**, 132-140.

CLOWES, F. A. L. (1972). *In* "The Dynamics of Meristem Cell Populations (M. W. Miller and C. C. Kuehnert, eds), pp 133-147. Plenum Press, New Yor

CONTIN, M. and MARX, G. A. (1974a). *Pisum Newsl.* **6**, 8.

CONTIN, M. and MARX, G. A. (1974b). *Pisum Newsl.* **6**, 9.

COOPER, D. C. (1938). *Bot. Gaz.* **100**, 123-132.

CROZIER, A. and REID, D. M. (1971). *Can. J. Bot.* **49**, 967-975.

ERIKSEN, E. N. (1974). *Physiologia Pl.* **30**, 163-167.

FRIES, N. (1954). *Symb. bot. upsal.* **13**, 1-83.

FURUYA, M. and TORREY, J. G. (1964). *Pl. Physiol.* **39**, 987-991.

GREENWOOD, M. S., HILLMAN, J. R., SHAW, S. and WILKINS, M. B. (1973 *Planta* **109**, 369-374.

HABERLANDT, G. (1902). *Sber. Akad. Wiss. Wien* **11**, 69-92.

HAYWARD, H. E. (1938). "The Structure of Economic Plants." Macmillan, Ne York.

HILLMAN, S. K. and PHILLIPS, I. D. J. (1970). *J. exp. Bot.* **21**, 959-967.

JANCZEWSKI, E. DE (1874). *Annls Sci. nat. bot.* **20**, 162-201.

KAMINEK, M. (1967). *Biologia Pl.* **9**, 86-91.

KAMINEK, M. (1969). *Biologia Pl.* **11**, 86-94.

KLASOVA, A., KOLEK, J. and KLAS, J. (1972). *Biologia Pl.* **14**, 249-253.

KONINGS, H. (1969). *Acta bot. neerl.* **18**, 528-537.

LIBBENGA, K. R. and HARKES, P. A. A. (1973). *Planta* **114**, 17-28.

LIBBENGA, K. R. and TORREY, J. G. (1973). *Am. J. Bot.* **60**, 293-299.

LIBBENGA, K. R., VAN IREN, F., BOGERS, R. J. and SCHRAAG-LAMERS, M. F (1973). *Planta* **114**, 29-39.

LIBBERT, E. (1957). *Z. Bot.* **45**, 57-76.

McDAVID, C. R., SAGAR, G. R. and MARSHALL, C. (1972). *New Phytol.* **7** 1027-1032.

McDAVID, C. R., SAGAR, G. R. and MARSHALL, C. (1973). *New Phytol.* **72** 465-470.

MOHAMMED, S. and ERIKSEN, E. N. (1974). *Physiologia Pl.* **32**, 94-96.

MORRIS, D. A., BRIANT, R. E. and THOMPSON, P. G. (1969). *Planta* **89**, 178-19

NAGAO, M. (1936). *Sci. Rep. Res. Insts Tôhoku Univ.* **10**, 721-731.

NAGAO, M. (1942). *Sci. Rep. Res. Inlts Tohoku Univ.* **17**, 137-158.

NAYLOR, A. W. and RAPPAPORT, B. N. (1950). *Physiologia Pl.* **3**, 315-333.

NEWCOMB, W. (1976). *Can. J. Bot.* **54**, 2163-2186.

PECKET, R. C. (1957a). *J. exp. Bot.* **8**, 172-180.

PECKET, R. C. (1957b). *J. exp. Bot.* **8**, 181-194.

PHILLIPS, D. A. (1971a). *Planta* **100**, 181-190.

PHILLIPS, D. A. (1971b). *Physiologia Pl.* **25**, 482–487.
PHILLIPS, D. A. and TORREY, J. G. (1970). *Physiologia Pl.* **23**, 1057–1063.
PHILLIPS, D. A. and TORREY, J. G. (1972). *Pl. Physiol.* **49**, 11–15.
POPHAM, R. A. (1955a). *Am. J. Bot.* **42**, 267–273.
POPHAM, R. A. (1955b). *Am. J. Bot.* **42**, 529–540.
RADLEY, M. (1961). *Nature, Lond.* **191**, 684–685.
REEVE, R. M. (1948). *Am. J. Bot.* **35**, 591–602.
REINHARD, E. (1954). *Z. Bot.* **42**, 353–376.
RIPPEL, K. (1937). *Ber. dt. bot. Ges.* **55**, 288–292.
SCHEITTERER, H. (1931). *Arch. exp. Zellforsch.* **12**, 141–176.
SCOTT, P. C. and NORRIS, L. A. (1970). *Nature, Lond.* **227**, 1366–1367.
SHORT, K. C. and TORREY, J. G. (1972a). *Pl. Physiol.* **49**, 155–160.
SHORT, K. C. and TORREY, J. G. (1972b). *J. exp. Bot.* **23**, 1099–1105.
SITTE, P. (1958). *Protoplasma* **49**, 447–522.
SKOOG, F. and MILLER, C. O. (1957). *Symp. Soc. exp. Biol.* **11**, 118–131.
SYŌNO, K., NEWCOMB, W. and TORREY, J. G. (1976). *Can. J. Bot.* **54**, 2155–2162.
SYŌNO, K. and TORREY, J. G. (1976). *Pl. Physiol.* **57**, 602–606.
TANG, V. W. and BONNER, J. (1948). *Am. J. Bot.* **35**, 570–578.
THIMANN, K. V. (1936). *Am. J. Bot.* **23**, 561–569.
THIMANN, K. V. and WENT, F. W. (1934). *Proc. K. ned. Akad. Wet.* **37**, 456–459.
TIETZ, A. (1971). *Planta* **96**, 93–96.
TIETZ, A. (1973). *Z. Pflanzenphysiol.* **68**, 382–384.
TORREY, J. G. (1949). PhD Thesis. Harvard University, Massachusetts, U.S.A.
TORREY, J. G. (1950). *Am. J. Bot.* **37**, 257–264.
TORREY, J. G. (1951). *Am. J. Bot.* **38**, 596–604.
TORREY, J. G. (1952). *Pl. Physiol.* **27**, 591–602.
TORREY, J. G. (1953). *Am. J. Bot.* **40**, 525–533.
TORREY, J. G. (1954). *Pl. Physiol.* **29**, 279–287.
TORREY, J. G. (1955). *Am. J. Bot.* **42**, 183–198.
TORREY, J. G. (1956). *Physiologia Pl.* **9**, 370–388.
TORREY, J. G. (1957). *Am. J. Bot.* **44**, 859–870.
TORREY, J. G. (1959). *Physiologia Pl.* **12**, 873–887.
TORREY, J. G. (1961). *Expl Cell Res.* **23**, 281–299.
TORREY, J. G. (1962). *Physiologia Pl.* **15**, 177–185.
TORREY, J. G. (1963). *Symp. Soc. exp. Biol.* **17**, 285–314.
TORREY, J. G. (1965). *Handb. Pfl Physiol.* **15**, 1256–1327. ✕
TORREY, J. G. (1967). *Physiologia Pl.* **20**, 265–275.
TORREY, J. G. (1976). *A. Rev. Pl. Physiol.* **27**, 435–459.
TORREY, J. G. and SHIGEMURA, Y. (1957). *Am. J. Bot.* **44**, 334–344.
TORREY, J. G. and WALLACE, W. D. (1975). *In* "The Development and Function of Roots" (J. G. Torrey and D. Clarkson, eds), pp 91–103. Academic Press, London and New York.
TULI, V. and MOYED, H. S. (1967). *Pl. Physiol.* **42**, 425–430.
VAN DER MAST, C. A. (1970). *Acta bot. neerl.* **19**, 363–372.
VAN OVERBEEK, J. (1939). *Bot. Gaz.* **101**, 450–456.
VAN'T HOF, J. (1966). *Am. J. Bot.* **53**, 970–976.
VAN'T HOF, J. and KOVACS, C. J. (1972). *In* "The Dynamics of Meristem Cell Populations" (M. W. Miller and C. C. Kuehnert, eds), pp 15–32. Plenum Press, New York.
WEBSTER, P. L. and LANGENAUER, H. D. (1973). *Planta* **112**, 91–100.
WENT, F. W. (1938). *Pl. Physiol.* **13**, 55–79.

WIGHTMAN, F. and BROWN, R. (1953). *J. exp. Bot.* **4,** 184–196.
WILSON, G. B., MORRISON, J. H. and KNOBLOCH, N. (1959). *J. biophys. biochem. Cytol.* **5,** 411–420.
YORKE, J. S. and SAGAR, G. R. (1970). *Can. J. Bot.* **48,** 699–704.
ZOBEL, R. W. (1975). *In* "The Development and Function of Roots" (J. G. Torrey and D. Clarkson, eds), pp. 261–275. Academic Press, London and New York.

6. Metabolic Aspects of Cell Growth and Development in the Root

J. K. HEYES

Department of Botany, Victoria University of Wellington, New Zealand

I. INTRODUCTION

Roots, particularly primary roots of seedlings, have long been a favourite experimental system for the study of cell growth and development in plants. The reasons for this are not difficult to find. The root is a geometrically simple organ with an apical meristem protected by the root cap, from which cells are produced basipetally which develop into the mature-root cells after a period of division and enlargement. As a consequence, serial or successive segments cut from the tip, towards the base, contain cells in progressively advanced stages of development (Heyes and Brown, 1956, 1965).

The analysis of such serial segments provides data on changes during the process of development. A similar approach has been made to the investigation of cell development in other tissues with distinctive meristems, such as the monocot leaf and coleoptile. Such tissues have the disadvantage that growth is limited in time and the meristematic region does not persist indefinitely, but the usefulness of the root is limited by the fact that, sooner or later, lateral meristems arise from the pericycle which complicate the geometrical symmetry and the developmental sequence.

Roots of *Pisum sativum* have been selected in many such studies for a number of reasons. For biochemical or chemical measurements, a reasonable

quantity of tissue is needed for measurements to be accurate and meaningf
and this makes the use of smaller roots more difficult and the majority of
studies using the serial-segment technique have been carried out using *Vic*
faba L., *Zea mays* L. or *P. sativum* L. The relatively large primary roots of
V. faba would seem to be ideal, but considerable difficulty is experience
whenever excised tissue fragments are studied due to the presence of a
active polyphenol oxidase system. The absence of any massive mucilag
secretion in *Pisum*, the ease with which sterile roots can be obtained for th
aseptic culture of root fragments and the ease with which consistent varietie
of seed can be obtained, have led to a large number of studies being mad
using *P. sativum* as a convenient experimental material rather than *Z. may*
although the two roots are similar in size.

It is, perhaps, less fortunate for those carrying out research on pea roo
that in the last 20 years there have been great changes in the commercia
varieties of peas available due to a change from selection for good canning c
"fresh-picked" qualities to selection for qualities suited to the frozen-vegetabl
industry. A second problem has been caused by the change to large-scal
mechanical harvesting which makes the supply of round pea seeds with fev
cracked testas extremely difficult. Large pea seeds with wrinkled coats, ric
in soluble sugars, are not only more difficult to surface-sterilize, but are sub
ject to considerable damage during harvesting (particularly when it is mech
anical). These changes provide seed samples from which consistently clear
disease-free primary roots are extremely difficult to obtain.

The selection of many new varieties in recent years has also made i
difficult to continue research with the same variety in some cases and there i
no doubt whatsoever that not only does the rate of root growth vary betwee
varieties, but the rate, extent and distribution of the basic growth-determinin,
processes (such as cell division, cell elongation and cell maturation) are no
invariable. As with much research in plant physiology, there are enormou
advantages in working with material from plants which are extensivel
cultivated as a horticultural crop with a defined, well-known and stable geneti
background, but these advantages tend to be negated when the character
selected for happen to change due to changing crop requirements. Despit
these difficulties—which the pea shares in common with most other moder
crop plants—the pea root remains entrenched as one of the most intensivel
studied tissue systems, particularly in studies of cellular aspects of growth anc
development.

II. THE SERIAL-SEGMENT TECHNIQUE

As stated above, at increasing distances from the root tip, cells are found to
be at a progressively more advanced stage of development. This fact has beer

exploited, and sufficient material has been obtained for biochemical and chemical analyses of such cells by cutting segments serially from the tip of seedling roots towards the base. Using a variety of simple sectioning devices, groups of roots can be sectioned at the same time and large numbers obtained for such analyses. By determining the quantity or activity of various parameters on such groups of segments, and by using a maceration technique followed by counts of cell numbers in suitably diluted aliquots (using a haemocytometer on further segment samples), data can be obtained which indicate the changes in the "average" quantity or activity per cell in each of the successive segments (Brown and Broadbent, 1950). These values have been interpreted as reflecting the progressive changes in a hypothetical average cell during cell growth and development.

There are two major restrictions on the usefulness or validity of this technique. Firstly, the thickness of each serial segment must be greater than a certain critical value which will depend upon the average cell length in that part of the root. Clearly, if cells are approximately 100 μm long, it is pointless attempting to cut segments even as thin as 200 μm, since 50% of the cells are likely to be cut, with a consequential loss of at least part of the cell contents. At a point where cell expansion has ceased in cortical cells of young seedling roots of *P. sativum* cv. 'Meteor' cells are about 150 μm long, and even at the commencement of the phase of maximum elongation, cells are still 20–30 μm long. Thus even with segments 1 mm thick, an appreciable proportion of cells are cut open when segments are cut. Another way of assessing the gravity of this source of error is to realize that about 2600 cells are cut when a root is sectioned transversely, and when the segment so obtained by two successive cuts contains fewer than a certain minimum number of cells (the 1 mm segment which is 10 mm from the root tip of cv. 'Meteor' contains only about 9000 cells), the proportion of damaged cells and hence cells with less than their original contents, becomes a factor worthy of consideration. Jensen (1955) fully realized this problem when attempting to use extremely thin segments of *V. faba* roots, and limited his measurements to the extreme tip of the root, before cell expansion, particularly in the axial direction, had begun.

The second weakness in the technique which has led to some criticism of the results obtained from its use, is that values obtained for the so-called average cell in each segment are unreal since the "average cell" does not exist, changes occurring in different types of cell at different positions within the root (Jensen and Kavaljian, 1958; see Figs 5.2A and 5.3B). The original intention in using this technique, and of the term, was to consider the hypothetical average cell as a concept and to treat the results as reflecting basic principles involved in cell development, rather than showing in precise detail changes occurring during development of any specific cell or cell type. The concept is still valid and even more valid are the consequences of a consideration of the primary results. A further justification, if such was needed,

rests with the fact which becomes apparent when a transverse section of a pe
root is studied. At 1·5 mm from the tip where the increase in average volum
of pea root cells begins, and at 7 mm from the tip, where the average volum
reaches a maximum, the proportion of cortical cells remains almost constan
comprising 66 and 70% respectively. Thus the majority of cells in each se
ment are cortical cells and it is changes in these cells which will contribute t
the greatest extent to the average cell changes obtained by the serial-segmer
technique. The technique was first developed to study the process of ce
expansion, including its contribution to root growth as a whole and it
control, and it was recognized that cell growth occurs to a greater extent i
certain root cells than in others, and in some cells, considerably more in on
dimension than another. Such differences between cells in different root tissue
need not invalidate the whole approach, since it was considered that ce
expansion occurred in all cell types, and what was being sought was th
underlying common denominator for all expanding cells, regardless of th
direction of their eventual differentiation.

As will be mentioned later, data obtained by the use of the serial-segmen
technique alone can be considerably amplified and made more meaningfu
by the use of other techniques in parallel studies. Studies of structural change
may be made using classical histochemical, autoradiographic and electron
microscopical techniques and these often amplify and extend findings obtaine
using the serial-segment and the "average cell" concept, but seldom (if ever
have they completely invalidated conclusions drawn from the use of th
basic method as outlined.

III. CHANGES IN THE AVERAGE CELL

One of the first descriptions which the use of the serial-segment techniqu
made possible, was that of the quantitative changes in cell contents from th
meristem to the stage of early maturity. Two sets of data are shown in Fig
6.1 and 6.2, both obtained using young pea-seedling roots, but obtaine
several years apart, in different laboratories and using two different varieties
Included in the figures are the changes in water content (fresh wt. — dry wt.
per cell, dry wt. per cell and protein per cell. Cell growth was completed i
one case at about 5 mm from the apex and in the later experiments at abou
8–9 mm from the apex, a situation also found (probably coincidentally) fo
cv. 'Alaska' by Sutcliffe and Sexton (1969). The final cell volume (as reflecte
by water content) was remarkably similar in the two cases, as were the dry
weight and protein content per cell, even though the duration and rate o
cell expansion were probably quite different, as was indicated by the differen
distance from the tip at which cell expansion was complete. Cell growth i
the pea seedling root has always been found to be accompanied by an increas

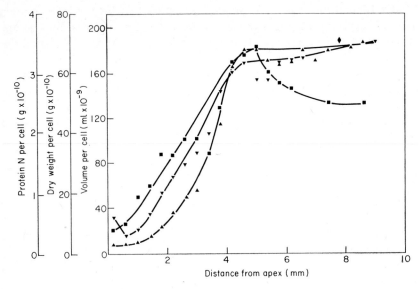

FIG. 6.1. Average water per cell, —▲—; dry weight per cell, —▼—; and protein N per cell, —■—; all measurements taken at increasing distances from the apex of pea-seedling roots. (From Brown and Broadbent, 1950)

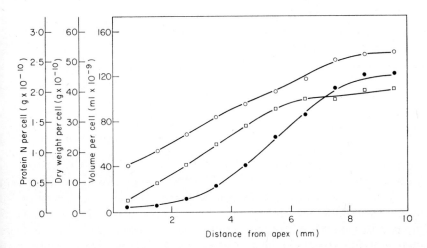

FIG. 6.2. Average water per cell, —●—; dry weight per cell, —□—; and protein N per cell, —○—; all measurements taken at increasing distances from the apex of pea-seedling roots. (Protein N values from Heyes, 1960)

F*

in both protein and total dry weight, indicating that, in the intact root at
least, cell development involves an increase in cellular material. Such simple
primary data show even more than the fact that cells increase in volume and
in cellular materials. At the apex, protein constitutes the major portion of the
total dry weight of the cell, while as cell development continues, it makes up
a progressively smaller proportion, reflecting the fact that non-protein
material must be laid down at a greater rate than protein as cells develop.

Other cellular constituents also increase, as can be seen from the data of
Fig. 6.3, in which the changes in inorganic phosphate, organic (alcohol-

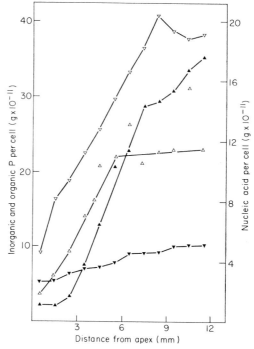

FIG. 6.3. Average DNA per cell, —▼—; RNA per cell, —▽—; inorganic phosphorus
per cell, —▲—; and organic (alcohol-soluble) phosphorus per cell, —△—; all measurements taken at increasing distances from the apex of pea-seedling roots. (RNA and
DNA values from Heyes, 1960)

soluble) phosphorus, ribonucleic acid (RNA) and deoxyribonucleic acid
(DNA) are shown. All of these constituents increase during this phase of cell
development, and they tend to do so to different extents. This again reflects
the fact that during cell expansion and early maturation, not only is there
more than just an inflation due to water uptake, but there is an increase in all
components—whether vacuolar, cytoplasmic or nuclear—and there is
evidence that the increase is not a uniform one. These differential changes

mply that the capacity of cells for accumulating or synthesizing cellular
constituents changes progressively during development, indicating that the
metabolic capacity of the cell is undergoing a progressive change.

V. CHANGES IN CELLULAR METABOLISM

The changing composition of cells indicated by the use of the serial-segment
technique was naturally extended to a study of the changes in enzyme activity
in preparations from groups of serial segments. Among the first studies was
that of Brown and Broadbent (1950), which showed that the respiratory
capacity of pea-root cells increased some six-fold during expansion, reached
a maximum at the same point at which cell volume and protein increase
ceased, and then declined. The use of a more sensitive technique (the oxygen
electrode), reported by Sutcliffe and Sexton (1974), showed even more clearly
that not only does respiratory activity increase during cell expansion by a
factor of more than eight, but in these roots the oxygen uptake per unit
protein increased four times. This indicates that while protein is increasing,
the ability of that protein to support aerobic oxidation is also increasing,
directly implying a change in the metabolic capacity of the cell.

Sutcliffe and Sexton (1969) also showed that the patterns of change in the
activity of invertase, β-glycerophosphatase and adenosine triphosphatase are
not the same and their data are presented in Fig. 6.4. This supports evidence
obtained from similar studies with roots of other species (e.g. *V. faba* as
reported by Brown and Robinson, 1955), that the activities of different
enzymes change to different extents during development.

Fowler and Ap Rees (1970) have shown that during the growth of pea-root
cells, there is a switch in the pathway of respiration from the use of the glycolytic to the pentose phosphate pathway, while Gibbs and Earl (1959)
had obtained evidence from the study of enzyme changes in young and old
pea roots that this change was a consequence of a change in the actual
enzyme complement. This question has been thoroughly re-investigated
using the serial-segment technique by Sutcliffe and Sexton (1974) who studied
the activities of four enzymes, these being: glucose-6-phosphate dehydrogenase, 6-phosphogluconate dehydrogenase (two pentose-phosphate-pathway
enzymes), the NAD-dependent glyceraldehyde-3-phosphate dehydrogenase
(from the glycolytic sequence) and malate dehydrogenase (from the Krebs
cycle). The activities of these enzymes, expressed per unit protein are shown
by the data of Fig. 6.5. It is clear that there are different patterns of change
in the activities of the pentose-phosphate-pathway dehydrogenases, the
glycolytic dehydrogenase and the Krebs cycle dehydrogenase, adequately
confirming the evidence obtained earlier.

The question of the precise meaning of changes in enzyme activity has been
raised by many workers. Changes in activity have frequently been interpreted

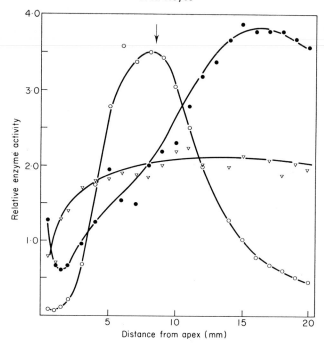

Fig. 6.4. Relative enzyme activity per cell of: invertase, —○—; β-glycerophosphatase —●—; and adenosine triphosphatase, —▽—; all measurements taken at increasing distance from the apex of pea-seedling roots. The arrow denotes the end of cell expansion. (Redrawn from Sutcliffe and Sexton, 1969)

as reflecting changes in the quantity of different enzyme proteins (Heyes and Brown, 1965), but a wide range of factors can influence enzyme activity apart from changes in the quantity of enzyme protein, such as activators or allosteric effectors (Truman, 1974), and many workers have attempted to prove unequivocally that there is a parallel change in enzyme protein. Sutcliffe and Sexton (1974) used polyacrylamide gel electrophoresis to separate proteins extracted from different regions of the pea root, but the changes observed were confined to changes in the intensity of the minor bands. Again, this rather inconclusive finding is not evidence against a change in the proteins, since each of the 20 protein bands observed probably contains many proteins and it is highly unlikely that they would all change in the same sense simultaneously. A further refinement of this technique, in which pea roots were incubated in a solution containing a ^{14}C-labelled amino acid for a period of time and then the protein was extracted from distinct regions after protein synthesis had occurred, was reported by Tulett (1963). Little difference was observed in the pattern of proteins as revealed by staining with amido-black, but the relative incorporation of the labelled amino acid in the different

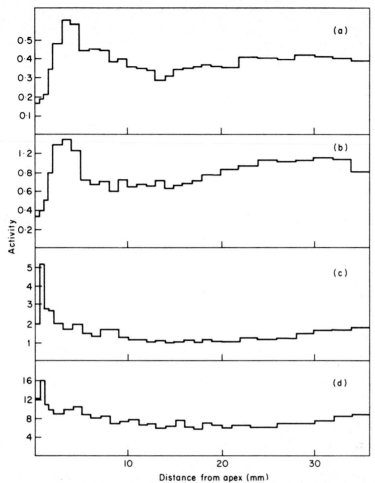

FIG. 6.5. The activity of (a) glucose-6-phosphate dehydrogenase, (b) 6-phosphogluconate dehydrogenase, (c) glyceraldehyde-3-phosphate dehydrogenase and (d) malate dehydrogenase in 0·5 and 1-mm zones cut at increasing distances from the apex of pea-seedling roots. All activities are expressed as μmol NADP or NAD reduced or NADH oxidized min^{-1} mg protein N^{-1}. (Data from Sutcliffe and Sexton, 1972)

protein bands was not constant in progressing from apical to elongating to mature cells. This result indicated not simply that the protein composition was changing during development, but that the protein being synthesized was changing as growth and development proceeded.

The nature of the relationship between the changing metabolic pattern and cell development has been further investigated by an extension of the serial-segment technique. Brown and Sutcliffe (1950) showed that segments excised from the growing zone of pea roots will continue expansion when transferred to a suitable medium, reflecting in their capacity for growth the stage of development attained by the cells at the time of excision from the intact root. Segments cut from the zone of cell expansion continue to expand when provided with an energy source such as sucrose, the segment 2–4 mm from the tip giving maximum growth. Segments cut closer to the apex exhibit a lag phase prior to the onset of the maximum rate of cell expansion, while segments cut further from the tip exhibit maximum rates of expansion immediately, but cease expansion sooner (Robinson and Brown, 1954).

It is considered that these observations reflect the fact that the protein or enzyme complement of the cells is initially in a state where it will not support growth, changes into a state which is capable of supporting growth and, at a later stage of development, changes into a state which no longer supports growth. Changes were observed in the catalytic properties of isolated segments similar to those found in the intact root; for example, respiration increased during expansion and decreased upon cessation. Similarly the pattern of change of invertase and phosphatase reflected the changes occurring in the intact root. Heyes and Vaughan (1967b) showed similar changes for invertase, phosphatase, ribonuclease and oxygen uptake, the initial increase in enzyme activity being completely dependent upon the presence of sucrose. In the absence of sucrose, not only did growth not occur, but the protein content, instead of remaining constant or decreasing slightly, underwent a considerable decrease (20% in 24 h). The implication is that in the presence of sucrose, the protein content is maintained due to protein synthesis almost balancing protein degradation, while in the absence of sucrose, no synthesis (or little) occurs. This conclusion is supported by the fact that even when sucrose is present, an inhibitor of protein synthesis blocks the transient increase and the enzyme change resembles that obtained in the absence of sucrose.

Therefore, it appears that there is a considerable body of evidence, from the use of the serial-segment technique, that cells undergo a change in their protein and enzyme complement during these early stages of development. The differentiation of distinct tissues and cell types has always carried with it the idea that the more defined structure and functions of the specialized, fully differentiated cells must be a consequence of the possession of different enzymic capabilities. The extension of this recognizable cyto-differentiation to include the less obvious change from the meristematic, through the expanding to the mature cell can be considered to change the whole concept of cellular differentiation in plants. Cell growth can be considered as but one aspect of cell differentiation; all cells become differentiated when they leave the meristem and lose some or all of the characteristics of such cells. They usually lose their isodiametric shape, the cell wall almost inevitably changes, vacuolation occurs to a variable extent—indeed, the cell becomes committed to a particular course of differentiation determined by its position and other micro-environmental factors. Furthermore, the more commonly accepted phenomena of differentiation—the acquisition of special shapes, thickenings, biosynthetic capabilities, etc.—only occur during or after expansion has occurred. Cell expansion is thus considered to be not only one aspect of cell differentiation but an essential vehicle for the more obvious types of cell differentiation.

V. CHANGES IN NUCLEIC ACIDS

As the data of Figs. 6.2 and 6.3 show, the average RNA and protein contents increase during cell growth to very similar extents, the RNA/protein ratio remaining remarkably constant. Such a parallel change in RNA and protein has been observed in other systems where the rate of protein synthesis could be considered to be limited by the amount of RNA. Evidence for the involvement of some or all of the cellular RNA in effecting the metabolic changes during cell development is still fragmentary, but comes from a number of lines of investigation. The major experimental approaches involve the use of analogues or inhibitors which interfere with the normal course of nucleic acid metabolism, and the more sensitive analysis of nucleic acid composition— mainly of the RNA—and of its changes during cell development.

Early attempts to fractionate total cellular RNA and to identify changes in composition similar to those found for enzymes and other proteins provided some evidence for there being such a change. The base composition (Heyes, 1960) of total RNA extracted from the initial 4 mm of the apex was only slightly different from that extracted from fully expanded cells at 8–12 mm from the apex, but the RNA changed considerably at 30 mm from the tip. A portion of cellular RNA could be extracted from the precipitated macromolecules after maceration of tissue using cold 1 N perchloric acid, and the proportion of the total RNA that was extractable was found to change progressively in successive 3 mm segments cut from seedling roots, 80% being extractable in the terminal 3 mm, but only 55% at 9–12 mm from the apex. The meaning of such changes remains unclear, as further attempts to fractionate total RNA using similar techniques yielded only slight additional information.

It is possible to reduce the "extractability" of RNA from acid-precipitated macromolecules from pea roots by the addition of 10^{-1}mM Fe^{3+} salts during the maceration process, but there is little effect of other ions. This ability to modify RNA stability or ease of acid extraction using Fe^{3+} salts may be reversed by EDTA and does not fit easily with a suggestion that the difference in composition of the extractable and non-extractable moiety could be due to partial digestion of RNA by RNase. Attempts to study changes in nucleic acid composition since the elucidation of the different classes of RNA has yielded little additional information. It is clear that in many tissues, especially leaves, the stability of ribosomal RNA changes, usually decreasing with age, particularly with regard to the stability of the chloroplast ribosomal RNA, but such changes which can be modified by the ionic environment do not appear to occur in pea roots.

Attempts have been made to detect changes in the biosynthesis of specific RNA molecules by using labelled precursors, interest naturally being centred upon the RNA fraction known as mRNA. Separation of total pea-root RNA using chromatography on MAK columns (methylated albumin-coated kieselguhr) has failed to provide an identifiable single fraction of D-RNA, but there is some evidence that a bound fraction contained this type of RNA together with another AMP-rich RNA (Jackson and Ingle, 1973a). The use of ^{32}P-orthophosphate in labelling newly synthesized RNA has enabled the pattern of nucleic acid biosynthesis to be determined after a 1 h labelling period. Jackson and Ingle (1973b) made the remarkable observation that the presence of ^{32}P in the incubation liquid bathing the roots severely modifies the growth pattern. Under the experimental circumstances, the rate of root extension remained constant for 16 h, but the addition of only 60 μCi ml^{-1} of ^{32}P to the incubation liquid, which already contained 0·4 μM carrier orthophosphate, resulted in a sudden cessation of growth after 5 h. The addition of 120 μCi ml^{-1} caused growth inhibition after 1·5 h and 160 μCi ml^{-1} caused cessation of growth after only 0·5 h. Analysis of the nucleic acids— particularly the rRNA and the rRNA precursors using electrophoresis on 2·2% polyacrylamide gels—showed that variation in the concentration of ^{32}P about the roots during the incubation period of 1 h changed not only the growth potential of the root but also the pattern of isotope incorporation. The data could be interpreted as indicating that the radio-isotope caused an accumulation of the RNA precursor (mol. wt. 2·5 × 10^6), and does not preclude an effect upon the synthesis of the non-ribosomal polydisperse RNA. These results make the use of the high concentrations of ^{32}P needed in short-term labelling experiments invalid. It is thus difficult to confirm the suggestion (Brown, 1963) that factors responsible for the changing protein synthesis during cell development in the pea root may include a changing informational RNA produced in the nucleus.

VI. ULTRASTRUCTURAL CHANGES

It is extremely difficult to separate structure and function at the sub-cellular level, but there is considerable evidence for a changing metabolic capacity that can be related not simply to an enzyme or to a type of molecule, but to one of the recognizable organelles. Cytological changes were observed using electron microscopy of isolated sub-cellular fractions from the roots of Z. *mays* by

Lund *et al.* (1958) in a parallel study of such changes and biochemical changes. Similar findings have been made for the pea root (Heyes, 1963b) where the average cellular mitochondrial RNA and protein content increases to a considerably greater extent than the content of microsomal RNA and protein. Loening (1961) demonstrated that the composition of the microsomal fraction isolated from serial segments of pea roots also changed. A free ribosomal fraction, with no membranous vesicles, was separated from the more typical vesicular fraction. The proportion of the membranous fraction increased with the age of the cell—a not surprising result when the increase in cell membranes concomitant with increased vacuolation is considered. During cell expansion the protein and RNA associated with the vesicular component increased considerably while the RNA and protein content per cell of the free ribosomes increased at first and then decreased.

The nucleus also changes in its composition and properties as Lyndon (1963) has shown with nuclei isolated from successive pea-root segments. The picture which emerges from all of these studies is one in which the pea-root cell is undergoing a progressive change in its metabolic capacity during growth, but the mechanism through which the change occurs has still to be clearly shown.

VII. CELL WALL CHANGES IN RELATION TO GROWTH

Studies on the cell walls of pea roots to determine the extent to which the wall alone may determine the growth potential of a cell have yielded considerable information. Although studies using tissue cultures of more or less identical cells have provided much data (e.g. suspension cultures of sycamore), the relevance of such studies to cell growth in an intact organized system is not always clear.

The cell wall has long been known to be anything but inert and a number of activities have been postulated to be associated with the cell wall and indeed to be involved in some part of the control mechanism. Dixon (1963) found that cell walls isolated from pea roots and thoroughly washed contained not only a hydroxyproline-rich protein, but that the enzyme invertase had a higher specific activity in the wall preparation than in the non-wall, cytoplasmic proteins. Other enzyme proteins have also been thought to be localized within the cell wall, presumably having been either secreted by the cell and incorporated along with the partially synthesized polysaccharide molecules, or entrapped from the plasma membrane as partially synthesized polysaccharide molecules are transported to the growing wall. The finding (Lamport and Miller, 1971) that 85% of the hydroxyproline is linked to arabinose, along with the suggestion (from work with other tissues) that the major structural polysaccharides are cross-linked through glycoproteins containing arabinose-linked hydroxyproline-rich peptides and galactose-linked serine residues (these serve to reduce wall extensibility, see Lamport, 1973), places cell-wall proteins in a rather unique position. Studies by Vaughan and Cusens (1973) and Vaughan *et al.* (1974) on the effects of hydroxyproline and other amino acid analogues on the growth of pea-root segments have supplemented results from research into the mechanism of action of ethylene on cell growth and cell wall changes in the pea epicotyl (Ridge and Osborne, 1970).

TABLE 6.I. Protein-bound hydroxyproline (Hyp) content of the cell walls of successive segments (2 mm) cut basipetally from the root tip of *Pisum sativum*. (Data calculated from Vaughan and Cusens, 1974)

Distance from root tip (mm)	Cell-wall protein (pg per cell)	Hyp (pg per cell)	Hyp content (µg of Hyp per mg of cell-wall protein)	No. of cells per segment
0–2	189	5·5	29·0	153 000
2–4	218	9·2	42·1	110 000
4–6	388	21·6	55·8	49 000
6–8	681	59·5	87·3	27 000
8–10	861	86·1	100·0	18 000
10–12	875	100·0	114·3	16 000

The data of Table 6.1 show quite clearly that in successive 2-mm segments cut from the root, there is a progressive increase in the hydroxyproline content of the cell-wall protein. The amount of wall protein per cell increases up to the time that cell expansion ceases, but the proportion of hydroxyproline increases markedly even when the total amount of cell wall protein is increasing only slightly.

The studies by Brown (1961, 1963a,b) on intercellular cohesion in the pea root implicated an acid-resistant component of the total intercellular cohesion. The data showed that intercellular cohesion is under quite close metabolic control, various metabolic inhibitors promoting the formation of firm, acid-resistant cohesion between cells particularly between the axial walls of adjacent cells. On the basis that this component involved protein linkages, for which some evidence had been obtained, Brown concluded that there was a connection between the role of cell wall protein bonds in mediating intercellular cohesion and in the control by such bonds of cell growth. These observations of changes in cell wall composition could be extended to include changes in the carbohydrate composition of cell walls, but these have not been studied in as comprehensive a manner as they have in the cell walls of *Allium cepa* roots (Jensen, 1960) or of the lupin hypocotyl (Monro *et al.*, 1972). In these tissues, it is clear that the average cell wall composition changes progressively during cell expansion and development with respect to the proportion of different classes of polysaccharide, and to the degree of association between polysaccharide and protein components.

The involvement of the plasmalemma in the control of growth of the cell wall with one of the following possibilities is likely to receive further attention in the near future. These include:

1. A hormone-mediated conformational change in the membrane leading to altered activities
2. A hormone-sensitive membrane-bound anisotropic ATPase or proton pump effecting wall softening
3. A transcriptional factor-release on hormone activation as mentioned by Roland (1973)

The mechanism of growth control in the normal, intact pea root under the influence of the endogenous hormone levels or as controlled primarily by changes in the cellular metabolic capability (with or without the involvement of hormones) is something that must be shown to be equivalent or at least comparable to the artificial test systems which the physiologists have developed.

VIII. STUDIES USING ISOLATED SEGMENTS

Data obtained showing the changes in the hypothetical average cell during cell development in the pea root have been considerably extended by studies which have determined quantitative and qualitative metabolic changes in isolated fragments of root tissue. Various approaches have been used and the conclusions drawn from the results obtained have a more or less limited applicability to the normal situation of cell development in the intact root, because of the degree of interference in normal processes that are imposed by the technique. The advantages and the disadvantages of using isolated root fragments stem from the fact that the pea seedling root is dependent upon the cotyledons for the nutrients upon which growth depends. Any attempts to modify the nutritional environment of root cells in attached roots is limited by the fact that normal nutrients reach the meristem and the cells in the early stages of development by conduction in the phloem and then by diffusion and cell-to-cell transport from the end of the provascular cylinder. Added nutrients or antimetabolites are imposed upon cells against a nutrient background which will vary according to the position of cells in the root tip and which is unknown. Excision of root tips (1 cm) and culture in the presence of an energy source (4% sucrose) plus mineral salts provides a defined nutrient medium which permits root growth to occur at a reasonable rate.

The disadvantage which accompanies the ability to define the nutrient conditions for the growth of the 1-cm root tips is of course that there is a considerable difference in the normal sequence of cellular changes which occurs when compared with the changes in the intact root due to the separation from the normal nutrient flow. These effects of excision upon 1-cm pea root tips which have been fully described by Abbott (1972b) and Heyes (1963a), considerably modify the pattern of protein and nucleic acid metabolism, presumably because of the interruption in the normal supply of precursors which the root-tip cells are apparently unable to provide in sufficient quantity for synthesis to continue at a normal rate. The data of

TABLE 6.II. Comparison between cells[a] of 1-cm root tips taken from roots grown for 6 days either attached or in culture. (From Abbott, 1972b)

	Cultured	Attached
Fresh weight (ng)	34·04	49·81
Dry weight (ng)	3·85	4·72
Protein N (pg)	82·33	201·80
Soluble N (pg)	36·95	162·26
RNA (pg)	38·37	98·30
RNA_1[b] (pg)	32·60	71·10
RNA_2[b] (pg)	5·80	27·3
DNA (pg)	6·41	11·47

[a] Values are means of four batches of 100 root tips and are expressed per mean cell.
[b] RNA_1 is that portion of the total RNA extracted by 1 N perchloric acid at 0–2°C during 24 h. RNA_2 represents the remaining RNA which cannot be extracted under these conditions.

Table 6.II show how the mean cell composition in 1-cm root tips, excised from roots grown in sterile culture for 6 days after a single sub-culture from a root tip grown for 4 days since removal from the 4-day-old seedling, compares with that of tips excised from sterile roots grown still attached to the seed for the same length of time. It is apparent that the cells are in a completely different state after adapting to the less-complete nutrient conditions of growth which prevail in culture. All components are reduced considerably, cell size and dry mass less than the others, which probably means that cell wall synthesis is not affected as much as the synthesis of other components.

Abbott (1972b) suggested that RNA_2 is associated with seed RNA but a closer investigation of the changes in the relative quantities of RNA_1 and RNA_2 in cultured pea roots, and particularly of the changes in the total quantity of RNA_2 makes it unlikely that RNA_2 is simply a seed-based RNA component. The total amount of this component increases from 21·5 μg to 23·0 μg after 2 days and to 25·8 μg per root after 4 days in culture, even though the amount per cell continues to decrease in the apex. Between 2 and 4 days in culture, the root shows a net gain in total nucleic acid content, and during this period the decline in the amount of RNA_1 per cell in the tip is halted. The significance of the changes in the relative quantities of these rather crudely separated fractions obviously requires an improved understanding of the reactions involved in cold acid extraction of RNA. Redundancy of components in "normal" attached root tips has been invoked to account for similar decreases in quantity while still permitting apparently adequate metabolism to be maintained, as for example, in the redundancy of DNA as envisaged by Wolff (1969). The ability of a cell to operate effectively with less RNA or protein is not surprising and does not detract from the usefulness of studies on isolated roots grown in culture.

Brown (1959) showed that despite the dramatic change consequent upon excision and growth in a defined medium, the rate of cell division which is very low initially increases steadily over the first 5 days and then declines. Figure 6.6 shows that the increasing rate of cell division occurs in a root at a time when the increase in cell number proceeds in an exponential manner. As Brown stated, one of the normal characteristics of a colony of unicellular organisms in the exponential phase is that all the cells or a constant proportion of them are capable of division and the constant relative rate of increase is thus a consequence of a constant rate of division in these cells. The rate of division in Fig. 6.6 was calculated on the basis that all non-vacuolated cells are capable of division, but the calculation does not take into account the

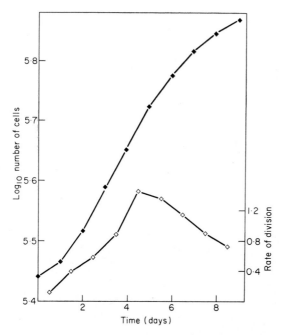

FIG. 6.6. The change with time in: the log[10] total number of cells, —◆—, and the rate of division, —◇—, in a cultured pea root. Rate of division calculated from the total number of non-vacuolated cells. (From Brown, 1959)

possibility that there may be changes in the so-called quiescent zone (Clowes, 1958). Abbott (1972b) reported that in the tips of roots cultured for 10 days, the quiescent zone could not be detected, indicating that there may well be an increasing proportion of meristematic cells dividing during the establishment of a root growing in culture, as well as an increasing rate of cell division as has been suggested by Brown.

Such an experimental situation has been used to study a wide range of requirements for root growth and to test the ability of roots to utilize a wide range of nutrients and metabolites, more so perhaps in the tomato (Street, 1957) than in the pea. Brown and Possingham (1957) and Abbott (1972a), however, have used pea roots growing in sterile culture to determine the cellular processes primarily affected by iron deficiency. Cultured roots have the advantage that secondary thickening is very much delayed or postponed indefinitely and there are relatively infrequent lateral roots, especially when a simple mineral salt plus sucrose medium is used. Studies of successive segments can thus be continued over a much more extended range of cell development and can provide further information on developmental changes in cellular metabolism. During the study of the effects of iron deficiency, changes in respiratory activity were observed, and Fig. 6.7 shows that the ratio of

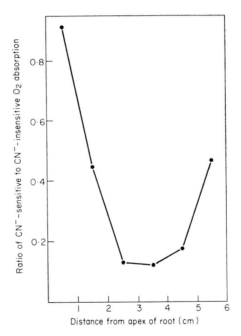

Fig. 6.7. Ratio of cyanide-sensitive to cyanide-insensitive oxygen absorption in successive 1-cm segments of 7-day-cultured pea roots. (From Brown and Possingham, 1957)

cyanide-sensitive to cyanide-insensitive oxygen uptake changes from the tip to the base of the cultured root, using successive 1-cm segments. Thus it appears that the changes observed during the early stages of cell development continue through the later stages.

The effects of added antimetabolites have also been studied using cultured roots in order to provide further information on the role of nucleic acids in the changing protein pattern. The purine analogue 8-azaguanine (Heyes, 1963a) was found to have several effects and revealed a fundamental weakness in the use of whole cultured pea roots in such a study. It was clear that 8-azaguanine inhibited cell division and the two populations of cells, that in the control and that in the treated roots, were no longer strictly equivalent. Comparison of the activities of a number of enzymes showed, however, that enzyme proteins synthesized in the presence of the base analogue had a lower specific activity and it was concluded that this was a consequence of incorporation of 8-azaguanine into the RNA. Nuti Ronchi et al. (1965), using intact pea roots, showed that 8-azaguanine prevented the entry of interphase nuclei into mitosis, reducing the mitotic index to zero in 12 h, and that it also inhibited DNA synthesis between 12 and 24 h. Incorporation of the analogue into RNA presumably could be reversed and DNA synthesis returned to normal in the presence of guanine; repetition of these treatments has led to endopolyploid mitosis. The unusual feature of these experiments was that such induction of endo-reduplication had been practically unsuccessful with roots of any other plant. Only with *Pisum* had this effect been obtained, but this consequence of 8-azaguanine treatment has not altered the significance of the observation that modification of nucleic acid composition resulted in the synthesis of aberrant proteins.

The use of even simpler fragments of pea roots than this has enabled the complications due to effects on mitosis to be avoided. Segments cut from the zone of cell enlargement, usually 2–4 mm from the apex, retain considerable potential for enlargement and the provision of 2% sucrose alone gives an increase of up to three-fold in segment length with no increase in cell number. The problem of separating the consequences of effects on cell division from those consequent upon changes in the growth and development of cells in the early stages of expansion is thus solved. The metabolism of such cells is still modified by the fact of excision, since during cell expansion in the isolated segment (Heyes and Vaughan, 1967) there is no increase in protein content and there is a considerable decrease in the nucleic acid content (Fig. 6.8), as compared with the average increase in both constituents in the intact root (Figs 6.2 and 6.3). The 2·4-fold increase in cell size is accompanied by an equivalent increase in dry weight, and this is not all due to uptake of sucrose or other low-molecular weight products, but is probably a real increase in cell wall material as suggested by Brown and Sutcliffe (1950).

Many treatments have been found to reduce cell growth in the isolated

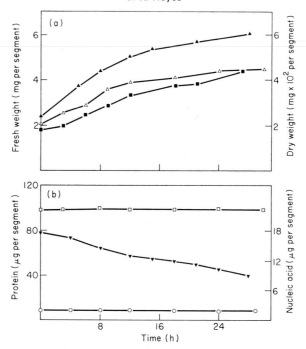

FIG. 6.8. Analysis of the contents of the sub-apical 2-mm segment during culture in 2% sucrose. Changes in: (a) fresh weight, —▲—; dry weight, —△—; and residual (alcohol-insoluble) dry weight, —■—; and (b) protein, —□—; RNA, —▼—; and DNA, —○—. (From Heyes and Vaughan, 1967)

root segment but a number of treatments have resulted in an increase in the rate or duration of cell growth (Fig. 6.9) (Heyes, 1963b; Yeoman, 1962; Vaughan *et al.*, 1974). Three of these treatments involve an interference with nucleic acid metabolism—8-azaguanine, 2-thiouracil and ribonuclease— while another, hydroxyproline, interferes with protein metabolism. The two base analogues are both incorporated into the RNA, 2-thiouracil replacing 3·4% of the RNA-uracil and 8-azaguanine replacing 7–9% of the guanine. Although Loening (1965) showed that in excised pea root segments (as in excised 1-cm root tips) there is a greatly reduced synthesis of RNA, there is every likelihood that the analogues are incorporated not only into mRNA but also into tRNA and even rRNA. Autoradiographic studies with both analogues show that the most rapid and most intensive incorporation is into RNase-sensitive material associated with the nucleus. In both cases greater labelling was found over sections of nuclei containing the nucleolus, but since this means that the section (8 μm thick) probably contained an entire nucleus or a median section through the nucleus, rather than a glancing section, the autoradiographic stripping film must have been exposed to β-particles

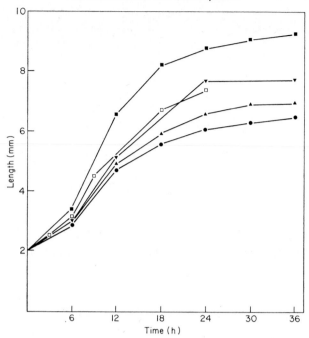

Length (mm)

Time (h)

FIG. 6.9. Growth of the sub-terminal 2-mm segment of pea-seedling roots in 2% sucrose only, —●—; and in 2% sucrose containing: 8-azaguanine, 5×10^{-6} M, —▲—; ribonuclease, 10 μM ml^{-1}, —▼—; 2-thiouracil, 10^{-3} M, —■—; and hydroxyproline, 3×10^{-3} M, —□—. Data for growth with hydroxyproline redrawn from Vaughan and Cusens (1973).

incorporated into a full 8 μm depth of nucleus, as compared with a segment from a sphere of probably less than 8 μm thickness. Thus the detection of greater labelling apparently associated with the nucleolus does not necessarily mean that the analogues are preferentially incorporated most rapidly into nucleolar RNA. The data so far obtained are insufficient to confirm the suggestion made by Brown (1963) and Heyes and Brown (1965) that the succession of protein states with their associated enzyme complements, so characteristic of cell development in the pea root, depends upon the production by the nucleus of a succession of changing mRNA molecules.

Attempts to study the effects of 2-thiouracil upon rate of biosynthesis of both proteins and RNA are subject to the problems associated with excision— these being that the normal precursor pools provided by the breakdown of reserve materials in the cotyledons have been removed—but it appears (Heyes and Vaughan, 1967b) that the incorporation of both ^{3}H-uridine and ^{14}C-leucine is increased slightly by the analogue. The use of differential centrifugation enables various sub-cellular fractions to be isolated but it is apparent that there is little evidence for a specific stimulation of biosynthesis of either macromolecule associated with any particular sub-cellular system. It is interesting to note that even in the isolated segment, cell growth and development are associated with the same sort of changes in sub-cellular components as was found in the intact root. There is a greater incorporation of ^{14}C-leucine into protein of the nuclear and soluble fractions, but there is real incorporation into the protein of all sub-cellular fractions, indicating that the constant protein content per segment reflects turnover. The incorporation of ^{3}H-uridine into the RNA of the nuclear and mitochondrial fractions increases during the first 6 h but decreases thereafter while activity of the RNA of the microsomal and soluble fractions declines gradually after a 1 h incorporation period. Clearly the rates of biosynthesis and the stablility of different sub-cellular RNA fractions vary, and cellular RNA metabolism is much more susceptible to interference due to excision than is the metabolism of cellular protein.

Hydroxyproline is not incorporated as such into cell wall protein, but is incorporated into a glycoprotein precursor within a membrane-bound vesicle as proline, and is subsequently hydroxylated and secreted into the cell wall (Sadava and Chrispeels, 1973). Stimulation of cell elongation by hydroxyproline was thought to be a consequence of an interference with normal proline incorporation into proteins, similar to the interference caused by azetidine-2-carboxylic acid. Vaughan and Cusens (1973), using the sub-apical 2-mm segment, have shown that azetidine-2-carboxylic acid inhibits elongation and the incorporation of both ^{14}C-leucine and ^{14}C-proline into the protein of all cellular fractions; but hydroxyproline enhances elongation while causing only a slight but significant reduction in the incorporation of both leucine and of proline into protein. It has been suggested that stimulation of growth by hydroxyproline results from an inhibition of the formation of hydroxyproline from protein-bound proline as was shown for *Avena* coleoptiles by Cleland (1967).

Cultured pea roots have also provided an experimental system for the study of changes associated with the later stages of cell development. Habeshaw and Heyes (1971) took advantage of the fact that the basal 1 cm of cultured pea roots does not produce lateral roots and no cell divisions appear to be completed in this segment at all once the apical meristem has grown away from the base. Changes in this segment could then be followed during an extended period of time. With three root tips in 20 ml of culture solution, the sucrose in the medium is completely utilized after 12 days, providing nutrient conditions less and less favourable for vigorous growth and increasingly conducive to the acceleration of senescence. Brown and Possingham (1957) studied changes in roots grown in culture for up to 12 days, but by selecting only the basal segment, root tissue can be studied during a culture period of 27 days. Total respiration and the activity of a number of individual enzymes (succinic dehydrogenase, β-glycerophosphatase, invertase, polyphenol oxidase, alanyl glycine dipeptidase and protease) continued to change in the basal segment throughout the whole period; no common pattern of change could be seen and the results indicated that the change in the pattern of metabolic activity supported by cell proteins continued throughout cell development, including senescence. Three enzymes showed a pronounced increase—transient only—in their activity at about 15 days, these being succinic dehydrogenase (but respiration as a whole did not), dipeptidase and protease. Electrophoresis of the buffer-soluble proteins on polyacrylamide gels also showed that the protein composition was undergoing considerable change including an increase, at quite a late stage of culture, in certain proteins.

Changes in the RNA and DNA content and of the buffer-soluble and buffer-insoluble proteins in the basal segment (Fig. 6.10) showed certain

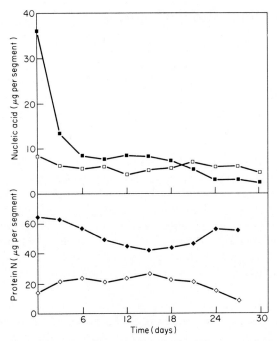

FIG. 6.10. Changes in: the RNA, —■—; DNA, —□—; buffer-soluble protein, —◇—; and buffer-insoluble protein, —◆—, in the basal 1-cm segment of cultured pea roots. Data recalculated from Habeshaw and Heyes (1971).

peculiarities of this ageing tissue. Changes during the first 6 days could not be related to changes in the same group of cells since it was only from the sixth day, when expansion had ceased in all cells of this basal segment, that changes could be associated with a fixed population of cells. Soluble protein increased up to 15 days and then declined considerably while the buffer-insoluble protein increased from 15 days. RNA content declined steadily with a slight increase just before 15 days with electrophoresis showing that both ribosomal RNAs declined at the same rate while low-molecular weight RNA remained constant until 21 days when there was a slight increase. The changes in DNA, whether measured by the Dische diphenylamine reaction, or by microdensitometry after Feulgen staining remained as shown, with a consistent transient increase at 9 days and again at 21 days. The significance of changes such as this could not be established from a biochemical analysis of the whole segment or of extracts of the segment and understanding of such transient changes was only obtained after supplementary experiments using histochemical techniques which enabled changes to be localized within the multicellular tissue.

IX. HISTOCHEMICAL AND CYTOCHEMICAL TECHNIQUES

One of the limitations of the serial-segment technique for the study of developmental changes in root cells is a consequence of the presence within the geometrically simple cylinder of root tissue of a number of distinct tissues, predominantly cortex, but the stele does undergo distinctive changes and significant changes in one minor cell type may well be lost completely against the background of change in other cell types. Jensen (1955) extended the serial-segment technique by combining with biochemical determinations a histochemical analysis of the distribution of substances or enzyme activities in particular cells or cell types in root tips of *V. faba.*

This approach has been utilized by Sutcliffe and Sexton (1969, 1974) and by Habeshaw and Heyes (1971) in extending the results of the biochemical analysis of developmental change. Sutcliffe and Sexton observed an increase in β-glycerophosphatase (βGPase) activity of "average" cell protein followed by a secondary rise about 10–15 mm from the tip of the root and a steady increase up to 40 mm. Activity also increased in the extreme 0·5 mm of the tip and agreed with the histochemical observation that root-cap cells showed βGPase activity increasing from the inner to the outer zone, with intense activity in the root-cap cells adhering to the cortex as far as 2 mm from the apex.

High concentrations of enzyme activity were also observed in differentiating xylem and phloem and in cortical cells bordering the emerging lateral-root initials. Comparison was made with other tissues in which increases in βGPase had been associated with senescent or dying cells. Habeshaw and Heyes (1971) observed a change in βGPase activity consistent with a localizationin lysosomes in living cortical parenchyma cells—the distribution of activity being evidenced by coloration in discrete granules. After 9 days of culture, cells of the piliferous layer become disorganized and activity was seen as diffused throughout the cytoplasm and to be more intense. By day 21 (at this stage cultured roots were 20 cm long), cells in the outer cortex began to show diffuse and intense activity and this necrotic or senescent change had spread to the inner cortex by day 30. Thus, the progression of enzyme changes in cells within a tissue could be clearly seen and correlates well with the increased βGPase activity consequent upon lysosome disintegration in other tissues after injury or during natural senescence.

Sutcliffe and Sexton (1974) reported similar studies of the localization of other enzymes in studying respiratory changes (cf. Fig. 6.5) and in this case, the data failed to show that the changing pattern of respiration was entirely due to the appearance of pentose phosphate pathway dehydrogenases in one specific cell type while declining glycolysis- and Krebs cycle-dehydrogenases were restricted to other cells. Thus the data in this case tended to confirm that the changes in the "average" cellular metabolism might be due to both a general decline in one respiratory pathway and a general increase in enzymes of the other pathway in individual cells, and also to a change in the relative contribution made by different tissues.

The cytochemical measurement of DNA using Feulgen staining and microspectrophotometry similarly enables changes in total DNA per segment to be visualized and localized. The changes in the DNA content shown in Fig. 6.10, when studied using this technique, showed clear differences between the change in the mean DNA content in the outer and the central cortex. The use of autoradiography after incubation of roots in ^{3}H-thymidine for a 6 h period provided further confirmation of the existence of two periods of DNA synthesis during culture (see Table 6.III).

TABLE 6.III. Changes in the DNA content of nuclei (microdensitometry after Feulgen staining) in different regions of the basal 1-cm segment of cultured pea roots and in DNA synthesis (intensity of labelling with ^{3}H-thymidine after a 6 h pulse). (From Habeshaw and Heyes, 1971)

Time in culture (days)	DNA (densitometer units)			^{3}H-Thymidine labelled nuclei No. per R.L.S.[a]
	Outer cortex	Mid cortex	Stele	
0	—	—	—	—
3	15·2	14·5	7·0	48
6	15·0	17·1	8·0	156
9	15·7	18·2	10·0	26
12	14·0	17·8	9·0	26
15	12·8	17·6	9·0	7
18	10·0	17·0	10·0	74
21	10·8	20·0	11·0	30
24	10·0	18·1	11·0	0
27	9·7	13·2	10·0	0
30	9·2	9·8	10·0	0

[a]R.L.S. = radial longitudinal section

The explanation put forward by Habeshaw and Heyes (1971) after consideration of the cytochemical demonstration of the localization of βGPase activity in senescing cells and in cells in which the DNA-increase was occurring was as follows. Senescence and the concomitant increase in hydrolytic enzymes proceeds from the outer layers of cortical cells inwards. The piliferous layer undergoes senescence changes first, producing soluble degradation products, some of which will leak from the root while others will pass through the cortex towards the stele. The transient change in concentration of substances including DNA precursors may well promote a wave of DNA synthesis. A subsequent gradual senescence of even more cells in the outer cortex with a loss in their average DNA content (some nuclei become Feulgen-negative) produces another wave of breakdown products/precursors and a transient DNA synthesis occurs in the mid-cortex to be followed in turn by senescence, hydrolysis and loss of DNA and other cellular material in these cells too.

The reversal of developmental processes—dedifferentiation—is of course limited to cells whose development has not involved such changes as loss of cytoplasmic or nuclear integrity and is considered by Torrey and Zobel in Chapter 5; it does involve, however, as shown by Phillips and Torrey (1973), a reversal of some of the changes involved in normal cell development to the mature state. Segments are cut from a region of the pea root containing cells in the state of early maturity and in order to study changes in the non-specialized cortical cells, the central vascular tissues are removed completely, and changes observed using both cytochemical and biochemical techniques. These experiments provide a system for the study of those aspects of cytodifferentiation concerned with the formation of tracheary elements from otherwise cortical cells. The question of whether the particular hormonal treatment (auxin plus 1·0 ppm kinetin) only brings about the differentiation of a proportion of the mature cortical cells after dedifferentiation and the induction of mitosis, has yet to be answered, and the results of further experiments in this direction are awaited with interest.

Few studies of pea roots have been made in which cell growth has been studied at all levels—biochemical and histochemical as well as at the ultrastructural level—as has been carried out for the changes accompanying the development of a callus (or the induction of cell division) in explants taken from the tuber of the Jerusalem artichoke (Yeoman and Aitchison, 1973), but there is no doubt that such a comprehensive study is necessary if the metabolic changes involved in normal cell growth and development are to be fully understood. Just as it is necessary to define an otherwise arbitrary subcellular fraction (isolated by techniques such as differential or gradient centrifugation) by electron microscopy before changes in composition or function of such fractions can be correlated with cellular changes (cf. Heyes, 1963), so it will be necessary to apply the full range of available techniques to the study of cell development in the pea root.

X. CONCLUSIONS

Heyes and Brown (1965) expanded an earlier concept (Brown and Robinson, 1955) in which the various stages of cell development were considered to be characterized by a series of protein states, each carrying different overall metabolic activities with a different set of catalytic capabilities. The change from one state to another must be a consequence of both the operation of specific enzymes and of the provision of specific templates. The experimental verification of all aspects of this concept has not yet been achieved, mainly because of the extreme sensitivity of normal nucleic acid metabolism to interruption by many manipulative and experimental procedures (radiation from ^{32}P, thermal shock, slight dehydration and even excessive handling). There are few who would disagree that cell development involves a change in the metabolic capacity from the state in which a cell near the meristem is unable to expand at maximum rate, through a state in which growth-dependent reactions can proceed, and that the cessation of growth is determined by a transition to a state in which the metabolic capacity will no longer support growth. The growth phase which is so amenable to experimental study is only one part of the whole course of development which terminates in senescence and cell death. The growth process is capable of variation, particularly with regard to the development of specific differences in the growth pattern, in forming characteristic cell shapes, and to the development of specific activities and hence constituents associated with a particular cell type and cell function (see Chapter 5).

The understanding of the control of cell growth in the pea root has been complicated at times by the emphasis upon the need to place the control at the level of phytohormones and systems responsive to changes in these regulators. The considerable body of research linking auxin-mediated inhibition of root growth with IAA-dependent ethylene production is exemplified by work using the pea root (Chadwick and Burg, 1970), one of the most significant features of which is the consideration of the growth response to IAA in different fluid volumes. The inference from the observations is that the IAA-dependent production of ethylene is correlated more closely with the effects of endogenous IAA than is the longer-term additional direct effect of IAA, particularly in any consideration of the involvement of IAA in the geotropic response mediated by this hormone. The reason for the difference in sensitivity of pea stem and root tissue to IAA concentration in the production of ethylene has still to be adequately explained, but may well relate to our complete lack of knowledge of the actual concentration of auxin and of the origin and route of transport of that auxin to the site of action of the hormone in the root.

The development of new techniques for the identification and isolation of specific "messenger" RNA molecules may well provide a useful experimental tool. The use of a specific binding technique to isolate polyadenylic-acid-rich RNA molecules has enabled mRNA to be isolated from *Lemna gibba* (Tobin and Klein, 1974). The ability to isolate mRNA through the immunological properties of the partially synthesized protein may not prove as useful as it already has in many animal systems (except perhaps in developing seeds; see Chapter 3), but a greater knowledge of the changing pattern of production of pea root mRNA must await further research.

REFERENCES

ABBOTT, A. J. (1972a). *New Phytol.* **71,** 85–92.

ABBOTT, A. J. (1972b). *J. exp. Bot.* **23,** 667–674.

BROWN, A. P. (1961). *J. exp. Bot.* **12,** 147–156.

BROWN, A. P. (1963a). *J. exp. Bot.* **14,** 114–119.

BROWN, A. P. (1963b). *J. theoret. Biol.* **5,** 372–388.

BROWN, R. (1959). *J. exp. Bot.* **10,** 169–177.

BROWN, R. (1963). *In* "Meristems and Differentiation". Brookhaven Symposia in Biology, No. 16, pp 157–169. Brookhaven National Laboratory, New York.

BROWN, R. and BROADBENT, D. (1950). *J. exp. Bot.* **1,** 249–263.

BROWN, R. and POSSINGHAM, J. V. (1957). *Proc. R. Soc.* B **147,** 145–166.

BROWN, R. and ROBINSON, E. (1955). *In* "Biological Specificity and Growth" (E. G. Butler, ed.), pp 99–118. University Press, Princeton.

BROWN, R. and SUTCLIFFE, J. F. (1950). *J. exp. Bot.* **1,** 88–113.

CHADWICK, A. V. and BURG, S. P. (1970). *Pl. Physiol.* **45,** 192–200.

CLELAND, R. (1967). *Pl. Physiol.* **42,** 271–274.

CLOWES, F. A. L. (1958). *J. exp. Bot.* **9,** 229–238.

DIXON, R. O. D. (1963). Unpublished data.

FOWLER, M. W. and AP REES, T. (1970). *Biochim. biophys. Acta* **201,** 33–44.

GIBBS, M. and EARL, J. M. (1959). *Pl. Physiol.* **34,** 529–532.

HABESHAW, D. and HEYES, J. K. (1971). *New Phytol.* **70,** 149–162.

HEYES, J. K. (1960). *Proc. R. Soc.* B. **152,** 218–230.

HEYES, J. K. (1963a). *Proc. R. Soc.* B. **158,** 208–221.

HEYES, J. K. (1963b). *Symp. Soc. exp. Biol.* **17,** 40–57.

HEYES, J. K. and BROWN, R. (1956). *In* "The Growth of Leaves" (F. L. Milthorpe, ed.), pp 31–49. Butterworths, London.

HEYES, J. K. and BROWN, R. (1965). *In* "Encyclopedia of Plant Physiology" (W. Ruhland, ed.), Vol. XV (1), pp. 189–212. Springer-Verlag, Berlin and Heidelberg and New York.

HEYES, J. K. and VAUGHAN, D. (1967a). *Proc. R. Soc.* B. **169,** 77–88.

HEYES, J. K. and VAUGHAN, D. (1967b). *Proc. R. Soc.* B. **169,** 89–105.

JACKSON, M. and INGLE, J. (1973a). *Biochem. J.* **131,** 523–533.

JACKSON, M. and INGLE, J. (1973b). *Pl. Physiol.* **51,** 412–414.

JENSEN, W. A. (1955). *Expl Cell Res.* **8,** 506–522.

JENSEN, W. A. (1960). *Am. J. Bot.* **47,** 287–295.

JENSEN, W. A. and KAVALJIAN, L. G. (1958). *Am. J. Bot.* **45,** 365–372.

LAMPORT, D. T. A. (1973). *In* "Biogenesis of Plant Cell Wall Polysaccharides" (F. Loewus, ed.), pp 149–164 Academic Press, New York and London.

LAMPORT, D. T. A. and MILLER, D. H. (1971). *Pl. Physiol.* **42,** 481–486.

ᴸOENING, U. E. (1961). *Biochem. J.* **81,** 254–260.
ᴸOENING, U. E. (1965). *Biochem. J.* **97,** 125–133.
ᴸUND, H. A., VATTER, A. E. and HANSON, J. B. (1958). *J. biophys. biochem. Cytol.* **4,** 87–98.
ᴸYNDON, R. F. (1963). *J. exp. Bot.* **14,** 419–430.
MONRO, J. A., BAILEY, R. W. and PENNY, D. (1972). *Phytochemistry* **11,** 1597–1602.
NUTI RONCHI, V., AVANZI, S. and D'AMATO, F. (1965). *Caryologia* **18,** 599–617.
PHILLIPS, R. and TORREY, J. G. (1973). *Devl Biol.* **31,** 336–347.
RIDGE, I. and OSBORNE, D. J. (1970). *J. exp. Bot.* **21,** 843–856.
ROBINSON, E. and BROWN, R. (1954). *J. exp. Bot.* **5,** 71–78.
ROLAND, J. C. (1973). *Int. Rev. Cytol.* **36,** 45–92.
SADAVA, D. and CHRISPEELS, M. J. (1973). *In* "Biogenesis of Plant Cell Wall Polysaccharides" (F. Loewus, ed.), pp 165–174. Academic Press, New York and London.
STREET, H. E. (1957). *Biol. Rev.* **32,** 117–155.
SUTCLIFFE, J. F. and SEXTON, R. (1969). *In* "Root Growth" (W. J. Whittington, ed.), pp 80–102. Butterworths, London.
SUTCLIFFE, J. F. and SEXTON, R. (1974). *In* "Structure and Function of Primary Root Tissues" (J. Kolek, ed.) pp 203–219. Veda, Bratislava, Czechoslovakia.
TOBIN, E. M. and KLEIN, A. O. (1974). *Pl. Physiol.* Annual supplement, 37.
TRUMAN, D. E. S. (1974). "The Biochemistry of Cyto-differentiation". Blackwell Scientific Publications, Oxford.
TULETT, M. H. (1967). Ph.D. Thesis, University of Edinburgh, Scotland.
VAUGHAN, D. and CUSENS, E. (1973). *Planta* **112,** 243–252.
VAUGHAN, D. and CUSENS, E. (1974). *Biochem. Soc. Trans.* **2,** 124–126.
VAUGHAN, D., DE KOCK, P. C. and CUSENS, E. (1974). *Physiologia Pl.* **30,** 255–259.
WOLFF, S. (1969). *Int. Rev. Cytol.* **25,** 279–296.
YEOMAN, M. M. (1962). *J. exp. Bot.* **13,** 390–396.
YEOMAN, M. M. and AITCHISON, P. A. (1973). *In* "Plant Tissue and Cell Culture" (H. E. Street, ed.), pp 240–268. Blackwell Scientific Publications, Oxford.

7. The Shoot Apical Meristem

R. F. LYNDON

Department of Botany, Edinburgh University, Scotland

I. MORPHOLOGY AND ANATOMY

A. Morphology

The external morphology of the pea shoot apex is shown in Fig. 7.1. Below the rounded apical dome the youngest leaf primordium is seen as a swelling on the side of the apex. As the primordium grows it extends around to the sides of the apex so forming a V-shaped protuberance, the lateral parts subsequently developing into the stipules. About one plastochron after its first appearance the primordium develops two lobes or protuberances about halfway between the axil and the tip. These are the first leaflet primordia. Subsequent pairs of leaflet primordia develop in acropetal succession as the primordium lengthens. Sometimes one or more pairs of leaflet primordia develop into tendrils or fail to develop at all. The stipules of the pea are characteristic in developing from the lateral parts of the primordium which gives rise to the leaf, and not from separate primordia of their own. The development and vasculature of the stipules have been described by Mitra (1950) and Nougarède

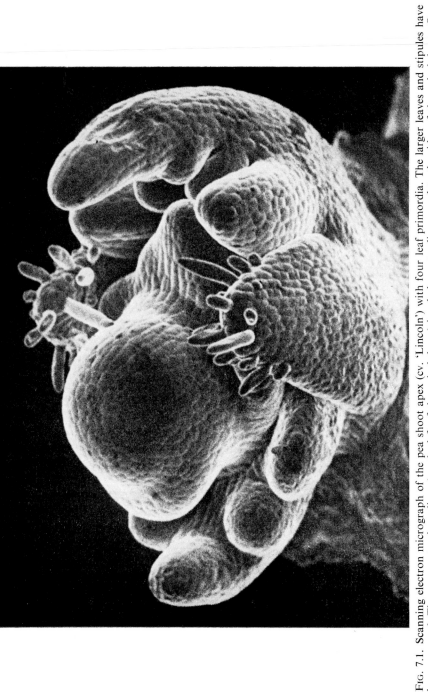

Fig. 7.1. Scanning electron micrograph of the pea shoot apex (cv. 'Lincoln') with four leaf primordia. The larger leaves and stipules have been removed. The youngest primordium (on the left of the apical dome) is a bulge extending round to the sides of the apical dome. One plastochron later (the next youngest primordium, on the right of the apical dome), one of the leaflet primordia and a developing stipule are

and Rondet (1973a). Successive leaves are formed alternatively on opposite sides of the apex so that the divergence angle between successive primordia is 180°. The phyllotaxis is therefore distichous, the leaves being in two vertical ranks.

B. Anatomy

The pea apex, as seen in a longitudinal section passing through the young leaves, has an epidermis in which all cell divisions are anticlinal (Thomson and Miller 1962; Lyndon 1971). The tunica (which includes the epidermis) is usually two-layered (Mitra 1950; Thomson and Miller, 1962; Nougarède and Rondet 1973b), although up to four layers have been observed (Reeve 1948). The tunica is distinguished from the corpus by the fact that it consists of cells which appear, from the arrangement of their cell walls, to have divided only anticlinally. If there are cells in the outer layer of the corpus which sometimes divide only anticlinally, while at other times they also divide periclinally, then these would not have only anticlinal walls as the tunica cells have. When the orientation of mitotic spindles was used as the sole criterion the tunica was found to be about seven-layered for one-half of the plastochron and to be two-layered for the other half of the plastochron (Lyndon 1970b, 1972a). Variation in the number of tunica layers during the course of the plastochron has been noted in other plants (Gifford, 1954) and is of possible significance in leaf initiation (Lyndon 1976; see Section VI). Variation is greatest at the point where a new leaf primordium is to be formed, for here the tunica becomes reduced to a single layer (the epidermis) by the periclinal divisions which occur in the sub-epidermal cell layers.

   The tunica–corpus arrangement of the pea shoot apex develops during late embryogeny (Reeve, 1948). The apex arises between the two cotyledons and at an early stage of its development shows two distinct outer cell layers. However, periclinal divisions are often observed in both layers during the early development of the apex, so there is no clear tunica at this stage. Even when the embryo apex has initiated the first primordium, occasional periclinal divisions may still be found in the epidermis. The apex does not show any appreciable change in the degree of organization until the tissues of the axis are well differentiated, and by the time the embryo is fully developed, two tunica layers have become sharply defined with sometimes a third, less discrete, layer also present in the mature embryo. On the flanks of the apex more than three tunica layers can sometimes be seen. The anatomy of the fully developed embryo apex differs very little from the apices of seedling and adult plants (Reeve, 1948); during development in embryogeny the apical dome increases slightly in size, and may continue to increase after germination and during vegetative growth of the plant.

The first sign of cell differentiation in the apex is a slight elongation of cells in the centre of the apex about level with the youngest primordium. These are the cells which are beginning to vacuolate and form the pith of the stem. In etiolated plants grown in complete darkness there are numerous starch grains in these cells which are not present in light-grown plants. Apart from this, the external and internal structures of the apex are exactly the same in plants grown in the light as those grown in the dark (Thomson and Miller, 1962). Procambium is first visible in the apex in relation to the development of the youngest leaf primordium, while the procambium in relation to the stipules develops later (Mitra, 1950).

The intensity of staining with pyronin or with general stains in longitudinal sections of the pea apex (Fig. 7.2) is greatest in the young leaf primordia and on the flanks of the apex, and is least in the enlarging cells of the pith and in the central zone (which consists of the cells at the extreme summit of the apical dome). The zonation pattern in the pea is not as marked as in most other plants (Nougarède and Rondet, 1973b). The lower intensity of staining in the pith is clearly because the cells are larger and more vacuolated than elsewhere in the apex. The lower staining intensity in the central zone is due to a slightly larger cell size than in the rest of the apical dome. The pattern of cell density, which is the inverse of cell size, is obtained by scoring the positions of individual cells in a median longitudinal section of the apex. By superimposing the drawings obtained from several apices, all at exactly the same stage of development, the mean cell density per unit area can be obtained for each position in a grid placed over the drawings (Lyndon, 1972a). From the values so obtained a diagram can be prepared in which the density of shading is directly proportional to the numbers of cells per unit area (Fig. 7.2B). The areas of densest shading are therefore the areas of smallest cell size and the areas in which the shading is lightest are those in which cells are largest. A comparison of the distribution of cell density with the distribution of staining intensity shows that they are very similar. The staining intensity is therefore a direct function of cell density and the zonation pattern seen in the pea is consistent with there being a constant amount of stainable material per cell throughout the apex. This has been confirmed by histochemical measurements (see Section III).

The central zone is also characterized by having larger nuclei than the cells in the surrounding tissues. Measurements of the sizes of the nuclei, in conjunction with microdensitometric measurements of the DNA value for each nucleus, showed that the 2C nuclei in the central zone were approximately twice the volume of 2C nuclei in the rest of the apex (Lyndon, 1973). The difference between 4C nuclei in the central zone and the rest of the apex was less marked.

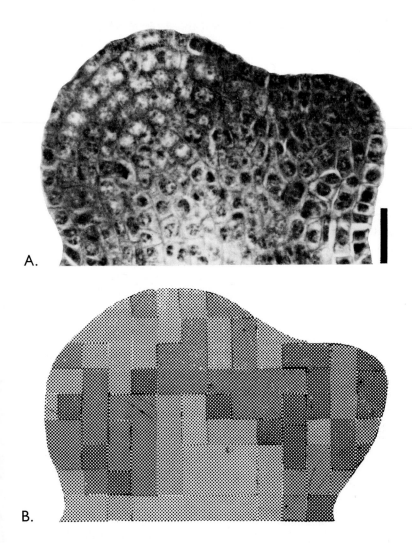

FIG. 7.2. (A) A longitudinal median section of the shoot apex stained with pyronin. (B) The relative densities of cells. In (A) the flanks of the apex and the young leaf primordium are darkly stained and the central zone and incipient pith are lightly stained, corresponding to the distribution of RNA. This is similar to the pattern of cell density (B) showing that the different concentrations of RNA throughout the apex are a function of cell density and inversely proportional to cell size. This is consistent with a constant amount of RNA per cell throughout the apex. Bar = 30 μm

C. Plastochronic Changes

As each new leaf primordium is formed the shape of the apex changes. The formation of a primordium on the flanks of the apical dome reduces the apical dome to minimal area. The apical dome then grows to maximal area just before the initiation of the next primordium. The interval between the formation of successive leaves is a plastochron. Associated with the formation of leaves are changes in the structure of the apex, these changes being repeated at each plastochron. From the time of its first emergence as a bump on the flanks of the apex the youngest primordium grows outwards about 60 μm before the next leaf is formed on the opposite flank of the apex. The development of the apex within a single plastochron has been followed by classifying apices according to the stage of development of the youngest primordium (Lyndon, 1968a). The base of the primordium was taken as the line joining the axil of the youngest primordium itself and the axil of the next primordium immediately below it. The length of the youngest primordium was taken as the length of the perpendicular from this base-line to the tip of the primordium. Since the primordium increased in length by 60 μm during the first plastochron of its existence, i.e. before the next primordium emerged, the plastochron could be divided into 10 distinct and morphologically recognizable stages, each represented by a 6 μm increment in the height of the youngest primordium.

A more sensitive method for measuring plastochron stage was devised by Hussey (1971a) and was used to follow the development of pea apices (Hussey, 1972). As soon as the youngest primordium is large enough to form an axil, a line joining the surface of the primordium to the surface of the apical dome no longer touches the surface of the apex in the region of the axil. The shortest distance from the axil to this line is the axillary distance. The value for the axillary distance at the time that the axil of the next leaf primordium becomes formed is the axillary distance increment during the course of a single plastochron. In this way it was possible to measure the developmental stage of an apex to within 1/100 of a plastochron.

When following changes in the morphology or physiology of a plant it is usually more useful to measure the time taken for the plant to reach a given developmental stage than to measure the average developmental stage reached by a certain time. By the latter method differences between developmental stages can be obscured because of the averaging of plants which are not exactly synchronous in their development. The former method, which is essentially a modification of the plastochron index (Erickson and Michelini, 1957), is to be preferred since plants at a given developmental stage can then be compared with plants at another clearly different and defined developmental stage.

The changes in structure of the pea shoot apex during the course of a plastochron have been described by Nougarède and Rondet (1973b) and Lyndon (1968a, 1970b). The formation of a new leaf primordium is accompanied by the occurrence of periclinal divisions in the underlying tunica and corpus. These are first seen as a change in orientation of the mitotic spindles in the I1 region* of the apex which occurs about half a plastochron before the new primordium emerges as a bump (see Section V). The incipient pith region is regenerated by cell division and growth at the base of the apical dome, the mitotic spindles being about 15° from the vertical so that the axis is inclined in successive plastochrons about 30° to the right or left of the previous orientation. Longitudinal mitotic spindles on the flanks of the apical dome are associated with the anticlinal divisions which result in the upward growth of the apical dome. In the incipient pith region there is a sudden increase in the proportion of plastids with starch which also occurs half a plastochron before the new primordium emerges and at exactly the same time as the change in orientation of the mitotic spindles in the I1 region (see Section II).

These changes in structure and shape of the apex are associated with changes in the overall dimensions. The height of the apical dome increases slowly just after a leaf has been initiated but increases much more rapidly during the second part of the plastochron, just before the initiation of the next leaf (Lyndon, 1968a). The increase in size of the apical dome was also measured by the increase in cell number during the plastochron. The increase in cell number was slow during the first part of the plastochron and rapid during the second part just before the initiation of the next leaf. Conversely, the increase in cell number in the incipient primordium and the associated axial tissue was more rapid during the first part of the plastochron, when the primordium had just been initiated, and was slower during the second part when the apical dome itself was growing more rapidly. The increase in cell number, and size, of the whole apex was nevertheless exponential, and the growth of the young primordium was also exponential (Lyndon, 1968a). The apparently discontinuous growth of the different regions of the apex is because the apex changes in shape as it grows.

The central zone at the summit of the apical dome appears to change in size during the plastochron, being at a maximum of about 100 cells in the middle of the plastochron and falling to a minimum at the times of emergence of a new leaf, i.e. at the beginning and end of a plastochron (Lyndon, 1968a). The rate of cell division also changes in the central zone, being lowest in the middle of the plastochron and reaching its highest value at the beginning and the end of the plastochron (Lyndon, 1970a).

* The I1 region is that part of the apical dome which will form the next primordium, and the I2, I3, etc. regions those which will form the next +1, next +2, etc. primordia.

G*

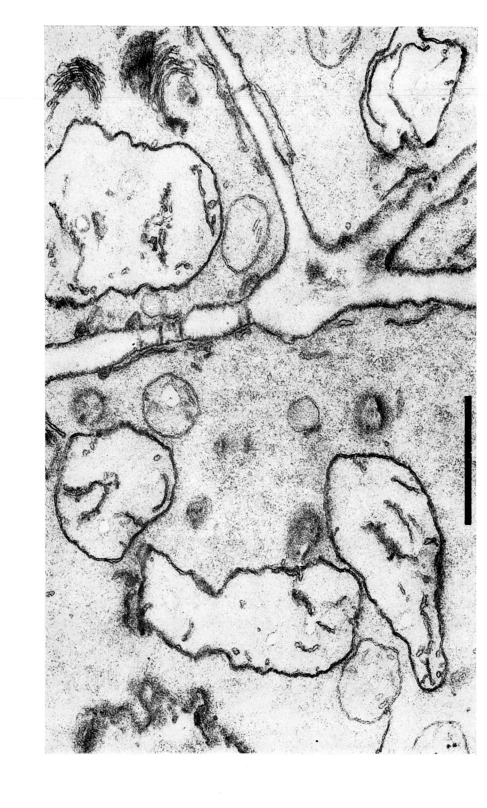

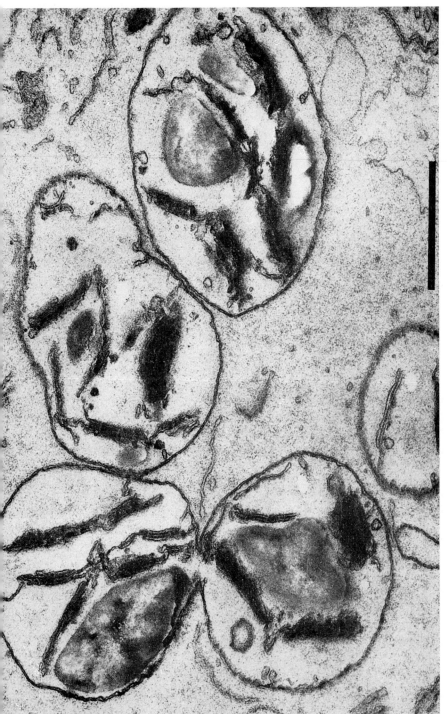

Fig. 7.3. B. Plastids from the incipient pith are larger, have better developed membrane systems and contain starch. One of the plastids (left) appears to be just completing division. Bar = 1 μm

II. ULTRASTRUCTURE

A systematic survey of the ultrastructure of the pea apex (Lyndon and Robertson, 1975) showed that most of the differences between cells in different parts of the apex were quantitative rather than qualitative. Observations were made on longitudinal sections which were cut near the median line of apices at six defined stages within a single plastochron. The sections were mounted on formvar/carbon films so that the whole of each section could be seen and eleven predetermined positions could be photographed. On each of the resulting electron micrographs the relative volumes of the different classes of organelles were measured by quantitative stereological methods (Weibel, 1969). The total numbers of profiles of each recognizable organelle were also counted. From these primary data the relative volume occupied by each class of organelle, the numbers of organelles per cell, and the sizes of the organelles were then calculated. The results were subjected to statistical analyses of variance.

Within the course of a single plastochron there were no changes in the volumes per cell of the different classes of organelles, or of the numbers of organelles per cell, or of the sizes of the organelles. The only change which could be linked to the plastochronic functioning of the apex and the regular initiation of leaves was an increase in the percentage of plastids with starch in the central, axial parts of the apex and in the developing leaf axils, but not in the apical dome and the youngest primordium. This increase occurred at plastochron stage 9·5, which is precisely the point in the plastochron at which the orientation of the mitotic spindles changes in the I1 region just prior to the emergence of a new leaf (Lyndon, 1970b). The cells in which the increase in starch was most marked (the developing leaf axils and the incipient pith) are cells which are not destined to play any part in the formation of the leaf itself. The starch was almost certainly synthesized from precursors already present in the apex and not as a result of photosynthesis because the membrane systems of the plastids were poorly developed (see Fig. 7.3 A, B).

The quantitative differences in ultrastructure between different regions of the apex were related to cell differentiation (Figs 7.4 and 7.5). In the course of development of cells from the summit of the meristem to the incipient pith the cells enlarged and the volume and numbers of most of the cytoplasmic components increased. The amount of endoplasmic reticulum and dictyosomal material per cell and the numbers of microbodies and vacuoles per cell increased. Plastids and mitochondria differed in that the number of plastids per cell remained constant though their size increased, whereas the number of mitochondria per cell increased although total mitochondrial volume did not increase and therefore the size of individual mitochondria decreased slightly.

This implies that as the cells enlarged, mitochondrial replication out-stripped cell replication whereas plastid replication kept in step with cell replication.

In the development of the leaf axils the number of plastids per cell remained the same from the axil of the eleventh (youngest) leaf to the axil of the ninth leaf, but the number of mitochondria per cell decreased. This was associated with a decrease in cell size in this axillary region during its development and again implies that the plastid replication has kept pace exactly with cell replication, whereas mitochondrial replication has not kept up with the rate of cell division.

In the development of cells which are displaced from the summit of the apex to the flanks and then incorporated into the developing leaf primordia, the numbers per cell of plastids, mitcohondria, dictyosomes, microbodies and vacuoles all increased. This is the only cell differentiation sequence in which the number of plastids per cell increased and it implies that in this case plastid replication, like mitochondrial replication, is faster than cell replication. This probably represents the beginning of the increase in plastid number per cell which is usually associated with leaf development.

Since mitochondrial replication occurred faster than plastid replication in developing pith cells, slower in developing axillary cells, and at the same rate in developing primordium cells, the implication is that the replication of mitochondria and plastids are not under a common control and that neither are under the same control as cell replication.

In all three sequences of cell differentiation—the formation of the incipient pith, the development of the axillary position and the development of the young leaf cells—there was an increase in the number of vacuoles per cell. In the pea apex therefore, vacuolation accompanies cell differentiation, even, as in the case of the axillary development, when the cells are becoming smaller as a consequence of not regaining their former size after division. Although there are differences in the rate of division of cells within the apical meristem of about three- or four-fold (Lyndon, 1973) there were no ultrastructural differences that could be correlated with the different rates of cell division.

The zonation patterns (Fig. 7.2) which are seen in stained sections of apices (Lyndon, 1972a) were not reflected in the quantitative distribution of any of the recognizable organelles in this study. However, ribosomes were not examined, and in other apices (Nougarède, 1967) the greater density of stain on the flanks of the apex and in the incipient primordium has been shown to be associated with a higher density of ribosomes in these regions. The zonation pattern in the pea is therefore presumably a reflection of the different concentrations of ribosomes and possibly soluble proteins in the apex but is not a reflection of the general composition of the cells in terms of other organelles.

The mean numbers of organelles per cell in the central zone of the pea apex were 11 plastids, 58 mitochondria, 15 dictyosomes, 4 microbodies and

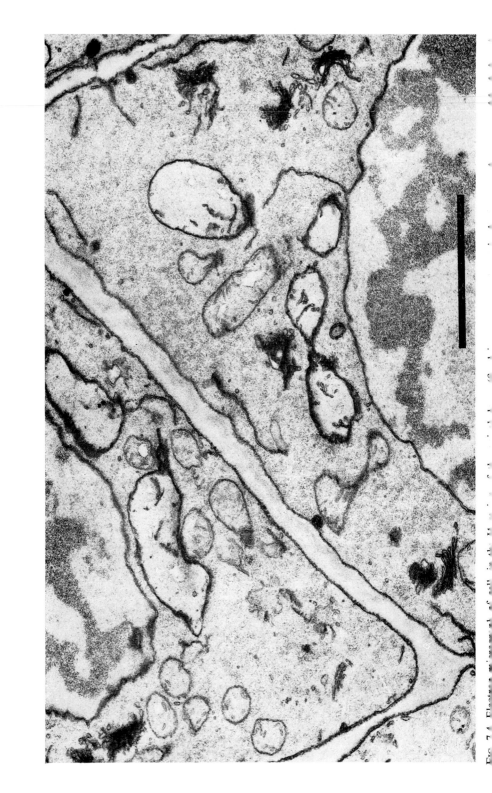

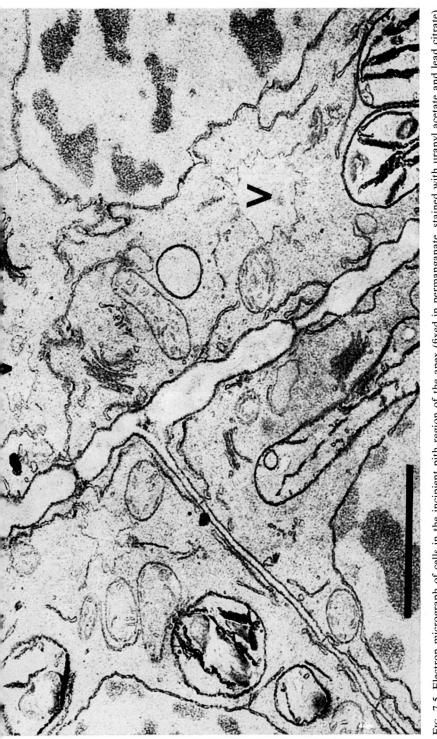

FIG. 7.5. Electron micrograph of cells in the incipient pith region of the apex (fixed in permanganate, stained with uranyl acetate and lead citrate). The internal membranes of the plastids are developing and often have starch associated with them. Vacuoles (V) are present. Bar = 2 μm

1 vacuole. This is comparable to the 11 plastids, 60 mitochondria and 24 dictyosomes in each telophase daughter cell in the *Epilobium* shoot meristem (Anton-Lamprecht, 1967) and the 12 plastids per cell in the shoot apex of spinach (Cran and Possingham, 1972). Although there was on average 1 vacuole per cell in the central region of the pea apex these were only small vacuoles and the apical dome has the general appearance of being non-vacuolated. This single average vacuole in the central zone was about half the volume of the typical plastid.

III. CELL GROWTH AND METABOLISM

The relative amounts of RNA, DNA and protein per cell in six regions of the apical meristem of the shoot were measured histochemically (Lyndon, 1970c) in sections stained with gallocyanin, Feulgen and dinitrofluorobenzene respectively. The absolute values for RNA and DNA per cell could be calculated because the 2C DNA amount was known to be 9·5 pg (Lyndon, 1967) and mitotic figures could therefore be used as an internal standard for the histochemical measurements. Independent chemical measurements indicated a protein:DNA ratio of approximately 6. The mean composition per cell in all regions of the shoot meristem was approximately 12 pg DNA, 9 pg RNA and 70 pg protein. The enlarging cells of the incipient pith were not measured but presumably would have greater amounts of RNA and protein per cell. The ratio of RNA:DNA is less than 1 in most of the apex and this is much lower than is usually found in plant cells, but is in fact comparable with the ratios of 1 or less recorded for the shoot apex of *Lolium* and for young wheat leaves (Rijven and Evans, 1967; Williams and Rijven, 1965). A low RNA:DNA ratio in the pea apex is also suggested by the RNA:DNA ratio in apical segments, each consisting of the apical dome plus the five youngest primordia and therefore containing a considerable number of enlarging and differentiating cells. Scans of the total nucleic acid extracted from these segments and subjected to electrophoresis on acrylamide gels shows that the RNA:DNA ratio was not more than about 2 (Fig. 7.6). The usual characteristic RNA components are present except for plastid RNA which would presumably occur only in small amounts, and is also easily degraded.

Since the gross composition of all the cells in the apex which were measured was the same it follows that all these cells have the same amounts of constituents to synthesize during the cell cycle. The rates of increase in the amounts of RNA and protein are therefore presumably a function of the rates of cell growth and division and will be less in the slowly dividing cells at the summit of the apex than in the more rapidly dividing cells. If the different rates of increase result from different rates of synthesis, and if the rate of incoproration of RNA and protein precursors is a measure of the rate of

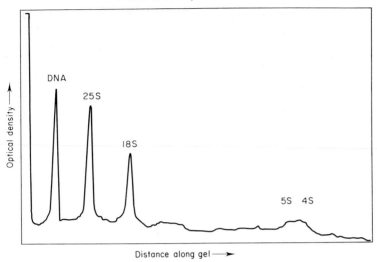

Fig. 7.6. Scan of nucleic acid components, separated by acrylamide gel-electrophoresis, from an extract of pea shoot apices, including the apical dome and the five youngest leaf primordia. RNA components are identified by their sedimentation constants.

synthesis, then the incorporation of ^{3}H-uridine and ^{3}H-leucine should be least at the summit of the apex and greatest on the flanks. However with both precursors the labelling was uniform throughout the pea meristem (Lyndon, 1972b). If the rates of incorporation of precursors were indicative of the rates of RNA and protein synthesis then this would mean that these rates were uniform throughout the apex. Were this so then the slower increase in the amounts of RNA and protein at the summit of the apex could result only from enhanced rates of RNA and protein breakdown.

A more probable and plausible explanation is that the rates of incorporation of precursor are not an indication of the rates of synthesis. This would only be so if the incorporation of the labelled precursor into the RNA or protein was itself the limiting step in the process. If some other step is limiting, such as the rate of uptake of precursor into the cells, or if there is an endogenous precursor pool already present in the cells, thus diluting the exogenous label to different extents in different parts of the apex, then the rate of incorporation will reflect these parameters rather than the rate of synthesis. Further experiments and kinetic data are necessary to determine whether or not these results can be explained, for example, by smaller pool sizes of precursors in the central zone rather than in the flank regions, or whether there is some other explanation. Rates of incorporation, in the absence of other data, obviously do not provide sufficient information from which to draw conclusions about the rates of synthesis of the compounds which become labelled.

The rate of DNA synthesis seems to be a function of the rate of cell division because the length of the S phase of the mitotic cycle is a function of the length of the whole cell cycle (Lyndon, 1973). In the cells on the flanks of the apex and in the incipient primordium the length of S was about 7 h, whereas in the more slowly dividing cells in the central zone it was about 11 h (Lyndon, 1972b, 1973; see also Table 7.I). Presumably the number of initiation sites for the replication of the DNA molecule is less in the central zone than in the rest of the apex, as it is in slowly dividing animal cells (Callan, 1972).

One aspect of the growth of the cell which can be measured visually is the growth of the nucleolus during the cell cycle. The average sizes of telophase and prophase nuclei in the pea shoot meristem are given in Fig. 7.7. Nuclei which are intermediate in size represent nuclei at some stage of interphase, progressively larger nuclei being progressively nearer prophase. Taking nuclear volume as an indicator of progress through the cell cycle (although the x axis may not necessarily be a linear time-scale) nucleolar volume per cell may be plotted against size of nucleus and so the increase in nucleolar volume per cell through the cell cycle from telophase to prophase may be inferred (Fig. 7.7). Nucleolar volume per cell increases from zero at telophase to 6·5 μm^3 in early prophase, just before the nucleolus disappears at mitosis.

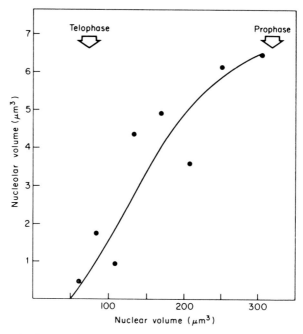

Fig. 7.7. Growth of nucleoli during the cell cycle in the pea shoot meristem. Mean volumes of 10 telophase and 10 prophase nuclei are indicated. Each point is the mean of 4–15 values, measured from sections.

This increase is slower than in the root meristem, in which the increase in nucleolar volume during the cell cycle is three times as great as in the shoot (Lyndon, 1968b). Since the cell cycle in the root meristem is probably shorter than in the shoot (Lyndon, 1973) this implies a rate of increase of nucleolar volume in the cell cycle in the shoot meristem of about one-sixth or less of that in the root.

The values in Fig. 7.7 refer to total nucleolar volume per cell. This is because there may be more than one nucleolus per cell and per nucleus. There are seven chromosomes in the haploid complement of *Pisum* and two of these have nucleolar organizers (Atakebowa, 1959). In the diploid cells of the shoot apical meristem one would therefore expect to find four nucleolar organizers per chromosome complement with a maximum number of four nucleoli per cell. In the pea shoot apical meristem four nucleoli are very rarely observed but three nucleoli can often be seen and most cells have one or two nucleoli. This suggests that fusion of the nucleoli occurs fairly quickly. Although different varieties of peas may have slightly different karyotypes, with differences in the position of the nucleolar organizing region, the total number of nucleolar organizers in the diploid cell is apparently always four (Sen and Tiwari, 1966).

IV. RATES OF CELL DIVISION

A. Mean Cell Generation Time

The overall rate of cell division for the meristem as a whole, the mean cell generation time (MCGT), can be calculated most readily from the rate of increase in cell number during a plastochron. Counts of the number of cells in the apical dome at different times during a single plastochron show that there is an increase from about 830 cells just after the formation of a leaf primordium to about 2600 just before the emergence of the next. Since the increase appears to be exponential (Lyndon, 1968a) the MCGT can be calculated as 28 h (Lyndon, 1970a) either graphically or from the formula:

$$\text{MCGT} = \frac{t \, log \, 2}{log \, y - log \, a}$$

where t = the plastochron (h); a = number of cells in the apical dome at the beginning of a plastochron; y = number of cells in the apical dome at the end of a plastochron.

B. Division Rates Measured by C-metaphase Accumulation

The rate of cell division has also been measured by the method of accumulation of colchicine-metaphases (Lyndon, 1970a). However, it was observed that the maximum rate of accumulation of C-metaphases which could be achieved was about $1 \cdot 5\%$ h^{-1}. This was less than the $2 \cdot 5\%$ h^{-1} which was theoretically necessary with a MCGT of 28 h. Numerous attempts with different methods of application of colchicine and with different batches of colchicine all failed to increase the rate of accumulation of C-metaphases above $1 \cdot 5\%$ h^{-1}. Had this been the only piece of data available the MCGT would have been calculated as about 45 h. The colchicine was apparently inhibiting not only exit from metaphase but also progress through interphase or prophase, i.e. at some point before entry into metaphase. The use of lower concentrations of colchicine might have resulted in the calculation of shorter MCGTs but since such lower concentrations did not eliminate anaphases and telophases their efficacy was not certain and the values for the MCGT would have been of doubtful validity. It is clearly necessary to corroborate MCGTs obtained from accumulation of C-metaphases with independent data for the MCGT in untreated plants. Only in this way can true cell cycle times be obtained rather than lengthened times which are in fact experimental artefacts.

Relatively high ($0 \cdot 5\%$) concentrations of colchicine were necessary to inhibit division in the shoot apex of the pea, and although lower concentrations can be used for roots, the pea does seem to be relatively insensitive to colchicine. Higher concentrations than those which are applied to the root must be applied to the shoot apex because the latter seems relatively impermeable to substances placed on it. The entry of substances into the shoot apex is facilitated by wounding or excising tissues near the apex (Bernier and Bronchart, 1963; Lyndon, 1973).

Although colchicine inhibited processes other than the exit of cells from metaphase, and although after 8 h of treatment with the $0 \cdot 5\%$ colchicine solutions the accumulation of metaphases ceased in the tissue altogether, the distribution of C-metaphases did not change. This showed that the degree of inhibition of entry into metaphase was the same in all parts of the apex and that the accumulation of C-metaphases did not cease in some regions sooner than others. Had this occurred then the pattern of C-metaphases would have changed with time. Since it did not, then the pattern of C-metaphases was always indicative of the pattern of rates of cell division, and it was valid to use apices which had been treated with colchicine for any length of time to observe the distribution of metaphases. Knowing the absolute mean rate of cell division for all cells in the apex (the MCGT) it was then possible to convert the relative rates of cell division obtained from the C-metaphase data into absolute rates. The absolute rate of division and the length of the cell cycle corresponding to a given density of C-metaphases could then be calculated so that lines joining cells having the same rate of cell division and the same cell cycle length could be superimposed on the outline of sections of the apex (Lyndon, 1973).

By recording the positions of all the C-metaphases in all the serial sections through pea apices it was possible to reconstruct sections in any plane and so examine the distribution of rates of cell division throughout the apex in three dimensions. The general structure of the apex in terms of cell division rates was most easily seen from median longitudinal sections (Lyndon, 1970a). At the beginning of a plastochron when the apical dome is at minimal area the whole of the dome is occupied by cells dividing relatively slowly. During the course of the plastochron a band of cells showing a more rapid rate of division develops at the base of the apical dome between the axil of the youngest leaf primordium which is just emerging and the axil of the next oldest leaf primordium. Reconstructions of sections in other planes show that this band of rapidly dividing cells is, in fact, a plate of cells. At the end of the plastochron the apex consists of a region of slowly-dividing cells which occupies most of the apical dome. This region is separated from that of the youngest primordium and the axial tissue which it subtends (and which has an intermediate rate of division) by a region of much more rapid divisions at the base of the apical dome. There does not seem to be any clear distinction in terms of rates of division between the central zone at the summit of the apex and the other tissues of the apical dome which surround it. The region of slow division rate in the apical dome is much larger than the central zone. Since the cytological characteristics of the central zone are not a function of a lower rate of division compared with the adjacent cells of the apical dome this suggests that the distinctiveness of the central zone has a physiological basis not related to division rate. When sections were reconstructed in other planes it was seen that the plate of rapidly dividing cells at the base of the apical dome was continuous with regions of rapid division lying in the positions of the incipient procambium at the stipular sides of the apex. At this stage of development this procambium is difficult to pick out by microscopical examination of anatomical preparations. It is notable that the procambium in the region of the young primordium, where the procambium is most distinct, does not have a higher rate of division than the cells around about it (Lyndon, 1970a). This may indicate that the high rate of division which is associated with the incipient procambial strands occurs only at the very earliest stage of procambium formation, and before the strands are anatomically distinct, the rate of division may diminish to that of the cells around them. The procambium in the region of the young primordium is seen to have a higher rate of division only at the stage before the primordium itself is formed, i.e. in the late I1 phase just after the orientations of the spindles have changed and before the primordium has begun to grow out as a bulge.

The recording of the C-metaphases in all the sections of the apical dome allowed reconstruction of the distribution of divisions over the surface of the apical dome. At the summit of the apex, in the region of the central zone and the tissues immediately adjacent to it, the rate of division was lower than in the rest of the apex, but divisions were present over the whole surface of the apex (Lyndon, 1970a).

The rate of division in the epidermis was in general about two-thirds of that of the underlying tissues (Lyndon, 1971). This is what one would anticipate on geometrical grounds because the surface of a solid body increases as the square of the linear dimensions, whereas the volume increases as the cube (see also p. 207).

C. Cell Cycle

The cells in different parts of the pea apex have different rates of division and it is clearly of interest to know whether certain parts of the cell cycle are extended or reduced at the expense of others. The only measurement of the lengths of the phases of the cell cycle in the pea (Lyndon 1973) have been done using the method of Mak (1965). The cells are labelled with ^{3}H-thymidine and the nuclei stained with Feulgen. The nuclei of the apex which are not labelled are therefore in parts of interphase other than S. From microdensitometric measurements of the DNA content of unlabelled nuclei the relative numbers of 2C and 4C nuclei can be obtained and from these values the relative lengths of the G_1 and G_2 phases of the cell cycle can be calculated. The relative length of the S phase is obtained from the proportion of labelled nuclei. The length of mitosis is obtained from the mitotic index. If the absolute length of the cell cycle is known then the absolute lengths of the different phases can easily be calculated. This technique was used by Lyndon (1973) to obtain the lengths of the phases of the cell cycle in the different regions of the pea apex. Very similar values were obtained by another method which depended on the measurement of average amounts of DNA in cells in different regions of the apex. Previous measurements of the DNA values (Lyndon, 1970c) had given mean C values for the DNA per cell in different regions of the apex. These values depend upon the relative numbers of nuclei in G_1, G_2, S and M and since the characteristic C amounts of DNA associated with each of these phases can be assumed it is possible to work out a formula from which the proportion of nuclei in all the four phases can be found if the proportions of nuclei in S and M are known (Lyndon, 1973). The values for the lengths of the phases of the cell cycle obtained by both methods agreed and the means are shown in Table 7.I. In the central zone, where the mitotic

TABLE 7.I. Lengths of the phases of the cell cycle (h) in four regions of the pea shoot meristem. (Data from Lyndon, 1973)

Region	Whole Cycle	G_1	S	G_2	M
Central Zone	69	37	13	18	1
Flanks of the } I2	30	14	8	7	1
Apical Dome } I1	28	15	8	4	1
Leaf Primordium	29	15	10	3	1

cycle is longer than in the rest of the apex, the lengths of S, G_2 and G_1 are all longer than elsewhere in the apex, S being extended by about 50% and G_1 and G_2 by 150%. The length of M (mitosis) remained the same irrespective of cycle length. The different rates of division in the pea apex therefore seem to be the result of an extension of all the phases of interphase.

Since the length of S is extended according to the length of the cell cycle then it follows that the proportion of cells labelled by ^{3}H-thymidine, i.e. the proportion of cells in S, will remain more or less constant irrespective of the rate of cell division. This is in fact what was found for Lincoln peas (Lyndon, 1972a), for apart from the incipient pith all regions of the apex showed a similar labelling index. The lesser extension of S, compared with G_1 and G_2, in the slowly dividing summit cells was shown more clearly by the lower labelling index for the summit than for the flanks of the apex in the peas used by Nougarède and Rondet (1973b).

V. PLANES OF CELL DIVISION AND GROWTH

The direction of growth and the plane of cell division are related in tissues in which the cells remain isodiametric, since the isodiametric shape can only be maintained if the axis of growth of the cell is the same as the axis of the mitotic spindle. If the axis of the mitotic spindle is at right angles to the direction of growth then the cells change in shape as they do in the procambium. The axis of the mitotic spindle in the procambium can be across the cell whereas the cell may grow in length and so become longer and thinner. Where the cell shape is isodiametric and remains so then the direction of growth can be inferred from the orientation of the mitotic spindles. The planes of growth throughout the pea apex have been measured and recorded by noting the orientation of all anaphase and telophase mitotic spindles (Lyndon, 1970b).

The distribution of spindles on the surface of the apical dome (i.e. in the epidermis) is predominantly radial, i.e. longitudinal. This, together with the gradient in the rate of growth and division from a minimum at the summit of the apex to a maximum on the flanks, results in the hemispherical shape of the apical dome in the pea (Lyndon, 1976).

In the sub-epidermal cells (i.e. all cells other than the epidermis) the orientation of growth is somewhat more complex and changes during the course of the plastochron. Median longitudinal sections of the apex show that the growth on the flanks of the apex where the primordium is about to appear is predominantly longitudinal and mitotic spindles are in the plane of the section (Lyndon, 1970b; Nougarède and Rondet, 1973b). This can be envisaged as giving rise to upward growth of the apex as a whole. In the central part of the apical dome the spindles (and hence the direction of growth) are perpendicular to the plane of the sections and are transverse, i.e. the spindles are aligned across the apex (Lyndon, 1970b). This is thought to be associated with the growth of the stipules which are formed at the sides of the pea apex. The proportions of spindles in the longitudinal and transverse orientations were essentially the same throughout the plastochron except in the region of the apical dome immediately above and adjacent to the emerging primordium. Here the mitotic spindles and the plane of growth are essentially transverse in the first part of the plastochron when the young leaf primordium which is immediately adjacent is becoming established, whereas in the second part of the plastochron the orientation is predominantly longitudinal during the upward growth of the apical dome once the axil of the young leaf primordium has become established (Lyndon, 1970b). This change in the orientation of many of the spindles, and in the direction of growth, in this part of the apical dome (the I2a region) is restricted to the sub-epidermal cells. Such a change is not found in the epidermal cells and is consistent with the hypothesis that the changes in the directions of growth originate in the interior of the apex rather than at the surface.

The orientation of the spindles and the plane of growth can also be classified according to whether the subsequent cell divisions are periclinal or anticlinal. Except for the region immediately above the site of the incipient primordium the growth of the apical dome during the first part of the plastochron was almost exclusively the result of anticlinal cell divisions associated with growth parallel to the surface of the apex (Lyndon, 1970b). Only in the second part of the plastochron were periclinal divisions observed in that part of the apical dome where the next primordium was to appear (the I1 region). These divisions would be associated with growth perpendicular to the surface, resulting in the subsequent bulging which is seen as the formation of a new primordium. In the second part of the plastochron the number of mitotic spindles resulting in periclinal divisions increased considerably in the I1 region (excluding the epidermis in which all divisions were anticlinal) so that the proportion of spindles orientated in the three planes periclinal, anticlinal (plane of section) and anticlinal (perpendicular to section), were in the ratio 1:1:1 (Lyndon, 1972a). The periclinal divisions were found throughout both the corpus and the second tunica layer in the I1 region. The equal distribution of divisions between the three planes of growth in the sub-epidermal cells suggested that the orientation of the spindle (and hence the direction of growth) at this time during the plastochron was random. This suggests that there is a constraint upon the planes of division in the first part of the plastochron, when growth results in anticlinal divisions, and that this constraint is lifted and that the cells can divide in any direction during the second part of the plastochron in the I1 part of the apical dome where the leaf is about to be formed. This has been interpreted as the primary event of leaf initiation and primordium formation (Lyndon, 1976). In other words, leaf formation appears to be associated with a loss of polarity in the apex rather than the imposition of a new polarity. The imposition of the polarity is in the maintenance of the anticlinal divisions and the planes of growth which allow the growth of the apical dome but not the formation of leaf primordia.

VI. MECHANISM OF LEAF INITIATION

The formation of a new leaf primordium at the shoot apex results from the outward growth of part of the apex to form a bulge. This change in shape could be accomplished by changes in the direction of growth, or by localized changes in the rates of growth with constraints on the directions in which the mass of dividing cells could grow, or it could result from a combination of both these factors.

Changes in the rates of growth in an apex with entirely anticlinal divisions would not result in a change in form of the apex and the formation of a new leaf. However, in an apex in which only the outer layers of cells are stratified and the inner cells divide and grow in all directions, i.e. there is a tunica and a corpus, then change in shape of the apex could occur if there were a change in the rate of division localized to one part of the apex. In this case one could expect to find the initiation of the leaf marked first of all by an increase in the rate of division localized to a particular part of the flank of the apex. If a change in the direction of growth were the main factor then the initiation of the primordium would be first marked by a change in orientation of growth and seen as a change in the orientation of mitotic spindles. Such a change in the orientation of mitotic spindles has been found in the pea shoot apex and occurs in the I1 region, where the new leaf primordium is about to be formed, about 16 h before it first emerges as a visible bump on the surface. The overall rate of division and growth in the I1 region of the apex was very much the same as in the I2 region in which a leaf was not to be immediately formed (Lyndon, 1970a). On the basis of this information Lyndon (1970a) argued that the formation of the leaf primordium depended primarily on changes in the directions of growth rather than changes in the rates of growth. However, Hussey (1972) pointed out that the region of fastest growth in the apex appeared to be the base of the I1 region and that Lyndon's data were inadequate in that they did not take into account the different rates of division in different parts of the I1 and I2 regions. Hussey (1971b) showed that in the tomato apex the formation of the leaf primordium was preceded by a localized increase in the rate of division of the cells in the corpus of the I1 region. He pointed out that in the pea apex there was a similar region at the base of the I1 region where divisions were faster than in any other part in the apex (Hussey, 1972). By placing marks of carbon on the apical dome of the pea and tracing their subsequent position after growth through a plastochron,

Hussey (1972) was able to show that it was the basal or abaxial part of th
developing primordium which grew fastest in an upwards direction and h
suggested that this followed from the establishment of the higher rate o
division at the base of the I1 region. He concluded that the primordium wa
formed primarily because of this localized rapid growth at the base of the I
region. Lyndon (1976) subsequently obtained data which supported Hussey'
contention. The distribution of C-metaphases in the epidermis of the pri
mordium indicated a greater division rate in the abaxial part of the primor
dium (Lyndon, 1976). Hussey (1972) also pointed out that the higher rate o
division at the base of the I1 region was a continuation of a higher rate o
division at the base of the I2 region. Hussey therefore concluded that th
primary event in the formation of a leaf primordium was the increase in th
rate of division which occurred on transition, or displacement, of a cell from
the I3 to the I2 region. A cell being displaced down the apical dome into th
leaf-forming region would therefore have its rate of division increased as i
entered the base of the I2 region approximately two plastochrons before i
would contribute to the initiation and formation of the bulge of the young
leaf primordium.

   The facts of the rates of division and planes of division within the pea ape
are not in dispute. What has been contested is the interpretation placed upo
these as to what is the primary event in leaf initiation. Lyndon (1976) arguec
that as the increase in the rate of division from I3 to I2 took place approxi
mately four cell cycles before the leaf primordium was formed, and that sinc
the I2 region itself did not grow out to form a leaf primordium, then thi
increase in the division rate was not an event associated with the formatior
of the bulge of the new primordium itself. He argued that the high rates o
division at the sides of the apex and in the I1 region seemed to be more
connected with the differentiation of the procambial strands, and that the
plate of rapidly dividing cells was concerned with the elevation of the apica
dome during the second part of the plastochron. The increased rate of division
at the base of the I1 region was interpreted by Lyndon as resulting in the
upward growth of the abaxial surface of the leaf primordium, as also claimed
by Hussey, but Lyndon pointed out that this was a factor affecting the shape
of the primordium rather than the emergence of the primordium itself. I
seems clear that the immediate event resulting in the formation of the bulge,
which is the new primordium, is the change in orientation of growth about
16 h before the primordium appears. Since each primordium (after the first)
is part of an integrated system in which it has been preceded by other pri
mordia and will be followed by further primordia it follows that the formation
of a primordium results from the organized growth of the apex, and since the
changes in rates of division are part of this organization it is difficult to point
to any one event in this continuum of events which is likely to be the starting
point for the formation of a new leaf primordium. Perhaps the question could

be resolved or at least clarified to a further extent by the examination of leafless mutants of peas or other plants.

The changes in the orientation of growth were in the first instance followed by comparing the rates of division with the rates of cell accumulation in arbitrary regions of the apex (Lyndon 1970a). This analysis was extended to the epidermis alone which could then be compared with the sub-epidermal regions of the apex (Lyndon, 1971). The bulging of tissues in one part of the apex was recorded as a displacement of cells from one arbitrary region to another. When the analysis was done for the epidermis it was found that the degree of displacement of cells from one region to another was less than it was for the underlying cells. This was interpreted as showing that the changes in the orientation of growth were predominantly internal phenomena and that the morphogenetic changes occurring in the apex and associated with leaf initiation were therefore events which were initiated deep within the apex rather than in the surface layers. The epidermis could be interpreted as behaving like a skin, growing locally in response to the bulging of the tissues within it.

Mechanical tensions do not seem to be involved in leaf initiation in the pea for all parts of the apex appear to be under slight compression, as shown by the lack of gaping (Gulline and Walker, 1957) or the closing up of incisions made in the apex (Hussey, 1973). Compression in the tissues of the young leaf primordium is consistent with the observation that here the rate of cell division is as high in the epidermal cells as in the underlying cells (Lyndon, 1971).

VII. RATE OF LEAF INITIATION

At the temperature at which peas are normally grown (15°–25°C) they appear to initiate leaves at the rate of about one leaf every 2 days. In *Pisum sativum* cv. 'Lincoln' no leaves were initiated during the first 3 days after sowing (Lyndon, 1968a). This is when the seed is imbibing water and the radicle is beginning to grow. Thereafter leaves were initiated quite rapidly from days 4 to 8 and the length of the plastochron during this stage of development was only about 23 h. Then the rate of leaf initiation settled down to a steady rate of about one leaf every second day, i.e. a plastochron of about 2 days. In *P. sativum* cv. 'Alaska' the rate of leaf initiation from days 2 to 16 after sowing seemed to be fairly constant and perhaps decreased with time, the plastochron again being just over 2 days (Thomson and Miller, 1961). The Alaska peas had six leaf primordia in the seed, and the Lincoln peas had seven primordia, the seventh primordium being very small.

Lincoln peas show strict apical dominance and flowering does not occur until about 19 leaves have been formed (Lyndon, 1968a). However in early varieties, such as 'Feltham First', flower buds may be present as soon as the seed germinates.

The length of the plastochron as determined by the rate of leaf initiation may not be the same as the length of the plastochron measured by the rate at which unfolded leaves are formed. Measurements on Lincoln peas (Fig. 7.8) show that the rate of formation of unfolded leaves is slower than the rate of leaf initiation so that the number of young leaf primordia on the shoot apex increases as it develops. The plastochron measured in these two different ways is therefore different.

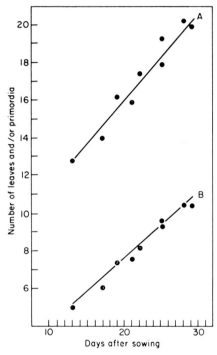

FIG. 7.8. The total number of leaves plus primordia (A) increases faster than the number of expanded leaves (B) so that the number of primordia plus folded leaves at the apex (the vertical difference between lines A and B) increases as the plant ages. Each value is a mean from 10 plants. The two small epicotyledonary leaves are included in both cases.

There is some effect of light on the rate of initiation of primordia in Alaska peas (Thompson and Miller, 1961). In the light the length of the plastochron was about 2·4 days compared with 2·9 days for plants kept in complete darkness. This represented growth over the first 16 days and during the formation of five or six new leaf primordia. In *P. sativum* cvs. 'Greenfeast' and 'Telephone' the rate of leaf initiation was the same, about one leaf every 2 days, in light or in darkness (Low, 1971). However, in darkness leaf initiation topped abruptly when the 12th leaf was formed. When light-grown seedlings were transferred into darkness, if leaf 12 had not yet been initiated at the time of transfer then leaf 11 or leaf 12 would be the last to be initiated in the dark. If leaf 12 had already been initiated before the transfer, then leaf initiation either stopped immediately or after the initiation of one more leaf. Apices which had stopped leaf initiation in the dark resumed initiation 2–3 days after being transferred back into the light, irrespective of whether this was done immediately or after the plants had been up to 10 days in the dark.

VIII. CULTURE OF EXCISED APICES

Shoot apices consisting of the meristem and the youngest leaf primordium together with 3 mm of stem have been successfully cultured for approximately one plastochron (Hussey, 1972). The apices were obtained from seedlings grown in sterile conditions for 7 days after sowing. The 3 mm piece of stem was embedded in agar and particular care was taken not to touch or damage the apical dome. The nutrient medium consisted of the Murashige and Skoog salt mixture and 2% sucrose together with 100 mg l^{-1} myo-inositol and 0·5 m gl^{-1} thiamine.HCl in 1% agar. The cultured apices, maintained at 25°C in a light intensity of 10 000 lx, grew for one plastochron, which was about 4 days. Since the plastochron in an intact seedling is about 2 days this means that before their growth eventually stopped these explanted apices were growing at only about half the rate of intact apices. Despite this slower growth rate their growth pattern was apparently the same as that of intact apices as far as could be judged from their anatomy.

Apical domes, with no primordia attached, have been cultured on agar with varying concentrations of benzyladenine (BA) and naphthalene acetic acid (NAA), both singly and together (Kartha et al., 1974). The apical cells formed a callus and if BA was present (with or without NAA) shoots were initiated but not roots. Only when 10^{-6} M NAA was supplied, in the absence of BA, were roots as well as shoots formed. With lower concentrations of NAA (10^{-7} and 10^{-8} M) no roots were formed and only shoots were initiated. Calluses which initiated shoots could also be obtained from macerates of apical meristems (Gamborg et al., 1974).

IX. GRAFTING OF THE APEX

Successful grafting of apical meristems back on to the parent stock can be achieved when special precautions are taken to prevent the apices drying out (Gulline and Walker, 1957). Young seedlings were grown until they were about 7 days old and their roots were about 3·8 cm (1·5 in) long The seedlings were then decapitated above the first node. After pruning away the bracts, and axillary buds which had more than one primordium, new axillary apices were formed and were readily accessible for experimentation. Using pieces of thin razor blade as knives, apical segments 200 μm or less were cut from the summit of the apex. The apical segments were then replaced on the stock from which they had been cut. If there were primordia present then the apical segment was rotated through approximately 180° to ensure the discontinuity of the original conducting paths. In the cases where successful grafting occurred this sometimes began as little as 1 or 2 days after cutting, and the grafts developed into small buds within a week. The smallest piece of apex which was successfully grafted was 50 μm in height. This presumably represented only the apical dome itself. The smallest apex which regenerated a bud was 150 μm high and consisted of about 500 cells.

The critical procedure for successful grafting was the performance of the cutting and grafting in a cabinet in which the atmosphere could be maintained saturated with a mist. The slightest drying out resulted in the graft being unsuccessful. Gulline and Walker concluded that the limitations to the size of the graft were imposed by the technical difficulty of cutting smaller pieces and transferring them back to the parent stock, and in doing so avoiding damage to the apex, for even touching the surface of the apex without causing any visible damage could result in its death. Unsuccessful grafts were usually caused by the cut surfaces not being in sufficiently close contact.

REFERENCES

ANTON-LAMPRECHT, I. (1967). *Ber. dt. bot. Ges.* **80**, 747–754.
ATAKEBOWA, A. J. (1959). *Biol. Zbl.* **78**, 424–438.
BERNIER, G. and BRONCHART, R. (1963). *Bull. Soc. r. Sci. Liége* **32**, 269–283.
CALLAN, H. G. (1972). *Proc. R. Soc.* B **181**, 19–41.
CRAN, D. G. and POSSINGHAM, J. V. (1972). *Protoplasma* **74**, 345–356.
ERICKSON, R. O. and MICHELINI, F. J. (1957). *Am. J. Bot.* **44**, 297–305.
GAMBORG, O. L., CONSTABEL, L. F. and SHYLUK, J. P. (1974). *Physiologia Pl.* **30**, 125–128.
GIFFORD, E. M. (1954). *Bot. Rev.* **20**, 477–529.
GULLINE, H. F. and WALKER, R. (1957). *Aust. J. Bot.* **5**, 129–136.
HUSSEY, G. (1971a). *J. exp. Bot.* **22**, 688–701.
HUSSEY, G. (1971b). *J. exp. Bot.* **22**, 702–714.
HUSSEY, G. (1972). *J. exp. Bot.* **23**, 675–682.
HUSSEY, G. (1973). *Ann. Bot.* **37**, 57–64.
KARTHA, K. K., GAMBORG, O. L. and CONSTABEL, L. F. (1974). *Z. Pflanzenphysiol.* **72**, 172–176.
LOW, V. H. K. (1971). *Aust. J. biol. Sci.* **24**, 187–195.
LYNDON, R. F. (1967). *Ann. Bot.* **31**, 133–146.
LYNDON, R. F. (1968a). *Ann. Bot.* **32**, 371–390.

LYNDON, R. F. (1968b). *In* "Plant Cell Organelles" (J.B. Pridham, ed.), pp 16–39. Academic Press, New York and London.
LYNDON, R. F. (1970a). *Ann. Bot.* **34**, 1–17.
LYNDON, R. F. (1970b). *Ann. Bot.* **34**, 19–28.
LYNDON, R. F. (1970c). *J. exp. Bot.* **21**, 286–291.
LYNDON, R. F. (1971). *Ann. Bot.* **35**, 263–270.
LYNDON, R. F. (1972a). *Physiol. Végétal.* **10**, 209–222.
LYNDON, R. F. (1972b). *Symp. Biol. Hung.* **13**, 345–353.
LYNDON, R. F. (1973). *In* "The Cell Cycle in Development and Differentiation" (M. Balls and F. S. Billett, eds), pp 167–183. Cambridge UniversityPress, London.
LYNDON, R. F. (1975). *In* "Cell Division in Higher Plants" (M. M. Yeoman, ed.), pp 285–314. Academic Press, London and New York.
LYNDON, R. F. and ROBERTSON, E. S. (1975). *Protoplasma* **87**, 387–402.
MAK, S. (1965). *Expl Cell Res.* **39**, 286–289.
MITRA, G. C. (1950). *Proc. Indian Acad. Sci.* **31**, 210–222.
NOUGARÈDE, A. (1967). *Int. Rev. Cytol.* **21**, 203–351.
NOUGARÈDE, A. and RONDET, P. (1973a). *C. r. hebd. Séanc. Acad. Sci., Paris* **277** D, 393–396.
NOUGARÈDE, A. and RONDET, P. (1973b). *C. r. hebd. Séanc. Acad. Sci., Paris* **277** D 997–1000.
REEVE, R. M. (1948). *Am. J. Bot.* **35**, 591–602.
RIJVEN, A. H. G. C. and EVANS, L. T. (1967). *Aust. J. biol. Sci.* **20**, 1–12.
SEN, S. K. and TIWARI, C. B. (1966). *Nucleus* **9**, 173–176.
THOMSON, B. F. and MILLER, P. M. (1961). *Am. J. Bot.* **48**, 256–261.
THOMSON, B. F. and MILLER, P. M. (1962). *Am. J. Bot.* **49**, 303–310.
WEIBEL, E. R. (1969). *Int. Rev. Cytol.* **26**, 235–302.
WILLIAMS, R. F. and RIJVEN, A. H. G. C. (1965). *Aust. J. biol. Sci.* **18**, 721–743.

8. Control of Vascular Differentiation

T. SACHS

Department of Botany, The Hebrew University, Jerusalem, Israel

I. INTRODUCTION

Pea seedlings offer a number of important advantages for experimental work on vascular differentiation:

1. It is easy to prepare a large number of genetically uniform seedlings for experimental purposes. The seedlings develop well in a wide variety of temperature, light and humidity conditions.

2. The seedlings are extremely vigorous and respond rapidly to various treatments even when they are supplied with nothing but water.

3. A wide range of varieties are readily available.

4. The plants can be operated on both in the apices and in the mature axis, with relative ease. The convenience of work on apices is due to the almost total absence of hairs.

5. Methods for culture of organs (roots) (see Chapter 5) and of callus are available (Torrey, 1968).

The phenomena of cell differentiation in peas and other organisms raise two problems which are of general relevance to biology. The first is the question of the nature of the cellular events which lead to the formation of different cells from the same genetic system. The answer to this question is expected in terms of the regulation of gene products. A second problem is the nature of the controls which cause the cells to differentiate in a definite relation to one another, forming organized tissues in defined locations relative to

H

the rest of the organism. These controls of patterns or of organization might be understood in terms of interactions between the differentiating cells and between these cells and the rest of the organism. This chapter is an attempt to review the available information concerning vascular differentiation in peas in relation to both of these problems. The main stress, however, will be on the problem of the nature of the interactions which control patterned differentiation. This emphasis is due both to the interests of the author and to the available information. Vascular differentiation always results in the formation of a number of cell types embedded in other tissues and the conversion of an entire tissue to one type of vascular element has not been found, either *in vivo* or *in vitro* (Torrey, 1971). Vascular differentiation, furthermore, is always a complex chain of processes leading to a final state which is totally different from the original meristematic cells. For these reasons vascular tissues are not a simple system for the identification of key regulatory mechanisms involved in differentiation. Studies of pattern formation, on the other hand, have been aided by the relative convenience of observing vascular differentiation and regeneration after wounds in large, vigorous seedlings, such as those of *Pisum sativum* L. They have also been stimulated by the available knowledge concerning the relation of vascular differentiation to the development of the rest of the plant and the chemical identification of one of the major differentiation signals involved in this interaction as auxin (see Section III).

II. DESCRIPTION OF VASCULAR DIFFERENTIATION

The vascular tissues of peas are organized in the leaves and reproductive organs as a network of strands which connect with a vascular cylinder in the stem and the roots. The vascular system, therefore, forms a connection between the photosynthetic and growing parts of the shoot and the root tips. This system is polar within the stem and the root in the sense that there are no individual vessels or sieve tubes which connect directly from one leaf to another or from one root to another. Passage of materials between organs of the same type must be made by transport from one vascular channel to another. The vascular system has a complex internal structure, as it is composed of three distinct tissues, the xylem, the phloem, and the cambium between them. In addition to the various types of cells which are always present in these tissues, the phloem includes phloem fibres, which are organized both in discrete regions outside the phloem, separated from the sieve tubes by a broad band of parenchyma, and also as completely separate cortical bundles in the stem (see below).

The arrangement of the vascular bundles in the pea is in general fairly usual for a dicotyledon; there are three leaf traces and three leaf gaps (see

Fig. 8.4a) at every node, and the root has a triarch pattern (see Chapter 5). There are, however, two unusual characteristics. The transition region, in terms of the arrangement of vascular tissues, between the root and the stem is found not only in the hypocotyl, which is very short, but also in the stem below the third internode. This transition is expressed in:

1. The location of the vascular cylinder relative to the organ as a whole (i.e. the presence and size of a pith).
2. The location of the phloem relative to the xylem.
3. The location of the protoxylem relative to the metaxylem.

For the details of this complex transition zone, see Gourley (1931) and Hayward (1938). The other special characteristic is the presence of cortical vascular bundles in the stem. Four such bundles may be seen readily in cross-sections of the lower parts of the stem (Fig. 8.1a); two of these, in the plane

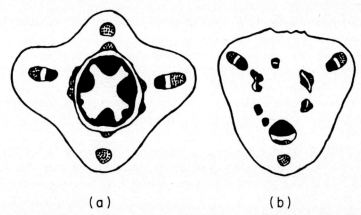

(a)   (b)

FIG. 8.1. Schematic representation of vascular tissues in cross-sections of pea stems, comparing an untreated stem (a) and one from which leaf primordia have been removed (b). Black areas mark the xylem, dotted areas the phloem fibres and clear areas next to the xylem mark the phloem and cambial regions. The sections were oriented so that the leaf petioles were above and below the vascular cylinder. Note the cortical bundles; the ones which connect directly to the petioles consist of fibres only. The treatment of (b) removed, at a very early stage of development, all but one leaf primordium from above the region of the stem which was later sectioned. Note that fibre and vascular strands which would have connected to the missing leaves did not differentiate. Drawn from Sachs (1972), with the kind permission of the Annals of Botany Co.

of the leaf petioles, and connecting directly to these petioles, consist only of phloem fibres. The two other cortical strands, in a plane perpendicular to that of the petioles, consist of all vascular tissues and connect to the stipules.

As in most other dicotyledons, peas have a vascular cambium which constantly adds new vascular tissues in the mature parts of the stems and the roots. Cambial activity is found very close to the shoot and root apices, the exact distance depending on growing conditions, and it does not cease

upon the onset of flowering. As new vascular tissues are added some of the
mature cells become non-functional; all of the primary phloem has been reported to be crushed at an early stage of shoot development (Wark and
Chambers, 1965). The actual size of the functional vascular system and the
changes in its capacity during growth are not known; however, a large
amount of vascular tissues is never formed, a characteristic which is probably
connected with the climbing habit. The exact course of cell divisions in the
cambial zone, which leads to the maturation of vascular elements, has been
described only for the phloem side of the cambium (Zee, 1968; Zee and
Chambers, 1969). The sequence of divisions is remarkably simple, and few
intermediate or transitional cells are formed. This simplicity is in contrast
to the situation in woody plants, making peas a convenient system for the
study of secondary phloem differentiation (Zee, 1968). In the stem, a cell
formed by the division of a cambial initial divides once or twice to form a
sieve element and a companion cell, or two parenchyma cells, or all three
types of elements (Zee, 1968). In the root, the sequence is even simpler, the
product of the cambium differentiating into a parenchyma cell or dividing
once to form a sieve element and a companion cell (Zee and Chambers, 1969).

A detailed account, at the electron-microscopic level, of the changes leading to the maturation of a vascular element from a meristematic cell is
available only for the phloem. The studies have included the sieve elements,
companion cells and phloem parenchyma cells, both primary and secondary,
in the young parts of the stem and the root (Bouck and Cronshaw, 1965;
Wark, 1965; Wark and Chambers, 1965; Zee, 1968; Zee and Chambers,
1968, 1969); fibre differentiation has apparently not been studied. The following description is of the major processes which have been inferred from the
EM pictures. Initially, the future sieve element develops as a typical vacuolated plant cell. The tonoplast then breaks down, followed by the dissolution
and disappearance of many of the cell organelles. In the final stages the sieve
element contains membranes, which have a bead-like appearance and are
presumably derived from the endoplasmic reticulum, as well as mitochondria
and plastids which have no developed lamellae and resemble proplastids.
"Fibrous bodies" and "amorphous material" are present in the early stages
of sieve element development and "slime" (p-protein) is present after the
nucleus and tonoplast have disappeared; the nature of, and relation between,
these materials are not clear. RNA synthesis, judged by ^{3}H-uridine incorporation (Zee and Chambers, 1969), ceases at the time the tonoplast breaks down.
The DNA of the nucleus, labelled with ^{3}H-thymidine (Zee and Chambers,
1969) becomes dispersed at the time the nucleus disappears, after the dissolution of the tonoplast.

The cell walls of the future sieve elements become specialized before any
changes are seen in the cytoplasm or the nucleus. A characteristic nacreous
layer develops on the lateral walls and callose deposits are found in regions

of future pores. The callose spreads as the pore is formed, and each pore develops from a number of plasmodesmata. Tubules of endoplasmic reticulum traverse the plasmodesmata and are connected with the cytoplasm of the two adjacent cells until a late stage of cell development; some fibrous material is always found in the pores.

A comparison of primary and secondary phloem development in the root and in the shoot (Wark and Chambers, 1965; Zee, 1968; Zee and Chambers, 1968, 1969) has shown that the processes of phloem differentiation are basically the same throughout the plant axis. In the secondary phloem of the stem, however, three types of inclusion bodies have been described which have not been seen in other pea phloem (Zee, 1968). The differentiation of companion cells and phloem parenchyma has also been studied and a normal structure for plant parenchyma was found (Wark, 1965; Zee, 1968; Zee and Chambers, 1968). No changes were seen in the cytoplasm or nucleus of these cells during the maturation of the sieve elements. The companion cells of the secondary phloem of the stem, however, have unusual cell wall ingrowths, or "trabeculae" (Wark, 1965).

III. POLAR CONTROL OF DIFFERENTIATION

The next group of problems to be discussed here is related to the pattern, or spatial arrangement, of the vascular elements relative to one another and to other parts of the plant. Specific phenomena to be accounted for are the arrangement of vascular tissues in elongated strands in which vascular elements are placed in vertical rows, the ending of these strands in the shoot and the roots and the orderly contacts between these strands. The vascular system will be treated here as a unit; the questions of the relation between the size of the vascular system, which increases constantly as a result of cambial activity, and the size of the plant as well as the relation between the various elements and tissues within a vascular strand will be taken up in the next section.

Many traits in the pattern of the vascular system can be either accounted for or investigated on the basis of the hypothesis that vascular differentiation occurs along the path of the movement of stimuli, one of which is auxin, which are produced in the young leaves and other parts of the shoot and are transported towards the root tips, where they are changed or destroyed. This working hypothesis is suggested by the following experimental results:

1. The removal of young leaves reduces vascular differentiation relative to untreated controls. This is true for the formation of a procambium, the differentiation of primary vascular tissues, the initiation and activity of the cambium and the differentiation of secondary vascular tissues. All aspects of this general effect, demonstrated very early for cambial activity in beans

by Jost (1893) and shown for primary vascular differentiation by Helm (1932), Young (1954) and Wangermann (1967), are readily repeatable with pea plants (Fig. 8.1b).

2. Promotion of vascular differentiation is not unique to young leaves; it is also produced by all other parts of the shoot. A mature leaf of a pea plant has been shown to promote the continued differentiation of the xylem in the internode below it (Benayoun et al., 1975). This effect of mature leaves on vascular differentiation is not as large as that found in other plants, such as beans (Hess and Sachs, 1972).

3. When vascular strands are cut, they regenerate readily (Simon, 1908). This regeneration often involves the redifferentiation of cortical parenchyma to phloem, xylem and a cambium (Sinnott and Bloch, 1945). It has been shown for a number of plants that this regeneration is very feeble when all leaves above the wounded region are removed (Jost, 1940; Jacobs, 1952). This effect of leaves on vascular regeneration is readily repeatable with peas.

4. The growth of buds released from dominance by the removal of the shoot above them is associated with the differentiation of vascular tissues which connect the buds with the rest of the plant. This process can include not only the normal aspects of vascular differentiation and redifferentiation of parenchyma cells, but also a reorientation of the axis of the new vascular elements so that they connect with the lateral bud rather than straight up the decapitated shoot. Growing buds include growing leaves, so that this is but another aspect of the effect of leaves on vascular differentiation. The relation between bud growth and vascular differentiation in peas has been studied by Sorokin and Thimann (1964). The reorientation of cambial elements towards the growing leaves can be seen to occur in peas, but has only been studied in woody plants, which are much more convenient for this purpose (Neeff, 1914; Kirschner, et al., 1971). Work which has not been repeated with peas also shows that buds cause vascular differentiation when they are grafted on a callus (Camus, 1949; Wetmore and Rier, 1963).

5. The effects of leaves on primary and secondary vascular differentiation, regeneration and reorientation occur along the axis connecting the respective leaf with the roots below. In this sense the effect of the leaves is strictly polar; although leaves supply photosynthetic products to the tissues above them, there is no evidence that they promote vascular differentiation in these tissues. Leaves have no effect other than an inhibitory one on vascular differentiation in branches which are lateral to the axis connecting the leaves to the roots. This generalization, though it has not been carefully documented, has been confirmed by many experimental observations on peas. In cleared pea plants it may be seen that vascular elements connecting to a leaf always lead only downwards, towards the roots.

6. The effects of the removal of young leaves on vascular differentiation can be partially replaced by auxin. The most common auxin used is a $0.001-1\%$

reparation of indol-3yl acetic acid (IAA) in lanolin. Auxin has been shown
n other plants to cause xylem and phloem differentiation (Jacobs and Morrow, 1957, 1958), cambial activity (Snow, 1935), vascular regeneration (Jost,
1942; Jacobs, 1952) and vascular reorientation (Kirschner *et al.*, 1971). In
peas, auxin has been shown to be one of the controls of cambial activity and
vascular differentiation in cultured roots (see Chapter 5) and to cause xylem
differentiation in cortical parenchyma (Fig. 8.2 and Sachs, 1968b, 1969);

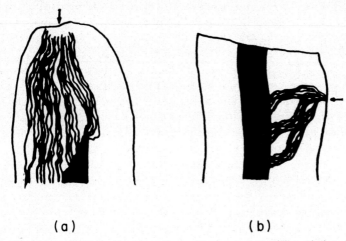

(a)   (b)

Fɪɢ. 8.2. Drawings of thick cleared sections showing xylem differentiation following
auxin treatments (1 % IAA in lanolin) of the regions marked by arrows. Black areas are
the vascular tissues present before treatment; lines mark the newly differentiated xylem.
Note that the new, induced xylem formed direct contacts between the auxin source and
the pre-existing vascular strands. (a) is a cut hypocotyl, split so the vascular tissues were
not included; (b) is half a root, split longitudinally. In both cases the treated tissues
remained in contact with the cotyledons and the root. (b) has been drawn from Sachs
(1968b) with the kind permission of the Annals of Botany Co.

other effects mentioned have also been observed by the author (unpublished
work). As in the case of leaves, all the effects of auxin on vascular differentiation are polar, and extend only in the direction of the roots. Auxin is especially
effective in terms of cambial activity if it is applied to a cut base of cultured
roots (Torrey, 1963 and Chapter 5).

   Other plant growth regulators have also been shown to promote vascular
differentiation; in peas, this is especially true of cytokinins when they are
added to cut sections with an inhibited bud on them (Sorokin and Thimann,
1964), to tissue cultures (Torrey, 1968) or together with auxin to wounded
plants (Sachs, unpublished). Cytokinins and other hormones might play
major roles in the regulation of vascular differentiation but there is no evidence at present that they can actually replace the effect of a part of the plant,
as does auxin. The results of the application of these hormones may depend
on their influence on other processes, such as bud growth which in turn

causes vascular differentiation, or on their being essential for an aspect of the
general metabolism which is necessary for the expression of differentiation
These hormones, therefore, may influence differentiation indirectly rather than
as specific "triggers" or inducing substances. Consequently, the role of
hormones other than auxin must be considered an open question, and certainly
a very important one, since auxin replaces the effect of leaves only partially
Auxin does not replace young leaves in causing procambium differentiation in
Lupinus (Young, 1954), an observation which has been repeated for peas
Exogenous auxin causes vascular differentiation in cortical and pith paren
chyma, but this effect is restricted to the vicinity of a wound (Jost, 1942), and
to parenchyma which is past the meristematic stage yet is not too mature
cortical cells at the stage of rapid elongation respond best (Sachs, unpub
lished). No auxin concentration reproduces the fibre-rich xylem which is
found in bean stems (Hess and Sachs, 1972) and in peas auxin does not re
place leaves in their effect on the differentiation of phloem fibres (Sachs, 1972)

7. Growing root tips, and roots as a whole, do not have the same effec
on vascular differentiation as do leaves. In peas, vascular differentiation was
found to continue even in the absence of all the roots (Sachs, 1968a). When
the roots are present their influence is complementary to that of the leaves, in
that they orient vascular differentiation in such a way that it connects into the
roots and reaches the root tips.

It follows from these results that leaves, and especially young leaves, pro
duce stimuli which control all the various aspects of vascular differentiation
The influence of those stimuli is not limited to tissues close to the leaves
their effects on cambial activity, secondary differentiation and vascular re
generation may be seen throughout the plant. The roots appear to act a
sinks for these stimuli; although roots produce stimuli essential for plan
integration, the fact that vascular differentiation occurs in the absence of the
roots indicates that these root stimuli have no directing effect on the location
of this differentiation (Sachs, 1968a). Hence the hypothesis stated above that
vascular differentiation occurs along the path of transport of stimuli from
the leaves to the roots. To account for the organized differentiation of th
various types of cells within the vascular system one must assume that the
leaf stimuli include more than one component (see Section IV). Auxin re
places many of the effects of leaves in terms of vascular differentiation; since
it is also known that auxin is produced by young leaves (Scott and Briggs
1960) it follows that auxin is one of the stimuli produced by the leaves which
control vascular differentiation in intact plants. Further evidence for this
conclusion comes from the correspondence between auxin synthesis and vas
cular differentiation in intact plants (Jacobs and Morrow, 1957, 1958). This
is, therefore, a case of induction of differentiation (Spemann, 1938) in which
part of the inducing substance or evocator is chemically known. The question of
the relation of xylem differentiation to auxin concentration gradients or auxin

ransport will be discussed in Section VI; in relation to the control of the
pattern of vascular differentiation a clear answer to this question is not essential.

V. THE CONTROL OF THE LONGITUDINAL PATTERN OF VASCULAR STRANDS

The advantages of the hypothesis of a control of vascular differentiation by
stimuli moving from the leaves to the root is that it accounts readily for many
known aspects of the pattern of vascular differentiation and suggests experiments concerning additional phenomena. The general pattern of vascular
tissues as connections between leaves and roots is readily understood. No
direct vessel or sieve tube connections can be expected between two leaves or
two roots, and none can be found in pea plants. The question of the control
of where contacts between vascular bundles do occur becomes amenable to
experimental work.

When an inhibited pea bud grows as a result of the removal of the shoot
above it the vascular strands which differentiate in the adjoining stem always
connect the bud to the pre-existing vascular tissue. The same phenomenon
may be seen when vascular differentiation is induced artificially on a cut epicotyl or root (Sachs, 1968b); differentiation always connects the source of
auxin with the nearest cut vascular strands (Fig. 8.2). Insight into the mechanism of this promotion of contacts between new and pre-existing vascular
strands and the role it plays in the plant is offered by the cases in which these
contacts do not form. The principal control of contact formation which has
been found experimentally is an inhibition of the formation of contacts with
existing strands which are supplied with auxin (Sachs, 1968b, 1969). The
system used to demonstrate this point was a cut pea epicotyl connected at its
base to the cotyledons and the roots. A cut was made separating a small part
of the epicotyl which consisted of parenchyma, fibres and epidermis but no
conducting vascular tissues (Fig. 8.3). This part of the epicotyl was treated
with auxin in lanolin and as a result strands were formed connecting on to
the pre-existing vascular cylinder below the cut (Fig. 8.3a). When the vascular
cylinder was also supplied with auxin, the formation of the contacts between
the old and the new strands was generally inhibited (Fig. 8.3b). The influence
of the auxin applied to the parenchyma was still evident, since the short cut
separating part of the epicotyl allows it to be expressed as a short strand
ending blindly. The formation of the contacts between the old and the new
strands was variable, but it was clearly under the control of the balance
between the concentrations of auxin applied to the parenchyma and the pre-
existing vascular cylinder (Sachs, 1969). The balance was not a simple one,
as the same concentration of auxin was more effective in preventing the formation of vascular tissues when it was applied to the vascular cylinder than
in promoting these contacts when it was applied to the separated cortex.

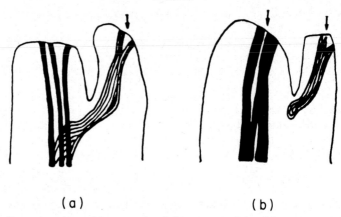

(a)   (b)

Fɪɢ. 8.3. Pea epicotyls used for a study of the control of the differentiation of contacts between vascular strands. Black areas mark the pre-existing vascular tissues of the epicotyl and lined areas mark xylem which differentiated after the application of auxin in lanolin to the locations marked by arrows. Note that contacts between the new and pre-existing vascular strands occurred when the pre-existing strands were not supplied with auxin in (a). The epicotyls used for this experiment remained in contact with intact cotyledons and roots. Drawn from Sachs (1969) with the kind permission of the Annals of Botany Co.

This control of contact formation is also evident in relatively intact plants, in which the pre-existing strands remain in contact with the young leaves above and new short strands are induced by local applications of auxin to wounds in the cortex. As expected from the fact that leaves supply auxin to the vascular tissues which connect them to the roots, removal of the shoot above a wound greatly increases the frequency of contacts between the new, artificially induced and the pre-existing vascular tissues. The formation of the contacts in these almost intact plants, however, is also dependent on a number of other factors (Sachs, unpublished). These contacts almost invariably form when the wounds to which the auxin is applied are deep, so that they damage the existing vascular system and disrupt the flow of auxin through it. The frequency of contact formation is also dependent on whether the auxin which induces the new strand is applied above or below the wound. Applications above the wound result in a very high proportion of new strands which make direct contacts with the existing vascular tissues, even when these tissues are in direct contact with an intact shoot above. This is presumably because auxin above the wound, unlike auxin below the wound, cannot be transported downwards through the cortical parenchyma and it therefore builds up to relatively high concentrations above the wound, finally forcing its way into the intact vascular tissues.

These results suggest a control for the formation of contacts between vascular strands in the intact, untreated plant. The pattern of the primary vascular tissues varies greatly depending on the species of plant, and the

ower part of the pea epicotyl is quite unusual in this sense, being a transition
petween the stem and the root (see above). The most repeatable aspect of this
pattern for all higher plants, however, is the presence of leaf gaps: regions
above the vascular strands which connect on to a leaf in which vascular
differentiation does not occur and the tissue becomes parenchyma. These
leaf gaps, therefore, are regions in which contacts are prevented between the
strands leading out to a leaf and new strands differentiating in connection
with the formation of new leaves above (Fig. 8.4a). The experimental results
described above suggest that these gaps are due to an inhibitory effect of the
leaves which supply the strands connecting them to the stem with auxin, and
presumably other stimuli, and thus prevent their being linked to new vascular
strands. An experimental test of this possibility is demonstrated in the
differentiation of vascular tissues which occurs in the region which would
normally be a leaf gap when the leaf, and thus the source of stimuli which
inhibit contact formation, is removed (Sachs, 1968b). Removing leaf primordia of pea seedlings very early in their development prevents the formation
of any vascular tissues leading to them, while removing the primordia later
on does not affect a pattern of vascular differentiation which has already been
established. If the primordia are removed when they are 100 μm long, however, cases can be found in which vascular differentiation occurs across leaf
gaps, and strands leading to the removed or greatly damaged leaves are connected directly to strands leading to young leaves above (Fig. 8.4b). This result,

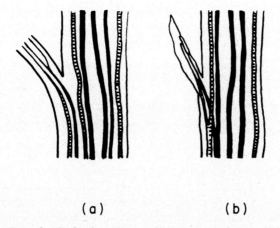

(a)   (b)

FIG. 8.4. The effect of a leaf primordium, which was allowed to develop in (a) and
removed at an early stage in (b), on the contacts between fibre and vascular strands.
Drawings of thick longitudinal sections through pea stems; dotted areas mark fibres
and black areas the conducting tissues. Note that the strands connecting to a normal
leaf in (a) are not joined directly by the strands leading to the leaves above them, so that
a leaf gap is present. When the leaf was removed at an early primordial stage in (b)
contacts were formed and so no leaf gap is present. Drawn from Sachs (1972) with the
kind permission of the Annals of Botany Co.

showing the influence of a leaf on the formation of leaf gaps, was found both
for the main traces of leaf petioles and for the traces entering the stem from
the stipules (Sachs, 1968b). The same control of the contacts made to a vascular strand by stimuli from the leaf associated with it was also found in the case
of a growing bud. When buds are released from dominance, their growth is
associated with the differentiation of vascular strands which connect with the
vascular system in the adjoining stem. The exact pattern of contacts is not
predetermined in the plant, and if one or more leaves are left intact their
vascular tissues leading to them are not contacted, or contacted to a much
lesser degree (Sachs, 1968b).

The results concerning the control of contacts between strands are relevant
not only to the general pattern of the vascular system, as described above, but
also to the organization of the vascular tissues in strands in which cells undergoing a similar differentiation, such as vessel elements, are arranged in
defined longitudinal rows. It was suggested above, on the basis of the effects
of leaves and roots on the various aspects of vascular differentiation, that
differentiation occurs along the channels of transport of stimuli, one of which
is auxin, from the leaves to the roots. The attractive influence of vascular
strands on the direction of new strand differentiation further suggests that
these strands are a sink for auxin, which they would be if they were the most
rapid or efficient channels for auxin transport. This suggestion is supported
by the effect of auxin, when supplied to existing vascular tissues, in inhibiting
their attractive influence on new strands which would otherwise form contacts
with them. It follows from these considerations that vascular tissues are the
most rapid channels for the transport of auxin, which is a stimulus for their
differentiation. This conclusion is supported by measurements of the movement of radioactive auxin in peas, which show that in intact plants its transport is most rapid in the vascular tissues, presumably in the phloem (Morris
et al., 1973). The following picture of the control of the differentiation of a
vascular strand may therefore be suggested: a gradient of the concentration
of stimuli, one of which is auxin, is set up between their source (leaves or an
artificial source of auxin) and the roots, or the vascular tissues which form
rapid channels of transport to the root. In response to this gradient auxin
movement occurs and the cells through which the auxin moves start to differentiate. At an early stage of this differentiation the cells become specialized
for a more rapid, oriented transport of auxin, and possibly other stimuli. The
cells which have become specialized will be the preferred channels of further
transport, and this transport will cause continued differentiation. Neighbouring cells will thus become drained of the stimuli necessary for differentiation
and will not become vascular elements even if their differentiation has started.
The cells directly below the ones which have started differentiation will be
subjected to a relatively large auxin gradient, and therefore also auxin
movement, and will be induced to start differentiating as well. The exact

determination of which cells become the preferred channels might sometimes depend on chance, at least in cases where exogenous auxin is added to cortical parenchyma (Fig. 8.2). Here the individual vessels may be seen to meander and their exact course differs for individual plants, apparently depending on chance factors. This coupling between the early stages of differentiation and the transport of stimuli, and its autocatalytic nature, therefore accounts fully for the arrangement of the vascular elements in longitudinal rows, and thus supplies a mechanism for the longitudinal interaction between cells within the vascular tissues.

V. THE CONTROL OF THE VOLUME AND CONSTITUTION OF THE VASCULAR TISSUES

The control of vascular differentiation by stimuli transported from leaves to the roots also suggests a consideration of the quantity, or volume, of vascular tissues in any part of the plant axis. As in other dicotyledons, the cambium of peas may be active throughout the life of the plant, so that the cross-section of the vascular tissues increases constantly. No quantitative results are available for peas, but it appears to be generally true that the larger the plant, and the more leaves the stem carries, the greater the cross-section of the vascular system which connects them to the roots. This is as would be expected, both from a functional point of view and from the evidence that leaves, especially when they are young, produce stimuli which promote vascular differentiation. A problem arises, however, when quantitative relations are considered. It is clear, even without exact measurement, that the cross-section of the vascular tissues at any given region of the axis is not a simple linear function of the number of leaves carried above this region. It is a general rule, which is especially evident in peas and other climbing plants, that as the number of leaves increases their effect in promoting additional vascular differentiation decreases. This is not due to a change in the leaves or their production of differentiation-inducing stimuli, since even leaves high up on a stem have a large effect on internodes directly below them, where no other leaves have had a previous effect. The decreasing effect of leaves as their number increases is also not likely to be due only to the increase in the distance between the new leaves and the observed region of the axis, since the promotive effect of leaves on vascular regeneration after wounding is found even when the leaves are at a considerable distance up the stem. The hypothesis that vascular differentiation is coupled to the transport of the stimuli for vascular differentiation, however, suggests a mechanism which could account for at least part of this complex, non-linear relation between the number of leaves and cross-sectional area of the vascular tissues which supply them. If, as suggested above, vascular differentiation is associated with an increase in the transport rate for auxin and other differentiation-inducing stimuli, then mature vascular tissues may inhibit the differentiation of additional tissues because the mature tissues would be the preferred channels for these stimuli. While this transport could not be expected to take place through the vessels, it could readily occur through the phloem (Morris and Kadir, 1972). This hypothesis is attractive because it suggests a dual control of the quantitative aspects of vascular differentiation which could be of obvious functional advantage; vascular differentiation would occur in response to stimuli from the leaves, and thus would depend on the size of the shoot which it supplied. It would be a response, however, not to the absolute quantity of stimuli produced but to the difference between this quantity and the transport capacity of the functional existing vascular tissues. Any stimulus which was not carried by the functional phloem would be transported by the cambial region; unlike the same stimulus when it is transported through the mature phloem, this transport through the cambial region would be associated with new vascular differentiation.

The effect of wounds on vascular differentiation could be a test of the possibility that mature vascular tissues carry stimuli for vascular differentiation. It is well known that the vascular tissues of dicotyledons regenerate readily when they are wounded (Simon, 1908), and this regeneration is easily observed in peas. Regeneration, however, would indicate that mature tissues influence the rate of the differentiation of new elements from the cambium only if the damaged mature tissues were replaced during vascular regeneration; it is also possible that what appears as a regeneration involves only a diversion of new tissues, which would have been formed regardless of the treatment, around the wound. The problem, therefore, is to detect whether a replacement of damaged tissues can occur. This replacement could be seen anatomically if regeneration reconstituted the continuity of individual vessels and sieve tubes, forming a bridge between the cut ends on the two sides of a wound. This possibility has been studied in a number of plants which have conveniently large vascular elements (Eschrich, 1953; Benayoun et al., 1975) and no such bridges were found; less thorough studies with peas have indicated that they are no exception to this rule. Replacement, however, would also be indicated if wounds increased the rate of vascular differentiation relative to unwounded plants, thus causing the replacement of the damaged vascular strands as a whole and not only the bridging of the cut ends on the two sides of the wound. Detecting such an increase in vascular differentiation as a result of wounds is complicated by the expected effects of wounds on the growth rate of the plant, which should also be reflected in the rate of vascular differentiation. Using pea plants from which all young leaves and shoot apices were removed, so that no effect on shoot growth was possible, Benayoun et al. (1975) found that wounds caused a reproducible increase in vascular differentiation. This differentiation was expressed in an increase in the cross-sectional area of the xylem in wounded relative to unwounded plants. This result indicates that wounding releases differentiation-inducing stimuli which are otherwise not expressed and that the effect of the wound is not limited to its immediate vicinity, in which case a bridging or repair of the cut ends of the sieve tubes, and possibly the vessels as well, could be expected. This conclusion fits in well with the results of auxin transport experiments, which show that transport can take place rapidly through the phloem, but only in intact, undamaged pea plants (Morris et al., 1973). Furthermore, the evidence that in pea plants auxin applied to very young leaves does not move through the phloem (Morris and Kadir, 1972), agrees with the suggestion that these leaves are inducing the differentiation of new vascular channels and that the transport through the procambium, the cambium and the differentiating elements is the relatively slow, polar transport. It may be concluded that vascular differentiation is controlled both by the leaves which are a source of stimuli and by mature vascular tissues, presumably the phloem, which can transport the stimuli away from the differentiating cells. These effects could supply a basis for understanding the control of the rate of vascular differentiation; this could only be a framework at best, however, and detailed, quantitative work is not available. Such quantitative work would have to take into account not only the vascular system as a whole but the control of the differentiation of its various components, which is the next subject to be discussed.

The patterns of differentiation discussed above are concerned with the vascular tissue as a whole and the relationships between its component tissues should now be considered. The specific questions, therefore, are both at a tissue level—how the relation between the phloem, the xylem and the cambium is controlled—and at the cellular level—how the relative location and quantity of the different cell types which compose these tissues is determined. The phenomena which are the basis of this general problem also appear in tissue cultures, where the tissues are much less organized than in intact plants. The most studied case of cell differentiation in plant tissue cultures is the formation of xylem. As yet, however, no conditions have been found in which more than 35% of the cells differentiate to tracheary elements, and this has been found in cultures of pea root callus (Torrey, 1971). There are various isolated descriptions of the differentiation of other cell types in culture, but no clear picture has emerged (Torrey, 1971). Unlike the problem of the control of the vascular system as a whole, there is no general hypothesis available concerning the control of the pattern of the component cells and tissues. The most studied case of a pattern within the vascular tissues is the arrangement of the phloem and the xylem in pea roots (see Chapter 5). This work does not, as yet, give a precise definition of the processes involved, but it suggests that the pattern of vascular tissues depends on hormonal levels and hormonal interactions.

In the case of the control of conditions promoting phloem differentiation as compared with xylem differentiation some information, which as yet doe not form any coherent picture, is available from work on plants other than peas. The ratio between the phloem and the xylem formed in *Syringa* callus has been found to depend on the concentration of sugar which is supplied (Wetmore and Rier, 1963). The cambium of some trees forms more tissue on the phloem side when gibberellins are applied (Wareing, 1958), although it is not clear that this phloem tissue actually includes mature sieve elements. The control of the relative amounts of fibres and vessels in the secondary xylem has been studied in *Phaseolus vulgaris* (Hess and Sachs, 1972). Mature leaves were found to promote the differentiation of a high proportion of fibres, an effect which could be replaced by a combination of gibberellins and auxin. Gibberellins alone did not cause xylem formation and auxin alone imitated young leaves in producing a xylem which was very rich in vessels.

The first vessel elements to mature in an elongating stem have annular secondary wall thickenings; afterwards elements with spiral thickenings are formed. Thompson and Miller (1963) made a careful study of the effect of etiolation on the structure of pea plants, and found that the ratio between the number of spiral and annular vessel elements is significantly increased by light (see Chapter 11). This effect was not due to any change in the cells which differentiate into annular wall thickenings but rather to the continued differentiation of new elements with spiral thickenings. This effect of light in causing continued differentiation as compared with etiolated tissues, is not specific in any way to vessel elements and is found in various aspects of shoot development. The change in the ratio between the two types of vessel elements, therefore, is an example of an external control of the duration of differentiation which has an indirect effect on the composition of the xylem.

Peas offer an advantageous system for comparing the control of the differentiation of different components of the vascular tissues because phloem fibres occur not only as part of the vascular system but also isolated in special strands in the cortex. Experiments involving leaf removal, hormone applications and regeneration after wounding, which have yielded useful information concerning the control of the differentiation of the vascular system as a whole, have been repeated in connection with the differentiation of these fibre strands (Sachs, 1972). Removing leaf primordia was found to prevent the differentiation of the fibres which would connect future leaves with the stem (Fig. 8.1), but this was true only when the primordia were removed before they reached the size of 25 μm, while an effect on xylem differentiation is apparent even when the primordia are removed at much later stages. Regeneration of fibre strands after wounding and through regions of a graft union were also found to occur, but only when the tissue treated was extremely young and had not yet started to elongate. The presence of a leaf gap in terms of fibre differentiation was also found to depend on the presence of the subtending leaf, and when this was removed at an early stage (but after the fibre strand had become determined) a direct contact was formed between the cortical fibre strands across the leaf gap (Fig. 8.4b). In all these details phloem fibre differentiation resembles the differentiation of the xylem, but appears to depend on a stimulus originating in a much earlier stage of leaf development. These results suggest, therefore, that fibre differentiation occurs along the channel of transport of a stimulus from the leaves towards the roots, in the same way as the vascular tissue as a whole does.

This conclusion led to experiments designed to identify the stimulus for phloem fibre differentiation in peas and to study its relation to the stimuli controlling the differentiation of phloem and xylem. No known hormones, however, were found to replace very young leaf primordia in their effect on the differentiation of phloem fibres; nor did the addition of any known hormones convert a differentiating fibre strand into conducting tissues or vice versa. Wounds were made to divert fibres and conducting tissues towards one another so as to see whether the stimuli responsible for the differentiation of one type of tissue will influence the other, but no such influence has as yet been found. The nature of the stimulus for phloem fibre differentiation in peas is therefore completely unknown. A general working hypothesis emerging from these results, however, is that the different types of vascular tissues are induced to differentiate by different stimuli originating in the leaves. If these stimuli are similar, or have components in common, it is possible that their first channels of transport will be the same cells which will only become specialized to transport the different stimuli at later stages of differentiation (Sachs, 1972; Hess and Sachs, 1972). This suggestion offers a possible mechanism for the joint pattern of differentiation of different cell types, such as fibres and vessel elements, and possibly other cases as well.

VI. THE CONTROL OF DIFFERENTIATION AT THE CELLULAR LEVEL

The final question to be discussed here is the nature of the first changes which lead to the regulation of cellular activities resulting in differentiation. As was explained in the Introduction, this is one of the two general problems concerning differentiation, the other being the nature of the control of patterned differentiation which was discussed above. Work related to the question of the regulation of cellular activities may be found in Chapters 5 and 6. This work, which will not be reviewed again here, shows that differentiation is related to metabolism in general and to changes involved in DNA synthesis and cell division in particular. Additional hints concerning the relation between tracheary element differentiation and metabolism in a wide variety of plants are available from work concerned with the exact culture conditions which lead to the formation of tracheary elements *in vitro* (Roberts, 1969) and such work has also been carried out with material taken from peas

(Torrey, 1968). Depending on conditions, different hormones or other factors may limit differentiation, but there is no way at present of separating the requirements for a metabolically active tissue, which is necessary for the expression of differentiation, and the specific signals which limit and control vascular differentiation in the plant.

Another approach to the question of the key controls of cellular differentiation is to study the determination of the axis of the vascular elements. This approach is obviously closely related to the other work discussed in this chapter and will therefore be considered in some detail.

Vascular elements do not differentiate as isolated cells but rather with a definite relation to surrounding tissues: sieve and tracheary elements have an axis (or, rarely, axes) defined by the structures which connect them to similar cells above and below. This axis generally corresponds to the axis of most rapid elongation of the cell and the organ in which it is found, but this correspondence does not hold for vessel elements in regions of regeneration (Neeff, 1914). The only factor which is known to control the determination of the axis of differentiating vascular elements is auxin: when IAA in lanolin is placed on cut pea epicotyls or roots, vascular differentiation occurs along the path of auxin movement from its lanolin source to the root tips (Fig. 8.2b) even if this path does not correspond to the original axis of elongation of the organ. Thus it has been shown that vascular differentiation can be induced transversely across epicotyl or root cortex (Sachs, 1968b). The question of the control of the axis of the differentiating tissues has advantages for experimental work concerned with the cellular regulation involved in differentiation relative to the control of vascular differentiation in general; this is because the process involved is specific and does not depend on all conditions which allow metabolic activity. It is a reasonable working hypothesis that in the plant the same control mechanism determines both the nature of a cell's differentiation and the axis of this differentiation, though this, of course, is only speculation.

There are two possible directional aspects of the presence of auxin which could determine the axis of differentiating cells: the gradient of auxin across the cells and the transport of the auxin through the cells. It is, of course, quite possible, or even likely, that the axis of transport is determined by the presence of an auxin gradient; the question, however, is which of these two aspects of the presence of auxin is the signal for determining the axis of the differentiating cells. Thompson (1970) showed that in pea stems (and in other plants) xylem differentiation occurs when auxin is transported in the basipetal direction but not when auxin is transported in the acropetal direction. This indicates an important function for auxin transport. Experiments designed to separate the possible effects of auxin gradients and auxin transport in pea epicotyls were carried out by Sachs (1974). Use was made of the fact that the direction of polar auxin transport is a determined characteristic of plant cells. Polar transport, therefore, may occur regardless of the direction of an auxin gradient. By applying local sources of auxin in lanolin to a cut surface of pea epicotyl it was possible to create conditions in which there was no fall in auxin concentration (or a gradient) from top to bottom and yet auxin transport could be expected. It was found that xylem differentiation occurred under these conditions. This result means that differentiation does not depend on the presence of a long-term auxin gradient; it might depend either on polar transport or on short-term local gradients which could be set up by auxin transport.

It was possible to distinguish between these two latter possibilities on the basis of the length of time auxin must be present for oriented differentiation to take place. This time could not be measured reliably by removing a source of auxin at different times, because there is no way of removing the auxin which has already entered the tissue. An estimation, however, was possible on the basis of the following experiment. When two spots of auxin in lanolin were placed one above the other on a cut pea epicotyl the xylem which differentiated connecting the upper source to the root did not pass in the immediate vicinity of the lower source (Fig. 8.5). This diverting effect of the lower source on the course of the xylem connecting to the upper source was present even when the application of the lower source was delayed for 1 or 2 days relative to the upper source. Since after periods longer than 2 days

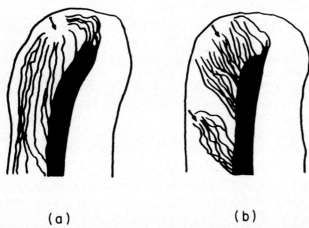

(a)   (b)

Fig. 8.5. Xylem differentiation in response to auxin in pea epicotyls. In the experiment the course of the xylem formed in response to one and to two auxin sources was studied. By separating the time of the application of the auxin sources and observing the course of the differentiation it was possible to measure the length of time necessary for xylem determination (see text). Black areas indicate the vascular cylinder, and lines indicate the xylem which differentiated in response to the application of auxin in lanolin, in the locations marked by arrows. Cut pea epicotyls which remained connected on to intact cotyledons and roots were used. Drawn from Sachs (1974), with the kind permission of the organizers of the Eighth International Conference on Plant Growth Substances.

mature new xylem could be observed, this result means that xylem differentiation occurs only when the inducing and orienting effect of auxin is present for
the entire or almost the entire period of differentiation. It follows that no
short-term gradients of auxin can determine the process in the cells in such a
way that they can continue even if conditions are changed. This rather complex reasoning concerning a simple set of experiments leads to the conclusion
that xylem differentiation, and presumably vascular differentiation in general,
is coupled in some way to auxin transport. In other words, a key controlling
process in vascular differentiation is the transport of auxin through the cells,
and this transport must continue throughout the period of differentiation.
Further evidence for this view of the control of differentiation comes from
considerations of the organized or patterned nature of vascular differentiation
in the previous sections.

VII. CONCLUSIONS

If our knowledge of vascular differentiation in peas is compared with our
knowledge of similar processes in other plants, it is found that many major
experiments have been either performed first or repeated on pea seedlings or
on organ and tissue cultures derived from them. A gap in our knowledge is the
description of the processes of xylem differentiation at the electron-microscopic
level, which is available for some plants (Torrey *et al.,* 1971) but
not for peas. A general question which appears to be open and offers promise
for future work is the nature of the controls which determine the relative
location and quantity of the different cell types within the vascular tissues. It
was shown above, in Section V, that these controls might depend on the
specific transport of differentiation-inducing stimuli by different cell types and
on a coupling between this transport and the processes of differentiation. This
suggested hypothesis is along the same lines as the major conclusion of the
work reviewed in this chapter: that a coupling between vascular differentiation
and the transport of stimuli for this differentiation, such as auxin, may
account for the relation between vascular differentiation and leaf growth. The
details of the pattern of the vascular system and the quantitative control of its
development may be the key cellular processes leading to differentiation.

REFERENCES

Aloni, R. and Sachs, T. (1973). *Planta* 113, 345–353.
Benayoun, J., Aloni, R. and Sachs, T. (1975). *Ann. Bot.* 39, 447–454.
Bouck, G. B. and Cronshaw, J. (1965). *J. Cell Biol.* 25, 79–95.
Camus, G. (1949). *Revue Cytol. Biol. vég.* 11, 1–199.
Eschrich, W. (1953). *Planta* 43, 37–74.

GOURLEY, J. H. (1931). *Bot. Gaz.* **92,** 367–383.
HAYWARD, H. E. (1938). *In* "The Structure of Economic Plants". MacMillan, New York.
HELM, J. (1932). *Planta* **16,** 607–621.
HESS, T. and SACHS, T. (1972). *New Phytol.* **71,** 903–914.
JACOBS, W. P. (1952). *Am. J. Bot.* **39,** 301–309.
JACOBS, W. P. and MORROW, I. B. (1957). *Am. J. Bot.* **44,** 823–842.
JACOBS, W. P. and MORROW, I. B. (1958). *Science, N.Y.* **128,** 1084.
JOST, L. (1893). *Bot. Ztg* **51,** 89–138.
JOST, L. (1940). *Z. Bot.* **35,** 114–148.
JOST, L. (1942). *Z. Bot.* **37,** 161–215.
KIRSCHNER, H., SACHS, T. and FAHN, A. (1971). *Israel J. Bot.* **21,** 129–134.
MORRIS, D. A. and KADIR, G. O. (1972). *Planta* **107,** 171–182.
MORRIS, D. A., KADIR, G. O. and BARRY, A. J. (1973). *Planta* **110,** 173–182.
NEEFF, F. (1914). *Z. Bot.* **6,** 465–547.
ROBERTS, L. W. (1969). *Bot. Rev.* **35,** 201–250.
SACHS, T. (1968a). *Ann. Bot.* **32,** 391–399.
SACHS, T. (1968b). *Ann. Bot.* **32,** 781–790.
SACHS, T. (1969). *Ann. Bot.* **33,** 263–275.
SACHS, T. (1972). *Ann. Bot.* **36,** 189–197.
SACHS, T. (1974). *In* "Plant Growth Substances 1973", pp 900–906. Hirokawa, Tokyo.
SCOTT, T. and BRIGGS, W. R. (1960). *Am. J. Bot.* **47,** 492–499.
SIMON, S. (1908). *Ber. dt. bot. Ges.* **26,** 364–396.
SINNOTT, E. W. and BLOCH, R. (1945). *Am. J. Bot.* **32,** 151–156.
SNOW, R. (1935). *New Phytol.* **34,** 347–360.
SOROKIN, H. P. and THIMANN, K. V. (1964). *Protoplasma* **59,** 326–350.
SPEMANN, H. (1938). "Embryonic Development and Induction". Yale University Press, New Haven, U.S.A.
THOMPSON, B. F. and MILLER, P. M. (1963). *Am. J. Bot.* **50,** 219–227.
THOMPSON, N. P. (1970). *Am. J. Bot.* **57,** 390–393.
TORREY, J. G. (1963). *Symp. Soc. exp. Biol.* **17,** 285–314.
TORREY, J. G. (1968). *In* "Biochemistry and Physiology of Plant Growth Substances" (F. Wightman and G. Setterfield, eds), pp 843–855. Runge Press, Ottawa.
TORREY, J. G. (1971). *In* "Les Cultures des Tissues des Plantes", pp 177–186. Editions du Centre Nat. Rech. Sci., Paris.
TORREY, J. G., FOSKET, D. E. and HEPLER, P. K. (1971). *Am. Scient.* **59,** 338–352.
WAREING, P. F. (1958). *Nature, Lond.* **181,** 1744–1745.
WANGERMANN, E. (1967). *New Phytol.* **66,** 747–754.
WARK, M. C. (1965). *Aust. J. Bot.* **13,** 185–193.
WARK, M. C. and CHAMBERS, T. C. (1965). *Aust. J. Bot.* **13,** 171–183.
WETMORE, R. H. and RIER, J. P. (1963). *Am. J. Bot.* **50,** 418–430.
YOUNG, B. S. (1954). *New Phytol.* **53,** 445–460.
ZEE, S. Y. (1968). *Aust. J. Bot.* **16,** 419–426.
ZEE, S. Y. and CHAMBERS, T. C. (1968). *Aust. J. Bot.* **16,** 37–48.
ZEE, S. Y. and CHAMBERS, T. C. (1969). *Aust. J. Bot.* **17,** 199–214.

9. Control of Root and Shoot Development

A. J. McCOMB

Department of Botany, University of Western Australia

I. INTRODUCTION

In a final analysis, the shape of a plant is determined by inherited information. But what controls the sequential expression of this inherited information as the plant develops, and how might a change in the environment affect this apparently well-ordered sequence? This complex question lies at the heart of an enormous body of experimental work on the development of many organisms, including animals, and it would be quite inappropriate to review this literature here. But we may touch upon several basic approaches. These are, firstly, to unravel the genetic complexity which underlies development, as a simple genetic control over a particular phase of development suggests a simple biochemical control over that phase. Secondly, we may try to isolate and identify specific chemicals from the organism, which affect particular developmental processes; if we are successful, a detailed understanding of the biochemical effects of these compounds might offer clues to normal control mechanisms in the plant. Thirdly, the effects on development of particular environmental factors can be analysed, and deductions made about the control mechanisms involved. Fourthly, cultures can be raised from isolated organs, tissues, cells and even protoplasts. The factors which trigger off new patterns of organization can be worked out, and inferences made about the formation of developmental patterns in the intact plant. Finally, a direct biochemical assault can be made on the changing patterns of enzymes (and other compounds), and even on the genetic material of the nucleus itself, as development proceeds.

The pea plant continues to be a most important experimental tool in the study of development, and in this chapter we will look briefly at its properties in relation to each of these five experimental approaches.

II. SIMPLE GENETIC CONTROLS OF DEVELOPMENTAL PATTERNS

A. Stem Growth

Growth of the stem is affected dramatically by simple genetic factors. Most familiar cultivars can be readily assigned to one of the two general categories of dwarf (characterized by short internodes) and tall (with long internodes). The first two internodes of tall plants are dwarf (though this is easily overlooked if seeds are planted below ground), but later internodes expand about twice as rapidly as those of the dwarf plants (Fig. 9.1). As Mendel found in his classic work on inheritance, the tall and dwarf varieties differ by a single factor, the tall allele (*Le*) being dominant over the dwarf (*le*).

It should be said straight away that the familiar tall and dwarf cultivars may differ genetically in a host of other ways, and some of the loci concerned, other than *Le*, may modify plant height. Node number is affected by at least three genes (*Mie, Miu, Min*), and other genes may modify internode length (eg. *Cot, Coe, Coh, Cona*) (Lamprecht, 1962). Flowering time affects stem length (Wellensiek, 1973). Nevertheless, it is a simple matter to demonstrate by crossing that the main difference in internode length between familiar cultivars—for example, the dwarf 'Progress No. 9' and the tall 'Alaska', which are often used in physiological experiments—can be attributed almost entirely to the *Le* locus (McComb and McComb, 1970). The effect of this single gene is so spectacular that, until we can explain biochemically the basis of the dominance of its effect, one wonders if there is any hope of explaining more subtle, modifying genetic controls of growth.

One may speculate that tall varieties synthesize a growth-stimulating factor at a higher rate than do dwarfs, and that the genetic block to growth operates in the pathway to the synthesis of this hypothetical compound (Brian, 1957). Reciprocal grafts between tall and dwarf varieties show that the effect of the

Le locus is not graft transmissible (Lockard and Grunwald, 1970; McComb and McComb, 1970), and so translocation of a growth-regulating compound from roots or mature leaves to the growing apex is not likely to be involved. When segments are cut from expanding internodes below the apex, and allowed to expand in a suitable medium, sections from tall plants grow at a higher rate than those from dwarfs (Stowe, 1960; Gorter, 1961), but it is not clear if this is a residual effect of substances translocated from the apex before excision, or due to the continued transcription and translation of genetic information within the expanding tissue segment after its isolation from the rest of the plant. If very short segments are used the growth of dwarf tissue may exceed that of tall (Tanimoto *et al.*, 1967), perhaps because of the relatively smaller number of cells in the excised tall tissue.

Another striking genetic effect on stem growth should be mentioned. Plants with very high growth rates may occur in the progeny of crosses between certain dwarf varieties, and these are designated "slender" (de Haan, 1927; Rasmusson, 1927; see Fig. 9.1). The height of these plants is due to more, as well as longer, rather thin internodes, and the increased internode length over dwarfs is due to increased cell number and length (de Haan and Gorter, 1936). Plants have a slender phenotype if they are carrying the homozygous recessive alleles *la* and *cry^s*, no matter which allele is carried at the *Le* locus. As it is the recessive which has such a very high rate of internode growth, one may speculate that slender plants lack an inhibitor, which is present in tall and dwarf varieties (Brian, 1957). Growth can thus be interpreted as a balance between growth stimulatory and growth inhibitory processes in the plant. Another allele has been described for the *Cry* locus, *cry^c*; plants carrying *le la cry^c*, designated "crypto-dwarfs", have growth rates comparable with slender plants during the expansion of the first few internodes, but growth then falls to about that of dwarfs (Rasmusson, 1927).

Fasciation of stems, which may lead to the clustering of flowers in a "pseudo-umbel" at the apex, constitutes another striking inherited characteristic. Mendel found that this appearance is conditioned by a single recessive factor (designated *fa*), but it emerges that the control is not straightforward. The gene shows incomplete penetrance and variable expression, so that in progeny of early generations there is a complete range of phenotypic expression from almost imperceptible to grotesque (Marx and Hagedorn, 1962). The effect is not graft transmissible (Scheibe and Wohrmann-Hillman, 1957).

Other genes affect stem growth, and genetic stocks are available with extremely short internodes, or with wavy "serpentine" stems, or in which all parameters of the shoot are reduced in size. While a thorough review of these controls would be out of place here, it is clear that the gene pool available in *Pisum* provides a most important potential source of material for general investigations of the physiological genetics of stem growth in plants (see also Chapter 2).

Fig. 9.1. The control of stem growth by known alleles. The seedlings are of the same age and were raised together. *Left*, a "dwarf" plant (carrying *le La cry*); *centre*, a "tall" plant (carrying *Le la Cry*); *right*, a "slender" plant (carrying *Le la cry*^{s}).
(Photograph: A. J. McComb)

. Leaf Shape

Much is known concerning the genetic control of leaf shape (e.g. Yarnell, 1962), and some aspects of the information on genetic control given below re expanded and illustrated in Chapters 2 and 12. In the familiar cultivars, leaves at nodes 2 and 3 (counting the node of the cotyledons as 1) are reduced to stipules and a vestigial axis—the so-called trifid bract or scale leaf. At higher nodes the axis also carries pairs of leaflets and (especially at later nodes) is terminated by one or more tendrils. This developmental pattern is modified by several genes. If *st* is carried, stipules are reduced in size. A maximum of one pair of leaflets is produced by each leaf if the genome includes *up*. Tendrils are converted to leaflets in 'Acacia' varieties, which carry *tl*, while leaflets are converted to tendrils by *af*. Other mutants are known which modify leaf shape, size and degree of indentation.

Clearly, patterns of leaf development are genetically controlled in a simple way, and one may speculate that lamina expansion is physiologically controlled independently of the branching system of the leaf axis and main veins. This concept fits in well with the theories of leaf structure homologies put forward by morphologists (Arber, 1950; and discussion by Audus, 1959). The lamina forms if *St* and *Af* are dominant, while *Tl* inhibits lamina formation; it is easy to suggest a balance between hypothetical stimulatory and inhibitory substances, the production of which are under genetic control. It seems likely that the patterns are controlled within the very young leaves; at least, effects of *tl* and *st* are not graft transmissible (Went, 1943).

C. Root Growth

As it is difficult to screen large numbers of plants for mutant forms of root growth, it is not surprising that there is little information in the literature about this topic, despite the importance of roots in the survival of plants. There is little information on the inheritance of morphological patterns in roots. Mean (1928) noted that although a dwarf ('Nott's Excelsior') had smaller roots than a tall ('Telephone'), shoot length/root length ratio of the dwarf was $1:1.4$, and of the tall $1:0.7$. When plants were crossed, the same ratios were maintained in tall and dwarf segregates of the F_2. Although this shows inheritance of root growth, it must be remembered that shoot growth affects root growth indirectly because of competition for a limited supply of assimilates (Lovell, 1971), so that for a convincing demonstration of direct genetic control over a root phenotype, efforts should be made, for example by reciprocal grafting or culture of excised roots, to unravel the complexities arising from the sink strength of shoots. Differences in rates of growth of roots excised from plants of different varieties and grown in culture do in fact occur (e.g. Pecket 1960). A discussion of the inheritance of root development is included in Chapter 5.

Of special interest in research on the control of gravitational responses is the ageotropic mutant of peas (Blixt et al., 1958), which is curiously unresponsive to gravitational fields. This is discussed below, in connection with auxin effects.

III. CHEMICAL CONTROL OF DEVELOPMENT

A. Gibberellins

These terpenoids are of classic interest to the student of plant growth, since the finding of Brian and Hemming (1955) that small amounts of gibberellic acid (GA_3), isolated from fungal cultures, were able to cause genetically dwarf pea seedlings to grow at the same rate as genetically tall pea seedlings. Thus a single chemical compound would overcome the effect of a single genetic factor, converting the dwarf genotype to a tall phenotype. Might an endogenous gibberellin be the hypothetical growth factor, the synthesis of which is controlled at the Le locus? This is a most important suggestion in view of the paucity of examples from higher plants in which the effect of a single gene is reversed by the application of a single chemical compound (Nelson and Burr, 1973).

Six gibberellins have been isolated from immature pea seeds and have been fully chemically characterized (Fig. 9.2). Of these, GA_{20} (III) and GA_{29} (IV) are the major components (Frydman and MacMillan, 1973; Frydman et al., 1974). Gibberellins are formed in excised pods (Baldev et al., 1965; see Chapter 15); chromatography and bioassay data show that gibberellins occur in the shoot (e.g. Radley, 1959; McComb and Carr, 1958; Kende and Lang, 1964; Köhler, 1965a; Railton and Reid, 1974). Two main fractions present on chromatograms, designated by Kende and Lang as GA_1-like and GA_5-like, may well be GA_{20} and GA_{29} (Frydman and MacMillan, 1973). Excised pea shoot apices release gibberellin, and are presumably a site of their synthesis (Jones and Lang, 1968), and gibberellins occur in xylem sap, where they are derived from the roots (Carr et al., 1964; Reid and Carr, 1967). Gibberellin also substitutes, in the short term, for the presence of the cotyledons in maintaining seedling shoot growth, so these organs may be a source of gibberellin for the seedling (Shininger, 1972).

Because of the dramatic effect of small amounts of gibberellin on the growth of the dwarf pea plant, it is a suitable organism for bioassays of gibberellin, and a convenient tool for examining the metabolic and biochemical effects of the gibberellins. After treatment, there is an increase in cell length (especially in internodes which are expanding at the time of treatment), cell number

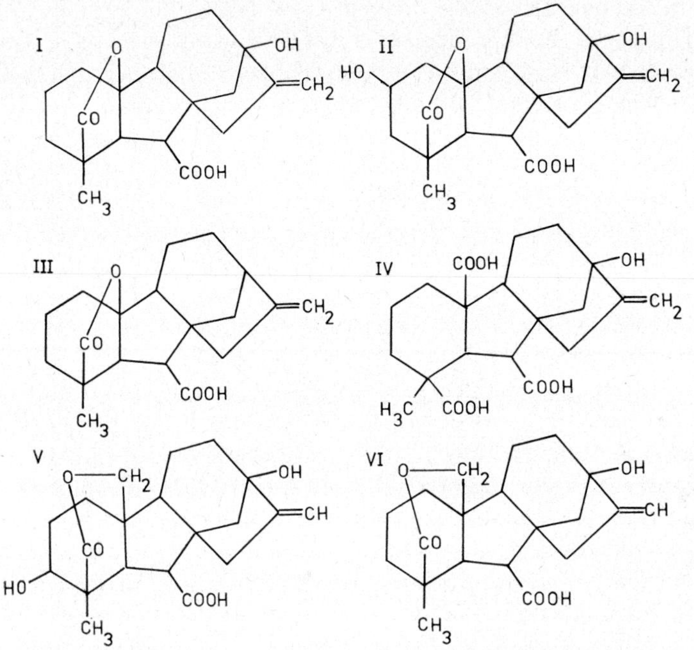

FIG. 9.2. Gibberellins from the pea which have been fully characterized. I, gibberellin A_{20}; II, gibberellin A_{29}; III, gibberellin A_9; IV, gibberellin A_{17}; V, gibberellin A_{38}; VI, gibberellin A_{44}. [Determinations were by combined gas chromatography and mass spectrometry (Frydman and MacMillan, 1973; Frydman et al., 1974.]

and DNA (especially in internodes close to the apex). There is a redistribution of dry matter from roots and leaves into the expanding stem, an increase in RNA and protein in the expanding internode, and an altered pattern of enzyme development, such as one might expect to accompany a developmental change of such magnitude and fundamental nature (McComb and Broughton, 1972).

Just which of these many changes are directly effected by the compound, and which are quite indirectly effected, are important and unresolved questions in understanding the mechanism of gibberellin action, and the final answers to them may well lie in the study of other metabolic systems such as the barley aleurone. For our purposes, it is important to note the considerable metabolic changes which can be triggered off by this single compound, and which might all be involved in the differences between tall and dwarf pea varieties; and further, to note how dangerous it would be to find a simple metabolic difference between two varieties, and from this finding to conclude that this difference alone accounts for the morphological difference between the two.

But more fundamentally, can the difference between tall and dwarf varieties

really be accounted for by differences in gibberellin content? Certainl
growth appears to be dependent on gibberellin, as the quaternary ammoniur
compound AMO1618, known to inhibit gibberellin synthesis in excised pe
fruits (Baldev *et al.*, 1965), greatly inhibits growth in shoots of both tall an
dwarf varieties, and its effect is reversed by gibberellin. (In fact, the growth c
tall varieties appears to be gibberellin limited; eg. Moore, 1967). If gibberellin
are extracted from tall and dwarf plants, and results are expressed on a "pe
plant" basis, the larger talls appear to have more gibberellin than the dwarf
(Köhler, 1965a,b). However, the level of gibberellin extractable from apica
regions, and which diffuses from excised apical buds, is the same for both tal
and dwarf plants (Kende and Lang, 1964; Jones and Lang, 1968). The possi
bility that the tall growth form is accounted for by high rates of gibberelli
production in roots or mature leaves is discounted by the results of the recipro
cal grafting experiments cited earlier. On the other hand, tall plants dwarfec
by AMO1618 respond more strongly to gibberellin than do dwarf plant
growing at comparable rates (Kende and Lang, 1964). We must reluctantl
conclude that the available evidence does not support the suggestion tha
differential rates of gibberellin synthesis explain the tall and dwarf phenotypes
but rather that tall plants are more "sensitive" to the gibberellins which the
contain. "Sensitive" in this context may cover a number of hypotheses—per
haps tall plants are able to convert gibberellin more efficiently to a mor
active form (Köhler, 1965b); perhaps there is a genetically controlled dif
ference between dwarf and tall at the site of gibberellin attachment; perhap
the difference between the two is, in a biochemical sense, far removec
from the mechanism of gibberellin action. But, despite all this, we migh
well bear in mind the simple genetic evidence that the *Le* gene acts tc
reverse an inhibitory process in the plant (see also Brian and Hemming
1958).

What of the "slender" plants, which appear to lack an inhibitory process
present in tall and dwarf varieties? If this hypothetical inhibition is normally
reversed by gibberellin, one might expect the presence of gibberellin to be
irrelevant to the slender phenotype. In fact, experimental findings agree with
this contention; AMO1618, at concentrations which greatly inhibit the
growth of tall and dwarf plants, does not affect the growth of slender plants
(McComb and McComb, 1970). When sufficient material is available, it will
be fruitful to assay the gibberellin content of these plants.

Other effects of gibberellins on pea plants might be noted in passing, particularly in view of what follows concerning other growth substances. Root
growth is inhibited by gibberellin application to the shoot, but this is probably
an indirect effect resulting from the extra deposition of assimilates in the
rapidly expanding stem (McComb, 1966); in cultures of excised pea roots,
Pecket (1960) records that gibberellic acid was without effect on one variety,
while in another, growth was enhanced. In both, formation of root hairs,

ormally suppressed in liquid culture, was greatly enhanced by the gibberel-
n. Senescence of the shoot may also be delayed by gibberellin (Lockhart
nd Gottschall, 1961).

. Auxins

uxins are readily detected in extracts from pea plants with biological assay
echniques, and are at rather higher concentrations in the shoot than the root
e.g. Cartwright *et al.*, 1956). Evidence for indol-3yl acetic acid itself (IAA;
ompound VII in Fig. 9.3) rests a little insecurely on bioassays of zones on

IG. 9.3. Auxins from the pea: VII, indol-3yl acetic acid (IAA); VIII, indol-3yl
cetaldehyde (IAA1), the biological activity of which is attributed to conversion to IAA;
X, methyl 4-chloroindol-3yl acetate; X, 4-chloroindol-3yl acetic acid. [VII detected
y paper chromatography and bioassay (van Overbeek, 1939; Bennet-Clark and Kefford,
953). VIII detected by paper chromatography, bioassay and similarity of u.v. spectrum
Rajagopal, 1967, 1968). IX and X detected by crystallization, i.r. spectra, and confirmed
by synthesis (Marumo *et al.*, 1968a, b).]

hromatograms corresponding in R_F with the known compound, together
vith visualization of chromatogram spots by indole reagents (van Overbeek,
939; Bennet-Clark and Kefford, 1953; Rajagopal, 1967). The aldehyde
VIII) is active in auxin assays, almost certainly because of its ready conversion
o the acid. Its occurrence in the pea plant, based on chromatography, bio-
ssay, colour reactions and u.v. spectrum (Rajagopal, 1967) has been con-
irmed with sterile material (Rajagopal, 1968). This latter step is important
)ecause of the synthesis of auxin by epiphytic bacteria in the pea plant
Wichner and Libbert, 1968), which can even have effects during the extrac-
ion process (Wichner, 1968). Marumo *et al.* (1968a,b) have found a unique
hlorinated auxin, methyl-4-chloroindol-3yl acetic acid (IX) and (in lower con-
entration) its free acid in extracts of immature pea seeds. Related chloro-
ryptophan derivatives have also been found (Marumo and Hattori, 1970).
    Auxins which cannot readily be attributed to any of the chemically defined
compounds have also been detected in extracts from the pea (e.g. Audus and

Gunning, 1958; Pate, 1958), and these include the widely distributed "accele
ator α" of Bennet-Clark and Kefford (1953); this may turn out to be a cor
plex involving chlorotryptophan derivatives (Marumo and Hattori, 197C
When labelled IAA is applied to pea plants a number of complexes ar
breakdown products are found (Schneider and Wightman, 1974). These san
compounds are, one presumes, present as part of the normal auxin metabolis
of the plant, but formal demonstration is always comforting. There is clear
scope for the more extensive application of modern analytical techniques i
the chemical definition of auxins and related compounds.

There are two important physiological differences between the effects c
auxins and gibberellins in pea plants. In contrast to the gibberellins, auxir
firstly inhibit growth at physiologically high concentrations; and secondl
they strongly promote the growth of excised segments of growing tissue. Pr
sumably these tissues, isolated from the source of auxin in the apical regio
of the plant, and containing oxidase enzymes which destroy auxin (e.
Galston and Dalberg, 1954), rapidly become auxin deficient and so respon
to an exogenous auxin supply. Segments of etiolated pea tissue have bee
widely used in the biological assay of auxins, and the "split pea epicotyl test"
in which segments from growing pea epicotyls, split for most of their lengtl
become contorted in the presence of auxin, has been used extensively i
searching for auxin activity among synthetic chemicals (e.g. Fawcett et al
1956).

Isolated segments of pea tissue have also been a favourite tool for examir
ing the metabolic effects of applied auxins. These effects are far reaching
and range from changes in wall plasticity (which accompany and presumabl
allow an increased rate of cell expansion) to effects on RNA and protei
metabolism, which may underlie the growth of the tissue. (For a review c
this general topic, see Audus, 1972.) As with the gibberellins, it is difficult t
unravel the primary events from the tangle of secondary effects, all of whicl
incidentally, might be affected by endogenous auxin levels in the intact plan
The mechanism of growth inhibition by high auxin concentrations remain
obscure.

It is generally acknowledged that auxin is required for cell growth, and her
we may ask if there is evidence that auxin has a regulatory role to play i
controlling growth of the intact plant. Stem growth is probably not auxi
limited, as the application of auxin to dwarf pea plants, while it may brin
about a small increase in growth, will not convert them to a tall phenotyp
in the way that gibberellin does (Brian and Hemming, 1955). The amount c
endogenous auxin might be controlled by regulation of synthesis, formatio
of complexes with other chemicals, or oxidation. Levels of IAA-oxidase ar
inversely related to growth (Galston and Dahlberg, 1954). Levels of fre
auxin appear to match growth rates, but this is different from saying that th
auxin controls growth. Perhaps it is not surprising to find that the auxi

:vels of tall pea plants are higher than dwarfs (e.g. McCune and Galston, 959). Similar levels of diffusible auxin occur in the apical regions of tall and warf plants, but the levels fall more rapidly in subtending internodes of the warf, where conversion to indol-acetyl aspartic acid is higher than in the alls (Lantican and Muir, 1969). When gibberellin is applied to dwarf plants he level of free auxin increases along with growth rate (Phillips *et al.*, 1959). .antican and Muir (1969) suggest that there is increased synthesis of auxin 1 the dwarf after gibberellin treatment. McCune and Galston (1959) report igher oxidase activity in internodes of dwarf plants, and there is a higher :vel of a phenolic inhibitor of the oxidase enzyme (Galston, 1959). Using ifferent assay methods, Ockerse and Weber (1970) found oxidase levels to •e higher in buds from tall plants than dwarfs, and to increase in the dwarf fter gibberellin treatment.

The observations of de Haan and Gorter (1936) suggest that in slender •lants auxin destruction is less effective than in dwarfs. Scheibe and Wohr-1ann-Hillman (1957) showed that application of the auxins IAA or naph-halene acetic acid to normal plants causes symptoms reminiscent of "fasci-ted" pea lines. They also found higher levels of four auxins in fasciated plants with the recessive *fa*) than normal, and suggest that this might be due to higher ryptophan levels in the mutant.

In isolated segments of tissue, incubated in sucrose, gibberellin will not timulate growth, but as we have seen, auxin does so dramatically; that is, rowth of the isolated segment is auxin limited. If auxin is present, gibberellin vill bring about further stimulation, and total growth of the segments is omparable with those of intact, gibberellin-treated dwarf plants (Brian and Hemming, 1958). Depending on the conditions of treatment (Purves and Hillman, 1958), the effects of the two compounds may range from less than dditive (Gorter, 1961) to synergistic (Ockerse and Galston, 1967). The ynergism is observed in segments from both tall and dwarf varieties (Tani-noto *et al.*, 1967). The actions of the two are different: gibberellin will stimu-ate growth at optimum concentrations of auxin, when any further increase n auxin would be inhibitory. If segments are cut from gibberellin-treated plants, they will subsequently grow more rapidly than segments from control plants only if auxin is present in the medium (Brian and Hemming, 1958). Thus, the effect of gibberellin is dependent on the presence of auxin. It re-mains possible that auxin-induced growth requires the presence of gibberellin Ockerse, 1970).

There are several other aspects of plant development in which auxin may aave a regulatory role. If the apex of the pea plant is removed, lateral buds it lower nodes begin to expand within 8 h, well before vascular connections o the bud have been established (Nakamura, 1964; Sorokin and Thimann, 1964; Wardlaw and Mortimer, 1970). Application of auxin to the cut stump, it a fairly high physiological concentration, suppresses the growth of these

I

laterals (Thimann and Skoog, 1933; Scott and Pritchard, 1968), and it
reasonable to suggest that auxin has a role to play in suppression of lateral
buds in the intact plant. However, van Overbeek (1938) could not detect
high levels of auxin in lateral buds of the pea plant and found that the auxin
content of these lateral buds increased when they were released from inhibition by removal of the apex. As we shall see below, application of cytokinin
will release dormant buds from inhibition. The situation is therefore a complex one, involving other growth substances, and also competition for nutrients; the whole field is summarized by Phillips (1969) and is dealt with in
detail for the pea plant in Chapter 10. A regulatory role for endogenous
auxin in bud suppression in the intact plant has not been firmly established.

Auxins are also implicated in the control of root development, where their
derivation from the shoot may be of importance. Shoots cut from the pea
plant will initiate roots, but this initiation is greatly reduced if the shoot is
disbudded (e.g. Eriksen, 1973). Auxin substitutes for the buds in the early
stages of initiation (Eriksen and Mohammed, 1974). Because of the effects of
auxin on growth of isolated roots in culture, they may well be involved in the
initiation of lateral roots and root expansion (Chapter 5). Decapitated pea
plants show a marked reduction in number and length of secondary roots,
but IAA applied to the cut stump largely compensates for the absence of the
shoot (McDavid et al., 1972). It may be concluded that the roots of the intact
plant are influenced by the supply of auxin from the shoot, although cultured
roots do synthesize some auxin on their own (e.g. van Overbeek, 1939).

The classical Cholodney–Went concept (Audus, 1959) links plant curvatures, both geotropic and phototropic, with a control over the transport of
auxin from meristematic regions to the subtending, expanding tissue. This
hypothesis appears to be quite tenable for the shoot, as it is consistent with
what is known about auxin metabolism, transport and sensitivity in this
region; the subject is discussed in Chapter 10. However, for the geotropic
response of roots the situation is far from clear, as transport of auxin in this
organ appears to take place from the base of the pea root towards the root
apex (Hillman and Phillips, 1970; Morris et al., 1969)—the reverse of the
situation in the shoot. Konings (1965) implicates auxin-oxidase activity in the
control of curvatures in this organ. For further details of this complex topic
see Chapter 5 and Audus (1969).

The ageotropic mutant of peas (Blixt et al., 1958) is of particular interest in
this context, and has been studied by Scholdeen and Burström (1960) and
Ekelund and Hemberg (1966). The roots of the mutant form grow upwards to
the soil surface, and then horizontally. The roots show no geotropic response
to gravitational fields of $2.7 \times g$, and the shoots are also in general ageotropic, though the response is less well defined. In their auxin response, the
roots behave in the same way as normal roots, showing growth inhibition
which suggests the presence of relatively high physiological auxin levels in

oth. If the root tip is placed eccentrically on its stump, the root bends in the
sual way. A "normal" tip placed on an "ageotropic" base brings about
ormal geotropic curvature in the base, while an ageotropic tip on a normal
ase leads to ageotropism in the base. The shoot of the ageotropic mutant
1ows a normal phototropic response.

. Abscisins

bscisic acid (ABA; Fig. 9.4) occurs in immature pea seeds and shoots
sogai et al., 1967; Kende and Kays, 1971; Komoto et al., 1972; Frydman
al., 1974). The closely related dihydrophaseic acid has also been isolated
om immature pea seeds (Frydman et al., 1974). The latter compound is
robably a metabolite of ABA, and its presence implies the presence of
haseic acid, an intermediate between ABA and dihydrophaseic acid (Fryd-
1an et al., 1974; Milborrow, 1974). The inhibitor xanthoxin (compounds
.III and XIV) is present in pea seedlings as two isomers (Burden et al., 1971),
nd is probably a precursor of ABA. ABA is formed in root systems (Tietz,
974), and presumably also in mature leaves, as the levels in the shoot
icrease dramatically when leaves are wilted (Simpson and Saunders, 1972).
his rise on wilting is presumably related, as in other species, to stomatal
losure (for review see Milborrow, 1974).

Because of its growth-inhibiting properties, it would be tempting to suggest
hat ABA is the hypothetical inhibitor present in tall and dwarf plants, the
ction of which is reversed by gibberellin (see above). It is certainly present in

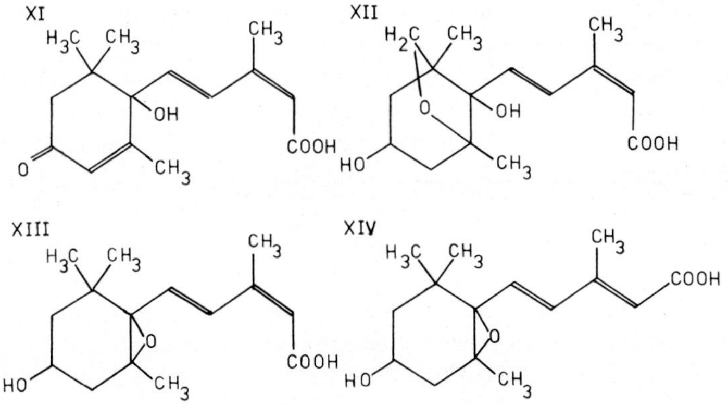

FIG. 9.4. Known abscisins from the pea: XI, (+)−abscisic acid; XII, dihydrophaseic
cid; XIII, (+)−xanthoxin (2-cis); XIV, (+)−xanthoxin (2-trans). Determinations
lepend essentially upon gas chromatography, typically of fractions of extracts showing
iological activity. [X reported, for example, by Isogai et al. (1967), Kende and Kays
1971), Tietz (1974) and Frydman et al. (1974); XII by Frydman et al. (1974); XIII and
XIV by Burden et al. (1971).]

tall and dwarf varieties, but no determinations are yet available for slender
plants. Simpson and Saunders (1972) conclude that differences in ABA level
do not account for differences between tall and dwarf plants, and this
consistent with the observation that dwarf mature leaves and root systems d
not inhibit the growth of tall scions.

ABA may also control some features of root growth. Tietz (1974) record
that it inhibits the growth of the main root, and so promotes the number o
lateral roots per unit of root length. Phillips (1974) found that ABA inhibit
nodule formation by inhibiting cortical cell divisions, at or adjacent to th
region of rhizobial infection.

D. Cytokinins

Cytokinins have been detected in extracts from pea seedlings (Zwar an
Skoog, 1963), roots (Short and Torrey, 1972b) and developing fruits in cultur
(Hahn et al., 1974). They occur both as free cytokinins and as constituent
of transfer RNA, from which they can be released by hydrolysis (e.g. Babcoc
and Morris, 1970). Cytokinins which have been fully characterized from th
pea plant are shown in Fig. 9.5. Most physiological work has been carrie
out with kinetin (6-furfuryl amino purine) and benzyl adenine, which, as fa
as one is aware, do not occur naturally in plants.

Because cytokinins are often needed as a supplement in plant tissue cul
tures, where they allow continued cell divisions, it is common to think c
them as concerned in regulating cell divisions in the intact plant; indeed, ce
divisions have been reported to follow the addition of kinetin to pea root
(Torrey, 1961). However, no regulatory role over rate of cell division ha
ever been clearly shown, and the elongation of roots and shoots is inhibite
by exogenous cytokinin (e.g. Sprent, 1968). In stem sections kinetin reduce
elongation but promotes an increase in stem diameter, acting synergisticall
with IAA. The effect is due to increased cell diameter (Thimann and Laloraya
1960; Hashimoto, 1961; Katsumi, 1962; 1963).

Cytokinin inhibits formation of roots on cuttings (Kaminek, 1966), thoug
at very low concentrations root initiation may be enhanced somewha
(Eriksen, 1974). The inhibition occurs only at an early stage of root initiation
In contrast to this inhibiting effect on the growth of stem and root, cyto
kinins stimulate the development of lateral buds, counteracting the inhibitor
effect of auxin. Once released from dormancy, early growth of the bud ca
be stimulated by auxin. Such observations lead to the tentative suggestio
that a balance between auxins and cytokinins is important in regulating apica
dominance (Wickson and Thimann, 1958; Sorokin and Thimann, 1964
Sachs and Thimann, 1967; and see Chapter 10).

Synthesis of cytokinins by roots could offer a mechanism for a hormona
form of "communication" from root to shoot. (As seen above, transport o

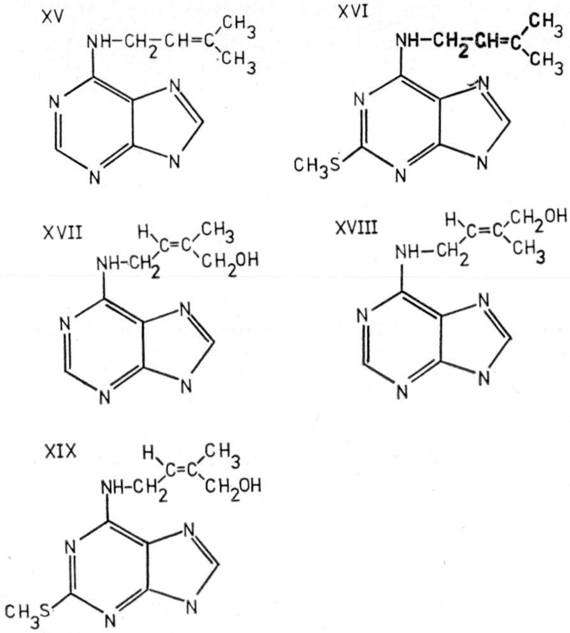

XV

NH—CH₂—CH=C⟨CH₃ / CH₃⟩

XVI

NH—CH₂—CH=C⟨CH₃ / CH₃⟩

CH₃S

XVII

H / CH₃   C=C   CH₂OH

NH—CH₂

XVIII

H / CH₂OH   C=C   CH₃

NH—CH₂

XIX

H / CH₃   C=C   CH₂OH

NH—CH₂

CH₃S

Fig. 9.5. Known cytokinins from the pea plant. Note that in each case identification was not made of the free molecule illustrated above, but of the corresponding riboside from hydrolysates of pea transfer RNA. Closely related or identical molecules, not fully characterized, occur freely in the plant (e.g. Short and Torrey, 1972b). Identifications of compounds with biological activity were by rigorous chemical analysis including (apart from XVI) u.v. spectroscopy and mass spectrometry.

XV. isopentenyl adenine (IPA), 6-(3-methyl-2-benzylamino) adenine; from roots and shoots (Hall *et al.*, 1967; Babcock and Morris, 1970; Vremen *et al.*, 1972).

XVI. MS-IPA, 6-(3-methyl-2-butenylamino)-2-methylthio adenine; identification based on chromatographic data, from shoot RNA (Vremen *et al.*, 1972).

XVII. *cis*-zeatin, 6-(*cis*-4-hydroxy-3-methyl-2-butenylamino) adenine; from shoots and roots, references as for XV.

XVIII. *trans*-zeatin, 6-(*trans*-4-hydroxy-3-methyl-2-butenylamino) adenine; from shoots (Vremen *et al.*, 1972).

XIX. MS-zeatin. 6-(4-hydroxy-3-methyl-2-butenylamino)-2 methyl thio adenine. The *cis* form is shown, but *cis* and *trans* forms could not be distinguished by the procedures used. From shoots (Vremen *et al.*, 1972)

auxin could be important in communication in the reverse direction.) In fact, the xylem sap of the field pea contains cytokinin (Carr and Burrows, 1966), apparently derived from the root system, and McDavid *et al.* (1973) found that the accelerated shoot senescence of a pea seedling, brought about by root pruning, could be offset by application of a cytokinin to the shoot.

There do not appear to be any reports of inherited abnormalities which involve cytokinins. It is possible that together with auxins, they could be involved in the syndrome of fasciation, especially since fasciation of pea plants brought about by *Corynebacterium fascians* is clearly due to the synthesis of

cytokinin by the bacterium. However, the symptoms of the disease involv
the development of clusters of laterals, rather than true fasciation (Thiman
and Sachs, 1966).

E. Ethylene

Pea seedlings respond dramatically to ethylene (Fig. 9.6), showing charac
teristic symptoms of stunting, stem swelling and prostration ("diageotropism"
of the shoot (e.g. Knight and Crocker, 1913). The plant is very sensitive t
the gas, responding, for example, to traces of ethylene produced by th
ballasts of fluorescent tubes, which were sometimes present in the earlier cor
trolled-environment cabinets designed for plant growth (Wills and Patterson
1970). The gas is also produced by the pea plant itself, and so it may b
involved in controlling growth in the intact plant.

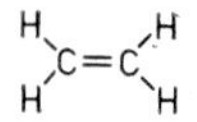

$$\begin{array}{c}\mathrm{H} \\ \diagdown \\ \mathrm{H}\diagup\end{array}C=C\begin{array}{c}\diagup\mathrm{H} \\ \\ \diagdown\mathrm{H}\end{array}$$

FIG. 9.6. Ethylene, the simplest organic substance from the pea plant which show
marked effects on plant growth. This gas has been detected by gas chromatography i
many investigations (e.g. Burg and Burg, 1968).

   The unfolding of the plumular hook, a characteristic response of etiolate
pea plants placed in the light (see Chapter 11), does not take place if ethylen
is administered. If ethylene produced by etiolated plants is removed by absorp
tion, the hook will open in the dark. Ethylene is produced by the hook region
in the dark at a rate sufficient to maintain the concentration of gas in th
tissue at a physiologically effective level, and ethylene production falls in th
light (Goeschl et al., 1967). As Audus (1972) points out, this is compelling
evidence for a regulatory role of ethylene in hook opening. Ethylene is pro
duced in response to injury, and also to simple physical constraint of the
growing shoot (Goeschl et al., 1966). It may therefore be of especial impor
tance in the regulation of early seedling growth, below the soil surface.
   The inhibiting effect of high auxin concentration can be simulated in roo
and shoot by ethylene, and indeed high doses of auxin result in ethylene pro
duction; it is reasonable therefore to suggest that the inhibiting effect o
auxin is mediated by ethylene (e.g. Chadwick and Burg, 1967; 1970). On
closer inspection, however, the effects of the two compounds can in part be
separated, though clearly the interaction between the two remains of impor
tance.
   Ethylene inhibits stem growth, an effect accompanied by reduced cel
division at the apex (Apelbaum and Burg, 1972). The inhibitory effect in
expanding tissue can be reversed by gibberellin, except in the expanding tissue

losest to the apex (Stewart *et al.,* 1974). IAA applied to the stump of a seedling from which the apex has been excised will bring about swelling, which reminiscent of swelling in intact seedlings due to ethylene. Fan and Mac-Lachlan (1966) showed the swelling effect of auxin to be associated with increased cellulase, but Ridge and Osborne (1969), while confirming this, found that the swelling due to ethylene was not associated with an increase in this enzyme. They suggested that ethylene might affect the mechanical properties of the wall in a different way, having noted increased hydroxy-proline and peroxidase in the walls (Ridge and Osborne, 1970; Sargent *et al.,* 1973).

In the roots as well, some of the effects of ethylene resemble those of auxin, but again separate effects of the two compounds can be distinguished (e.g. Andreae *et al.,* 1968; Scott and Norris, 1970). The role of ethylene in root growth is explored further in Chapter 5.

IAA stimulates ethylene production in the nodal regions of pea stems, suggesting that ethylene might also be involved as a mediator in bud inhibition by auxin; however, the level of ethylene production at the nodes does not fall after decapitation of the apex (Burg and Burg, 1968).

Information on possible interactions between ethylene formation and genotype appear to be lacking for the pea plant. In *Phaseolus vulgaris* dwarf plants have been found to produce more ethylene (expressed per unit dry weight) than those of a tall variety (Aharoni *et al.,* 1973). Gibberellin is without effect on ethylene production in the pea (Fuchs and Lieberman, 1968).

V. ENVIRONMENTAL CONTROLS OF DEVELOPMENT: ETIOLATION

Temperature, water availability, mineral nutrition, light, mechanical damage and other factors all affect morphology, and we have already touched upon some of these factors—physical constraint might affect development through ethylene levels, and wilting causes an increase in the level of ABA. Low temperatures reduce growth rates, and a short period of cooling inhibits apical dominance and causes the plant to throw laterals; the effect is not reversed when the plants are returned to higher temperatures, and auxin transport may be involved (Husain and Linck, 1967). Water stress can also reduce lateral bud growth and thereby enhance apical dominance (McIntyre, 1971). Effects of low temperature and photoperiod on flowering, which in turn affects the duration of vegetative growth, are discussed in Chapter 14.

Undoubtedly the most striking effect on development is that brought about by light. We have mentioned some effects of light in passing, and the subject is explored in depth in Chapter 11. Here we want to draw attention especially to the phenomenon of etiolation, i.e. the rapid growth, suppression of leaf expansion, and other changes which occur when plants are maintained in absolute darkness. How does more rapid growth in the dark relate to tall and dwarf genotypes, and to gibberellin treatment?

Both tall and dwarf plants show more rapid internode elongation in the dark (including, significantly, the first two internodes which are usually dwarf in both types of plants), but the effect is more striking in the dwarfs, and to such an extent that the differences between the two varieties under conditions of etiolation are almost eliminated (e.g. Lockhart, 1956; Gorter, 1961). To put it another way, dwarf plants react more strongly to light than tall ones, and one can think of the dwarf habit as therefore being induced by light. Might increased gibberellin synthesis in the dark be involved? Dark-grown plants of tall and dwarf cultivars respond slightly to gibberellin, so that the level may still be limiting in the dark. Consistently with this, AMO1618 reduces growth in the dark and the effect is reversed by gibberellin. Köhler (1965b, 1966) found more gibberellin in dark-grown than in light-grown plants when whole plants were extracted, but Kende and Lang (1964) were unable to detect differences in the gibberellin contents of shoots from seedlings of dark- and light-grown plants. Their results, and those of Jones and Lang (1968), favour the suggestion that light increases the sensitivity of the plant to endogenous gibberellin; if AMO1618 is used to dwarf the dark-grown plants to an extent comparable with that seen in the light, the dark-grown plants then respond more readily to applied gibberellin than do plants in the light. Thus the difference in stem growth between dwarf plants growing in light and dark may prove to be similar to the difference between dwarf and tall plants growing in the light. All of this has interesting implications concerning the action of light in relation to the expression of the *Le* locus.

It would be useful to have further information concerning the physiological control of slender plants, for these grow even more rapidly than tall plants do in the light. Might it turn out that the slender phenotype lacks the mechanism for light inhibition of stem growth which is present in tall and dwarf plants?

It is pertinent to note that light is not believed to bring about its inhibiting effects on growth of pea shoots by controlling ABA levels (Kende and Kays, 1971; Simpson and Saunders, 1972; Dörffling, 1973), though light has certainly been shown to affect the ABA levels of roots (Tietz, 1974). Burden *et al.* (1971) reported that light does depress the level of the related inhibitor xanthoxin in the shoot, and as we have seen, ethylene levels also fall in the light.

We should remind ourselves that the effect of light in stimulating leaf expansion is quite another matter, as the differences in leaf areas between tall, dwarf and gibberellin-treated plants grown in the light are trivial in comparison with the overall effect of reducing leaf area in all by plunging the plants into darkness.

V. FORMATION OF CALLUS AND REGENERATION OF PLANTS

A. Formation of Callus

We have now seen something of the factors which control development, and judging from the genetic information and the effects of growth substances and light, this control may turn out to have an underlying simplicity. But it must be admitted that in most of what we have been considering, even though the effects may be dramatic, the controls involved are only effecting changes in an existing pattern of some sort—for example, a modification of rate of stem growth, or the expansion or not of lateral buds. An exception is found in the formation of roots as new organs on pieces of stem, an effect which is so effectively enhanced, and presumably controlled, by auxin.

In this section we will briefly explore the formation and culture of disorganized pea callus, and ways in which the developmental pattern of shoot and root might be imposed on this undifferentiated tissue. Apart from intrinsic interest in understanding the factors controlling development, the possibility of cloning plant material for genetic purposes and of using callus and isolated protoplasts in studies on the uptake of macromolecules and bacteria, needs no emphasis (e.g. Cocking, 1972; Melchers and Labil, 1973; McComb, 1974). As we shall see, regeneration of pea plants from protoplasts and disorganized tissue is proving to be difficult, but the comparative ease with which intact plants can be regenerated from protoplasts of other species, such as *Nicotiana* spp. and *Petunia* spp., suggests that the problems to be overcome with pea plants are due to technique, rather than to some fundamental lack of totipotency in the cells. Regeneration of intact plants from callus in another legume genus, *Medicago*, has been achieved (Saunders and Bingham, 1972).

Spontaneous calluses occur on the pods of certain lines, where they arise from the subsidiary cells of the stomata, thus constituting an interesting example of dedifferentiation (Burgess and Fleming, 1973). For culture work, callus is usually obtained by placing segments of tissue in an appropriate medium containing chemicals which act as suitable agents for dedifferentiation. For the pea plant, calluses have been obtained from stem, leaf and root tissue, using an auxin such as 2,4-dichlorophenoxyacetic acid, a cytokinin such as kinetin, and a suitable basal medium (Fig. 9.7). Among a number of suitable media are those of Torrey (1961), Schenk and Hilderbrandt (1972), and the B_5 medium of Gamborg et al. (1968). Part of this primary callus can be sub-cultured on to a new batch of medium, and will grow indefinitely in a disorganized way. The calluses produce gibberellin (Nickell, 1958), cytokinins (Short and Torrey, 1972a), ethylene (Atkins and McComb, unpublished) and presumably other growth factors. Initially callus from different

I*

organs is different in appearance, but after several sub-cultures all calluses appear to be the same so that it is not possible to discern the organ of origin. Calluses of peas can be shaken apart and used as a source of material for suspension cell cultures (e.g. Torrey and Shigemura, 1957; Veliky and Martin, 1970). There are no reports for the pea of haploid callus formation from developing pollen or, for that matter, of formation of haploid plantlets from developing pollen.

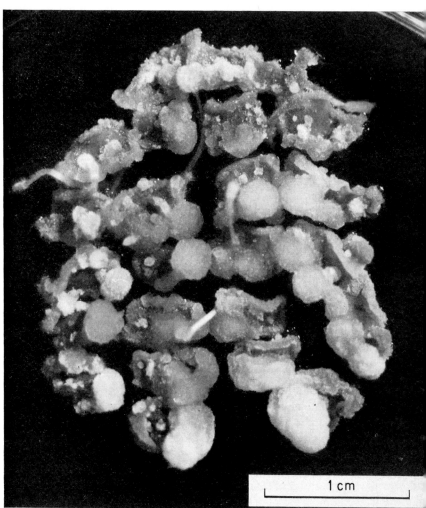

1 cm

FIG. 9.7. The induction of callus on pea leaf tissue. A leaflet of the dwarf cultivar 'Meteor' has been surface-sterilized, cut into pieces, and placed for 5 weeks on the medium of Shenk and Hilderbrandt (1972) containing the auxin 2,4-dichlorophenoxy-acetic acid (1.0 mg^{-1}) and the cytokinin benzyladenine (0.1 mg^{-1}). Note callus and some root formation. (Photograph: J. Atkins and A. J. McComb)

B. Regeneration of Plants

Regeneration of intact plants might occur following simultaneous root and shoot formation, or formation of one structure followed by the other. If clonal plants are required, regeneration should follow soon after callus formation, as prolonged culture leads to chromosomal abnormalities (e.g. Torrey, 1967). Root initiation by pieces of stem or leaf in culture follows high auxin treatment, but the roots in these instances are almost certainly derived from the excised parental tissue. However, Torrey and Shigemura (1957) and Torrey (1967) have recorded root formation on pea callus, some of which had been cultured for long periods; shoot formation did not follow. Hilderbrandt et al. (1963) noted formation of leaves and shoots, as well as callus, on pieces of pea stem placed on a medium containing coconut milk, but no roots were reported.

The most hopeful report is that of Gamborg et al. (1974), who found that minced apical tissue, when spread on to a suitable medium, will form many calluses; some of these produce shoots, and (somewhat unpredictably) roots may also form, so that intact plants can be obtained. It seems likely that the shoots are formed de novo, and not as a result of preservation of some of the original geometry of the apex. Although this approach is promising, yields of intact shoots are at present low.

C. Isolation of Single Cells and Protoplasts

Large numbers of calluses could be derived from a single genetic source if pea mesophyll cells or protoplasts could be cultured, and the use of protoplasts would open the possibilities of fusion and the uptake of macromolecules. Protoplasts have been isolated from pea leaves (Constabel et al., 1973) and roots (Landgren and Torrey, 1973) by means of suitable cellulase enzymes; some 90% of leaf cells can be converted to living protoplasts (Constabel et al., 1973). The protoplasts have intact membranes, show protoplasmic movements and regenerate cell walls. Before wall formation the protoplasts may fuse together, especially after treatment with polyethylene glycol, and Vicia–Pisum heterokaryons have been formed (Kao and Michayluk, 1974). Bacteria will pass into the protoplasts (Davey and Cocking, 1972). The formation of callus from protoplasts has been reported, but only in yields of about 0·5% of the isolated protoplasts (Constabel et al., 1973); much higher yields have been obtained with certain other species. Intact pea plants have not yet been regenerated from isolated protoplasts, but it is presumably just a matter of time before this is accomplished.

In conclusion, it is certainly possible to destroy the geometry of a piece of pea plant, and grow disorganized tissue, using known growth regulators as stimulants. Reliable plantlet formation will probably be achieved from secondary callus, but studies on the pea plant in this area of research have not yet extended the conclusions which can be drawn from experiments with more accommodating species.

VI. CONTROL OF GENETIC MATERIAL IN THE NUCLEUS

The effects and interactions of the growth-regulating compounds are complex and we have pointed out in passing how they may set in train fundamental changes in the metabolism of plant tissue, changes which are presumably directly or indirectly, related to effects on plant morphology. As morphology is the final expression of genetic information, we may well ask whether changes in morphology are accompanied by changes at the level of the genetic material in the nucleus. Further, might the primary action of the growth-regulating compounds be to control gene expression in a very real biochemical sense, at the level of transcription of genetic information? These are simple questions and a clear affirmative answer to each would have considerable appeal. But answering the questions presents many difficulties, and the pea plant is playing a part which has to be viewed in relation to the literature on control of gene action in other eukaryotes. Here we want only to touch upon work of direct relevance to the pea and to the substances we have talked about. We are also consciously bypassing all that is known about changes in biochemical machinery during development, including nucleic acids and proteins (some of which are dealt with in other chapters, for example, Chapters 3, 6 and 15, in this book), ignoring fine structural studies, and concentrating upon the properties of chromatin. In so doing we feel obliged to offer the warning that there are many points at which controls may operate in the sequence between genetic information and morphological expression, including feed-back controls, pool sizes of substrates, enzyme activation and inhibition, and so on.

Chromatin can be extracted from the pea plant by simple methods maceration, coarse filtration and gentle centrifugation, followed by recovery of the pellet and its sedimentation through a sucrose gradient (Huang and Bonner, 1962). The resulting gelatinous pellet of chromatin contains DNA protein and RNA in the ratio (by mass) $1:1:0.25$. [These results and much of the following data are summarized in Bonner et al. (1968).] It also contains a small amount of the enzyme RNA polymerase, and in an appropriate incubation medium this will synthesize RNA in the test tube, using DNA of the chromatin as a template. This extracted enzyme can be supplemented with RNA polymerase from another source, such as a micro-organism, and the yield of RNA from the chromatin can thus be increased. Immediately we see that two important properties can be assayed and expressed per unit of DNA: the level of endogenous RNA polymerase activity associated with the chromatin, and the rate at which RNA can be synthesized at saturating levels of

RNA polymerase. Pure DNA can be extracted from the chromatin, and the level of RNA synthesis achieved using this pure DNA can be compared with that for chromatin; chromatin supports far less RNA synthesis than does the pure DNA. This is due to a restriction in the amount of DNA "template" transcribed by RNA polymerase, as shown, for example, by studies in which the amount of hybrid which forms between purified DNA and the RNA transcribed from it is compared with that which forms between purified DNA and RNA transcribed from chromatin. In the pea bud, the template available in chromatin represents some 6% of the whole DNA, in the embryonic axis 12%, and in the growing cotyledon 32%; the amount of DNA template available therefore differs in different tissues.

Evidence that the "genetic repression" observed in these investigations is similar to that *in vivo* comes from DNA/RNA hybridization experiments, and from the observation that, in an appropriate incubation medium, chromatin from pea cotyledons will suppport the synthesis of seed globulins, while chromatin from the buds will not.

It is also possible to demonstrate that much of the restriction of template activity is due to a "blocking" action by the major protein fraction of the chromatin, the histones, which are proteins characterized by a relatively negative charge. When they are removed selectively, the level of template available for transcription rises dramatically towards that of pure DNA. There are some six different histone fractions, which can be separated by gel filtration and electrophoresis, and that they have a fundamental role to play in the organization of the chromosomal material is emphasized by the fact that very similar fractions have been isolated from a number of eukaryotes. They turn out to be, in an evolutionary sense, the most conservative proteins known; for example, histone fraction 4 from the pea plant differs by only two amino acid residues in 102 from the histone fraction 4 of calf thymus (De Lange *et al.*, 1969).

Clearly, the six histone fractions could not in themselves offer the sophisticated control which is required to modulate gene expression, and it is to the two non-histone proteins, or to the RNA of the chromatin ("chromosomal RNA"; Holnes *et al.*, 1972), that we might look for compounds of sufficient diversity to either select the sites at which histones are deposited (and thus become repressed) or, alternatively, which select the sites which are available for transcription (that is, are specifically activated: e.g. Britten and Davidson, 1969). Details of mechanisms for gene regulation in eukaryotes are fields of considerable experimentation and speculation at the present time, and other factors—for example, the structure of the RNA polymerase itself—could prove to be important in selecting sites for transcription.

We can now return to the possible role of growth-regulating compounds, and inquire about their ability to affect the measurable properties of chromatin. Although the RNA content of pea tissue is enhanced by gibberellic acid, the compound does not enhance RNA synthesis when it is added to isolated pea chromatin, nor will it enhance RNA production by isolated pea nuclei. However, if the nuclei are isolated from gibberellin-treated tissue, they will form more RNA than do nuclei from controls (Johri and Varner, 1968), and chromatin isolated from gibberellin-treated pea plants shows an enhanced level of RNA polymerase activity (McComb et al., 1970). Clearly, something mediates between the hormone and this effect on the nuclear material. In this regard a search for gibberellin-binding proteins in the pea plant, which could represent sites of gibberellin action, is of considerable interest (Musgrave et al., 1969; Stoddart et al., 1974), as is the occurrence in pea plants of adenosine 3′,4′-cyclic monophosphate (Kessler and Levinstein, 1974), a compound known to mediate in certain hormone responses of animal tissues. However, the effect of the hormone on RNA synthesis is not accompanied in any obvious way by the exposure of parts of the genome, which were previously repressed. There is no increase in template as detected by direct assay (McComb et al., 1970), such as is observed when dormancy is broken in other plants by gibberellin treatment. Gibberellin does not change the distribution of histones from pea chromatin (Spiker and Chalkley, 1971), and hybridization experiments fail to disclose a difference in the RNA present after gibberellin treatment as compared with that synthesized before treatment (Thompson and Cleland, 1972). Thus, although gibberellin might yet turn out to have an effect on a small part of the genome which is too subtle to be detected by these rather blunt techniques, the prospects for further breakthrough are not encouraging. Surely in view of these results one could not attribute the bulk of the increase in RNA and protein (with its altered balance of enzymes) to the expression of new, previously repressed genetic information?

While changes in RNA synthesis undoubtedly accompany and are important for growth, another reason for doubting that formation of new species of RNA is the key to the primary action of gibberellin comes from investigations on the rapidity with which the growth regulator exerts its effect in the pea plant: changes in growth rate can be detected within 15 min of treatment, and one wonders if this is sufficient time to allow the synthesis of new RNA and the formation and expression of new enzymes (McComb and Broughton, 1972).

For auxins and cytokinins, a somewhat similar picture emerges. Addition of these compounds to isolated pea chromatin does not result in increased RNA synthesis, but chromatin isolated from treated tissue does form more RNA in vitro. Addition of the growth factors along with a protein fraction from the tissue also gives enhanced RNA synthesis, and there is some evidence that in the case of auxin the protein is organ specific (Matthyse, 1970; Matthyse and Abrams, 1970). Again, however, the stimulation of growth of pea tissue can be so rapid—Muryama and Ueda (1973) claim an effect within 1 min of auxin treatment at elevated temperatures—that it is difficult to see how an effect on the nuclear material is likely to be the cause of the initial effect of auxin on growth (Evans, 1974), important no doubt as it must be in relation to a sustained growth response.

In summary, we can see that techniques are becoming available for examining the properties of nuclear material during ontogeny, and that there is now quite direct evidence for differences in the amount of DNA template available in different tissues. Preliminary work suggests that while changes can be observed in the ability of chromatin to support RNA synthesis when the chromatin is isolated from plant material treated with growth-regulating chemicals, the effects are mediated via other compounds in the plant. The effects are no doubt important in maintaining growth responses, but at present one may feel sceptical as to whether they represent the primary action of these chemicals in initiating short-term responses.

VII. CONCLUDING REMARKS

We have seen how the genetics of developmental patterns suggest quite simple biochemical controls, which may operate, along with light, through the effects of known and unknown growth-regulating chemicals. These may be "stimulatory" or "inhibitory", and the balance between them is clearly important. They may have relatively simple effects on growth (such as controlling rate of tissue elongation) or trigger off new patterns of development (as in the formation of callus, and the initiation of roots and shoots). It is sometimes hard to decide if a compound, which may affect growth and even be necessary for the growth of isolated organs, really has a regulatory role in the intact plant.

These chemicals may have complex effects, some of which apparently do not involve the nucleus, but others which may, via intermediate compounds, affect RNA transcription. In no single case can we trace the sequence from gene, to growth-regulating chemical, to expression of an effect, but we are probably closest to doing so in the case of the gibberellins.

The interesting fields of research touched on here are all advancing and are, to some extent, interdependent; but anything approaching a fully meaningful synthesis depends upon further information about the mechanisms modulating gene expression in the nucleus during development. There is no doubt that *Pisum* will continue to prove an invaluable experimental organism in all of these areas.

REFERENCES

AHARONI, Y., PHAN, C. T. and SPENCER, M. (1973). *Can. J. Bot.* **51**, 2243–2246.
ANDREAE, W. A., VENIS, M. A., JURSIC, F. and DUMAS, T. (1968). *Pl. Physiol.* **43**, 1375–1379.

APELBAUM, A. and BURG, S. P. (1972). *Pl. Physiol.* **50**, 117–124.

ARBER, A. (1950). "The Natural Philosophy of Plant Form". Cambridge University Press, London.

AUDUS, L. J. (1959). "Plant Growth Substances", 2nd edition. Leonard Hill London.

AUDUS, L. J. (1969). *In* "Physiology of Plant Growth and Development" (M. B Wilkins, ed.), pp 205–242. McGraw-Hill, London.

AUDUS, L. J. (1972). "Plant Growth Substances", 3rd edition. Vol. I. Chemistry and Physiology. Leonard Hill, London.

AUDUS, L. J. and GUNNING, B. E. S. (1958). *Physiologia Pl.* **11**, 685–687.

BABCOCK, D. F. and MORRIS, R. O. (1970). *Biochemistry, N.Y.* **9**, 3701–3705.

BALDEV, B., LANG, A. and AGATEP, A. O. (1965). *Science, N.Y.* **147**, 155–157.

BENNET-CLARK, T. A. and KEFFORD, N. P. (1953). *Nature, Lond.* **171**, 645–647.

BLIXT, S., EHRENBERG, L. and GELIN, D. (1958). *Agri Hort. Genet.* **16**, 238–250.

BONNER, J., DAHMAS, M. E., FAMBROUGH, D., HUANG, R., MARUSHIGE, K. and TUAN, D. Y. H. (1968). *Science, N.Y.* **159**, 47–56.

BRIAN, P. W. (1957). *Symp. Soc. exp. Biol.* **11**, 166–181.

BRIAN, P. W. and HEMMING, H. G. (1955). *Physiologia Pl.* **8**, 669–681.

BRIAN, P. W. and HEMMING, H. G. (1958). *Ann. Bot.* **22**, 1–17.

BRITTEN, R. J. and DAVIDSON, E. H. (1969). *Science, N.Y.* **165**, 349–357.

BURDEN, R. S., FIRN, R. D., HIRON, R. W. P., TAYLOR, H. F. and WRIGHT, S. T. C. (1971). *Nature, New Biology,* **234**, 95–96.

BURG, S. P. and BURG, E. A. (1968). *Pl. Physiol.* **43**, 1069–1074.

BURGESS, J. and FLEMING, E. N. (1973). *Protoplasma* **76**, 315–325.

CARR, D. J. and BURROWS, W. J. (1966). *Life Sci.* **5**, 2061–2077.

CARR, D. J., REID, D. M. and SKENE, K. G. M. (1964). *Planta* **63**, 382–392.

CARTWRIGHT, P. M., SYKES, J. T. and WAIN, R. L. (1956). *In* "Chemistry and Mode of Action of Plant Growth Substances" (R. L. Wain and F. Wightman, eds), pp 32–39. Butterworths Scientific Publications, London.

CHADWICK, A. V. and BURG, S. P. (1967). *Pl. Physiol.* **42**, 415–420.

CHADWICK, A. V. and BURG, S. P. (1970). *Pl. Physiol.* **45**, 192–200.

COCKING, E. C. (1972). *A. Rev. Pl. Physiol.* **23**, 29–50.

CONSTABEL, F., KIRKPATRICK, J. W. and GAMBORG, O. L. (1973). *Can. J. Bot.* **51**, 2105–2107.

DAVEY, M. R. and COCKING, E. C. (1972). *Nature, Lond.* **239**, 455–456.

DE HAAN, H. (1927). *Genetica* **9**, 481–498.

DE HAAN, I. and GORTER, C. J. (1936). *Recl. Trav. bot. néerl.* **33**, 434–446.

DE LANGE, R. J., FAMBROUGH, D. M., SMITH, E. L. and BONNER, J. (1969). *J. biol. Chem.* **244**, 5669–5679.

DÖRFFLING, K. (1973). *Z. Pflanzenphysiol.* **70**, 131–137.

EKELUND, R. and HEMBERG, T. (1966). *Physiologia Pl.* **19**, 1120–1124.

ERIKSEN, E. N. (1973). *Physiologia Pl.* **28**, 503–506.

ERIKSEN, E. N. (1974). *Physiologia Pl.* **30**, 163–167.

ERIKSEN, E. N. and MOHAMMED, S. (1974). *Physiologia Pl.* **30**, 158–162.

EVANS, M. L. (1974). *A. Rev. Pl. Physiol.* **25**, 195–223.

FAN, D. F. and MACLACHLAN, G. A. (1966). *Can. J. Bot.* **44**, 1025–1034.

FAWCETT, C. H., TAYLOR, H. F., WAIN, R. L. and WIGHTMAN, F. (1956). *In* "The Chemistry and Mode of Action of Plant Growth Substances" (R. L. Wain and F. Wightman, eds), pp 187–194. Butterworths Scientific Publications, London.

FRYDMAN, V. M. and MACMILLAN, J. (1973). *Planta* **115**, 11–15.

FRYDMAN, V. M., GASKIN, P. and MacMILLAN, J. (1974). *Planta* **118**, 123–132.
FUCHS, Y. and LIEBERMAN, M. (1968). *Pl. Physiol.* **43**, 2029–2036.
GALSTON, A. W. (1959). *In* "Photoperiodism and Related Phenomena in Plants and Animals" (R. B. Withrow, ed.), pp 137–157. American Association for the Advancement of Science, Washington, D.C.
GALSTON, A. W. and DALBERG, L. Y. (1954). *Am. J. Bot.* **41**, 373–380.
GAMBORG, O. L., MILLER, R. A. and OJIMA, K. (1968). *Expl Cell Res.* **50**, 151–158.
GAMBORG, O. L., CONSTABEL, F. and SHYLUCK, J. P. (1974). *Physiologia Pl.* **30**, 125–128.
GOESCHL, J. D., PRATT, H. K. and BONNER, B. A. (1967). *Pl. Physiol.* **42**, 1077–1080.
GOESCHL, J. D., RAPPAPORT, L. and PRATT, H. K. (1966). *Pl. Physiol.* **41**, 877–884.
GORTER, C. J. (1961). *Physiologia Pl.* **14**, 332–343.
HAHN, H., DE ZACHS, R. and KENDE, H. (1974). *Naturwissenschaften* **61**, 170.
HALL, R. H., CSONKA, L., DAVID, H. and McLENNAN, B. (1967). *Science, N.Y.* **156**, 69–71.
HASHIMOTO, T. (1961). *Bot. Mag., Tokyo* **74**, 110–117.
HILDERBRANDT, A. C., WILMAR, J. C., JOHNS, H. and RIKEN, A. J. (1963). *Am. J. Bot.* **50**, 248–252.
HILLMAN, S. K. and PHILLIPS, I. D. J. (1970). *J. exp. Biol.* **21**, 959–967.
HOLNES, D. S., MAYFIELD, J. E., SANDERS, G. and BONNER, J. (1972). *Science, N.Y.* **177**, 72–74.
HUANG, R. and BONNER, J. (1962). *Proc. natn. Acad. Sci. U.S.A.* **48**, 1216–1222.
HUSAIN, S. M. and LINCK, A. J. (1967). *Physiologia Pl.* **20**, 48–56.
ISOGAI, K., OKAMOTO, T. and KOMODA, Y. (1967). *Chem. pharm. Bull, Tokyo* **15**, 1256–1257.
JEAN, F. C. (1928). *Bot. Gaz.* **86**, 318–329.
JOHRI, M. M. and VARNER, J. E. (1968). *Proc. natn. Acad. Sci. U.S.A.* **59**, 269–276.
JONES, R. L. and LANG, A. (1968). *Pl. Physiol.* **43**, 629–634.
KAMINEK, M. (1966). *Biologia Pl.* **9**, 86–91.
KATSUMI, M. (1962). *Physiologia Pl.* **15**, 115–121.
KATSUMI, M. (1963). *Physiologia Pl.* **16**, 66–72.
KAO, K. N. and MICHAYLUK, M. R. (1974). *Planta* **115**, 355–367.
KENDE, H. and KAYS, S. E. (1971). *Naturwissenschaften* **58**, 524–525.
KENDE, H. and LANG, A. (1964). *Pl. Physiol.* **39**, 435–440.
KESSLER, B. and LEVINSTEIN, R. (1974). *Biochim. biophys. Acta* **343**, 156–166.
KNIGHT, L. I. and CROCKER, W. (1913). *Bot. Gaz.* **55**, 337–371.
KÖHLER, D. (1965a). *Planta* **65**, 218–224.
KÖHLER, D. (1965b). *Planta* **67**, 44–54.
KÖHLER, D. (1966). *Planta* **69**, 27–33.
KOMOTO, N., IKEGAMI, S. and TAMURA, S. (1972). *Agric biol. Chem.* **36**, 2547–2553.
KONINGS, H. (1965). *Acta. bot. neerl.* **18**, 528–537.
LAMPRECHT, H. (1962). *Agri hort. Genet.* **20**, 23–62.
LANDGREN, C. R. and TORREY, J. G. (1973). *In* "Protoplastes et Fusion de Cellules Somatiques Végétales", pp 207–213. Coll. Int. CNRS No. 212, Paris.
LANTICAN, B. P. and MUIR, R. M. (1969). *Physiologia Pl.* **22**, 412–423.
LOCKARD, R. G. and GRUNWALD, C. (1970). *Pl. Physiol.* **45**, 160–162.
LOCKHART, J. A. (1956). *Proc. natn. Acad. Sci. U.S.A.* **42**, 841–848.
LOCKHART, J. A. and GOTTSCHALL, V. (1961). *Pl. Physiol.* **36**, 389–393.

LOVELL, P. H. (1971). *Physiologia Pl.,* **25**, 382–385.

MARUMO, S. and HATTORI, H. (1970). *Planta* **90**, 208–211.

MARUMO, S., ABE, H., HATTORI, H. and MUNAKATA, K. (1968a). *Agric biol. Chem.* **32**, 117–118.

MARUMO, S., ABE, H. and MUNAKATA, K. (1968b). *Nature, Lond.* **219**, 959–960.

MARX, G. A. and HAGEDORN, D. J. (1962). *J. Hered.* **53**, 31–43.

MATTHYSE, A. G. (1970). *Biochim. biophys. Acta* **199**, 519–521.

MATTHYSE, A. G. and ABRAMS, M. (1970). *Biochim. biophys. Acta* **199**, 511–518.

MCCOMB, A. J. (1966). *Ann. Bot.* **30**, 155–163.

MCCOMB, A. J. and BROUGHTON, W. J. (1972). *In* "Plant Growth Substances, 1970" (D. J. Carr, ed.), pp 407–413. Springer Verlag, Berlin.

MCCOMB, A. J. and CARR, D. J. (1958). *Nature, Lond.* **181**, 1548.

MCCOMB, A. J. and MCCOMB, J. A. (1970). *Planta* **91**, 235–245.

MCCOMB, A. J., MCCOMB, J. A. and DUDA, C. T. (1970). *Pl. Physiol.* **46**, 221–223.

MCCOMB, J. A. (1974). *J. Aust. Inst. agric. Sci.* **40**, 3–10.

MCCUNE, D. C. and GALSTON, A. W. (1959). *Pl. Physiol.* **34**, 416–418.

MCDAVID, C. R., SAGAR, G. R. and MARSHALL, C. (1972). *New Phytol.* **71**, 1027–1032.

MCDAVID, C. R., SAGAR, G. R. and MARSHALL, C. (1973). *New Phytol.* **72**, 465–470.

MCINTYRE, G. I. (1971). *Nature, New Biology* **230**, 87–88.

MELCHERS, G. and LABIL, G. (1973). *In* "Protoplastes et Fusion de Cellules Somatiques Végétales", pp 367–372. Coll. int. C.N.R.S. No. 212, Paris.

MILBORROW, B. V. (1974). *A. Rev. Pl. Physiol.* **25**, 259–307.

MOORE, T. C. (1967). *Am. J. Bot.* **54**, 262–269.

MORRIS, D. A., BRIANT, R. E. and THOMPSON, P. G. (1969). *Planta* **89**, 178–197.

MURYAMA, K. and UEDA, K. (1973). *Pl. Cell Physiol.* **14**, 937–979.

MUSGRAVE, A., KAYS, S. E. and KENDE, H. (1969). *Planta* **89**, 165–177.

NAKAMURA, E. (1964). *Pl. Cell Physiol.* **5**, 521–524.

NELSON, O. E. Jr. and BURR, B. (1973). *A. Rev. Pl. Physiol.* **24**, 493–518.

NICKELL, L. G. (1958). *Science, N.Y.* **28**, 88.

OCKERSE, R. (1970). *Bot. Gaz.* **131**, 95–97.

OCKERSE, R. and GALSTON, A. W. (1967). *Pl. Physiol.* **42**, 47–54.

OCKERSE, R. and WEBER, J. (1970). *Pl. Physiol.* **56**, 821–824.

PATE, J. S. (1958). *Aust. J. biol. Sci.* **11**, 516–528.

PECKET, R. R. (1960). *Nature, Lond.* **185**, 114–115.

PHILLIPS, D. A. (1974). *Planta* **100**, 181–190.

PHILLIPS, I. D. J. (1969). *In* "Physiology of Plant Growth and Development" (M. B. Wilkins, ed.), pp 163–202. McGraw-Hill, London.

PHILLIPS, I. D. J., VLITOS, A. J. and CUTLER, H. (1959). *Contr. Boyce Thompson Inst. Pl. Res.* **20**, 111–120.

PURVES, W. K. and HILLMAN, W. S. (1958). *Physiologia Pl.* **11**, 29–35.

RAJAGOPAL, R. (1967). *Physiologia Pl.* **20**, 655–660.

RAJAGOPAL, R. (1968). *Physiologia Pl.* **21**, 378–385.

RADLEY, M. (1959). *Nature, Lond.* **178**, 1070–1071.

RAILTON, I. D. and REID, D. M. (1974). *Pl. Sci. Lett.* **2**, 157–163.

RASMUSSON, J. (1927). *Hereditas* **10**, 1–152.

REID, D. M. and CARR, D. J. (1967). *Planta* **73**, 1–11.

RIDGE, I. and OSBORNE, D. (1969). *Nature, Lond.* **223**, 318–319.

RIDGE, I. and OSBORNE, D. (1970). *J. exp. Bot.* **21**, 843–856.

SACHS, T. and THIMANN, K. V. (1967). *Am. J. Bot.* **54**, 136–141.

SARGENT, J. A., ATACK, A. V. and OSBORNE, D. J. (1973). *Planta* **109**, 185–192.
SAUNDERS, J. W. and BINGHAM, E. T. (1972). *Crop Sci.* **12**, 804–808.
SCHEIBE, A. and WOHRMANN-HILLMAN, B. (1957). *Z. Bot.* **45**, 97–121.
SCHNEIDER, E. A. and WIGHTMAN, F. (1974). *A. Rev. Pl. Physiol.* **25**, 487–513.
SCHOLDEEN, L. and BURSTRÖM, H. (1960). *Physiologia Pl.* **13**, 831–837.
SCOTT, P. C. and NORRIS, L. A. (1970). *Nature, Lond.* **227**, 1366–1367.
SCOTT, T. K. and PRITCHARD, J. B. (1968). *In* "Transport of Plant Hormones" (Y. Vardar, ed.), pp 309–319. North-Holland, Amsterdam.
SCHENK, R. V. and HILDERBRANDT, A. C. (1972). *Can. J. Bot.* **50**, 199–204.
SHININGER, T. L. (1972). *Pl. Physiol.* **99**, 341–344.
SHORT, K. C. and TORREY, J. G. (1972a). *J. exp. Bot.* **23**, 1099–1105.
SHORT, K. C. and TORREY, J. G. (1972b). *Pl. Physiol.* **49**, 155–160.
SIMPSON, G. M. and SAUNDERS, P. F. (1972). *Planta* **102**, 272–276.
SOROKIN, H. P. and THIMANN, K. V. (1964). *Protoplasma* **59**, 326–350.
SPIKER, S. and CHALKLEY, R. (1971). *Pl. Physiol.* **47**, 342–345.
SPRENT, J. I. (1968). *Planta* **81**, 80–87.
STEWART, R. N., LIEBERMAN, M. and KUNISHI, A. T. (1974). *Pl. Physiol.* **54**, 1–5.
STODDART, J., BREIDENBACK, W., NADEAN, R. and RAPPAPORT, L. (1974). *Proc. natn. Acad. Sci. U.S.A.* **71**, 3255–3259.
STOWE, B. B. (1960). *Pl. Physiol.* **35**, 262–269.
TANIMOTO, E., YANAGISHIMA, N. and MASUDA, Y. (1967). *Physiologia Pl.* **20**, 291–298.
THIMANN, K. V. and LALORAYA, M. M. (1960). *Physiologia Pl.* **13**, 165–178.
THIMANN, K. V. and SACHS, T. (1966). *Am. J. Bot.* **53**, 731–739.
THIMANN, K. V. and SKOOG, F. (1933). *Proc. natn. Acad. Sci. U.S.A.* **19**, 714–716.
THOMPSON, W. F. and CLELAND, R. (1972). *Pl. Physiol.* **50**, 289–292.
TIETZ, A. (1974). *Biochem. Physiol. Pfl.* **165**, 387–392.
TORREY, J. G. (1961). *Expl Cell Res.* **23**, 281–299.
TORREY, J. G. (1967). *Physiologia Pl.* **20**, 265–275.
TORREY, J. G. and SHIGEMURA, Y. (1957). *Am. J. Bot.* **44**, 334–344.
VAN OBERBEEK, J. (1938). *Bot. Gaz.* **100**, 133–166.
VAN OBERBEEK, J. (1939). *Bot. Gaz.* **101**, 450–456.
VELIKY, I. A. and MARTIN, S. M. (1970). *Can. J. Microbiol.* **6**, 223–226.
VREMEN, H. J., SKOOG, F., FRIHART, C. R. and LEONARD, N. J. (1972). *Pl. Physiol.* **49**, 848–851.
WARDLAW, I. F. and MORTIMER, D. C. (1970). *Can. J. Bot.* **48**, 229–237.
WELLENSIEK, S. J. (1973). *Scient. Hort.* **1**, 77–84.
WENT, F. W. (1943). *Bot. Gaz.* **104**, 460–474.
WICHNER, S. (1968). *Physiologia Pl.* **21**, 1356–1362.
WICHNER, S. and LIBBERT, E. (1968). *Physiologia Pl.* **21**, 227–241.
WICKSON, M. and THIMANN, K. V. (1958). *Physiologia Pl.* **11**, 62–74.
WILLS, R. B. H. and PATTERSON, B. D. (1970). *Nature, Lond.* **225**, 119.
YARNELL, S. H. (1962). *Bot. Rev.* **28**, 465–537.
ZWAR, J. A. and SKOOG, F. (1963). *Aust. J. biol. Sci.* **16**, 129–139.

10. Correlative Influences in Seedling Growth

P. H. LOVELL

Department of Botany, University of Auckland, New Zealand

I. GENERAL SURVEY

Germination and the ensuing growth of the pea seedling are highly organized processes which are under tight control. No one part of the seedling develops independently and at all stages each part relies upon and is affected by other parts. These interrelations may be called "correlative influences".

A. Influence of the Cotyledons

The garden pea (*Pisum sativum* L.) is a hypogeal species, that is the cotyledons stay below ground unless they are accidentally uncovered. They are very inefficient photosynthetic organs (Lovell and Moore, 1970), their major function being to act as a store of reserves to support the axis during early seedling growth. This function is usually almost completed after about a fortnight. For example, in cv. 'Alaska' the cotyledons lost 88% of their initial dry weight after 15 days' germination and cotyledon excision on or after day

13 had no effect on the rate of seedling growth (Killeen and Larson, 1968). Cotyledon removal earlier than day 13 caused a reduced rate of axis growth. The presence of the axis is necessary, especially in the first 2–3 days of germination, for the normal development of enzyme systems in the pea cotyledon (Varner et al., 1963; Guardiola and Sutcliffe, 1971a, b) and for the proper development of cotyledon ultrastructure (Bain and Mercer, 1966a,b). There is thus a marked degree of interdependence between cotyledons and axis (see Chapter 3 for a detailed treatment of the subject).

Apart from modifying growth rates, cotyledon excision can also change the timing of flowering. Moore (1964), Sprent (1966) and Johnston and Crowden (1967) found that cotyledon removal reduced the node at which the flowers first formed in late flowering pea varieties. However, in the early flowering cultivars 'Massey' and 'Alaska', the number of nodes produced before flowering is increased by 1–2 nodes when the cotyledons are excised at an early stage (Moore, 1964). Other correlative effects of pea cotyledons are the production of an inhibitor of root nodulation (Phillips, 1971) and an inhibitor of cotyledonary bud growth (Dostál, 1908).

B. Shoot–Root Interrelationships

During seedling development a wide range of growth correlations may be observed. For example, during seedling emergence the plumule may encounter resistance from the soil, particularly if the soil is compressed. When elongation of the epicotyl (cv. 'Alaska') is restricted there is an increase in ethylene production giving rise to reduced epicotyl length and an increased internode diameter (Goeschl et al., 1966). This in turn may increase the capacity of the seedling to overcome the resistance of the soil during the pre-emergence phase.

The interdependence of shoot and root during seedling development has been studied in cv. 'Kelvedon Wonder' by McDavid et al. (1972, 1973a,b). Cotyledon removal at day 15 and root pruning at day 15 and every subsequent 4 days resulted in accelerated senescence of mature leaves. This could be overcome to a large extent by dipping the shoot system in a 20 mg l^{-1} solution of 6-benzylaminopurine (BAP). McDavid et al. (1973b) suggest that cytokinins from the root system are important in the maintenance of photosynthetic systems and the delaying of leaf senescence. Supporting evidence for this comes from the work of Short and Torrey (1972) who have shown that cytokinins are present in pea roots. McDavid and co-workers also found that shoot removal after 2 days' germination substantially reduced lateral root production but that a 1% indol-3yl acetic acid (IAA) lanolin paste applied to the cut stump compensated to some extent for the absence of the shoot system. Ringing the base of the epicotyl of intact seedlings with 2, 3, 5, tri-iodobenzoic acid (TIBA), an inhibitor of auxin transport, also considerably reduced lateral root development. They concluded that auxin from the shoot system is important in initiation and growth of lateral roots. Peckett (1957) also suggested that lateral root initiation in peas is controlled by factors moving from or through the older root tissue to the areas of lateral root primordia initiation. However, Torrey (1952) in cv. 'Alaska' and Šebánek and Kopecký (1967) in cv. 'Raman' found that root growth was unaffected by removal of the epicotyl. In cv. 'Alaska' however, excision of the cotyledons completely prevented lateral root formation.

Obviously major surgery of this sort can give rise to big differences in growth pattern from those present in the intact seedling. For example, cuttings taken from pea plants by excising the epicotyls will readily form roots (Eriksen, 1973) and isolated shoot apical meristems will give rise to complete plants given appropriate conditions of culture (Kartha et al., 1974). Thus, if the integrity of the seedling is not maintained major changes in development may ensue. This is in contrast to the well ordered growth of the intact system where, for example, Klasová et al. (1971, 1972) showed that in cv. 'Pyram' the formation of lateral root primordia takes place at $54 \cdot 2 \pm 2 \cdot 9$ h after the start of imbibition and that by $73 \cdot 5 \pm 5 \cdot 5$ h laterals were apparent on the surfaces of the primary root. These times were not affected even when the growth of the primary root was retarded.

C. Source–Sink Relationships

After emergence of the epicotyl the importance of the cotyledon reserves declines and the leaves become the main suppliers of carbohydrate. The rate of export of photosynthate from the leaves is determined by the activity of shoot and root apices, the expanding leaves and by the number of mature exporting leaves on the plant. Removal of mature leaves tends to increase the rate of export from the remaining leaves, whereas removal of major sinks tends to reduce such export (Lovell et al., 1972). In general, the lower leaves export to the root system and the upper leaves supply the main stem apex and the expanding leaves, although in a vegetative plant the root system is the biggest single sink. The general topic of changing translocation patterns in the ageing plant is considered in further detail in Chapter 12.

The correlative influence which has excited by far the most attention is the effect of the main shoot apex on the lateral shoot buds (apical dominance). The remainder of this chapter will therefore be devoted to an analysis of apical dominance and the control mechanisms involved in this process in P. sativum.

II. CORRELATIVE INHIBITION: SEEDLING MORPHOLOGY

A. Growth of the Main Axis in Intact Plants

In most cultivars of pea a single unbranched shoot is produced during vegetative growth. The buds in the axils of the cotyledons and leaves do not grow
out but remain inhibited at least until quite late in the growth cycle. This
growth pattern is reinforced in plants grown in the dark or in low light conditions. The phenomenon by which the control of axillary bud growth is
determined by the main shoot apex is an example of "correlative inhibition".

Severe damage to the main shoot apex, which could occur in nature during
emergence or by trampling or other physical factors, releases the lateral buds
from inhibition. Shoot growth is then taken over by one or more of the
lateral buds. Although several lateral buds may commence growth, usually
only one continues to grow and the other buds are in turn correlatively inhibited by the new dominant shoot apex. The bud which becomes dominant is
determined to a large extent by the number of nodes present on the seedling
at the time of decapitation and by the amount of epicotyl removed (Husain
and Linck, 1966). Release of buds from dominance in the intact plant usually
occurs at about the time of flowering.

B. Axillary Bud Growth Following Decapitation

1. Decapitation Immediately Above the Cotyledons

If the epicotyl is removed from seedlings the only buds present are those in the
axils of the cotyledons. In each of these there is a main bud and a number of
subsidiary buds ("serials"; Šebánek, 1966a; Dostál, 1967). Epicotyl removal
after 48 h imbibition results in the outgrowth of the two main buds. These may
both grow out equally (Snow, 1931), as is generally the case with dwarf
varieties, or unequally, as in tall varieties (Dostál, 1968). Sachs (1966) also
regularly obtained two unequal cotyledonary shoots in a semi-dwarf (cv.
'Radio') after decapitation; the smaller shoot showing a little growth initially,
but subsequently becoming completely inhibited and dying. When the two
cotyledonary shoots grow out equally both survive and continue to grow but
when the initial growth is unequal the larger bud rapidly assumes dominance
and the smaller shoot is inhibited. Dostál (1968) found that in cv. 'Liblický
Bastard' the number of individuals in which the right cotyledonary bud
assumed dominance equalled the number in which the left bud was dominant
and that a correlation existed between the side of the pod the seed occupied
during development and the cotyledonary bud which subsequently assumed
dominance. In the tall cv. 'Alaska', however, Husain and Linck (1966) found
that the right lateral bud developed more quickly in most cases.

These two shoot systems produced by severing the internode just above the cotyledons have been used by a number of workers in studies of apical dominance and further reference to this system will be made later in the chapter.

2. Decapitation at Later Stages

Husain and Linck (1966), using tall cv. 'Alaska', decapitated seedlings at a series of ages (Table 10.I) and found that after decapitation below node 1 (in the internode just above the cotyledons) both cotyledonary buds grew out, but one rapidly assumed dominance. (The system of node numbering used in this chapter is given in Fig. 10.1.) Decapitation at internodes 3, 4, or 5 led to the axillary shoot which developed at node 2 assuming dominance, although

TABLE 10.I. The effect of the position of decapitation and plant age on lateral shoot outgrowth in cv. 'Alaska'. The buds were measured 21 days after the decapitation treatment. Modified from Husain and Linck (1966)

Site of Decapi- ation (internode)	Age at Decapi- tation (days)	coty- ledon node[a]	Length of Lateral Shoot (mm)						
			node 1	node 2	node 3	node 4	node 5	node 6	node 7
1	3	147	–	–	–	–	–	–	–
3	6	0	0	182	–	–	–	–	–
4	9	0	56	176	0	–	–	–	–
5	12	0	8	97	14	14	–	–	–
6	15	0	9	10	24	24	75	–	–
7	18	0	0	0	0	0	50	131	–
8	21	0	0	0	0	0	0	55	107

[a] Values for the cotyledonary node are for the longer of the two buds. Bud dead or completely inhibited (0); node removed as a result of treatment (–).

some growth of the lateral at node 1 also took place. In plants decapitated at internode 6 or at higher points the shoot emerging at the uppermost intact node dominated and the shoot immediately beneath it was the main competitor. In these experiments the decapitation treatments were carried out at different plant ages and therefore, for example, the bud at node 2 was much older in plants decapitated at internode 8 (21-day-old plants) than in 6-day-old plants decapitated at the third internode. There may therefore be an age as well as a positional effect determining which bud assumes dominance.

3. Time of Lateral Bud Outgrowth

When a lateral bud has been released from inhibition it commences growth very soon after the decapitation treatment and grows rapidly if it is to become the dominant shoot. Wardlaw and Mortimer (1970) measured the rate of growth of lateral buds by means of a microscope with a calibrated stage.

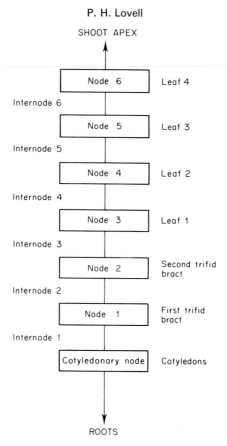

FIG. 10.1. Numbering of nodes and internodes in the pea plant

They found that the bud in the axil of the fourth expanded leaf (node 6) in cv. 'Alaska' started to grow within a few hours of decapitation of the main shoot immediately above leaf 4 (Fig. 10.2). There was a definite stimulation of growth after 8 h and the data suggest that the initiation of growth was probably as early as 4 h after decapitation. The lateral shoot continued to elongate and reached a length of 90 mm after 5 days. The corresponding bud on intact plants did not grow.

Both dwarf and tall peas respond in similar ways to decapitation treatments. Phillips (1969a) decapitated 14-day-old light-grown peas immediately below the fourth node and measured the lengths of the buds developing at nodes 1, 2 and 3 (Fig. 10.3). There was a rapid and continued increase in lateral shoot lengths in decapitated plants of both the tall ('Alaska') and the dwarf ('Meteor') cultivars. Buds in intact control plants grew to a very limited extent.

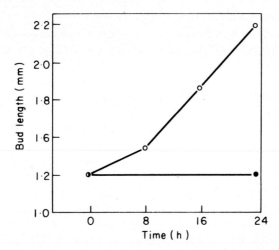

FIG. 10.2. The elongation of the lateral bud in the axil of leaf 4 (at node 6) in intact (●) and decapitated (○) light-grown plants. The plants were decapitated 2 cm above the sixth node. Figure redrawn from Wardlaw and Mortimer (1970)

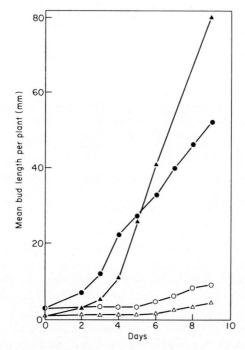

FIG. 10.3. The elongation of lateral buds after decapitation of the main shoot in light-grown plants. Values are totals for the buds developing at nodes 1, 2 and 3. Dwarf peas, cv. 'Meteor': ○, intact; ●, decapitated. Tall peas, cv. 'Alaska': △, intact; ▲, decapitated. Redrawn from Phillips (1969a)

4. Bud Anatomy and Vascular Connections

There are several buds in the axil of a cotyledon, trifid bract, or leaf although one bud is usually much further developed than the others at a node. For example, there are three buds in the axil of the trifid bract at node 2 in cv 'Alaska' and usually only the central and largest one undergoes development (Sorokin and Thimann, 1964), and four buds of different sizes in the axil of the lowest trifid bract in cv. 'Dun' (Sachs, 1968). There can be considerable variation in size of the inhibited buds and also in the amount of xylem development in the traces leading to them (Sorokin and Thimann, 1964). In the cv. 'Champion of England' studied by Gourley (1931) there was only a small trace linking the inhibited lateral bud to the vascular elements in the stele of the main axis. Similarly, Sorokin and Thimann (1964) suggest that the vascular connections to inhibited lateral buds are poor and consist of two traces for each bud, with a branch trace extending from the bud into the primary tissues of the stele. Whilst a bud remains inhibited the strands differentiating in the bud do not make contact with those developing acropetally in the stem. Connections develop only after the bud is released from inhibition, either by removal of the upper parts of the stem (Sachs, 1968) or by supplying kinetin to the bud (Sorokin and Thimann, 1964). Contact between the stem and bud strands takes place after 55–70 h followed by a rapid widening of the strands (Sorokin and Thimann, 1964). After the release from inhibition of the bud in the axil of the fourth expanded leaf in cv. 'Alaska', Wardlaw and Mortimer (1970) did not observe any increase in vascular connections between the bud and stem in the first 24 h but after 48 h additional xylem elements were evident and autoradiography showed a localization of photosynthate in the outer part of the trace at the site of phloem differentiation.

III. CORRELATIVE INHIBITION: FACTORS INVOLVED IN APICAL DOMINANCE

A. The Complexity of Apical Dominance

Having established that the main shoot apex plays a major role in the control of lateral bud growth the question that arises is "What are the factors involved in this control?" Reviews by Phillips (1969b) and Guern and Usciati (1972) and the paper by Tucker and Mansfield (1973) on apical dominance give much essential information in relation to this question for a range of species. It is clear that the control mechanisms involve complex interactions between endogenous hormones, nutrients and environmental factors. The situation is further complicated by the fact that all of the major groups of plant hormones have been implicated in apical dominance. The garden pea has been one of the most extensively used plants in apical dominance studies and many of the critical experiments have been carried out on this species. The reasons for this are probably that the seed is readily available, germination is rapid and relatively uniform, the plants are able to withstand a wide range of surgical treatments, and fairly absolute apical dominance is present during vegetative growth.

Apical dominance is an extremely complex process and the interactions between the controlling factors are critically important. Since there is great variation in the light and nutrient status, the method and place of hormone applications, the age and cultivar of the plant and the site of manipulative treatment from experiment to experiment major difficulties arise in interpreting and correlating all the relevant information.

B. The Inhibition of Lateral Bud Growth by Auxin

1. Evidence for Auxin Involvement

Almost all the available evidence suggests that IAA is either directly or indirectly involved in apical dominance. Auxin is produced by the apex and young leaves giving rise to a high auxin content in the terminal shoot bud (van Overbeek, 1938). Decapitation, which leads to the outgrowth of the axillary buds, is associated with a sudden decline in auxin moving down the stem. Other treatments which lead to a reduction in auxin movement down the main stem have similar effects. Thus, interruption of the basipetal movement of auxin in the stem by the application of TIBA just below the first node of 8-day-old seedlings (cv. 'Kelvedon Wonder') caused the release of the cotyledonary buds from apical dominance (McDavid et al., 1972).

If auxin is applied to the cut shoot immediately following decapitation, the axillary buds remain inhibited (Skoog and Thimann, 1934). However, the longer the delay between decapitation and auxin application, the less effect IAA has on lateral bud inhibition in both tall (cv. 'Alaska') (van Overbeek, 1938) and dwarf (cv. 'Meteor') peas (MacQuarrie, 1965). Not only may the time and site of application of the hormone be important but also the concentration and frequency of its application. Clearly, exogenous hormone treatments may give rise to uptake problems and also lead to changes in the pattern of endogenous hormone distribution. The application of high levels of IAA to plants can also cause ethylene to be evolved which in turn has been shown to inhibit lateral bud growth (Burg and Burg, 1968). Application of IAA dispersed in lanolin (MacQuarrie, 1965) or as an aqueous solution (Scott and Pritchard, 1968) have both been shown to be effective. In general, rather high levels of externally applied IAA (e.g. 100 mg l^{-1}, Scott and Pritchard, 1968) are necessary to give a substantial inhibitory effect on bud outgrowth following decapitation. In 35-mm-long stem segments with a single bud (cv. 'Alaska'), concentrations of 10 mg l^{-1} IAA were sufficient to inhibit completely lateral bud outgrowth (Wickson and Thimann, 1958). The effect of externally applied auxin is not one of poisoning since the buds are released from inhibition very rapidly after the source of auxin is removed.

Auxin action in apical dominance could be direct in that auxin from the shoot apex could move basipetally in the stem and acropetally into the lateral buds and in some way inhibit their growth. Alternatively, the auxin could induce the production of an inhibitor during its passage down the stem; this inhibitor could then inhibit bud growth. There has been heated discussion as to whether the role of auxin in apical dominance is a direct or an indirect one ever since the experiments carried out by Snow (1931, 1937). He showed that, in a two-shoot pea system, the inhibitory influence on bud growth must be able to move substantial distances in an acropetal direction (whereas auxin movement is predominantly basipetal) suggesting an indirect role of auxin. A third possibility is that the polar movement of auxin has an effect on the movement of other hormones and nutrients—diverting them from the lateral buds. These are not the only possibilities.

If inhibition of lateral buds is due solely to the relative difference in auxin level between the bud and the associated main stem region then application of auxin directly to the lateral bud should modify this balance and release the bud from inhibition. However, when Thimann (1937) applied auxin to small, fully inhibited lateral buds of peas the buds remained inhibited. Further work by Sachs and Thimann (1964, 1967) showed that auxin could stimulate elongation of a bud only when some growth had already taken place. Similarly it had no effect on the completely inhibited shoots on the two-shoot system used by Sachs (1966).

2. Movement of Auxin

There is some evidence that auxin movement from the shoot into a lateral bud can occur (van Overbeek, 1938; Wickson and Thimann, 1960) but it may be that it is not the absolute level of IAA in the axillary bud that is important in bud inhibition but rather the relative levels in bud and stem. Perhaps it is when the level in the bud is lower than the level in the stem that the bud remains inhibited.

Auxin applied to a mature foliage leaf moves in a fast, non-polar manner together with assimilates in the phloem, but when applied to the apical bud of an intact plant it moves in a polar manner at a rate similar to the rate of basipetal movement of endogenous auxin (Morris and Thomas, 1974).

3. Auxin Effects on Nutrient Movement

An auxin source, either the main shoot apex, or exogenous auxin applied to the surface of a shoot after decapitation, results in nutrient movement to the point of auxin application. This has been shown both for the movement of ^{14}C-labelled sucrose from leaves to the uppermost stem internode (Booth et al., 1962; Morris and Thomas, 1968) and for ^{32}P-orthophosphate applied o the roots accumulating in the upper stem below the point of auxin application (Nakamura, 1964; Davies and Wareing, 1965; Bowen and Wareing, 1971). When IAA is not applied after decapitation nutrients do not accumulate in the uppermost internode, for example ^{32}P applied to the roots moves nto the lower stem and the developing lateral shoots (Nakamura, 1964, Husain and Linck, 1966). The directional effects of auxin on nutrient movement are obviously important in main shoot/lateral bud relationships.

C. Less Complex Systems

Since whole plants and decapitated plants possessing a number of stem nodes are very complex, simpler systems have been used in attempts to clarify the mechanisms operating in apical dominance. The two most commonly used systems are stem sections consisting of a single node and the two shoot system obtained by decapitating recently germinated seedlings just above the cotyledons.

1. Stem segments

The use of stem segments possessing a single node has the obvious advantage of simplicity. Correlative effects of roots, buds on other nodes and of the main shoot apex are avoided and it is possible to look more directly and under more uniform conditions at the effect of different treatments on the lateral bud. Wickson and Thimann (1958) used 35-mm-long stem sections (cv. 'Alaska') and showed that culture in 10 mg l^{-1} IAA completely inhibited bud outgrowth but that kinetin applied directly to the lateral bud released it from the inhibition imposed by IAA. Kinetin promoted the growth of buds on two-node stem sections of cv. 'Lincoln' even when it was not applied directly to the buds (Kamínek, 1965). It was inferred that acropetal transport of kinetin had taken place through the stem sections and into the lateral buds.

The movement of ^{14}C-labelled IAA into the stem segment was predominantly basipetal but some acropetal movement did take place, and Wickson and Thimann (1960) showed that the level of radioactivity in the lateral bud was directly proportional to the degree of inhibition shown by the bud. These experiments reinforce the view that auxin plays a major role in the inhibition of lateral buds and that cytokinins can release buds from inhibition.

2. The "Two-Shoot" System

The two-shoot system has been used to clarify the factors involved in the control of the interrelationships of the two shoots and to elucidate the role of the cotyledon in growth inhibition of cotyledonary buds.

When the epicotyl is removed and two unequal cotyledonary shoots are obtained the larger (dominant) shoot inhibits the smaller (inhibited) shoot and the smaller shoot eventually dies (Sachs, 1966). If the dominant shoot is weakened, for example by decapitation, then the smaller shoot is released from inhibition and commences growth (Šebánek, 1966a). By feeding $^{14}CO_2$ to a fully expanded leaf on decapitated and intact cotyledonary shoots, Lovell (1969) showed that in intact dominant shoots virtually all the radio-carbon remained in the fed shoot but that in decapitated dominant shoots there was a substantial transfer of labelled photosynthate to the other cotyledonary shoot. Thus, treatments that reduce the competitive ability of one shoot reduce the levels of nutrients received by that shoot and can result in transfer of materials from it to the other shoot. However, the inhibited shoot can be released from inhibition by the application of kinetin (Sachs, 1966). The application of IAA to the inhibited shoot cannot relieve the inhibition if the shoot is totally inhibited at the time of application. However, if auxin is applied to a dominant shoot that has been weakened by decapitation, shading or by removal of the young leaves then the IAA can reimpose the dominance originally shown by the shoot (Snow, 1931, 1932, 1937).

The relationship between the cotyledon and the bud in its axil is extremely fascinating. If one of the two cotyledons from a dark-grown seedling is amputated after decapitation of its epicotyl, only the cotyledonary bud in the axil of the amputated cotyledon grows out (Dostál, 1908). Thus the cotyledon which remains inhibits the development of the bud in its axil (Fig. 10.4). This inhibition can be broken by growth under high light intensity (Dostál, 1939, 1967) or by the application of BAP to the remaining cotyledon (Šebánek, 1965a). There is evidence to suggest that the inhibitory effect of the cotyledon may be due to auxin since if, after decapitation of the shoot, both cotyledons are cut in half perpendicularly and IAA paste applied to the cut surface of one of them, inhibition of the axillary bud of the cotyledon to which IAA was applied always occurs (Dostál, 1939).

In further experiments on dark-grown plants, Šebánek and Zelinka (1971) found that the application of abscisic acid (ABA) to one of the cotyledonary buds inhibited its growth. In the experimental system described above, in which one of the cotyledons was amputated, treatment of the cotyledonary bud in the axil of the amputated cotyledon with 10 mg l^{-1} ABA inhibited it and the lateral in the axil of the attached cotyledon grew out. Thus ABA acts in opposition to the cytokinins.

Amputation of one of the cotyledons of dark-grown plants gives rise to a rapid increase (within 48 h) in the level of endogenous gibberellin in the axillary bud of the amputated cotyledon. An increased uptake of ^{32}P after treatment of the root system with labelled phosphate occurs before any morphological differences become apparent between the cotyledonary shoots (Šebánek, 1965b,c). Šebánek's later work (1966b, 1972) supports the possibility of synthesis of gibberellins by the roots and their involvement in the correlations between cotyledons and bud growth.

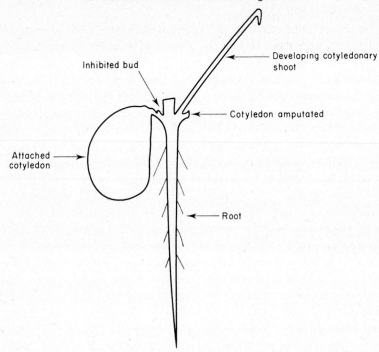

Developing cotyledonary
shoot

Inhibited bud

Cotyledon amputated

Attached
cotyledon

Root

FIG. 10.4. A diagram of a dark-grown pea seedling decapitated just above the cotyledonary node. When one of the cotyledons is amputated the remaining cotyledon inhibits
the growth of the cotyledonary bud in its axil.

Thus, when only one cotyledon is present in a two-shoot system, it imposes
inhibition on the bud in its axil and this inhibition can be simulated by IAA
applied to the cotyledon or ABA applied to the bud. Kinetin overcomes the
inhibition and an increase in gibberellin level and a more rapid uptake of
mineral nutrients to the bud released from inhibition occurs before any
growth is observed.

To what extent can these data be used to interpret the situation present in
light-grown decapitated and intact plants?

D. The Role of Gibberellins

1. Movement of Gibberellins

Gibberellins are synthesized in the apical regions of the shoot (Jones and
Phillips, 1966, in sunflower) and in the roots (Šebánek, 1966b, in pea). They
move freely within plants in a non-polar manner. Gibberellins move both
acropetally from the cotyledons and basipetally from the shoot apex of pea
(Moore, 1967), and McComb (1964), using intact dwarf pea seedlings, found
movement of labelled gibberellic acid (GA_3) from mature to young leaves.

K

2. Action in Apical Dominance

Most reports of gibberellin action on decapitated plants have shown that th
application of gibberellin to the cut stem surface causes an increase in th
rate of lateral bud elongation (Brian *et al.*, 1955; Wickson and Thimann
1958; Scott and Pritchard, 1968) and that when applied at the same time a
IAA to the cut surface of the stem in decapitated plants, it acts in oppositio
to IAA in both tall and dwarf varieties (Phillips, 1969a). However, Scott *et a.*
(1967) found that GA_3 increased the inhibitory effects of IAA when applie
to the cut surface of the shoot of decapitated peas grown in vermiculite
Phillips (1971a,b) using *Phaseolus multiflorus*, showed that treatment of de
capitated plants with GA_3 antagonized the inhibition of axillary bud growt
by IAA when applied to a fully elongated internode, but treatment of a
elongating internode with these hormones after decapitation gave rise to a
increase in inhibition of bud growth by IAA in the presence of GA_3. Thi
compensatory growth effect could explain the different results obtained b
Scott *et al.* (1967) and Phillips (1969a) although nutrient factors could also b
involved since Scott *et al.* (1967) mentioned that under high light and favour
able nutrient conditions, GA_3 promoted lateral outgrowth even in th
presence of IAA.

When gibberellin application promotes lateral bud growth it is probabl
acting on the extension processes following release from inhibition. There i
no evidence that GA_3 application to a bud can release it from inhibition.

E. Cytokinin Involvement

1. Point of Application

(a) *Application to lateral buds.* Interest in the involvement of cytokinins ir
the control of lateral bud growth has run high ever since Wickson and Thimann (1958) showed that kinetin applied directly to an axillary bud on a stem
segment partially immersed in IAA released it from inhibition. Kinetin
applied as a drop in 50% ethanol and polyethylene glycol at a final concentration of 500 mg l^{-1}, also caused the release from inhibition of the lateral bud
to which it was applied on intact pea plants (Sachs and Thimann, 1964). The
pattern of release of bud inhibition by kinetin was the same whether the original inhibition was due to the intact apex or to IAA. Sachs and Thimann
(1964) suggest that the action of kinetin is on the release phenomenon itself
However, for the continued elongation of lateral shoots released from inhibition by kinetin it was necessary to supply auxin to the apex of the lateral
bud (Sachs and Thimann, 1967). They suggested that the development of a
lateral is a two-stage process, the first "release from inhibition" stage being
under cytokinin control and the second "elongation" phase being governed
by auxin from the lateral shoot apex.

(b) *Application to the cut stump.* In the experiments cited above Thimann
and his co-workers applied kinetin directly to a lateral bud and effected its
release from inhibition. Morris and Thomas (1968) applied kinetin and IAA
together in lanolin paste to the cut stump of 14-day-old peas decapitated
above the third node and found that the growth of the lateral buds remained
suppressed—presumably because insufficient cytokinin reached the inhibited
buds.

One of the features of cytokinin application is that if sufficient movement
occurs within the plant the inhibition of many buds may be released. Thus,
Scott and Pritchard (1968) found that kinetin applied in solution to the cut
stump of an epicotyl decapitated at the top of the fifth internode stimulated
more buds to grow out than in the decapitated controls to which only water
had been applied. However, in plants treated with kinetin all the buds were
rather short and of similar length whereas in the water controls usually one
or two buds developed (one large and one small) and the others remained
inhibited. It is possible that poor nutrition of the plants retarded the growth
of the kinetin-released buds.

(c) *Application to seeds and roots.* Sprent (1968a) found that the application
of 10 mg l^{-1} or higher of BAP to the seed of the tall cultivar 'Improved Pilot'
and the dwarf cultivar 'Supreme' gave rise to extensive branching in intact
seedlings. The cotyledonary buds and sometimes buds at higher nodes grew
out. When she treated one of the cotyledons with BAP only the shoot in the
axil of that cotyledon continued to develop, although initially axillaries of
both of the cotyledons were released from dominance (Sprent, 1968b). Treatment of cotyledons with GA$_3$ and BAP increased the rate of elongation of all
growing shoots but did not alter the pattern of distribution of growth between
main and axillary shoots. The cytokinins initiated growth in buds that were
completely inhibited. The application of kinetin to a totally inhibited shoot
on a two-shoot system (cv. 'Radio') also relieved the inhibition (Sachs, 1966).

Sprent (1968a,b) observed that BAP treatment of the cotyledons gave rise
to a retardation of cotyledon senescence and an overall inhibition of root
and shoot growth. Wittwer and Dedolph (1963) found no effect of kinetin or
BAP on shoot branching (cvs. 'Alaska' and 'Little Marvel') when applied to
the root system although reductions in both shoot and root growth occurred
as a result of treatment. Cytokinins can thus radically alter the morphology
of intact plants when applied to the cotyledons or to the root system.

2. Movement of Cytokinins

Experiments involving cytokinin, especially of the relatively immobile substance kinetin, give different effects depending upon the point of application.

Morris and Winfield (1972), using kinetin-8-^{14}C, found that in intact plants
there was no export of kinetin from the apical regions to which it was applied.
In decapitated plants most of the kinetin still remained in the uppermost

internode but substantial levels were present in the lateral buds and roots. Th
youngest bud had the highest specific activity but the largest bud (at node 2
had the greatest amount of radioactivity. When IAA was added at the sam
time as the kinetin there was very little movement into the buds. However
application of IAA at the same time as kinetin to the cut stump of a decapi
tated plant has been shown to increase the basipetal movement of kineti
(Seth *et al.*, 1966). Woolley and Wareing (1972a) found that in *Solanum
andigena*, IAA inhibited the action of root-produced cytokinins in releasing
lateral buds by inhibiting the accumulation of cytokinins in the buds. Clearl
the root-produced cytokinins must be much more mobile than the non
endogenous kinetin.

3. Events taking place in the Lateral Bud

Morris and Thomas (1968) observed that when kinetin was applied to cu
stumps of pea shoots in the absence of IAA a marked accumulation of radio
active material occurred in lateral buds at nodes 1 and 2 in plants fed wit
^{14}C-sucrose on the adaxial surface of leaf 3, even though the growth of th
lateral buds was no greater than in lanolin-treated controls. It was also note
that, 24 h after treatment with kinetin, mitotic indices were significantl
higher in the buds of kinetin-treated plants than in the buds of decapitate
controls and that a higher frequency of division was maintained for at leas
72 h after treatment. However, in kinetin-treated plants given a single dose o
19 μg of IAA, mitotic figures remained at a low level for at least 48 h. Th
failure of lateral buds on kinetin-treated decapitated plants to elongate an
more rapidly than buds on decapitated controls may have been due to
shortage of auxin in the lateral buds.

Other events which occur soon after bud release as a result of BAP appli
cation to inhibited buds on intact plants include a stimulation of DNA
synthesis within 24 h (Schaeffer and Sharpe, 1969, in tobacco) and a rapi
increase in RNA level in the treated buds between 16 and 27 h after BAI
application (Schaeffer and Sharpe, 1970, in tobacco). Such changes in nuclei
acid levels as a result of cytokinin application have not yet been reported i
peas. Significant growth of the lateral buds after decapitation of the mai
shoot has been detected after very short time periods. Nagao and Rubinstei
(1974), using a dissecting microscope, found that growth of the lateral bu
in cv. 'Alaska' was stimulated 6–10 h after decapitation of the main shoot
The growth of the lateral buds over the 24 h period following decapitatio
was the same in plants with roots as in de-rooted plants and they suggest tha
the supply of cytokinin from the roots may not play a major role during th
release of lateral buds from apical dominance. It is possible, however, tha
sufficient cytokinin which originated from the roots was present in the ste
at the time of root removal and that this cytokinin was involved in the releas
of the lateral buds.

. Is Abscisic Acid a Major Controlling Influence?

. *Application of Abscisic Acid*

he involvement of ABA in correlative inhibition is suggested by several lines
f evidence. Firstly, Arney and Mitchell (1969) using tall ('Sutton's Achieve-
ient') and dwarf ('Kelvedon Wonder') cultivars showed that 100 μg of ABA
pplied as an ethanolic drop to the axils of the disbudded trifid bracts sub-
antially inhibited the outgrowth of lateral buds in the axil of the first true
oliage leaf in decapitated tall and dwarf plants. Interestingly, ABA was effec-
ve in the control of lateral shoot growth only in plants grown under low
ght conditions prior to decapitation. Plants under high light and high nutrient
onditions grew vigorously and produced laterals on intact plants. The appli-
ation of ABA to these plants following decapitation had no effect on lateral
utgrowth. Recent work by Bellandi and Dörffling (1974a,b) showed that
'hen ABA at a level of 0·1–5·0 μg was applied directly to a lateral bud of de-
apitated pea seedlings (cv. 'Senator'), the bud was inhibited. The magnitude
f the inhibition was related to the concentration of ABA used (Fig. 10.5).

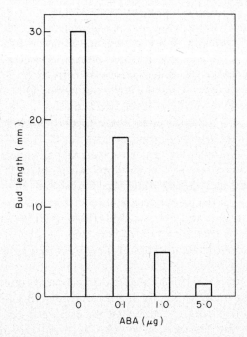

FIG. 10.5. The effect of ABA on the elongation of the lateral bud at node 2 of decapi-
ated pea seedlings. The ABA was applied to the lateral bud and measurements were
aken 16 days after treatment. The initial bud length was 1·5 mm. Data from Bellandi
and Dörffling (1974a)

However, when the ABA was applied to the intact main shoot apex, it inh
bited the growth of the main shoot, and the lateral buds were released fro
inhibition. Application of ABA to the cut surface of seedlings decapitate
about 5 mm above the second node resulted in a slight inhibition of the bu
at the second node and in growth promotion of the bud at the first node. The
results strongly suggest that the effect of ABA is strictly local when applie
to stems, inhibiting only at the point of its application. It was found th.
ABA-2-^{14}C applied to the intact main shoot apex did not move down tl
shoot to any great extent but movement was more pronounced when it wa
injected into the stem tissue at the apical bud (Bellandi and Dörffling, 1974b
When labelled ABA was applied to an expanded leaf there was a substanti
accumulation of radioactivity in the lateral roots and in the growing apic
regions. The ABA did not affect the pattern or the rate of distribution
radioactive IAA applied as an ethanolic drop to the main shoot apex. Whe
GA$_3$ was added together with ABA to the apical bud of intact pea seedlin;
it was found that the GA$_3$ almost overcame the inhibitory effect of ABA c
the apical bud and as evidence of this no growth of lateral buds occurred.

2. Moisture Stress, Abscisic Acid and Apical Dominance

Another avenue of exploration arises from McIntyre's (1971) study of tl
relationship between water stress and apical dominance. He used a cultiva
('Alaska') which normally shows strong apical dominance, lateral shoo
growing only when the main shoot is decapitated. However, when the tran
piration rate and hence the water stress was reduced by growing seedling
at a lower temperature (15°C) and a high relative humidity (85%), he foun
a very marked increase in length of the lateral buds at nodes 1 and 2. A tot.
length of 76 mm was obtained under these conditions whereas only a 3 m
increase in length of lateral shoots occurred at 25°C and 40% r.h. Rem
(1968), using cv. 'Volontaire', has also shown that treatments which modi
the pattern of water stress also modify the pattern of bud inhibition. Sim
larly, Husain and Linck (1966) found that lateral buds (cv. 'Alaska') whic
did not continue to develop after decapitation of the main shoot often showe
signs of loss of turgor, and Sorokin and Thimann (1964) showed that th
vascular connections between inhibited buds and the stem are poor and the
attributed some of the characteristics of these buds to a poor water or solut
supply.

Simpson and Saunders (1972) noted that intact plants of cv. 'Alaska', sut
jected to moisture stress, contained higher levels of ABA than those grow
under non-stressed conditions. Experiments using excised epicotyls of c
'Meteor' showed that the increase in ABA occurred very rapidly after th
onset of wilting. An 8 h period of wilting gave rise to a seven-fold increase
ABA in the epicotyl. The degree of moisture stress in these experiments wa
not especially severe. Excised epicotyls were placed beneath a fan at 20°

until a 6% loss in fresh weight occurred and the epicotyls were then maintained at this moisture content. The levels of ABA present in cv. 'Alaska' and cv. 'Meteor' in Simpson and Saunder's experiments are of the same order as those found for dwarf peas by Kende and Kays (1971) and for 'Gradus' and 'Progress No. 9', a tall and a dwarf cultivar respectively, by Barnes (1972) suggesting that ABA levels in tall and dwarf cultivars are similar.

From these experiments it is reasonable to suggest that the following series of events takes place:

$$\text{moisture stress} \xrightarrow{\text{rapid increase in ABA level}} \text{inhibition of bud growth}$$

However, although the overall level of ABA has been shown to increase in the plant in these experiments, data on ABA level for the lateral buds are not available. It is of course quite possible that wilting has its most dramatic effect on increasing levels of ABA in the lateral buds and that the real magnitude of this effect is obscured if the whole epicotyl is analysed. Certainly, Tucker and Mansfield (1973) found massive levels of ABA in inhibited buds of *Xanthium* following a far-red treatment which caused lateral bud inhibition. The levels of ABA were far higher in the inhibited buds than in the other parts of the stem.

Further support for the involvement of ABA can be inferred from the ideas put forward by Snow (1937) who suggested that the auxin control of apical dominance must operate indirectly, possibly involving an inhibitory substance which must, amongst other things, exhibit non-polar movement. Abscisic acid is an obvious candidate for Snow's hypothetical substance.

3. Phytochrome-Mediated Responses

Phytochrome is synthesized in the embryonic axis of young (cv. 'Alaska') seedlings (McArthur and Briggs, 1970) and phytochrome-mediated responses are involved in the control of bud growth and the directional movement and accumulation patterns of sucrose in the epicotyl (Goren and Galston, 1966; Anand and Galston, 1972). Khudairi et al. (1971) using light-grown dwarf peas (cv. 'Little Marvel') removed the apical bud and all the axillary buds except the one at node 2. They then transferred the seedlings to darkness at 24°C and gave short bursts (one per day) of either far-red (FR) or red (R) irradiation. Internode elongation was enhanced by FR and inhibited by R, but the buds exposed to R showed increased DNA content, RNA content and cell divisions, and growth of the axillary bud continued at a faster rate than in seedlings exposed to FR light. The inhibitory effect of FR on lateral bud growth is also shown by Tucker and Mansfield (1972, 1973) in *Xanthium* where the lateral buds of plants given an end-of-day FR irradiation were strongly inhibited. The inhibited buds contained very high levels of ABA.

G. Integration and Speculation

This section pulls together the data on apical dominance in pea plants by means of two flow diagrams. The first of these (Fig. 10.6) is concerned primarily with factors involved in the maintenance of lateral bud inhibition and the initial release mechanisms and the second (Fig. 10.7) with the events taking place after release from inhibition. Experimental evidence is not available for all the stages shown in the figures.

1. Release Mechanisms

(a) *Intact plants.* In intact untreated plants (identified in Fig. 10.6 by double boxes) IAA produced by the main shoot apex and young leaves is envisaged as moving down the stem in a polar manner (1) and causing the continued production of ABA which then accumulates in the lateral buds (2) (Fig. 10.6). It is suggested that the level of ABA present in these buds is sufficient to maintain bud inhibition, especially if the light intensity and nutritional levels are not high. Under conditions in which the lateral buds are inhibited there is a strong directional flow to the shoot apex of photosynthates from the

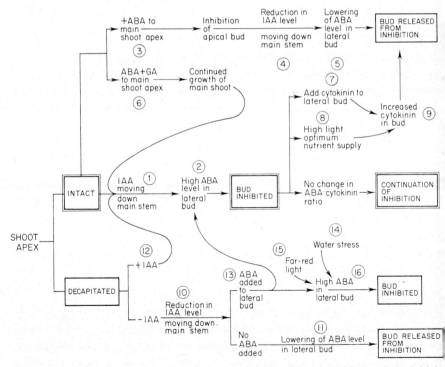

Fig. 10.6. Factors involved in the maintenance of lateral bud inhibition and the initial release mechanisms.

mature leaves, and of mineral nutrients, various elaborated substances and cytokinins from the root. Little movement occurs, however, into the lateral buds. Vascular connections between the main stem vascular system and the inhibited buds are poor.

If ABA is added to the apical bud (3) its growth is inhibited and presumably this results in a reduction of the level of IAA moving down the stem (4), which in turn is associated with a lowering of the ABA level in the lateral buds (5). These are then released from inhibition. If the continued presence of IAA in the stem is necessary to maintain ABA levels in the bud, either the IAA causes continued synthesis of ABA at a faster rate than ABA breakdown or it protects the ABA which is already present from being broken down. Presumably in the absence of IAA or when IAA levels in the stem are low, ABA breakdown predominates, giving rise to a sufficient reduction in ABA level in the lateral buds to enable their dormancy to be broken. When ABA and GA_3 are added together to the main shoot apex (6) the growth of the main shoot is not affected and it is reasonable to suppose that IAA movement down the stem (1) is again responsible for the observed inhibition of lateral bud growth.

Any factor that modifies the ABA:cytokinin ratio in a lateral bud appears to modify the degree of inhibition present in that bud. Thus, the addition of cytokinin directly to an inhibited lateral bud (7) releases it from inhibition. Similarly, environmental factors such as high light intensity and high nitrogen status (8) also tend to reduce the degree of apical dominance, presumably by increasing cytokinin production (9) in the roots (Woolley and Wareing, 1972b, in *Solanum andigena*) or possibly by the buds themselves. The evidence suggests that nutritional factors are operating through changes in hormone balance rather than directly since inhibited buds of *Phaseolus vulgaris* (Phillips, 1968) do not appear to be deficient in nitrogen, phosphorus or potassium.

(*b*) *Decapitated plants.* When the main shoot apex is decapitated there is a reduction in the level of IAA moving down the stem (10) which presumably results in a decrease in ABA level in the lateral buds (11) (Fig. 10.6). The buds are then released from inhibition. If IAA is added to the cut stem immediately after decapitation (12) then the lateral buds remain inhibited. Delay in the application of IAA reduces its inhibitory effect possibly due to a reduction in the efficiency of the IAA transport system or to the resumption of active growth by the lateral buds in the intervening period.

After decapitation of a plant, inhibition of its lateral buds can be reimposed either by adding ABA directly to the bud (13) or by giving treatment such as water stress (14) or far-red light (15) which give rise to increased levels of ABA in the lateral buds (16). If ABA is added to the cut shoot of a decapitated plant the lateral buds will be inhibited only if the ABA moves into them, and since its movement is limited, inhibition will only occur if the bud is close to the point of ABA application. It is possible that ABA application is more effective when the endogenous cytokinin levels in the plant are low.

(c) *Gibberellin effects.* Work by Catalano and Hill (1969) on IAA-inhibited buds of tomato showed that growth promotion by kinetin and GA_3 applied together to a bud was more effective if the buds were treated first with kinetin and then with GA_3, rather than if the two were applied simultaneously. This can be explained most readily by suggesting that the gibberellins are acting only on elongation and not on the processes of bud release. Thus, if the upper internodes on the main shoot below the point of decapitation have not completed elongation the GA_3 applied to the lateral bud may move out into the upper stem internodes and increase the inhibitory effect of the IAA (see Phillips, 1971a,b). If, however, the GA_3 was applied to the bud after its release then the gibberellin would be expected to cause a localized increase in elongation of that bud.

2. Events Following Bud Release

Once the lateral bud has been released from inhibition a very rapid series of changes takes place leading to bud elongation within only a few hours. In these very early stages increases in DNA, RNA, cell divisions and small but measurable length increases are recorded and these may be accomplished largely from reserve materials and hormones already present in the bud itself or in stem tissues close to it (Fig. 10.7). Although under certain circumstances exogenous sucrose has been shown to be necessary to stimulate mitoses in the

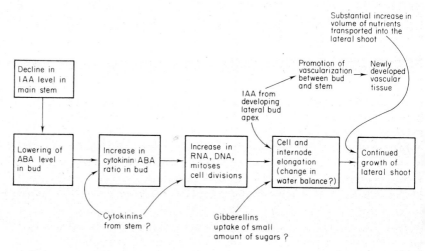

FIG. 10.7. Events taking place in the lateral bud after release from inhibition.

lateral buds (Ballard and Wildman, 1964, in dark-grown sunflower plants) limitation of carbohydrate does not seem to be the cause of bud inhibition in light-grown peas. Connections between bud and stem vascular bundles are poor at this stage and show no significant development prior to the first observable bud elongation, but it is possible that there is an immediate improvement of the water status of the buds at the time of its release from dormancy.

The rapid increase in gibberellins in buds released from inhibition and their promotive effects in the elongation of intact shoots suggest that gibberellins are directly involved in bud extension. Once bud growth commences auxin produced by the bud apex is important for the continued elongation of the lateral shoot and also for the differentiation of the vascular tissue linking the bud and stem bundles. These connections are made after about 2 days and substantial increases in the traces subsequently take place. This increase in vascularization is essential because the developing bud rapidly becomes a major sink for nutrients and the poor vascular connections present initially would obviously be quite inadequate for transport of such substances in sufficient quantities. Thus, the improvement of vascular connections should be seen as an integral part of the syndrome of bud growth rather than as a reason for the release of bud inhibition.

IV. CONCLUSIONS

Although only one example of a correlative growth process, "Apical Dominance", has been discussed in detail in this chapter, all of the correlations mentioned above are under the control of complex interactions between hormones and nutrients, the levels of which are affected by environmental conditions. Much valuable information has been gained by exogenous growth-regulator applications and by surgical treatments, but it is obvious that the responses of plants treated in this way may differ fundamentally from those of untreated plants.

A substantial body of information is now available indicating which hormones are involved and some knowledge of their roles in various correlative processes has now come to light. The links between light, nutrient level and endogenous hormone level and distribution are much less clear, but the information gained from plants subjected to hormonal and surgical treatments makes it possible to design more critical experiments on intact plants under different environmental regimes. Information is needed particularly with regard to the effects of environmental factors on hormone levels in intact plants and concerted efforts are necessary in this connection if we are to achieve a fuller understanding of the biology of correlative processes in intact plants.

REFERENCES

ANAND, R. and GALSTON, A. W. (1972). *Am. J. Bot.* **59**, 327-336.
ARNEY, S. E. and MITCHELL, D. L. (1969). *New Phytol.* **68**, 1001–1015.
BAIN, J. M. and MERCER, F. V. (1966a). *Aust. J. biol. Sci.* **19**, 69–84.
BAIN, J. M. and MERCER, F. V. (1966b). *Aust. J. biol. Sci.* **19**, 85–96.
BALLARD, L. A. T. and WILDMAN, S. G. (1964). *Aust. J. biol. Sci.* **17**, 36–43.
BARNES, M. F. (1972). *Planta* **104**, 182–184.
BELLANDI, D. M. and DÖRFFLING, K. (1974a). *Physiologia Pl.* **32**, 369–372.
BELLANDI, D. M. and DÖRFFLING, K. (1974b). *Physiologia Pl.* **32**, 365–368.
BOOTH, A., MOORBY, J., DAVIES, C. R., JONES, H. and WAREING, P. F. (1962). *Nature, Lond.* **194**, 204–205.
BOWEN, M. R. and WAREING, P. F. (1971). *Planta* **99**, 120–132.
BRIAN, P. W., HEMMING, H. G. and RADLEY, M. (1955). *Physiologia Pl.* **8**, 899–912.
BURG, S. P. and BURG, E. A. (1968). *Pl. Physiol.* **43**, 1069–1074.
CATALANO, M. and HILL, T. A. (1969). *Nature, Lond.* **222**, 985–986.
DAVIES, C. R. and WAREING, P. F. (1965). *Planta* **65**, 139–156.
DOSTÁL, R. (1908). *Rozpravy Čes. Akad.* **17**, 1–44.
DOSTÁL, R. (1939). *Ber. dt. bot. Ges.* **57**, 204–230.
DOSTÁL, R. (1967). *Biologia Pl.* **9**, 330–339.
DOSTÁL, R. (1968). *Flora, Jena* **159**, 274–276.
ERIKSEN, E. N. (1973). *Physiologia Pl.* **28**, 503–506.
GOESCHL, J. D., RAPPAPORT, L. and PRATT, H. K. (1966). *Pl. Physiol.* **41**, 877–884.
GOREN, R. and GALSTON, A. W. (1966). *Pl. Physiol.* **41**, 1055–1064.
GOURLEY, J. H. (1931). *Bot. Gaz.* **92**, 367–383.
GUARDIOLA, J. L. and SUTCLIFFE, J. F. (1971a). *Ann. Bot.* **35**, 791–807.
GUARDIOLA, J. L. and SUTCLIFFE, J. F. (1971b). *Ann. Bot.* **35**, 809–823.
GUERN, J. and USCIATI, M. (1972). *In* "Hormonal Regulation in Plant Growth and Development" (H. Kaldewey and Y. Vardar, eds), pp 383–400. Proc. Adv. Study Inst. Izmir 1971. Verlag Chemie, Weinheim.
HUSAIN, S. M. and LINCK, A. J. (1966). *Physiologia Pl.* **19**, 992–1010.
JOHNSTON, M. J. and CROWDEN, R. K. (1967). *Aust. J. biol. Sci.* **20**, 461–463.
JONES, R. L. and PHILLIPS, I. D. J. (1966). *Pl. Physiol.* **41**, 1381–1386.
KAMÍNEK, M. (1965). *Biologia Pl.* **7**, 394–396.
KARTHA, K. K., GAMBORG, O. L. and CONSTABEL, F. (1974). *Z. Pflanzenphysiol.* **72**, 172–176.
KENDE, H. and KAYS, E. (1971). *Naturwissenschaften* **58**, 524–525.
KHUDAIRI, A. K., JOHNNYKUTTY, A. T. and AGARWAL, S. (1971). *Planta* **101**, 185–188.
KILLEEN, L. A. and LARSON, L. A. (1968). *Am. J. Bot.* **55**, 961–965.
KLASOVÁ, A., KOLEK, J. and KLAS, J. (1971). *Biologia Pl.* **13**, 209–215.
KLASOVÁ, A., KOLEK, J. and KLAS, J. (1972). *Biologia Pl.* **14**, 249–253.
LOVELL, P. H. (1969). *Physiologia Pl.* **22**, 506–515.
LOVELL, P. H. and MOORE, K. G. (1970). *J. exp. Bot.* **21**, 1017–1030.
LOVELL, P. H., OO, H. T. and SAGAR, G. R. (1972). *J. exp. Bot.* **23**, 255–266.
MACQUARRIE, I. G. (1965). *Can. J. Bot.* **43**, 29–38.
MCARTHUR, J. A. and BRIGGS, W. R. (1970). *Planta* **91**, 146–154.

McComb, A. J. (1964). *Ann. Bot.* **28**, 669–687.

McDavid, C. R., Sagar, G. R. and Marshall, C. (1972). *New Phytol.* **71**, 1027–1032.

McDavid, C. R., Sagar, G. R. and Marshall, C. (1973a). *New Phytol.* **72**, 269–275.

McDavid, C. R., Sagar, G. R. and Marshall, C. (1973b). *New Phytol.* **72**, 465–470.

McIntyre, G. I. (1971). *Nature, Lond.* **230**, 87–88.

Moore, T. C. (1964). *Pl. Physiol.* **39**, 924–927.

Moore, T. C. (1967). *Am. J. Bot.* **54**, 262–269.

Morris, D. A. and Thomas, E. E. (1968). *Planta* **83**, 276–281.

Morris, D. A. and Thomas, A. G. (1974). *Planta* **118**, 225–234.

Morris, D. A. and Winfield, P. J. (1972). *J. exp. Bot.* **23**, 346–355.

Nagao, M. and Rubinstein, B. (1974). *Pl. Physiol. Suppl. Rep. No.* 50.

Nakamura, E. (1964). *Pl. Cell Physiol.* **5**, 521–524.

Peckett, R. C. (1957). *J. exp. Bot.* **8**, 172–180.

Phillips, D. A. (1971). *Physiologia Pl.* **25**, 482–487.

Phillips, I. D. J. (1968). *J. exp. Bot.* **19**, 617–627.

Phillips, I. D. J. (1969a). *Planta* **86**, 315–323.

Phillips, I. D. J. (1969b). *In* "The Physiology of Plant Growth and Development" (M. B. Wilkins, ed.), pp 163–202. McGraw-Hill, London.

Phillips, I. D. J. (1971a). *Planta* **96**, 27–34.

Phillips, I. D. J. (1971b). *J. exp. Bot.* **22**, 465–471.

Remy, M. (1968). *C. r. Acad. Sci., D,* **266**, 676–679.

Sachs, T. (1966). *Ann. Bot.* **30**, 447–456.

Sachs, T. (1968). *Ann. Bot.* **32**, 781–790.

Sachs, T. and Thimann, K. V. (1964). *Nature, Lond.* **201**, 939–940.

Sachs, T. and Thimann, K. V. (1967). *Am. J. Bot.* **54**, 136–144.

Schaeffer, G. W. and Sharpe, F. T. (1969). *Bot. Gaz.* **130**, 107–110.

Schaeffer, G. W. and Sharpe, F. T. (1970). *Ann. Bot.* **34**, 707–719.

Scott, T. K., Case, D. B. and Jacobs, W. P. (1967). *Pl. Physiol.* **42**, 1329–1333.

Scott, T. K. and Pritchard, J. B. (1968). *In* "The Transport of Plant Hormones" (Y. Vardar, ed.), pp 309–319. North-Holland, Amsterdam.

Šebánek, J. (1965a). *Acta Univ. Agric. Brno* **3**, 363–370.

Šebánek, J. (1965b). *Biologia Pl.* **7**, 380–386.

Šebánek, J. (1965c). *Biologia Pl.* **7**, 194–198.

Šebánek, J. (1966a). *Acta Univ. Agric. Brno* **4**, 587–596.

Šebánek, J. (1966b). *Biologia Pl.* **8**, 470–475.

Šebánek, J. (1972). *Biologia Pl.* **14**, 337–342.

Šebánek, J. and Kopecký, F. (1967). *Rostl. výroba* **13**, 843–848.

Šebánek, J. and Zelinka, B. (1971). *Rostl. výroba* **17**, 451–456.

Seth, A. K., Davies, C. R. and Wareing, P. F. (1966). *Science, N.Y.* **151**, 587–588.

Short, K. C. and Torrey, J. G. (1972). *Pl. Physiol.* **49**, 155–160.

Simpson, G. M. and Saunders, P. F. (1972). *Planta* **102**, 272–276.

Skoog, F. and Thimann, K. V. (1934). *Proc. natn. Acad. Sci. U.S.A.* **20**, 480–485.

Snow, R. (1931). *Proc. R. Soc.* B **108**, 305–316.

Snow, R. (1932). *Proc. R. Soc.* B **111**, 86–105.

Snow, R. (1937). *New Phytol.* **36**, 283–300.

Sorokin, H. P. and Thimann, K. V. (1964). *Protoplasma* **59**, 326–350.

SPRENT, J. I. (1966). *Nature, Lond.* **209**, 1043–1044.
SPRENT, J. I. (1968a). *Planta* **78**, 17–24.
SPRENT, J. I. (1968b). *Planta* **81**, 80–87.
THIMANN, K. V. (1937). *Am. J. Bot.* **24**, 407–412.
TORREY, J. G. (1952). *Pl. Physiol.* **27**, 591–602.
TUCKER, D. J. and MANSFIELD, T. A. (1972). *Planta* **102**, 140–151.
TUCKER, D. J. and MANSFIELD, T. A. (1973). *J. exp. Bot.* **24**, 731–740.
VAN OVERBEEK, J. (1938). *Bot. Gaz.* **100**, 133–166.
VARNER, J. E., BALCE, L. V. and HUANG, R. C. (1963). *Pl. Physiol.* **38**, 89–92.
WARDLAW, I. F. and MORTIMER, D. C. (1970). *Can. J. Bot.* **48**, 229–237.
WICKSON, M. and THIMANN, K. V. (1958). *Physiologia Pl.* **11**, 62–74.
WICKSON, M. and THIMANN, K. V. (1960). *Physiologia Pl.* **13**, 539–554.
WITTWER, S. H. and DEDOLPH, R. R. (1963). *Am. J. Bot.* **50**, 330–336.
WOOLLEY, D. J. and WAREING, P. F. (1972a). *New Phytol.* **71**, 781–793.
WOOLLEY, D. J. and WAREING, P. F. (1972b). *New Phytol.* **71**, 1015–1025.

11. The Physiology of Photomorphogenesis and of Tendril Response

A. W. GALSTON

School of Biological and Environmental Studies, Yale University, New Haven, U.S.A.

I. INTRODUCTION

For at least four decades, the garden pea (*Pisum sativum* L.) has been one of the favourite tools of the experimental plant physiologist. I first started to work with etiolated pea seedlings in 1947, and for many years, these slender and pale yet vigorous young plants constituted my main experimental "guinea plant" in investigations of hormonal physiology and metabolism. Curiosity later led me to explore physiological differences between totally etiolated, partially de-etiolated and fully green plants; this in turn brought me into such varied worlds as phytochrome physiology, flavonoid biosynthesis and tendril movements. Now that my interest in phytochrome-controlled leaf movements has led me away from the pea plant, I welcome the opportunity to summarize the results of my own investigations on varied aspects of the physiology and biochemistry of the garden pea, while drawing appropriate attention to the work of others. Much relevant material, especially in the field of hormonal response, is found in Chapter 9 and is not repeated here.

A. Advantages of Etiolated Pea Seedlings in Physiological Research

Let us begin by noting some of the advantages of the etiolated pea seedling as an experimental tool.

1. Large quantities of genetically identical seeds may be obtained from commercial seedsmen; we have used mainly Asgrow Incorporated, formerly of New Haven and Orange, Connecticut, U.S.A. With some personal intervention from time to time, one can ensure the delivery of virtually identical and disease-free seed batches from year to year. This is important in facilitating long-range experimental comparisons.

2. There is sufficient food reserve within the pea seed to ensure rapid growth and development of the seedling under completely heterotrophic (dark) conditions for about 2 weeks, which is long enough for most physiological experiments.

3. The pea seed may be conveniently imbibed in total darkness, planted densely in pre-washed, water-saturated, fine vermiculite, and the seedlings harvested some days later without additional feeding of water or mineral nutrients. If kept at about 25°C in a humid (> 70% r.h.) chamber in total darkness, such seeds produce predictable crops of sturdy, rapidly growing and virtually identical seedlings within 3 days (Fig. 11.1A) although most of our experiments have called for 6 to 7-day-old seedlings (Fig. 11.1B). More than 300 such seedlings can conveniently be harvested from a single polyethylene tray about 30 cm long × 20 cm wide × 16 cm deep.

4. Genetically tall and dwarf varieties are available, facilitating experiments in gibberellin physiology, phytochrome-controlled responses and growth kinetics.

5. Excised portions of the epicotyl, apical hook, root and plumule respond well to growth regulatory stimuli, facilitating the analysis of responses of individual organs (Fig. 11.2).

6. The different regions of the aerial portion of the seedling, e.g. epicotyl, apical hook and terminal bud, respond differently to such growth regulatory stimuli as light and hormones, thus permitting analysis of complicated interorgan interactions.

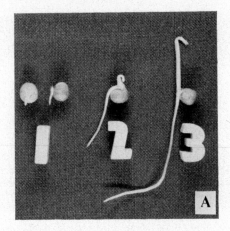

FIG. 11.1A. Seedlings of 'Alaska' peas 1, 2 and 3 days after imbibition and growth in total darkness at 26°C.

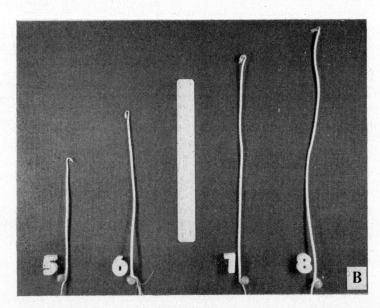

FIG. 11.1B. Seedlings after totally etiolated growth for 5, 6, 7 and 8 days. Note the rapid epicotyl elongation, the small size of the leaves and the apical hook. The epicotyl bears only tiny scale leaves at nodes 1 and 2. The scale indicates a length of 15 cm.

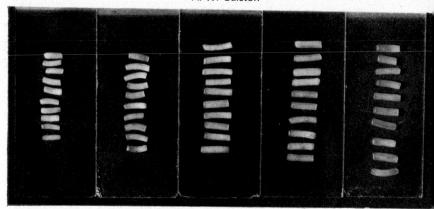

FIG. 11.2. Response of 5-mm long sub-apical etiolated 'Alaska' pea epicotyl sections to varying levels of IAA in the presence of sucrose and phosphate buffer. Left to right 0, 5×10^{-5}, 5×10^{-4}, 5×10^{-3} and 5×10^{-2} mm IAA; 18 h growth period.
(From Galston, 1950)

B. Experimental Advantages of the Green Pea Seedling (or of a Similar Plant)

1. When sown in vermiculite, watered daily with a salt mixture such as "Hyponex", and exposed to 1000–2000 ft candles (10 750–21 500 lx) of mixed fluorescent and incandescent light, the seeds produce sturdy green plants, suitable for many physiological experiments, within 12–14 days (Fig. 11.3).

2. The sub-apical portions of the stem respond well to such hormones as auxins and gibberellins. These responses differ from those of analogous etiolated tissues, providing a convenient material for study of the comparative hormonal physiology of etiolated and green tissues.

3. Unlike stem sections of etiolated seedlings, the growth of green sections is promoted, rather than inhibited, by light. They thus furnish a yet un-exploited material for the study of comparative photophysiology of etiolated and green tissues (Fig. 11.4).

4. The tendrils formed from leaves above the third node are rapidly grow-ing and extremely sensitive to tactile stimuli. This provides an unusually convenient material for the study of rapid plant movements.

Fig. 11.3. 'Alaska' pea plant grown in about 8600 lx mixed fluorescent–incandescent light. Note the successively larger green leaves at each higher node, the increasingly larger tendrils, the absence of an apical hook and the short internodes compared with etiolated plants. (From Galston and Baker, 1951)

II. GROWTH PATTERNS OF ETIOLATED PLANTS AND THEIR ALTERATION BY LIGHT

A. Epicotyl Growth

In the cultivar 'Little Marvel', grown in the dark at 25°C, the second internode begins elongation at day 4 after imbibition, reaches a length of about 40 mm by day 8, then elongates very little in the succeeding 4 days (Thomson, 1954). By contrast, the third internode begins elongating on day 5, accelerates its growth greatly on day 7 (when the previous internode is decelerating its growth), and grows rapidly for 2 days at a rate of about 30 mm day^{-1}, reaching a final length in total darkness of as much as 90 mm. The fourth internode, in turn, starts elongating slowly on day 8 and more rapidly on day 9. Thus the upward thrust of the etiolated epicotyl resembles the extension of a

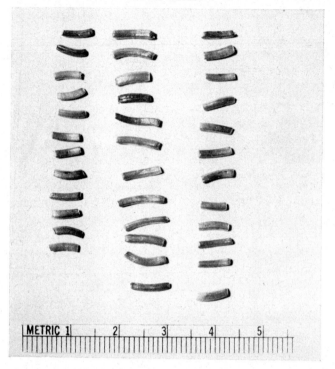

FIG. 11.4. The response of 5-mm sub-apical 'Alaska' green stem sections to (left to right) light minus auxin, light and auxin, auxin minus light; after 18 h of growth. (From Galston and Baker, 1951)

jointed antenna, with each internode starting to expand as the one below it slows down, reaching a maximum rate of extension as the one below it stops, and slowing down as the one above it starts. While all this stem growth activity proceeds, little more than scale leaves are being formed at nodes 1 and 2 and only slightly expanded, pale leaves at nodes 3 and 4.

Should a little light fall on this etiolated system, growth patterns change dramatically. In an overall sense, light inhibits the elongation of stems, promotes the expansion of leaves, and differentially affects the tissues of the hook so as to promote opening and erection of the terminal bud. But a detailed analysis shows that the situation is much more complex than this. Thomson (1954) showed that 3 h of weak white light administered on day 4 inhibited cell extension in the already expanding second internode relative to the dark control, but promoted the early extension rate of cells of internodes 3 and 4, which were still enclosed in the terminal bud. Similarly, light given on day 6 inhibited the extension of cells in the already expanding third internode, but promoted those of the fourth internode, still enclosed in the bud. A useful generalization is that light promotes subsequent elongation when

,iven to young cells, but inhibits subsequent elongation when given to older ells.

With regard to cell division, light given only on day 4 diminishes the rate t which new cells are formed in the third internode (enclosed in the bud) but loes not affect the ultimate number of cells contained in the internode. This ame light treatment increases the early rate of production of new cells in the till younger fourth internode above, but this light-induced difference also)ecomes much smaller with time. By contrast, light given every day inhibits mitotic activity in the older internodes (2 and 3) throughout their life history, vhile producing a temporary increase in the rate of cell production in inter-•ode 4. This detailed study should teach us that in interpreting the effects of ight on any plant, especially the pea, we must consider not only the kind of issue (i.e. stem *vs* leaf) but also the age and physiological status of the tissue)r organ in question.

Parker *et al.* (1949) showed that the action spectrum for the de-etiolation •ffects, especially the promotion of leaf growth (length of the first and second eaves) of 'Little Marvel' peas resembles that for photocontrol of floral nitiation in long-day and short-day plants. It has a single peak at about 570 nm in an actinic zone extending from 610–710 nm. We now know that his action spectrum is characteristic of absorption by phytochrome, and the system has been shown to be reversed by far-red light (FR). This subject will)e discussed further later in this chapter.

Epicotyls of seedlings of the tall 'Alaska' cultivars grown in a dark humid room at *c.* $26 \pm 1°C$ elongate at a rate of about $1·35$ mm h^{-1} for the first 2 days after emergence from the substratum and then at about $1·80$ mm h^{-1} for the subsequent 2 days. A 2 to 4-h exposure to photomorphogenetically active red light ($1·4$ mJ cm^{-2}) lowers the subsequent elongation rate by one-third to one-half, but this is corrected back to the control rate within 18–20 h after the light has been turned off (Fig. 11.5). If the light is administered at the end of the first 2-day period of growth above the substratum (i.e. $3·5$–$5·5$ days after imbibition), then the subsequent rise in elongation rate is completely prevented (Galston *et al.,* 1964). Excised sub-apical sections of the epicotyl can also be inhibited by red light, but only if their growth is first promoted by sucrose (Bertsch and Hillman, 1961); auxin- or gibberellin-induced growth is not similarly sensitive to phytochrome transformation.

B. Hook Opening

Irradiation with red light also leads to opening of the apical hook (Fig. 11.6). A 2 h irradiation period results in an increasing rate of hook opening up to 10–12 h after the onset of illumination; the rate then declines and hook opening is complete within about 18 h after the onset of irradiation (Galston *et al.,* 1964). Hook opening is probably related to decline in ethylene production in the hook region (Goeschl and Pratt, 1968), as discussed in Chapter 9.

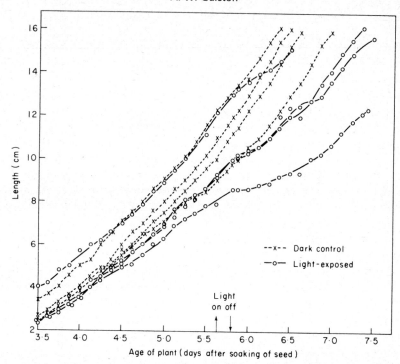

FIG. 11.5. Growth of 'Alaska' pea seedlings in total darkness and after exposure to red light for the indicated period. Note altered growth rates after a lag, and correction of the altered growth rate after a further lag period. (From Galston *et al.*, 1964)

C. Leaf Growth

Morphogenetically active light causes leaf growth rate to increase (Parker *et al.*, 1949; Bottomley *et al.*, 1966; Furuya and Thomas, 1964). Fresh weight doubles within 12 h after *c.* 200 mJ cm^{-2} of red light. This represents about a 30% rate increase over the dark controls. The effects of red light are essentially reversed by subsequently administered far-red light, and can be attributed to absorption by phytochrome. White light produces greater effects than red light alone, and one must presume the operation of at least one other photoreaction, possibly a blue-light mediated reaction.

D. Circumnutational Movements

Pea epicotyls show marked circumnutational patterns during growth in darkness (Galston *et al.*, 1964). When viewed from one side as a projection on to a single plane, the stem shows a *c.* 15° angular displacement from the vertical, oscillating from side to side with a period of *c.* 86 min at 26°C. Exposure to

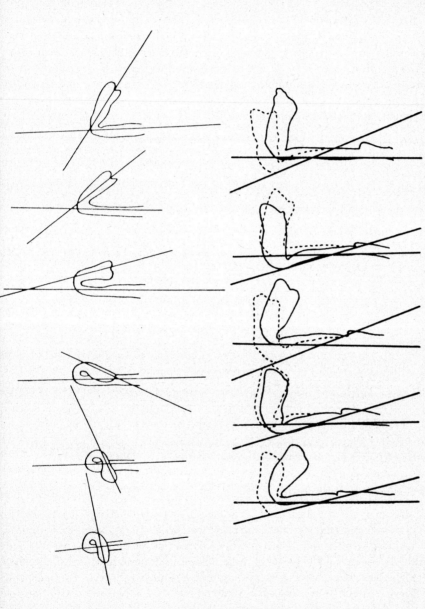

FIG. 11.6. Above. Stages in the opening of an etiolated apical hook (left) exposed to red irradiation. The straight lines indicate how angles were measured. Below. The "bobbing movement" shown by buds during seedling growth. (From Galston et al., 1964)

red light results, about 15–22 h later, in an increase of angular displacement to about 40° and a decrease in period of oscillation to about 76 min (Fig. 11.7).

Both auxins (Arnal, 1953) and gibberellins (McComb, 1962) have been linked to circumnutation. While it is clear that hormonal deprivation diminishes and hormonal supplementation increases such movement, no data convincingly separate these effects from direct promotion of growth. In effect, since nutations result from temporary and alternating inequalities in elongation rates on opposite sides of an organ, factors affecting growth rate would be expected to affect circumnutation.

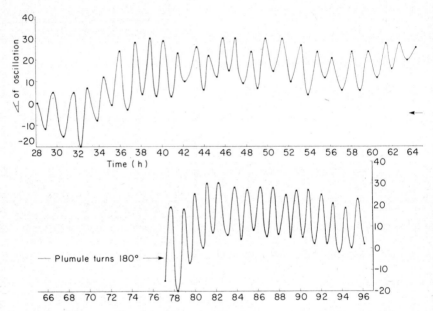

FIG. 11.7. Circumnutational patterns described by etiolated peas during growth, see text for details. (From Galston *et al.*, 1964)

E. Response to Hormones

1. Auxins

Completely etiolated sub-apical epicotyl sections, 5–10 mm long, grow rapidly in solutions containing 55 mM sucrose, 16 mM phosphate buffer (pH 6·1) and indol-3yl acetic acid (IAA) or other auxin. Longitudinally split etiolated pea epicotyls have been shown to curve in response to auxin (Thimann and Schneider, 1938, 1939; Galston and Hand, 1949). Endogenous auxin induces up to 30% elongation, while optimal auxin more than doubles the length within 18 h. The optimum IAA concentration for growth of etiolated pea stem sections is *c.* $5·7 \times 10^{-3}$ mM. This response is dependent on the presence

of sucrose in the medium (Purves and Galston, 1960). Prior irradiation with
red light shifts the IAA optimum in the presence of sucrose more than two
orders of magnitude to about 10^{-1} mM, but does not alter the maximum
growth obtained. An energy of c. 200 mJ cm^{-2} produces the maximal decrease in sensitivity to auxin if given about 18 h prior to the excision of tissue
for the growth test; about 0·1 maximal effect (optimum shifted to about
10^{-2} mM) if given 6–8 h before tissue excision; and no effect at all if given
24 h before tissue excision. Thus, it appears that the effects of morphogenetically active light are progressively magnified for about 18 h, then suddenly
decay to zero at 24 h (Galston and Baker, 1953).

Pretreatment with red light also greatly decreases the rate of removal of
exogenous IAA from the incubation medium, as determined by the Salkowski
test of the incubation solution (Galston and Baker, 1953). Homogenates of
entire etiolated epicotyls, fortified by the known cofactors 10^{-1} mM Mn^{2+} +
2,4-dichlorophenol (optimal for IAA oxidase activity, see below), show the
same effect, i.e. red light administered to the tissue approximately 16 h prior
to harvest depresses the activity of the homogenate. The effect of red light
is reversible by immediately applied FR, indicating light reception by phytochrome. The difference in IAA oxidase activities between R and FR treatments is annulled if the epicotyl homogenates are dialyzed (Hillman and
Galston, 1957). It was therefore inferred that R pretreatment leads to the
formation of a dialyzable inhibitor of IAA oxidase.

The inhibitor is most concentrated in the bud tissue, and is especially
abundant in green buds, young leaves and young internodes. The longer the
photoperiod under which the plant is grown, the higher is the titre of the
inhibitor. The inhibitor is also more abundant in tall than in dwarf pea
tissue, and its level in both is reduced by gibberellin application (Galston,
1959). The IAA oxidase inhibitor was later shown to be one or more flavonoids (Mumford et al., 1961; Furuya et al., 1962; Furuya and Galston, 1965).
This is discussed later below.

(a) *Phytochrome and flavonoid biosynthesis in peas.* In etiolated peas,
kaempferol-3-triglucoside (KG) and its p-coumaric acid ester (KGC) abound;
when phytochrome is transformed to the P_{fr} form by brief exposure to red
light, the terminal buds begin to make the analogous quercetin-3-triglucoside
(QG) and its p-coumaroyl ester (QGC) (Furuya and Galston, 1965; Bottomley
et al., 1965, 1966) while the internode tissue increases in KG and KGC
(Furuya and Thomas, 1964; Russell and Galston, 1967; see Fig. 11.8). The
changes in flavonoid content following P_r transformation to P_{fr} may possibly
be causally related to both the promotion of bud growth and the inhibition
of stem growth (Galston, 1969), although the available comparative kinetic
data on the observed increases in flavonoids and growth rate tend to diminish this as a likely possibility (Bottomley et al., 1966; Russell and Galston,
1967).

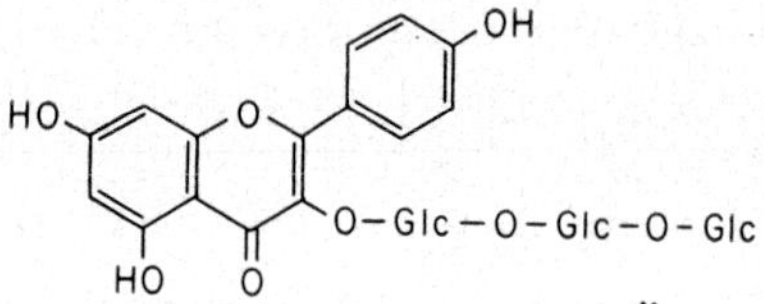

Kaempferol 3-triglucoside (I)*

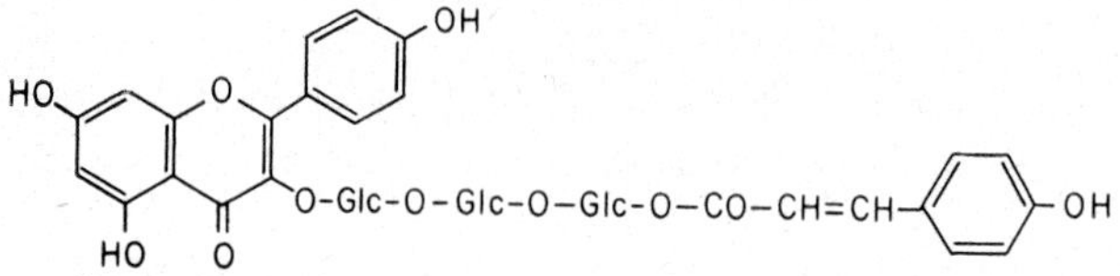

Kaempferol 3-(*p*-coumaroyltriglucoside) (II)

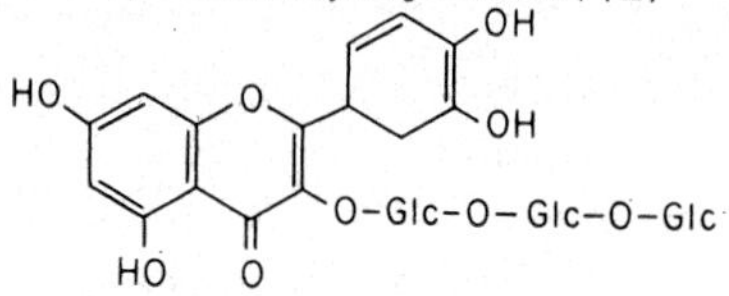

Quercetin 3-triglucoside (III)

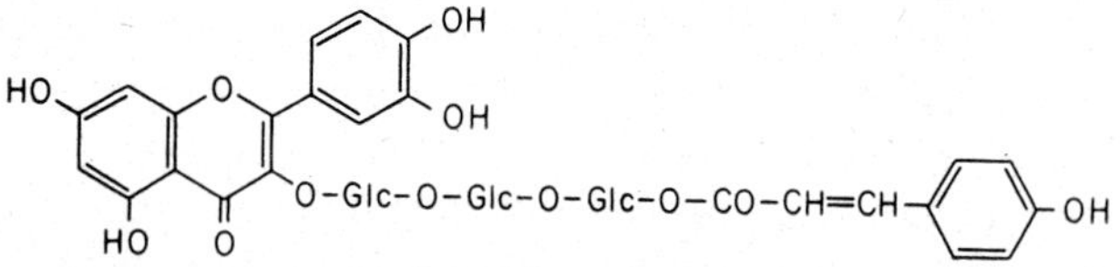

Quercetin 3-(*p*-coumaroyltriglucoside) (IV)

*Glc = glucose; linkage in trisaccharide probably β, 1→2; position of acyl linkage in II and IV still undetermined

FIG. 11.8. The structures of kaempferol and quercetin derivatives found in peas.

Later studies (Attridge and Smith, 1967) have shown that $P_r \longrightarrow P_{fr}$ conversion causes the appearance of considerably elevated activities of the enzyme phenylalanine ammonia lyase. Whether this is due to *de novo* synthesis or to activation of a pre-existing zymogen has been much debated, but the bulk of the evidence now favours the activation mechanism. Exposure of seedlings to light is also known to change some characteristics of the extracted enzyme (sensitivity to quercetin, decreased stability in the absence of sulfhydryl reagents) while not affecting the apparent molecular weight (Attridge and Smith, 1973).

(*b*) *IAA oxidase and peroxidase.* Etiolated pea tissue and its homogenates destroy IAA readily (Tang and Bonner, 1947, 1948). The destruction system is enzymatic, i.e. it is heat-labile and inhibited by cyanide, and is duplicated, at least in part, by crystalline horse-radish peroxidase (Goldacre, 1951; Galston *et al.*, 1953). The activity of the enzyme can be promoted by 10^{-1} mM Mn^{2+} (Wagenknecht and Burris, 1950) in the presence of *o*- and *p*-substituted monophenols (Goldacre *et al.*, 1953; Hillman and Galston, 1956), and inhibited by 10^{-1} mM Mn^{2+} alone (Hillman and Galston, 1956), by Co^{2+} (Galston and Siegel, 1954) and by o- and p-diphenols. Green tissues, originally reported to have no IAA oxidase activity at all (Tang and Bonner, 1948), do show such activity after scrupulous removal of polyphenolic inhibitors (Galston, 1969).

In etiolated tissue, the older the tissue, the higher is the IAA oxidase activity (Galston and Dalberg, 1954). Decapitation and IAA application may both, paradoxically, increase IAA oxidase activity in young, but not in old tissue (Birecka and Galston, 1970). Peroxidase, and thus presumably IAA oxidase as well, exists as numerous isozymes in the tissue (Siegel and Galston, 1967). In excised green pea stem sections, IAA represses the appearance of one of the isozymes of peroxidase (Ockerse *et al.*, 1966). The various organs of etiolated peas have multiple isoperoxidases, each organ showing its own pattern (Siegel and Galston, 1967).

When an etiolated pea homogenate is mixed with substrate-level IAA, visible light increases the rate at which IAA disappears. The action spectrum for this effect resembles that of riboflavin (Galston and Baker, 1949), and in fact free riboflavin can sensitize the photo-oxidation of IAA (Galston, 1949) and of various enzymes (Galston, 1950). Evidence was later adduced for photoactivation of a flavoprotein enzyme in the pea homogenate (Galston and Baker, 1951), and for the possible participation of such an enzyme in phototropism.

When dwarf pea plants ('Progress No. 9') with unusually high peroxidase activity are treated with a gibberellin, their peroxidase activity falls (with respect to controls) at about the time their growth rate rises; genetically tall ('Alaska') peas show no such effects (McCune and Galston, 1959; Galston and McCune, 1961). This diminution of peroxidase synthesis might be causally related to growth rate increase. It should be noted that GA_3 causes no qualitative change in the isoperoxidase pattern (Birecka and Galston, 1970).

2. Gibberellins

As discussed more fully in Chapter 9, totally etiolated peas are virtually insensitive to gibberellin application, but when seedling growth is altered by exposure to visible radiation, the plants become sensitive to gibberellin (Lockhart, 1956; Lockhart and Gottschall, 1959). Many dwarf types are in fact "photodwarfs", showing their dwarf character only after exposure to light. The metabolic basis for this effect of light is still not known, but it may involve photodestruction or photoinhibition of the synthesis of gibberellin, as well as altered sensitivity to gibberellins.

Excised, etiolated sub-apical pea epicotyl sections respond to auxins and gibberellins (Galston and Warburg, 1959; Purves and Hillman, 1958, 1959), and this interaction is not synergistic. In analogous green tissue, however, the reaction can be synergistic (Brian and Hemming, 1958; Galston and Kaur, 1961; Ockerse and Galston, 1967), and the degree of synergism is dependent on the photoperiod under which the plant has been grown.

3. Ethylene

We have already remarked that etiolated pea seedlings produce prodigious quantities of ethylene, especially in their apical regions (Goeschl and Pratt, 1968), and that phytochrome affects the rate of this process in the region of the hook (Kang *et al.,* 1967). Etiolated peas have long been known as a sensitive detector of ethylene, manifesting a "triple response" (inhibition of elongation, swelling and ageotropic behaviour) in response to minute quantities of the gas. Ethylene production is greatly increased by supra-optimal auxin levels (Burg and Burg, 1966, 1968) and the growth "inhibitions" beyond the optimum concentrations correlate well with hormone-induced ethylene production. It is interesting that green pea stem sections, which show no auxin optimum up to 10^{-1} mM IAA also produce no ethylene in the presence of elevated auxin levels (Burg and Burg, 1966).

F. Effects of Phytochrome on Transport and Utilization of Sucrose and Other Substances

Because P_{fr} promotes flavonoid synthesis, it is reasonable to expect it to promote the incorporation of precursors into flavonoids. In experiments designed to test this possibility, Goren and Galston (1966) excised epicotyls just above the cotyledons, dipped the cut ends into ^{14}C-sucrose, and studied the effect of red and far-red light on sucrose uptake and transformation. Red light (R) promoted both bud growth and sucrose uptake into the bud; both phenomena were completely reversible by far-red (FR) if applied immediately, half-reversible by FR if applied 4 h after R, and essentially irreversible by FR if applied after 8 h. The promotive effect by R was dependent on the length of epicotyl between site of sucrose application and the terminal bud. Thus, with sufficiently truncated epicotyls, R promoted neither bud growth nor sucrose uptake into the terminal bud, but with increasing length of epicotyl down to about the first node, the promotive effect of R was augmented. These effects were both promoted when *c.* 50–100 mM carrier sucrose was present in the medium.

The effect of R in promoting sucrose uptake and transport to the bud was more sensitive to low irradiances and occurred more quickly than growth effects, indicating a possible causal role in regulation of growth and flavonoid synthesis. The effect was also shown, though to lesser extent, with maltose, fructose and glucose, but not with ribose; phenylalanine and valine uptake and transport were also promoted by R, but not significantly reversed by FR, while tyrosine and cinnamic acid were taken up too slowly to show significant effects.

Both R-promoted growth and sucrose uptake by buds are inhibited by 10^{-3} mM gibberellic acid (GA_3) applied before or at irradiation time and reversed by GA_3 applied no more than 2 h after irradiation (Goren and Galston, 1967). GA_3 also reverses the effects of R on internode elongation and KGC synthesis, and alters the optimum of applied auxin (Russell and Galston, 1969).

The movement of ^{14}C-sucrose up the pea epicotyl requires active metabolism, being inhibited by anaerobiosis, low temperature and 2,4-dinitrophenol (Anand and Galston, 1972). Increase in bud labelling is detectable within 15–30 min and precedes the first detectable increase in growth rate; about 60–70% of the bud label remains as carbohydrate. In growing internodes, where P_{fr} decreases growth, it also decreases ^{14}C accumulation, lending support to the theory that growth control by P_{fr} is linked with its effects on uptake and transport of solutes. Investigations by Köhler et al. (1968) showed similar effects for ^{86}Rb-labelled K^+ solutions applied to pea roots. Thus, the effects of P_{fr} on membrane properties, which would be a logical control locus for both uptake and transport, must be fairly general, rather than specific for sucrose or carbohydrates.

In a comparative study of tall ('Alaska') and dwarf ('Progress No. 9') cultivars, Russell and Galston (1968) found identical effects and dose response curves for R and similar reversals by FR; however, decay of FR reversibility was more rapid in tall than in dwarf plants. The significance of this difference for the dwarf habit, if any, is not clear.

G. The Physiology of Pea Phytochrome

Using a differential two-wavelength "Ratiospect" photometer, and expressing their results as $\Delta(\Delta OD \text{ g f wt.}^{-1})$, Furuya and Hillman (1964) and Briggs and Siegelman (1965) found abundant phytochrome in the terminal bud and apical portions of the pea epicotyl, but relatively little in the remainder of 7-day-old and 5-day-old etiolated 'Alaska' pea seedlings. Phytochrome:protein ratios in peas, as in other etiolated seedlings, were invariably highest in meristematic regions where phytochrome content itself was highest. Phytochrome appearance and distribution in the embryonic axis of Alaska peas and in the young seedling have been chronicled by McArthur and Briggs (1970). Some microspectrophotometric evidence has been presented for phytochrome on the nuclear envelope of etiolated pea epicotyl cells (Galston, 1968).

The demonstrated presence of phytochrome in etiolated pea seedlings, however, does not automatically provide convincing proof of phytochrome participation in photophysiological events. Hillman (1965) showed that when excised etiolated pea stem sections are given varying light treatments, all P_{fr} levels inhibit their elongation, 40–60% P_{fr} producing maximal inhibition.

However, if the sections are excised from plants given 15 min red light 8–12 h before excision, then 5–20% P_{fr} promotes elongation, and only 20% P_{fr} inhibits. This suggests that there is a residue of P_{fr} in the pretreated tissue despite the fact that spectrophotometric assays show none to be present. This "*Pisum* paradox" (Hillman, 1967), like phytochrome paradoxes from other plant tissues, has been apparently resolved by the demonstration (Rubinstein *et al.*, 1969; Quail *et al.*, 1973) that P_{fr} but not P_r phytochrome binds to pelletable membranes which constitute only a small fraction of the total cellular phytochrome. Under such conditions, and if each fraction behaves independently and differently, no necessary correlation between spectrophotometric and physiological effects of phytochrome should be expected.

Another complexity in the phytochrome physiology of etiolated peas is the presence of a "killer factor" (Furuya and Hillman, 1966). This methanol- and butanol-soluble material interacts with P_{fr} to cause permanent losses in photoreversibility in *in vitro* phytochrome preparations. The titre of the factor is not affected by de-etiolation treatments and it is apparently absent from green pea tissue (Fox, 1975). Pea phytochrome is stabilized *in vitro* by metal-chelating agents including azide (Furuya *et al.*, 1965); this is understandable in terms of heavy-metal-promoted destruction of phytochrome *in vitro* (Lisansky and Galston, 1974). Brief R irradiation of etiolated pea tissues leads to the formation of rapidly reverting P_{fr}, with no destruction during the first 30 min. Later on, or after prolonged R irradiation, destruction begins (McArthur and Briggs, 1971). The *in vivo* significance of irreversible destruction, as distinct from reversion, is not at all clear, but it is instructive that assayable phytochrome levels in green tissues seem to be much lower than in etiolated tissues. This may be related to the greater synthesis of membranes, especially in such bodies as plastids, following illumination.

The chemistry of pea phytochrome has not been as extensively investigated as that of cereal phytochromes, but both Pratt (1973) and Rice and Briggs (1973) showed antigenic differences between pea and oat phytochrome.

III. GROWTH PATTERNS OF GREEN PEA PLANTS AND THEIR RESPONSE TO LIGHT

A. Growth Patterns of Entire Plants

A study of the growth pattern of 'Progress No. 9' peas has been made in connection with investigations on isoperoxidase ontogeny (Birecka and Galston, 1970). The plants were grown in vermiculite, watered with a solution of "Hyponex" mixed salts, and exposed to 16 h photoperiods at 21 500 lx at 23°C. Some plants were treated with 7 μg GA_3 in 5 μl of 3% ethanol. Fourteen days after planting, four nodes were in evidence, approximately 10, 19, 26

nd 31 mm above the ground level. The first two internodes had ceased
longating, but the remaining two continued growth for another 2 and 5 days,
espectively. By day 16, the fifth node was clearly apparent, by day 19 the
ixth could be discerned. Each successive node until the tenth appeared at
–3 day intervals, elongated vigorously for about one week, then ceased. Thus,
.t any one time, the two youngest internodes were vigorously elongating,
vhile the third youngest was in its last stages of growth in length. The pattern
s shown in Fig. 11.9. Concomitant with the increase in height was a decrease
n stem diameter; from the first internode, which weighed 8·5 mg mm^{-1}, down
o 4·3 mg mm^{-1} for the ninth internode.

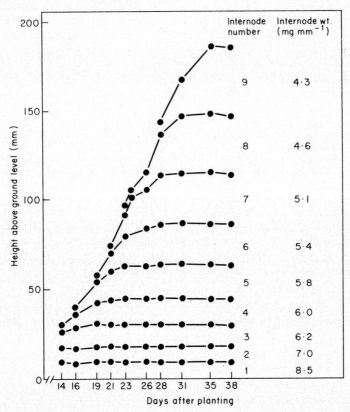

FIG. 11.9. Growth pattern of green 'Progress No. 9' pea plants (from Birecka and
Galston, 1970). Note the wavelike progression of growth from internode to internode.

The highest content of soluble protein per unit fresh weight was noted in
the youngest (3–4 mm long) internodes. A doubling of internode length was
accompanied by a halving of the protein content, but during subsequent
elongation, protein content decreased much less proportionally. Once an

internode ceased elongating, its protein content remained rather stable. Application of GA_3 caused an increase in elongation of growing internodes and a decrease in protein content and peroxidase activity relative to the control.

While decapitation inhibits the elongation of young internodes in genetically tall pea plants, it is without such effect in 'Progress No. 9' dwarf plants. Decapitation does, however, result in an increase in peroxidase activity in growing internodes and some change in the isoperoxidase pattern (Birecka and Galston, 1970).

B. Green Pea Stem Sections

Unlike sub-apical etiolated stem sections, analogous sections from light grown peas show no auxin optimum, at least up to $c.$ 10^{-1} mM IAA (Galston and Baker, 1951). Also, unlike etiolated peas, these sections are promoted rather than inhibited, by white light (mixed fluorescent–incandescent) o $c.$ 8600 lx. This light effect may be interpreted as due mainly to photosynthesis, since:

1. Its effect on growth continues to increase with increasing duration of exposure,

2. Sucrose does much to make up the difference in growth between darkened and illuminated green sections, and

3. The promotive effect of light is annulled by CMU and other substituted urea herbicides.

Paradoxically, $1–5 \times 10^{-1}$ mM CMU actually promotes the growth of etiolated sections in the presence of sucrose, or especially sucrose + IAA. The IAA optimum is unaffected by the inclusion of CMU. At 110 mM sucrose, optimal for growth of green sections in darkness, there is virtually no difference in growth attributable to light; however, at this concentration of sucrose, growth of light-grown sections is depressed about 25% below the optimum, at 0–25 mM sucrose (Galston and Kaur, 1961).

Light promotes the uptake of exogenous IAA by green pea stem sections (Thimann and Wardlaw, 1963). This may explain part of the acceleration of growth by light, although as previously mentioned, photosynthetic production of sucrose is also a factor.

The longer the light exposure per day of the parent plant, the better is the growth of excised green stem sections in the light and the worse is their growth in the dark period. These differences are especially manifest in the presence of sucrose, although the photoperiodic treatment produces no shift in the optimum for either sucrose or IAA. It thus appears that light-grown sections are "light adapted" and grow best in the light while etiolated sections may be said to be "dark adapted", since they grow best in the dark and are inhi-

ited by even low irradiances of visible light. The basis for such differences
unknown (Galston and Kaur, 1961).

In the light-grown pea, the apical bud is the only source of freely diffusible
uxin. The quantity of this auxin decreases progressively down to the base of
he epicotyl. Extractable auxin also declines, albeit less dramatically, and in
he growing internode, it does not decline at all. Removal of the apex results
n a disappearance of that part of the free auxin that is not mobile; ulti-
nately all the free auxin finds its way into the transport system and "dis-
ppears" into the base of the plant (Scott and Briggs, 1960). Subsequently, it
vas found that the youngest stem tissue has the greatest capacity for IAA
ptake, as measured by extraction, and that after uptake, IAA is transported
t about 10–12 mm h^{-1}. The IAA content in young, but not old stem tissue
an be increased by application of IAA in lanolin paste, and in this case, the
sual gradient of diffusible IAA is obliterated (Scott and Briggs, 1962). Dark-
rown peas contain only one-half the extractable auxin of light-grown peas,
nd show no diffusible auxin at all (Scott and Briggs, 1963). There is evidence
or auxin movement, however, since decapitation results in a slow disap-
earance of auxin only in apical regions (Scott and Briggs, 1963). Dis-
ppearance of auxin may result either from oxidative inactivation, as dis-
ussed previously, or by conjugation to form inactive derivatives (Andreae
nd van Ysselstein, 1956).

The growth of etiolated epicotyls is much more restricted topographically
han that of green pea stems. Thus, sub-apical etiolated sections show
reatest elongation and fresh weight gain in the presence of auxin when they
re about 3 mm long; sections longer or shorter than this length grow less in
esponse to auxin. By contrast, green sections also show an optimum response
t 3 mm length, but this optimum persists up to 20 mm in length. Excised
-mm long etiolated sections show maximum growth when derived from the
oungest internodes when they are 15–30 mm long, while green sections do
etter when excised from youngest internodes 25–50 mm long (Galston and
Kaur, 1961). The photoperiodic exposure of the plant from which the green
ections were excised does not affect the auxin optimum, but does affect the
pparent GA_3 optimum. Thus, plants grown at 8 or 16 h photoperiods furnish sections whose GA_3 optimum is around 10^{-2} mM, while those derived
from 24-h photoperiod plants show no optimum concentration even up to
3×10^{-1} mM.

Green sections of appropriate length and photoperiodic history show
marked $IAA–GA_3$ synergism (Galston and Kaur, 1961), while etiolated
sections do not (Purves and Hillman, 1958). Exposure of as little as 1 mm of
a section to GA_3 suffices to induce the synergistic interaction. The cobaltous
ion, which strongly promotes the elongation of etiolated pea sections in the
presence of sucrose (Miller, 1954), does not similarly promote the growth of
green stem sections.

L

C. Tendrils

Tendrils are produced by the third and subsequent leaves of light-grown 'Alaska' peas. At nodes 3, 4 and 5, these tendrils are unbranched and successively longer; at nodes after 6, the tendrils are branched once or several times (Jaffe and Galston, 1966a). The unbranched tendrils of node 5 reach maturity at a seedling age of about 10 days and retain maximum sensitivity to touch for about 3 days. When stroked on their ventral surface either on the plant or when excised, these tendrils coil rapidly, reaching a maximum of curvature within about 30 min to 3 h. Both elongation and coiling are sensitive to light, temperature and various hormones.

Tendrils are maximally sensitive when the plants bearing them are exposed to bright light after 8 h darkness (Jaffe and Galston, 1966b). Curvature increased almost linearly with logarithmic increases in white light intensity; since this light effect is sharply inhibited by DCMU and shows relative spectral efficiencies resembling chlorophyll absorption, it is assumed to be due to photosynthetic activity. Inhibitors of oxidative ATP formation such as dinitrophenol, CCP and anaerobiosis also reduced growth and/or curvature of tendrils. This led to the hypothesis that ATP might limit growth and curvature. Apparent confirmation was obtained when the addition of about 1 mM ATP to excised tendrils in the dark facilitated curvature, raising it almost to the level of the light control; 10 mM ATP raised growth to the light control value and caused a further increase in curvature. Further confirmation was obtained when illumination increased the ATP level of tendrils (determined by the firefly luciferase system) and mechanical stimulation sharply decreased the ATP titre, while increasing the level of inorganic phosphate. The activity of ATPase extractable from tendrils declines progressively after stimulation, reaching a minimum at about the time (60–90 min) the excised tendrils attain their maximum curvature. This ATPase also shows changes in specific viscosity in an Ostwald viscometer when ATP is added (Jaffe and Galston, 1967a) indicating a possible role of contractile proteins in the coiling phenomenon. An ATPase has been localized on the chloroplast envelope of tendrils (Sabnis, Gordon and Galston, 1970).

The flavonoid QGC (Fig. 11.8) is abundant in tendrils, especially near the most responsive tip. The titre of this flavonoid drops to less than half its former value when tendrils are stimulated to coil, and the kinetics of the decrease parallel the kinetics of the curvature itself. The flavonoid QGC, as well as aqueous extracts of unstimulated tendrils, represses curvature of excised tendrils. Thus, QGC titre either controls or is controlled by the state of coiling (Jaffe and Galston, 1967b). Auxin also induces curvature in unstimulated pea tendrils (Galston, 1971) as previously reported for other species by Reinhold (1967).

The cut bases of excised tendrils excrete more electrolytes (mainly H^+) and

eviously absorbed sucrose or acetate than do tendrils at rest (Jaffe and
ralston, 1968). There is also some evidence for greater loss of previously
bsorbed tritiated water from the ventral side (which contracts most) than
om the dorsal side.

Tendrils are so beautifully reactive that they must have much more to
each us than we have yet learned about them. This brief survey of photo-
orphogenesis and tendril response represents only "the tip of the iceberg",
nd despite the vast literature on the physiology and biochemistry of growth
nd development in the garden pea, much still remains to be explained.

EFERENCES

NAND, R. and GALSTON, A. W. (1972). *Am. J. Bot.* **59**, 327–336.
NDREAE, W. A. and VAN YSSELSTEIN, M. W. H. (1956). *Pl. Physiol.* **31**, 235–240.
RNAL, C. (1953). *Ann. Univ. Saraviensis* (Sci.) **2**, 186–203.
TTRIDGE, T. H. and SMITH, H. (1967). *Biochim. biophys. Acta* **148**, 805–807.
TTRIDGE, T. H. and SMITH, H. (1973). *Pl. Sci. Lett.* **1**, 247–252.
ERTSCH, W. F. and HILLMAN, W. S. (1961). *Am. J. Bot.* **48**, 504–511.
IRECKA, H. and GALSTON, A. W. (1970). *J. exp. Bot.* **21**, 735–745.
OTTOMLEY, W., SMITH, H. and GALSTON, A. W. (1965). *Nature, Lond.* **207**, 1311–1312.
OTTOMLEY, W., SMITH, H. and GALSTON, A. W. (1966). *Phytochemistry* **5**, 117–123.
RIAN, P. W. and HEMMING, H. G. (1958). *Ann. Bot.* **22**, 1–17.
RIGGS, W. R. and SIEGELMAN, H. W. (1965). *Pl. Physiol.* **40**, 934–941.
URG, S. P. and BURG, E. A. (1966). *Proc. natn. Acad. Sci. U.S.A.* **35**, 10–17.
URG, S. P. and BURG, E. A. (1968). *Pl. Physiol.* **43**, 1069–1074.
OX, L. R. (1975). *Pl. Physiol.* **55**, 386–389.
URUYA, M. and GALSTON, A. W. (1965). *Phytochemistry* **4**, 285–296.
URUYA, M. and HILLMAN, W. S. (1964). *Planta* **63**, 31–42.
URUYA, M. and HILLMAN, W. S. (1966). *Pl. Physiol.* **41**, 1242–1244.
URUYA, M. and THOMAS, R. G. (1964). *Pl. Physiol.* **39**, 634–642.
URUYA, M., GALSTON, A. W. and STOWE, B. B. (1962). *Nature, Lond.* **193**, 456–457.
URUYA, M., HOPKINS, W. G. and HILLMAN, W. S. (1965). *Archs Biochem. Biophys.* **112**, 180–186.
ALSTON, A. W. (1949). *Proc. natn. Acad. Sci. U.S.A.* **35**, 10–17.
ALSTON, A. W. (1950). *Science, N.Y.* **111**, 619–624.
ALSTON, A. W. (1959). *In* "Photoperiodism and Related Phenomena in Plants and Animals" (R. B. Withrow, ed.), pp 137–157. American Association for the Advancement of Science, Washington, D.C.
ALSTON, A. W. (1968). *Proc. natn. Acad. Sci. U.S.A.* **61**, 454–460.
ALSTON, A. W. (1969). *In* "Perspectives in Phytochemistry" (J. B. Harborne and T. Swain, eds), pp 193–204. Academic Press, New York and London.
ALSTON, A. W. (1971). *In* "Gravity and the Organism" (S. A. Gordon and M. J. Cohen, eds), pp 453–467. University of Chicago Press.

GALSTON, A. W. and BAKER, R. S. (1949). *Am. J. Bot.* **36**, 773–780.

GALSTON, A. W. and BAKER, R. S. (1951). *Pl. Physiol.* **26**, 311–317.

GALSTON, A. W. and BAKER, R. S. (1953). *Am. J. Bot.* **40**, 512–516.

GALSTON, A. W. and DALBERG, L. Y. (1954). *Am. J. Bot.* **41**, 373–380.

GALSTON, A. W. and HAND, M. E. (1949). *Am. J. Bot.* **36**, 85–94.

GALSTON, A. W. and KAUR, R. (1961). *In* "Light and Life" (W. D. McElroy ar H. B. Glass, eds), pp 687–705. John Hopkins Press, Baltimore, U.S.A.

GALSTON, A. W. and McCUNE, D. C. (1961). *In* "Plant Growth Regulation" pp 611–626. Iowa State College Press, U.S.A.

GALSTON, A. W. and SIEGEL, S. M. (1954). *Science, N.Y.* **120**, 1070–1071.

GALSTON, A. W. and WARBURG, H. (1959). *Pl. Physiol.* **34**, 16–22.

GALSTON, A. W., BONNER, J. and BAKER, R. S. (1953). *Archs Biochem. Biophy* **42**, 456–470.

GALSTON, A. W., TUTTLE, A. A. and PENNY, P. J. (1964). *Am. J. Bot.* **51**, 853 858.

GOESCHL, J. D. and PRATT, H. K. (1968). *In* "Biochemistry and Physiology Plant Growth Substances" (F. Wightman and G. Setterfield, eds), pp 1229 1242. The Runge Press, Ottawa, Canada.

GOLDACRE, P. L. (1951). *Aust. J. sci. Res. Ser. B.* **4**, 293–302.

GOLDACRE, P. L., GALSTON, A. W. and WEINTRAUB, R. L. (1953). *Archs Biochem Biophys.* **43**, 358–373.

GOREN, R. and GALSTON, A. W. (1966). *Pl. Physiol.* **41**, 1055–1064.

GOREN, R. and GALSTON, A. W. (1967). *Pl. Physiol.* **42**, 1087–1090.

HILLMAN, W. S. (1965). *Physiologia Pl.* **18**, 346–358.

HILLMAN, W. S. (1967). *A. Rev. Pl. Physiol.* **18**, 301–324.

HILLMAN, W. S. and GALSTON, A. W. (1956). *Physiologia Pl.* **9**, 230–235.

HILLMAN, W. S. and GALSTON, A. W. (1957). *Pl. Physiol.* **32**, 129–135.

JAFFE, M. J. and GALSTON, A. W. (1966a). *Pl. Physiol.* **41**, 1014–1025.

JAFFE, M. J. and GALSTON, A. W. (1966b). *Pl. Physiol.* **41**, 1152–1158.

JAFFE, M. J. and GALSTON, A. W. (1967a). *Pl. Physiol.* **42**, 845–847.

JAFFE, M. J. and GALSTON, A. W. (1967b). *Pl. Physiol.* **42**, 848–850.

JAFFE, M. J. and GALSTON, A. W. (1968). *Pl. Physiol.* **43**, 537–542.

KANG, B. G., YOCUM, S., BURG, P. and RAY, P. M. (1967). *Science, N.Y.* **156** 958–959.

KÖHLER, D., WILLERT, K. V. and LÜTTGE, U. (1968). *Planta* **83**, 35–48.

LISANSKY, S. G. and GALSTON, A. W. (1974). *Pl. Physiol.* **53**, 352–359.

LOCKHART, J. A. (1956). *Proc. natn. Acad. Sci. U.S.A.* **42**, 841–848.

LOCKHART, J. A. and GOTTSCHALL, V. (1959). *Pl. Physiol.* **34**, 460–465.

McARTHUR, J. A. and BRIGGS, W. R. (1970). *Planta* **91**, 146–154.

McARTHUR, J. A. and BRIGGS, W. R. (1971). *Pl. Physiol.* **48**, 46–49.

McCOMB, A. J. (1962). *New Phytol.* **61**, 128–131.

McCUNE, D. C. and GALSTON, A. W. (1959). *Pl. Physiol.* **34**, 416–418.

MILLER, C. O. (1954). *Pl. Physiol., Lancaster* **29**, 79–82.

MUMFORD, F. E., SMITH, D. H. and CASTLE, J. E. (1961). *Pl. Physiol.* **36**, 752–75

OCKERSE, R. and GALSTON, A. W. (1967). *Pl. Physiol.* **42**, 47–54.

OCKERSE, R., SIEGEL, B. Z. and GALSTON, A. W. (1966). *Science, N.Y.* **151** 452–453.

PARKER, M. W., HENDRICKS, S. B., BORTHWICK, H. A. and WENT, F. W. (1949) *Am. J. Bot.* **36**, 194–204.

PRATT, L. H. (1973). *Pl. Physiol.* **51**, 203–209.

PURVES, W. K. and GALSTON, A. W. (1960). *Am. J. Bot.* **47**, 665–669.

ЈRVES, W. K. and HILLMAN, W. S. (1958). *Physiologia Pl.* **11**, 29–35

ЈRVES, W. K. and HILLMAN, W. S. (1959). *Physiologia Pl.* **12**, 786–798.

UAIL, P. H., MARMÉ, D. and SCHÄFER, E. (1973). *Nature, New Biology* **245**, 189–190.

EINHOLD, L. (1967). *Science, N.Y.* **158**, 791–793.

ICE, H. V. and BRIGGS, W. R. (1973). *Pl. Physiol.* **51**, 939–945.

UBINSTEIN, B., DRURY, K. S. and PARK, R. B. (1969). *Pl. Physiol.* **44**, 105–109.

USSELL, D. W. and GALSTON, A. W. (1967). *Phytochemistry* **6**, 791–797.

USSELL, D. W. and GALSTON, A. W. (1968). *Planta* **78**, 1–10.

USSELL, D. W. and GALSTON, A. W. (1969). *Pl. Physiol.* **44**, 1211–1216.

ABNIS, D. D., GORDON, M. and GALSTON, A. W. (1970). *Pl. Physiol.* **45**, 25–32.

COTT, T. K. and BRIGGS, W. R. (1960). *Am. J. Bot.* **47**, 492–499.

COTT, T. K. and BRIGGS, W. R. (1962). *Am. J. Bot.* **49**, 1056–1063.

COTT, T. K. and BRIGGS, W. R. (1963). *Am. J. Bot.* **50**, 652–657.

EGEL, B. Z. and GALSTON, A. W. (1967). *Pl. Physiol.* **42**, 221–226.

ANG, Y. W. and BONNER, J. (1947). *Archs Biochem.* **13**, 11–25.

ANG, Y. W. and BONNER, J. (1948). *Am. J. Bot.* **35**, 570–578.

HIMANN, K. V. and SCHNEIDER, C. L. (1938). *Am. J. Bot.* **25**, 627–641.

HIMANN, K. V. and SCHNEIDER, C. L. (1939). *Am. J. Bot.* **26**, 328–333.

HIMANN, K. V. and WARDLAW, I. F. (1963). *Physiologia. Pl.* **16**, 368–377.

HOMSON, B. F. (1954). *Am. J. Bot.* **41**, 326–332.

ЈAGENKNECHT, A. C. and BURRIS, R. H. (1950). *Archs Biochem.* **25**, 30–53.

12. Photosynthesis and Translocation

D. M. HARVEY

John Innes Institute, Norwich, England

I. INTRODUCTION

The primary aim of this chapter is to bring together some of the diverse information on those aspects of the physiology of transport which impinge on the agronomic problems associated with the yield and nutritional value of *Pisum sativum* as a crop plant. A case study is presented of the integrated relationship between photosynthesis and assimilate translocation and the culmination of these activities as expressed by the production and filling of seeds. In attempting to broaden our understanding of the physiological basis of the effects of these processes on yield it is hoped that subsequent breeding and selection work may benefit. Where possible attention is drawn to those areas requiring further research and the problems relating thereto are defined and discussed.

This chapter places especial emphasis on the quantitative aspects of carbon and nitrogen assimilation in peas, and the translocatory fate of these elements is discussed with respect to both vegetative and reproductive growth. The effects of temperature, illumination and nutritional history on genotypic expression of the plant are considered in relation to the net gain of carbon and nitrogen made by a plant during a specific period of its life. Growth and seed production will be seen as the outcome of a programmed sequence of physiological events within a system which shows considerable powers of accommodation in response to constraints imposed by the plant's environ ment.

II. PHOTOSYNTHESIS: A NET CARBON GAIN

A. The Measurement of Net Photosynthetic Rate

One basic purpose of photosynthetic studies on higher plants is to enable advances to be made in our general understanding of the interaction between the genotype and the environment on plant yield. As far as *Pisum* is con cerned, studies of this nature have received little attention in comparison with other agriculturally useful species, despite the increasing demand for *Pisum* as a seed protein crop for human and animal consumption.

Carbon dioxide photoassimilation (CO_2 uptake) by a leaf can be described in physiological terms as the net flux of CO_2 (P_n) which takes place across the gradient between an external (C_o) and an internal (C_i) CO_2 concentration existing across the diffusive resistance (r_d) of the leaf. Thus:

$$P_n = \frac{C_o - C_i}{r_d} \qquad (1$$

The resistance term (r_d) represents the summation of separate and definable resistances, i.e. stomatal, mesophyll and chloroplast resistance.

The derivation and development of Equation (1) giving consideration to factors such as CO_2 solubility, light intensity, respiration and cytoplasmic reaction kinetics is set out by Acock *et al.* (1971). The applicability of the equation to CO_2 uptake measurement in typical open-circuit assimilation chamber studies is defended by Bravdo (1971), and a consideration of the same equation in relation to glasshouse studies as compared with studies in the field has recently been given by Bierhuizen (1973).

Harvey and Hedley (1974) state that to determine in practice the rate of net CO_2 uptake by illuminated leaves, five basic requirements must be satisfied:

1. The selection and design of a suitable assimilation chamber (cuvette).
2. The attainment of light saturation at the leaf surface.
3. The control and monitoring of leaf temperatures.
4. The control and monitoring of the CO_2 concentration, relative humidity and flow rate of the gas supply to the enclosed leaf.
5. The attainment of steady state conditions with respect to the leaf.

The many available methods for meeting these objectives cannot be
escribed adequately here, and the reader is referred to the review of metho-
ology in photosynthetic studies by Sestack *et al.* (1971). This section will
oncentrate specifically on aspects relating to the measurement of net CO_2
ptake rate in *Pisum* leaves using open circuit (i.e. differential mode) infra-
ed gas analysis (IRGA). The leaf response to changes in CO_2 availability,
ight intensity and temperature will then be examined.

Studies of net CO_2 uptake by *Pisum* leaves reveal that the rate of gas flow
o the cuvette is a major source of error (Flinn 1969; Hellmuth 1971). If
lepletion of the content of CO_2 in the cuvette exceeds its replenishment the
hotosynthetic rate will obviously be underestimated. Similarly, as Bravdo
1971) observes, if effective gas mixing does take place within the cuvette,
CO_2 concentrations at the cuvette entrance and exit may be almost equal,
but the net CO_2 uptake rate will still be consistently underestimated due to
boundary layer effects at the leaf surfaces.

Hellmuth (1971) demonstrated that shoots of 14-day-old plants of *P. sativum* cv. 'Meteor' exposed to saturating light intensity showed greatest response
o CO_2 concentration over the range 100–450 ppm, the rate of net CO_2 uptake
ncreasing almost linearly by a factor of 6·5 from top to bottom of the range
Fig. 12.1).

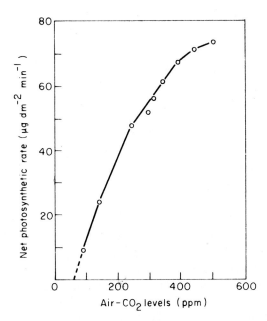

FIG. 12.1. Relationship between net photosynthesis and air-CO_2 concentration for 14-
day-old *Pisum sativum* cv. 'Meteor'. Leaf temperature 27°C, illuminance 17·6 klx.
(From Hellmuth, 1971)

L*

Hellmuth (1971) also showed that the optimum temperature for net CO_2 uptake at a light intensity of 17·6 klx was almost independent of the available CO_2 concentration within the range 140–520 ppm (Fig. 12.2). Fig. 12.2 also

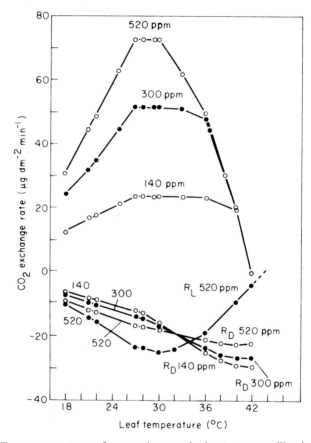

FIG. 12.2. Temperature curves for net photosynthesis at constant illuminance of 17·6 klx, and dark respiration (R_D) at various air-CO_2 concentrations, and the temperature curve for light respiration (R_L) at 520 ppm for 14-day-old *P. sativum* cv. 'Meteor'. (From Hellmuth, 1971)

shows that at 17·6 klx and 520 ppm CO_2 there was a 150% increase in net CO_2 uptake as the temperature was raised from 18 to 27°C. The response to a similar temperature increase at 140 and 300 ppm CO_2 was only a 100% increase. The response curve obtained at 300 ppm CO_2 (Fig. 12.2) is obviously the most appropriate to the effects to be expected under normal atmospheric conditions in the field.

Further studies by Hellmuth (1971) indicated that leaf temperature markedly influenced the magnitude of the maximum rate of net CO_2 uptake in lation to light intensity. The compensation point and saturation value for ght were found to be both markedly dependent on leaf temperature; the ght flux density required for saturation at a leaf temperature of 27–30°C as 17·5 klx, whereas exposure to 14·5 klx produced saturating conditions at 8 or 40°C (Fig. 12.3).

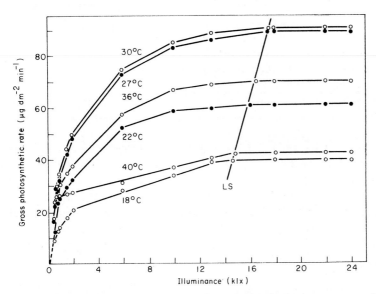

IG. 12.3. Light curves for gross photosynthesis at varying leaf temperatures and at an ir-CO_2 concentration of 520 ppm for 14-day-old *P. sativum* cv. 'Meteor'. LS = saturation illuminance. (From Hellmuth, 1971)

Further evidence is awaited before it can be decided whether the CO_2, light nd temperature responses reported by Hellmuth (1971) are general for the enus *Pisum* or whether they are prone to marked variation amongst cultiars. Additional work is also required to elucidate developmental and environmental effects on the responses.

B. Leaf Respiration

Respiratory processes in the leaf have received much attention in a wide ange of species possessing either the C_3 or the C_4 pattern of photosynthetic xation (see reviews: Jackson and Volk, 1970; Goldsworthy, 1970). An mportant conclusion for a C_3 species is that, for a given temperature, hotorespiration is quantitatively greater in its effect in reducing the net aily CO_2 uptake from the atmosphere than is the respiration that occurs uring darkness. All evidence suggests that the *Pisum* leaf exhibits a normal C_3 pattern of photosynthesis, the most obvious fact in support of this being the high CO_2 compensation point (60 ppm) recorded for this genus (Hellmut 1971).

In a recent study of *Phaseolus* leaf respiration, Mangat *et al.* (1974) hav utilized $^{12}CO_2$ and $^{14}CO_2$ to follow, over short time intervals, influx an efflux of CO_2 during light and darkness. Their results indicated that CO_2 evo lution by dark respiration was suppressed by 75% in the light, and it wa inferred that dark respiration and photorespiration utilized different sut strates and chemical routes. Little or no CO_2 derived from photorespirator processes could be detected in the dark, although dark respiration was show to be controlled in the illuminated leaf by photosynthetically produced ATF The possibility of similar responses in *Pisum* awaits confirmation.

Earlier work by Krotkov *et al.* (1958) on $^{12}CO_2$–$^{14}CO_2$ exchange in detache leaves of 22-day-old *P. sativum* (cv. 'Laxton's Progress'), indicated that in ai containing 5% CO_2 (at 20° $\pm$ 2°C) the illuminated leaf (*c.* 8·6 klx) produce less labelled CO_2 than did the corresponding leaf respiring in darkness Irrespective of whether a leaf has been previously experiencing light or dark its output of CO_2 in a subsequent period of illumination was about 0·73 m CO_2 h^{-1} g f. wt.$^{-1}$. In a corresponding period of dark a higher output o 2·45 mg CO_2 h^{-1} g f. wt.$^{-1}$ was recorded. Unfortunately these workers did no describe exchange response of this nature using the atmospheric CO_2 leve of 0·03%. So it cannot be ascertained whether the response obtained reall indicates a naturally low rate of photorespiration in the species or simply a suppression of photorespiration (and/or an increase in dark respiration) in the presence of high levels of CO_2.

Smillie (1962) studied dark respiration and photosynthesis in seedlings o *P. sativum* (cv. 'Laxton's Progress') throughout the first 17 days from ger mination, and, from study of O_2 exchange in intact leaves maintained a 13·5°C in 1·0% CO_2, indicated that the intensity of dark respiration of the first formed leaf on a unit fresh weight basis was at a maximum (600 μl O h^{-1} g f. wt.$^{-1}$) 5 days from germination. During the next 12 days of growth the respiration rate declined to 25% of this value. Over the same study period three respiratory enzymes—enolase (D-2-phosphoglycerate hydrolyase), 6 phosphogluconic dehydrogenase (6-phospho D-gluconate NAD oxido reductase), and aconitase (citrate (isocitrate) hydrolyase)—also showed a decline in activity. In contrast, maximum photosynthetic rate was attained in the first leaf 9 days after germination of the seed.

The rapid rise and fall in respiratory activity reported for seedling leave by Smillie (1962) contrasts with the almost stable rate of dark respiration recorded for the leaf subtending a pod of *P. arvense*. During a period of 2 days of fruit development, dark respiration of the adjacent blossom leaf wa maintained within the range 0·4–0·9 mg CO_2 leaf^{-1} h^{-1} (Flinn and Pate, 1970) In an attempt to study the interaction of environmental variables on the gas exchange of pea leaves, Hellmuth (1971) studied the rates of dark respira on and photorespiration in leaves of 14-day-old shoots of *P. sativum* cv. Meteor' over temperatures in the range 18–40°C and at CO_2 concentrations f 140, 300 or 520 ppm. Dark respiration rate was found to depend on both emperature and the CO_2 concentration (Fig. 12.2). At the near atmospheric alue of 300 ppm CO_2, an increase of temperature of from 18 to 40°C caused ae dark respiration rate to increase almost linearly from 8 to 27 μg CO_2 m^{-2} min^{-1}.

Hellmuth (1971) recorded that the interaction between CO_2 supply and leaf emperature was complex. At 32·5°C the dark respiration rate was similar at ach of the CO_2 concentrations employed, but CO_2 enrichment to 520 ppm acreased the dark respiration rate if temperatures below 32·5°C were used, nd decreased if the shoots were exposed to higher temperatures. Depletion f the CO_2 level to 140 ppm had opposite effects on dark respiration rate as hown in Fig. 12.2.

Hellmuth (1971) derived an estimate of the photorespiratory rate by xtrapolation of the plot for the net CO_2 exchange rate *vs* illuminance (see 'ig. 12.4). The indication was that at 520 ppm CO_2, photorespiration always xceeded dark respiration throughout the temperature range of 18–32°C, but hat beyond this point, dark respiration dominated. The highest rate of hotorespiration in the presence of 520 ppm CO_2 was 23·0 μg CO_2 dm^{-2} min^{-1}.

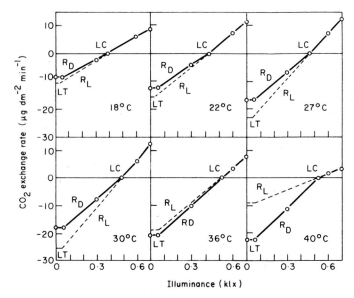

FIG. 12.4. Carbon dioxide exchange at very low illuminances and varying leaf tempera-ires and the evaluation of light respiration (R_L) for 14-day-old *P. sativum* cv. 'Meteor'. he air-CO_2 concentration was 520 ppm. R_D = dark respiration; LT = light threshold; LC = light compensation. (From Hellmuth, 1971)

This was achieved at 30°C and 17·6 klx. This was approximately 30 % greate than the corresponding value for dark respiration. Again it was unfortuna that photorespiration rate was not studied using this method in a norm; atmosphere. Attention should be drawn to the questions raised concernin the validity of extrapolation methods (Jackson and Volk, 1970).

As with other crop species the environment in which peas are grown ca markedly influence the rate of respiration of their leaves. For exampl Harvey (1972a,b) demonstrated that during the normal growing seaso of *P. sativum* (in the U.K.) the field-grown plants of cultivars 'Chemi Long', 'Orfac' and 'Sultan' each had a significantly higher dark respiratio rate per unit area of leaf than they did when grown under glasshous conditions. Clearly, considerable work remains to be done to evaluat the interplay of genetical and environmental effects on CO_2 loss by *Pisur* leaves.

C. Genotypic Variation in the Photosynthetic Rate

1. The Underlying Problems

The concept is now widely accepted that there may be within a genus highl significant variations in photosynthetic potential. For example, Cooper an Wilson (1970) have recorded an approximately two-fold difference in rate c apparent photosynthesis between two genotypes of *Lolium perenne* an Harvey (1972a,b) has indicated that a similar situation exists in cultivars o *P. sativum*. There is therefore the attractive possibility of selecting genotype which are photosynthetically more efficient and possibly of greater yiel potential than their parents. However, it must be borne in mind that infor mation regarding relative rates of CO_2 uptake is only one of several factor likely to influence yield. In particular, knowledge must also be available i relation to the following four factors:

1. The subsequent fate of the photoassimilated carbon, especially in term of fostering new growth above and below ground.

2. The effects on CO_2 fixation potential of factors such as temperature illumination and nutritional history.

3. The extent to which CO_2 evolution in respiration, or dark fixatio of CO_2, might influence the carbon balance of the plant and hence of it leaves.

4. The effects of ontogeny and leaf age on the photosynthetic performanc of the leaf.

Notwithstanding these complications, it can still be said that net CO_2 up take data for genotypes of pea may well help to indicate the extent to whic genetic variability might aid the plant breeder in his never-ending search fo better yielding cultivars.

2. The Chloroplast

Earlier work on the physiology of *P. sativum* chloroplasts was summarized by Smillie and Krotkov (1959) and in this summary they established a number of features. Firstly, non-oxidative photosynthetic phosphorylation was found to be responsible for ATP synthesis in the plastid, this being in clear contrast with the oxidative phosphorylation conducted by mitochondria. Secondly, it was shown that isolated *P. sativum* chloroplasts were capable of photoassimilation of CO_2 and of most of the other reactions characteristic of photosynthesis in intact leaf photosynthesis. Finally, and of central interest, was the finding that when expressed on the basis of a unit chlorophyll and an $ATP:O_2$ ratio of 3, the plastids capacity for ATP production was similar to the ATP requirement for photosynthesis. For example, Smillie and Krotkov (1959) showed that a photosynthetic oxygen output of 89 μM O_2 h^{-1} per mg chlorophyll was apparently associated with an ATP consumption of 267 μM ATP h^{-1} per mg chlorophyll and an ATP production of 200–300 μM ATP h^{-1} per mg chlorophyll.

A study of the chlorophyll content in *P. sativum* cv. 'Laxton's Progress' by Smillie and Krotkov (1961) indicated that a maximum concentration of 3·0 mg chlorophyll per g leaf was reached in the first seedling leaf 12 days after germination (Smillie and Krotkov, 1961), but further work has shown that the total chlorophyll content of the leaf can vary markedly from one genotype to another, in some cases apparently in excess of requirement. Leaves of normal plants of *P. sativum* (cv. 'Greenfeast') possess a chlorophyll a:chlorophyll b ratio of approximately 3, but Highkin *et al.* (1969) report on a spontaneously occurring mutant deficient in chlorophyll in which the corresponding ratio was about 19. However, the ratio became reduced to 12 in successive generations developed from the mutant. Total chlorophyll estimates revealed that the normal and mutant leaves respectively had 2150 and 1020 μg chlorophyll g f. wt.$^{-1}$, or if expressed in terms of leaf the mutant had only 40% of the chlorophyll content of the normal leaf; respectively, 17·4 and 45·1 μg chlorophyll cm^{-2}.

Studying photosynthesis, these workers showed that the rate of CO_2 uptake by the normal plant was light saturated at only 60 klx, whereas the mutant required 113 klx for saturation. At 113 klx the respective CO_2 uptake rates were similar on a unit area basis (about 20 mg CO_2 dm^{-2} h^{-1}) but on a chlorophyll basis the mutant had twice the activity of the normal plant. The Hill reaction (i.e. photolysis of water) by isolated chloroplasts was, in the case of the mutant, saturated at higher than normal light intensity and exhibited at least twice the normal rate on a chlorophyll basis. In *P. sativum* (cv. 'Alaska') the Hill reaction rate has been shown to depend on the temperature and illumination history of the chloroplast (Miller, 1960). For a wider assessment of chlorophyll levels, chloroplast structure and function in plants, the reader is referred to Park and Sane (1971).

Dowdell and Dodge (1970) conducted an IRGA assessment of CO_2 up take by intact leaves of *P. sativum* (cv. 'Meteor') after the rate of chlorophy synthesis had been artificially regulated by treatment either with an antibioti (chloramphenicol or terramycin) or by low light intensity. They conclude that there was no consistent relationship between chlorophyll content an the photosynthetic rate, except under conditions of very low chlorophyll. A Highkin *et al.* (1969) demonstrated, the chlorophyll content of chloroplast may vary over a wide range before appearing to be a limiting factor in CO uptake, so that it would clearly be unwise to attempt to select for bette photosynthetic potential by comparison of activity based on unit chlorophyll

3. Normal and Mutant Leaf Structure

The normal (or wild type) *Pisum* leaf is denoted in genetic terms as *Af/Af Tl/Tl* and it normally has one to three pairs of leaflets per petiole and a dista tendril. Mutants of markedly different leaf morphology to this have been recorded and held to be due to the activity of corresponding recessive genes In the case of the genotype *af/af, Tl/Tl* the leaflets are converted into much branched tendrils. Conversely, *Af/Af, tl/tl* has no tendrils but five or six pair of apparently normal leaflets. The homozygous double recessive (*af/af, tl/tl* has a much branched petiole and numerous relatively minute leaflets. Th literature relating to the origin of these mutations has been reviewed by Harvey (1972a).

Harvey (1972a) observed that the chloroplast-containing palisade an mesophyll tissues present in normal *Pisum* leaflets, were absent from norma tendrils and the tendrils of *af/af, Tl/Tl*. The chloroplasts of tendrils wer confined to four or five consecutive sub-epidermal layers of cells, thes layers occupying the entire circumference of the tendril. This type of chloro plast distribution is also characteristic of stems, petioles and tendrils o normal-leaved plants.

Stomata were observed to be present at four times the frequency per uni area on the lower than on the upper epidermis of mature leaves of *P. sativun* (cv. 'Chemin Long'). The stomatal openings were randomly orientated in the plane of the lamina except in the regions of leaf venation where the slit were orientated along the vein axis. The epidermis of normal tendrils and th tendrils of *af/af, Tl/Tl* had about 50% fewer stomata than the mean for both surfaces of the normal leaf, and the stomatal slits were orientated along th tendril axis.

The replacement of leaflets by tendrils carries the advantages of improvin the standing ability (i.e. lodging resistance) of the sward, and the rate of cro drying at harvest (Harvey, 1972a; Snoad 1974), but it has to be borne in mind that any reduction in the number of leaflets might adversely influenc the plant's photosynthetic potential. In a study of genotypic variation in *Pisum*, Harvey (1972a) estimated the photosynthetic rates attainable by th foliar mutations described above and normal-leaved cultivars. Estimates for the net CO_2 uptake rate were obtained using an IRGA for the youngest fully expanded leaves of vegetative plants grown in a glasshouse during April to June in England. At light saturation, 330 ppm CO_2, and at a leaf temperature of about 19°C, the data (Fig. 12.5) indicated that when compared on a unit surface area basis, the photosynthetic activity of the mutant forms lay within the range of values recorded for normal-leaved genotypes. Nevertheless, the mutant with tendrils instead of leaves was the least effective of those assayed in utilizing light intensities of less than two-thirds light saturation, and if its photosynthetic activity was expressed on a unit dry weight basis it was only 18% as effective as a normal leaf, presumably because of a higher proportion of non-photosynthetic tissue.

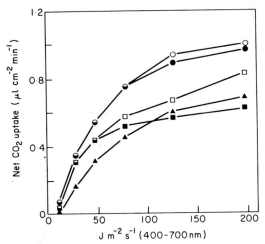

FIG. 12.5. The CO_2 photoassimilation response to different light intensities by two normal and three foliar mutant pea genotypes. (From Harvey, 1972) ○ *Af/Af, tl/tl,* ● 'Chemin Long', □ *af/af, tl/tl,* ▲ *af/af, Tl/Tl,* ■ 'Thomas Laxton', ◒ ◓ values overlap.

Attention is drawn to the distinction between leaflets and stipules. The normal stipule (*St/St*) can be reduced to a vestigal form (*st/st*) in a manner independent of the genes modifying leaflet specification (Snoad, 1974). Effects of stipule size on photosynthetic activity and efficiency have not yet been investigated, although according to Flinn (1969) the photosynthetic potential of a stipule is equivalent to that of a leaflet of the same area.

4. The Vegetative Phase

Smillie (1962) observed that during the early vegetative growth of *P. sativum* (cv. 'Laxton's Progress') the first-formed leaves each established a period of approximately 5 days when they maintained a near maximal activity in photosynthesis. Correlated changes in the activity of ribulose 1,5-diphosphate carboxylase and oxygen output were recorded, the corresponding maximum rates being $9.0\,\mu$mol substrate min^{-1} g f. wt^{-1} and 5000 μl O_2 h^{-1} g f. wt.$^{-1}$ respectively. Each leaf showed a well defined rise and fall in photosynthetic capacity. The attainment of the maximum rate of net CO_2 uptake often coincided with the completion of leaf expansion. The longevity of the optimum period for later leaves is variable and is affected by genetic and environmental factors.

5. The Reproductive Phase

It is quite conceivable that during the photosynthetic life of later-formed leaves, flower and pod production might influence photoassimilation by increasing the general demand for assimilates. This concept gained support when Flinn and Pate (1970) recorded for the leaf subtending a pod of *P. arvense* near maximal net CO_2 uptake rates of between 4·0 and 6·5 mg CO_2 h^{-1} per leaf sustained over a period of 22 (16 h) days in a controlled environment. It was further supported when Flinn (1974) showed that removal of a flower committed the subtending blossom leaf to a pattern of photosynthetic behaviour akin to that of the topmost vegetative leaf, and very different from the pattern expressed if the subtended fruit had developed. Removal of fruits also had the effect of allowing extra assimilates to be diverted to the root system and nodules (see Lawrie and Wheeler, 1974).

Nevertheless, as the results of Harvey (1972a) have implied, the maximum rate of CO_2 uptake attained by each successive leaf of *P. sativum* (cvs. 'Sultan' and 'Dark Skinned Perfection') appears to be a basic characteristic of the genotype and not markedly influenced by the transition from vegetative to reproductive phase. But this does not exclude the possibility that developing pods can increase the overall output of adjacent leaves by increasing the period of time during which they function at near maximum activity.

More recent evidence has indicated that in *P. sativum* (cv. 'Onward') the growth of the fruit influences markedly the photosynthetic potential of the subtending leaf. There are two phases of markedly increased rate of net CO_2 uptake, one corresponding with the attainment of maximum elongation of the pod, the other with the main period of swelling of the seeds (Flinn, 1974). However, this situation is not observed in blossom leaves of *P. sativum* cv. 'Orfac' (Harvey, 1974b) nor of *P. arvense* (Flinn and Pate, 1970). A further discussion of the inter-relationships of leaflet photosynthesis and fruit growth is to be found in Chapter 15.

III. TRANSLOCATION: A STUDY IN INTEGRATION

A. The Pathway of Translocation

Translocation is best defined for present purposes in terms of a developmental process within the whole plant—a mass flow or transfer, mediated by the regulated and simultaneous movement of different solutes via a vascular system which interconnects regions of solute origin (sources) and regions of solute utilization (sinks). In a recent succession of literature reviews [Milthorpe and Moorby (1969), Eschrich (1970), MacRobbie (1971), and Canny (1971 and 1973)] the respective authors have clearly indicated the present inconclusive state of knowledge and wide degree of conjecture as to the nature of the translocation mechanism.

Many hypothetical mechanisms for phloem transport have been forwarded but no proposed mechanism has gained general acceptance or been ascribed specifically to *Pisum*. Electron-microscopic studies of the fine structure of the phloem of *P. sativum* in relation to ontogeny and function have been conducted by Bouck and Cronshaw (1965), Wark and Chambers (1965) and by Wark (1965), who suggest that the sieve elements, companion cells and phloem parenchyma operate essentially as a co-ordinated structural and functional unit.

Pate (1975) observes that the pattern of interconnections within the vascular system of the *Pisum* stem corresponds to that previously described for *Trifolium* by Devadas and Beck (1972). In this system four axial bundles (designated A1, A2, A3, A4) run the length of the stem and in a defined sequence give rise to lateral traces supplying the organs of specific nodes. Two of these axial bundles connect to leaves in an alternate manner: bundle A1 links with the leaf at nodes 2, 4, 6, etc. while A4 links with the leaf at nodes, 1, 3, 5, etc. Bundles A2 and A3 are not directly linked to leaves. Thus there exist direct but unilateral vascular connections between specific leaves on the alternately branched stem (i.e. $\frac{1}{2}$ phyllotaxy).

The vascular supply to the stipule is more complex: at any node the stipule is linked to each of the three axial bundles which do not connect directly to the subtending leaf. For example the stipule at node 6 connects to A2, A3 and A4, whereas the leaf subtending node 6 is linked to A1. With regard to an axillary shoot (at node 6, for example) supply is via lateral traces which arise from A1 and A2, whereas the shoot at node 5 is linked to A3 and A4. Fruit vascular connections to the axial bundles are not indicated, although it could be inferred from studies of ^{14}C-assimilate translocation that at least two axial bundles might be involved, possibly with the same type of organization as described above for an axillary shoot.

In *Pisum* the output and solute composition of sap collected from severed elements of the vascular system has been much studied in relation to plant development. The xylem and phloem exhibit fundamental differences in terms of the range and levels of the individual solutes which they carry. These differences are primarily a function of the assimilatory activities of the respective metabolic centres from which the solutes come. The existence within the xylem and phloem sap of significantly different levels of solutes suggests that there may be little lateral exchange between adjacent channels within a vascular bundle. In terms of bulk transport the xylem and phloem are therefore likely to function very differently, not only in the spectrum of solutes which they convey but also in the extent of their acropetal or basipetal commitments in transport.

With respect to organic nitrogen movement, the xylem is almost exclusively concerned with the acropetal transport to the shoot of amides and amino acids synthesized within the roots. Pate and Wallace (1964) demonstrated that the balance of the various nitrogen compounds present in the xylem sap is dependent on, though not markedly altered by, the particular nitrogen source available to the roots, eg. $NO_3,^-$ NH_4^+, urea or symbiotically fixed nitrogen. In contrast, the phloem undertakes basipetal and acropetal transport of photosynthetically derived carbohydrate such as sucrose, and, in addition, carries amino acids synthesized in the leaf from some of the combined nitrogen which the leaf has received via the xylem. Lewis and Pate (1973) indicated that some 94% by weight of the phloem sap solutes is carbohydrate and only 6% is amino acids or amides and other constituents.

The overall complexities of the integrated transport system represented by the situation in the intact plant during its active growth have been summarized by Carr and Pate (1967) (Fig. 12.6). In interpreting the figure, attention must be drawn to the existence in *Pisum* of a gross mobilization of previously accumulated solutes after flowering. In this respect Fig. 12.6 may be viewed as a composite of the events throughout the growth cycle which lead ultimately to seed nutrition.

B. Lamina Vein Loading

Loading of the minor veins of the lamina of the leaf is essentially a process by which photosynthetic products are withdrawn from the mesophyll cells and accumulated in the parenchyma of minor veins prior to their export via the petiole phloem. It is quite conceivable that this loading process, not the rate of photosynthesis, might often be the factor limiting the rate of export of foliar assimilates.

A mechanism for the process of lamina vein loading has been proposed for *Pisum* by invoking a solute uptake and transfer function for specific cells that are found adjacent to the vascular elements of the minor veins of the

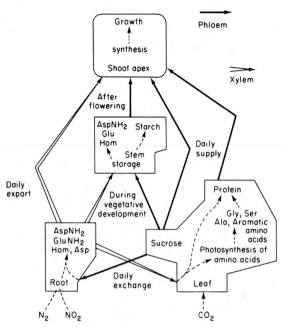

FIG. 12.6. Circulation of metabolites in the field pea (*P. arvense* L.) AspNH₂, asparagine; GluNH₂, glutamine; Hom, homoserine; Glu, glutamic acid; Gly, glycine; Ser, serine; Ala, alanine. (From Carr and Pate, 1967)

leaf (Gunning *et al.*, 1968; Pate and Gunning, 1972). A specialized structural characteristic of walls of these ancillary cells ("transfer cells") is the presence of abundant and highly convoluted ingrowths occupying a substantial proportion of peripheral region of the cell's volume. In some instances the wall ingrowths support a surface area of the plasma membrane some ten times higher than in a smooth-walled cell of similar size. This increase in surface area could be of some significance in a situation in which the transfer cells were able to draw upon solutes of the free space (apoplast) of the leaf, the very route, in fact, used in dispensing solutes to the leaf via the xylem. A high density of mitochondria occurs in transfer cells especially in the vicinity of their wall ingrowths, thus implicating them in having a special facility for active uptake against concentration gradients. This is the very situation that must exist when transferring solutes from a relatively dilute xylem to the much more concentrated fluids of the phloem. Plaut and Reinhold (1969), have obtained evidence that the ATP formed in non-cyclic photosynthetic phosphorylation may be expended during vein loading of *Phaseolus* leaf, and that the rate of translocation is influenced accordingly by light intensity. The existence of any such system in *Pisum* has not been confirmed.

An alternative explanation for the regulation of foliar export in *Pisum* has been proposed by Lovell *et al.* (1972), their main thesis being that the active and continued removal of translocated photosynthate by the sinks of the plant is a principal factor in controlling the export rate from a foliar surface. This conclusion was drawn from the following study conducted on vegetative plants of *P. sativum* cv. 'Kelvedon Wonder'. At a time 20 h prior to a 10-min period of photosynthesis in $^{14}CO_2$ by a single attached leaf, replicate plants were subjected to one of four different sink-removal treatments. One intact group of plants served as a control group. After the exposure to $^{14}CO_2$ the labelled source leaf was monitored for 3 h with a Geiger-Müller tube located below the leaf. The effect of each treatment was then assessed by measuring any subsequent changes in the rate of ^{14}C-export and expressing such changes as a percentage loss of radioactivity relative to the leaf's initial level of labelling. Results showed that defoliation of shoots (except for the source leaf) significantly increased the percentage of ^{14}C exported by the source leaf, but that removal of the shoot apex decreased greatly the percentage of the ^{14}C that was exported. A further significant decrease in export was observed when the shoot apex and the roots were removed, or when the apex, roots and leaves (other than the source leaf) were excised. Lovell *et al.* (1972) thus concluded that the progressive removal of available sinks for the ^{14}C-assimilate was positively associated with a reduction in the export activity of the source leaf. Unfortunately these authors did not measure changes in photosynthetic rate throughout the time course of their experiments, but, judging from results with other species, it is highly likely that sink removal would have ultimately depressed photosynthesis as well as translocation.

C. The Significance of Leaf Morphology

Genetically induced modifications of the morphology of the *Pisum* leaf, especially those concerned with leaflet and stipule area, might be expected to have quite profound influences on the translocation potential of the shoot, either by altering the quantity of assimilate exported, or by affecting the pattern with which this assimilate is distributed. A number of these morphological mutants are now available to test these suggestions and, as an added impetus to study, there is the possibility that some new foliar mutant might turn out to exhibit agronomically desirable features with regard to its pattern of interception of light and its ability to partition assimilates. These aspects have been investigated in some detail by Harvey (1974a) using two foliar mutants and comparing their performance with that of normal leaved *P. sativum* cv. 'Orfac' and cv. 'Dark-skinned Perfection'. It should be noted that the comparisons were carried out under field sward conditions.

Morphologically, the mutations studied involved either the conversion of leaflets to tendrils (genotype *af/af, Tl/Tl*) or the production of numerous minute leaflets in place of tendrils (*af/af, tl/tl*) (see Section II, part *3*). The particular leaf and growth stage selected for assay under field conditions corresponded to those previously shown to yield the most useful information on ^{14}C-distribution patterns in the normal leaved *P. sativum* cv. 'Orfac' (Harvey, 1973). Using a method of administration adapted for field use (Harvey, 1974a,b), the $^{14}CO_2$ was fed to specific leaves subtending either a vegetative or a reproductive node, and, after a 48-h translocation period the labelled plants were divided into haulm, roots and pods. A special ^{14}C assay procedure was adopted in which tissue fragments of dry material were rendered colourless and translucent prior to direct counting of the ^{14}C in the sample using scintillometry.

The ^{14}C data (Table 12.I) indicated that on a basis of specific activity of the respective sinks for ^{14}C, the mutant genotypes assayed were not significantly different from normal *P. sativum*. Specific activity was in this instance defined as the c.p.m./mg dry wt./10^6 c.p.m. exported from a fed leaf. The data showed that the distribution of ^{14}C amongst the various fruits on a shoot is by no means uniform. It was always found that the positions of those fruits acting as major sinks for labelled assimilate varied predictably with the positions of the leaf which had been fed with $^{14}CO_2$. A leaf at a fruiting node exported principally to its subtended fruit, but a leaf at a vegetative node exported mainly to the nearest fruit above it on the same side of the haulm, i.e. to the fruit most directly connected to it in terms of vasculature. In consequence the ^{14}C-distribution pattern obtained from the leaf subtending a pod at the second fruiting node was essentially similar to the pattern derived from the leaf of the last vegetative node.

The general conclusion from these studies was that genetically induced changes in leaf morphology did not markedly affect the translocation potential or pattern of partitioning of assimilates in the plant, implying that new foliage forms would be unlikely to exhibit any large scale imbalance in dry matter accumulation in vegetative or reproductive organs.

D. The Effect of Leaf Age on Export during Vegetative Growth

Carr and Pate (1967) studied the changing pattern of assimilate export activity during the productive life span of specified leaves of *P. arvense*. Developmentally related changes in the quantity and direction of exports were also determined. An initial study concentrated on the leaf at the third node of quite young vegetative plants and the standardized procedure was to measure the distribution of ^{14}C from this leaf after 1 h $^{14}CO_2$-photoassimilation in mid-photoperiod. The percentage of the ^{14}C fixed that was exported from the assayed leaf was recorded at growth stages ranging from the initial unfolding of the leaf at the plant apex through to leaf senescence. During the period studied a further seven nodes were expanded by the plant. The ^{14}C export

TABLE 12.I. The specific activities of the respective ^{14}C-sinks of two normal and two mutant genotypes of *Pisum sativum* that have been fed with $^{14}CO_2$ at specifed leaves prior to ^{14}C-assay and approximately 22 days from anthesis. Values are given as the mean ($\pm$ s.e.) c.p.m./mg dry matter/10^6 c.p.m. exported from the fed leaf

The Node Subtended by the $^{14}CO_2$ fed Leaf	Genotype	Cultivar or Mutant	Pods at First Fruiting Node		Pods at Second Fruiting Node		Pods at Third Fruiting Node	Distal Pods	Haulm plus Roots
			Wall	Seed	Wall	Seed	Wall plus Seed		
Last Vegetative	Af/Af, Tl/Tl	DSP[a]	32±6	124±35	173±56	657±274	59±9	55±5	20±8
	af/af, Tl/Tl	Tendrils	10±6	12±4	362±86	744±198	58±15	61±301	22±5
	af/af, tl/tl	Minute Leaflets	8±2	17±4	287±107	728±190	64±22	121±30	27±7
First Fruiting	Af/Af, Tl/Tl	'Orfac'	291±55	1030±46	4±1	14±6	32±10	14±6	10±3
	Af/Af, Tl/Tl	DSP	169±67	388±113	11±9	9±7	13±7	16±14	21±7
	af/af, Tl/Tl	Tendrils only	271±106	957±37	1±1	2±1	7±1	14±2	4±2
	af/af, tl/tl	Minute Leaflets	236±64	810±196	3±1	11±1	40±18	41±23	5±1
Second Fruiting	Af/Af, Tl/Tl	'Orfac'	5±2	22±18	334±39	740±373	42±18	22±6	10±6
	Af/Af, Tl/Tl	DSP	15±12	29±21	285±59	824±165	41±21	52±11	6±1
	af/af, Tl/Tl	Tendrils only	4±2	7±5	646±199	1167±436	6±2	22±11	5±2
	af/af, tl/tl	Minute Leaflets	6±1	10±4	323±80	1093±121	46±2	153±32	7±2

[a] DSP = 'Dark Skinned Perfection'.

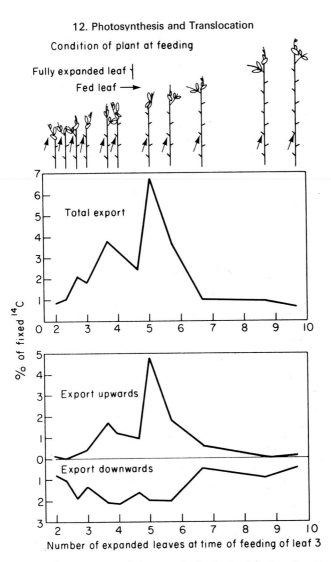

FIG. 12.7. The pattern of export of photosynthetically fixed carbon from leaf 3 of the field pea *P. arvense* L. $^{14}CO_2$ was fed to leaf 3 for 1 h and the percentage of the total fixed radio-carbon transported upwards and downwards from the leaf was determined. The age of leaf 3 at the time of feeding is measured with reference to the number of fully expanded leaves on the shoot. (From Carr and Pate, 1967)

data (Fig. 12.7) indicated that in each case the maxima for both upward and downward ^{14}C-assimilate movement from the leaf at the third node coincided with the expansion of the leaf subtending the fourth and fifth nodes respectively. Output from the third leaf had markedly declined before the expansion of the seventh leaf.

Due to the different physiological ages and photosynthetic conditions of successive leaves on the shoot not all will contribute with equal effectiveness at any one time, and the pattern of these levels of export will change continuously as lower leaves senesce and new ones expand at the apex. In a further study intended to examine these aspects Carr and Pate (1967) recorded the percentage of ^{14}C exported by each leaf on vegetative plants which had, at the time of study, three, five, seven or twelve leaves present on their main shoots. The ^{14}C distribution data (Fig. 12.8) indicated that during the ageing of

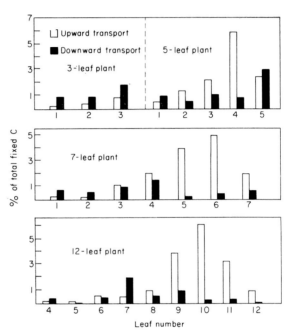

FIG. 12.8. The pattern of export from leaves at four times during the vegetative life of the field pea *P. arvense* L. Export upwards and downwards from each leaf is measured as percentage of total fixed carbon exported after 1 h of photosynthesis in ^{14}CO$_2$. The data are derived from individual measurements of export from single leaves of three replicates. (From Carr and Pate, 1967)

a leaf, significant and reproducible changes occur in the proportions of upward and downward export from the leaf. In the examples quoted, downward export dominated during the three-leaf stage, upward export from the leaf at the fourth node dominated the distribution pattern of the five-leaf plant, while at the seven- and twelve-leaf stages, the leaves at the sixth and tenth nodes respectively were those dominant in upward export.

Pate (1966) studied the contribution of photosynthetic carbon to current protein synthesis in the shoot apex of nodulated vegetative plants of *P. arvense* and observed that each mature leaf (at the eight-leaf stage) was observed to be effective in donating carbon to protein, the upper leaves feeding the apex directly with photosynthetically produced amino acids, while the lower leaves contributed carbon which had cycled through the roots. Sucrose accounted for more than 90% of the ^{14}C exported from a leaf of any age, and upper leaves mainly exported to the roots and nodules.

In a review of the effects of leaf age on translocation from leaves, Thrower (1967) emphasized the distinction between changes in distribution and changes in total quantity of assimilate exported, irrespective of destination. Both distribution and quantity change as the leaf ages. The explanation put forward (for the case of the vegetative plant) was that as further leaves unfold, so an older leaf becomes further and further removed from the influence of an apically derived auxin, and therefore less likely to respond to demands from the apical sinks. This suggestion is explored further in the next section.

E. Auxin Regulation of Assimilate Movement

Scott and Briggs (1960) demonstrated that a concentration gradient of apically derived endogenous auxin exists in the stem of *P. sativum* cv. 'Alaska', the diffusible auxin concentration decreasing with distance from the apex of a vegetative plant. Booth *et al.* (1962) observed the effect of 0·1 % indol-3yl acetic acid (IAA) in lanolin applied to the cut surface of decapitated plants of *P. sativum* cv. 'Meteor'. These workers measured the internodal distribution of ^{14}C label after export from a mature leaf to which ^{14}C sucrose had been applied. The ^{14}C-distribution corresponded in magnitude to that attained in the intact control plants, but decapitated control plants (which had received ^{14}C-sucrose but not IAA) accumulated only one-third as much ^{14}C at the internodes. The conclusion reached was that the IAA had stimulated ^{14}C-assimilate movement to, and accumulation by, those tissues from which it was diffusing.

Further to the latter concept, Davies and Wareing (1965) demonstrated that for vegetative plants of *P. sativum* cv. 'Meteor' an apical application of 0·1 % IAA induced the acropetal transport of ^{32}P injected as sodium orthophosphate at the first internode. Decapitated plants treated with IAA behaved in the same manner. In parallel experiments the stem was steam-girdled at the third internode (prior to the application of IAA) in order to kill locally the phloem tissues. Xylem appeared to remain functional as indicated by the maintenance of shoot turgidity. This treatment strongly inhibited the acropetal movement of ^{32}P.

In another approach Davies and Wareing (1965) employed a distal or midstem application of 1 % 2, 3, 5-tri-iodobenzoic acid (TIBA), a treatment previously known to block the polar movement of auxin in pea stems. They then applied IAA and ^{32}P to the TIBA-treated plants (also cv. 'Meteor') employing the same method as described above, and the result was that acropetal movement of ^{32}P was reduced by more than 90% compared with that in TIBA-free controls. The implication was that auxin-directed transport involved unimpeded basipetal movement of auxin and was not merely a localized effect of auxin at the site of its application. Further study by these workers suggested that from amongst a range of growth substances tested only IAA and naphthoxyacetic acid were effective in directing assimilate movement in *Pisum*. (Indol-acetonitrile; 2,4-dichlorophenoxyacetic acid, 6-furfuryl amino purine (kinetin) and gibberellic acid (GA$_3$) were the substances shown to be ineffective.)

Morris and Thomas (1968) presented evidence that the apex and root system of intact vegetative *Pisum* plants might be in active competition for translocated assimilates from leaves. In 14-day-old seedlings of *P. sativum* cv. 'Meteor', 53·1% of the total labelled translocate derived from ^{14}C-sucrose applied to the third leaf, was distributed to the shoot apex, and only 34·7% to the roots. The response to decapitation at the third internode was a readjustment of these allocations, lateral buds receiving an aggregate of 15·0%, roots 61·6%. Extending the work of Booth *et al.* (1962), Morris and Thomas (1968) also observed that ^{14}C-translocation patterns could be modified by the application of supplementary auxin (10 μg in lanolin) to the cut stump. Applications of either IAA, kinetin, or IAA plus kinetin restored capacity to attract ^{14}C assimilates towards the apex to within 4% of the value recorded for non-auxin-treated control plants. The response to kinetin application was to increase the share of label supplied to the lateral buds from a mere 0·2% to 30·9% of the total ^{14}C-translocate. In contrast to the latter response, IAA and IAA plus kinetin respectively increased internodal accumulation from 2·6% to 43·5 and 45·3% of the total ^{14}C available, without markedly increasing the amount of label reaching the lateral buds.

Of relevance to our study here of *Pisum* is the observation by Seth and Wareing (1967) that in *Phaseolus vulgaris* the translocation of assimilates to developing fruits might be regulated, or at least modulated, by the presence of auxin. Application of IAA to a peduncle from which the fruit had been previously removed had the significant effect of enhancing the accumulation of ^{32}P by this peduncle after the isotope had been injected through the stem-base. Furthermore, in comparison with their individual performances there was a strong synergistic effect between IAA, kinetin and GA$_3$, the effect of the synergism being to enhance even further acropetal transport of ^{32}P towards the site of auxin application. Seth and Wareing (1967) also demonstrated a somewhat similar effect of auxin in directing translocation of ^{14}C-photosynthate from the leaves to the peduncle. This suggested that auxin might have a key role in fostering assimilate flow to seeds. Further indirect evidence of this comes from the finding that seeds are rich sources of endogenous growth substances (see Chapter 15).

As summarized by Skoog (1973), cytokinins are now known to be involved in directing movement of assimilates and in the regulation of assimilate incorporation into apices. Possibly of greatest importance is the finding that cytokinins are capable also of determining the longevity of vegetative plant parts and therefore of regulating their facility to release nutrients for reproductive growth. The mechanism of cytokinin induced regulation is not yet fully understood. However, changes in the physical properties of cell membranes and an indirect influence on enzyme production and other biosyntheses are clearly implicated. For instance, in *Pisum*, Davidson (1971) has demonstrated that the induction of growth of a lateral bud treated with cytokinin is preceded by endogenous auxin synthesis coupled with a correlative decrease in auxin production by the terminal bud. The correlative influences in seedling growth dealt with in Chapter 10 are likely to have long-term implications in respect of transport similar to those described here.

F. The Supply of Photosynthate to Root Nodules

1. Development and Synchrony

The onset of effective nodulation in *Pisum* usually occurs at a relatively early stage of plant development and certainly usually well in advance of exhaustion of cotyledonary nitrogen. Subsequently there is a characteristic synchrony between the symbiotic activity and physiological events of the life cycle (see Chapter 13).

Pate and Wallace (1964) recorded that in *P. arvense* plants inoculated at sowing with an effective race of *Rhizobium*, root nodulation is first visible at 13 days. The haemoglobin pigment associated with effective functioning of the nodule in nitrogen fixation is formed 3 days later, and the first nitrogen fixation that could be detected by an increase in plant nitrogen occurred between 16 and 19 days from germination. By this time the third foliar leaf had expanded. According to Carr and Pate (1967) it is at this stage of plant development that the third leaf becomes capable of exporting ^{14}C-photosynthate to the roots. At a later stage, when eight mature leaves are present, Pate (1966) observed that only the lower leaves were really actively engaged in direct export of current assimilates to the roots and nodules; ^{14}C-labelled sugars transported to the roots after photosynthetic activity by the shoot were shown to be the major source of the carbon metabolized during the formation of amino acids in nitrogen fixation (see Chapter 13).

Lawrie and Wheeler (1973) have investigated in further detail the relationship between nitrogen fixation and photosynthate supply to the nodule using *P. sativum* cv. 'Alaska' plants inoculated with an effective race of *Rhizobium* at planting, and supplied with a nitrogen-free nutrient solution. The growth cabinet conditions used were a 16 h photoperiod of 22 000 lx at 25°C. To estimate nitrogenase activity, the acetylene reduction technique was employed in view of the supposed quantitative relationship of this assay with nitrogen fixation (Hardy *et al.,* 1973). To estimate the quantity of photosynthetic products that had accumulated within the nodules, Lawrie and Wheeler (1973) measured nodule radioactivity at a stage 6 h or 25 h after the intact plant shoot had photoassimilated $^{14}CO_2$ for 30 min. Correlative changes were observed (Fig. 12.9) between the rate of acetylene reduction, the accumulation

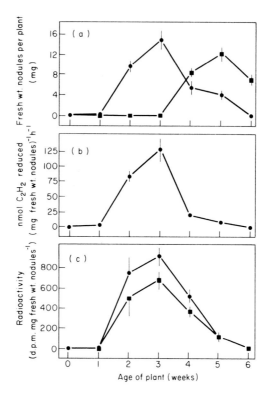

FIG. 12.9. Changes during growth in: (a) the distribution of active pink nodules (●) and inactive green nodules (■); (b) the acetylene reduction by nodules; (c) the accumulation of labelled photosynthates by nodules 6 h (●) and 25 h (■) after feeding $^{14}CO_2$ to the shoot for 15 min. (From Lawrie and Wheeler, 1973)

of ^{14}C-photosynthate by the nodules, and the population ratio of pink (active) and green (inactive) nodules. This was in close accord with earlier observations by Pate (1958a) who found that functional nodule activity in *P. arvense* attained a maximum performance at a plant growth stage coinciding with anthesis. This latter feature has been again confirmed for *P. sativum* cv. 'Laxton's Progress' by La Rue and Kurz (1973) who conducted acetylene reduction assays for nitrogen fixation non-destructively on the intact plant.

2. Regulation

As mentioned in Chapter 13 there is still no really satisfactory explanation as to why relatively high levels of combined nitrogen should inhibit nitrogen fixation. Certainly its effect could operate through translocation channels and to resolve this further Small and Leonard (1969) investigated whether combined nitrogen applied after nodules had already commenced nitrogen fixation could influence specifically the translocation of current photosynthates to the nodules. Their procedure consisted of growing *P. sativum* cv. 'Progress' under sterile conditions and then treating the roots with a commercially available root inoculum. Uninoculated plants served as nitrogen controls and these were given sodium nitrate at the time when the companion inoculated plants had initiated nitrogen fixation. Once the nodules were actively fixing nitrogen, the roots of groups of inoculated plants were subjected for 5 days to a nutrient solution containing 0, 28 or 200 mg l^{-1} Na NO_3 as the nitrogen supply. On completion of this conditioning period the attached shoot of each plant was exposed to $^{14}CO_2$ in the light for 15 min and after an interval of either 0, 1, 4 or 20 h the plants were harvested and dried. The radioactivity of the shoot, root and nodules was then determined and the percentage distribution of ^{14}C and its specific activity (c.p.m. mg^{-1}) were derived. Their data indicated that in plants which had received 200 mg l^{-1} of combined nitrogen, a higher proportion of basipetally translocated photosynthate was accumulated in the root apices than in the nodules. Small and Leonard (1969) also noted that after a 4 h or 20 h period of translocation of ^{14}C the percentage of radioactivity in the nodules of plants which received no nitrate was significantly higher than that in nodules of nitrate-treated plants. These findings were supported by autoradiographic evidence. Their conclusion was that nitrate application to effectively nodulated plants results in a significant shift in the pattern of ^{14}C-photosynthate translocation; a greater proportion of the labelled translocate being then received by the root apices than by the nodules.

Significant growth differences between *Pisum* plants utilizing nitrate-N or relying solely on N-fixation, have been recorded. For instance Minchin and Pate (1973) demonstrated that nitrate-fed plants of *P. sativum* cv. 'Meteor' produced a greater leaf area, chlorophyll content and root dry weight, but a lower shoot dry weight than did the plants dependent on nitrogen fixation. Despite these differences, the two treatments resulted in similar amounts of carbon fixed photosynthetically being respired by the root and apparent efficiency of usage of carbon in nitrogen assimilation was also similar: 6·2 mg C were respired for each mg nitrate-N assimilated, and 5·9 mg C for each mg N_2 fixed from the atmosphere.

The photosynthate supply to nodules may be under auxin regulation; Pate (1958b) reported that the nodules of *P. arvense* were intensive centres of auxin activity and were particularly rich in IAA. Furthermore, the concentration of auxin-like substances was significantly higher in the nodule than in the parent root. As pointed out in Section III (E), assimilates could be preferen tially translocated to regions of high IAA concentration. Therefore it i feasible that nodules might regulate their own import of photosynthate by means of control over the production and diffusion of endogenous auxins.

3. Root and Nodule Carbon Economy

Minchin and Pate (1973) made quantitative measurements of the carbon exchange of the root and nodules of young vegetative plants of *P. sativum* cv. 'Meteor' over the period 21 to 29 days from sowing. Their data showed that some 26% of the net gain of photosynthate was incorporated directly into the growing shoot. The remaining 74% travelled downwards via the phloem, 42% to the roots, and 32% to the nodules. Losses due to roo respiration accounted for 35% of the photosynthate gain of carbon wherea root growth incorporated only 7%. Of importance was the finding that for an expenditure of 12% of the photosynthetically fixed carbon on nodule respira tion and 5% on nodule growth, the substantial sum of 15% of the carbon wa combined with atmospheric nitrogen and cycled back to the shoot via the xylem attached to amino acids and amides.

G. The Transfer of Carbon and Nitrogen to the Developing Seed

1. Carbon and Nitrogen History

The long-term movement of carbon and nitrogen from the vegetative region to the developing seed of *P. arvense* has been studied by Pate and Flinn (1973) in an attempt to understand the significance of past and present assi milation of these elements in relation to seed nutrition. They investigated the fate of carbon from photoassimilated $^{14}CO_2$ and the nitrogen from root-fed ^{15}N-nitrate. A time course study of changes in isotope content was conducted after a prolonged (7 days) application of ^{15}N and ^{14}C that was given either during mid-vegetative growth, or much later, at the mid-fruiting period. The assay data (Fig. 12.10) indicated that 90% of the ^{15}N assimilated during early vegetative growth was still present at plant maturity (i.e. when all pods had ripened), whereas only 23% of the ^{14}C survived to plant maturity, probably because large losses of carbon had occurred in respiration. When expressed in relation to the initial amount of each isotope assimilated, the seed received 51% of the early fed ^{15}N but only 2% of the early fed ^{14}C. However, a quite different pattern emerged when the isotopes were applied during the pod production growth phase. By plant maturity 75% of the ^{14}C and 70% of the ^{15}N had been translocated to the seed (Fig. 12.11).

2. Nitrogen Transport

Lewis and Pate (1973) have demonstrated how effective the transpiration stream is in the distribution of ^{15}N-solutes directly or indirectly to all parts

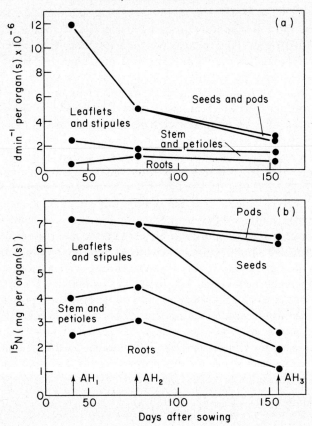

Fɪɢ. 12.10. Fate of (a) ^{14}C and (b) ^{15}N fed to field pea plants 34–41 days from sowing. Plants harvested immediately after feeding (41 days, AH$_1$), at 78 days (AH$_2$) and at plant maturity (155 days, AH$_3$). Nine plants were used for each harvest from each isotope treatment. Note the large loss of ^{14}C from treated plants between AH$_1$ and AH$_2$ and the low efficiency of transfer of ^{14}C to seeds in comparison with ^{15}N. (From Pate and Flinn, 1973)

of a fruiting plant. Fruiting plants of *P. sativum* cv. 'Meteor' were detached from their roots and stood continuously in a solution containing ^{15}N-nitrate, and a 48-h time course study of ^{15}N distribution showed that the ^{15}N was incorporated into leaves, pod and seeds at substantially different rates. There was a 5 h lag before the seeds became labelled but after this labelling increased in an exponential fashion. In contrast an almost immediate and linear incorporation of ^{15}N was observed for the leaves and pod walls. Lewis and Pate (1973) suggested that the lag in ^{15}N supply to seeds represented the period required for nitrogen to be processed in the vegetative regions and loaded on to the translocatory channel serving the seed. Flinn (1969) has observed that there is a poor xylem connection to the individual seeds within a pod of

Pisum and this might explain their extreme dependence on phloem for nourish
ment. Furthermore, a significant grouping of specialized transfer cells (define
in Section III(B)) was found to be associated with both the xylem and phloem
in the vascular bundles supplying the seeds, and Flinn (1969) has deduce
that these cells might aid in transfer of materials from xylem to the phloem
outside the pod. Autoradiographic evidence (Gunning *et al.*, 1968) has indi
cated that transfer cells which are associated with the *Pisum* vascular system
are able to incorporate radioactivity from ³H-amino acid supplied via the
transpiration stream, and if this passage of amino acids through the transfer
cell system is relatively slow it might be held responsible for the observed la
in incorporation of labelled solutes into the seed.

During pod production the remobilization of nitrogen assimilated prior to
anthesis is likely to supplement the diminishing supplies of nitrogen com
pounds received via the root system. The need for regulatory controls in such
a system can be envisaged in order to ensure that the first formed pods do no
sequester excessive amounts of nitrogen and thereby deprive subsequent pods
forming on the supplies of the shoot system. Strong evidence for the existence
of a closely regulated system for dispensing assimilates to seeds can be
inferred from the work of Ali-Khan and Youngs (1973) on protein levels in
seeds of differently placed pods on the shoots of 20 different varieties of
P. sativum. The assays indicated that a small, but significant, intra-plant
difference in seed protein content was attributable to the pod position on
the haulm. The mean percentage protein values for seed taken from
proximal, intermediate or distal pod positions were 24·1, 23·9 and 23·7
respectively.

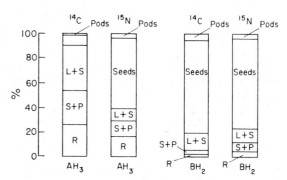

Fig. 12.11. Percentage distribution of ¹⁴C and ¹⁵N in field pea plants harvested at
maturity (155 days after sowing) after being fed for a week either in early vegetative
growth (34–41 days after sowing, AH₃), or later during fruiting (112–119 days, BH₂)
Note the high efficiency of transfer of ¹⁴C and ¹⁵N to seeds after the later period of feeding. The values refer to isotope surviving in the plant to its maturity and do not take
into account losses of isotope during the interval between feeding and harvest. L + S,
leaflets + stipules; S + P, stem segments + petioles; R, root + nodules. (From Pate
and Flinn, 1973)

3. *Carbon Transport*

Movement of carbon in the short term (24 h) to seeds is an essential background to understanding the role in fruit nutrition played by the different organs of a fruiting node. Flinn and Pate (1970) conducted such a study on *P. arvense* cv. 'Black-eyed Susan', studying ^{14}C distribution patterns at 4-day intervals over a 3-week period, from anthesis through to seed maturation. Plants were allowed to photoassimilate $^{14}CO_2$ for 1 h at either the leaf, the stipule or the pod of the first fruiting node, and 23 h later ^{14}C distribution assays were conducted, using a modified Schöniger combustion technique. The ^{14}C data for seeds borne at the first fruiting node showed that they exhibit a strong nutritional dependence on the pod stipule and leaf (Figs. 12.12 and 12.13). Considerable changes were observed in the proportion of ^{14}C derived by the seeds from each of these sources during the course of development. The pod never exported to the rest of the plant but it did receive ^{14}C assimilates from foliar organs especially early in its life before the seeds started to grow.

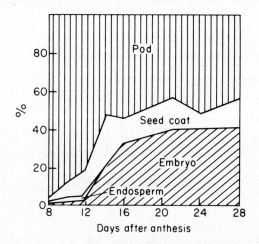

FIG. 12.12. Percentage distribution of radio-carbon remaining in pods and transferred to the various parts of the seeds 24 h after their pods had been fed in the light for 1 h with $^{14}CO_2$ in the middle of their 16 h photoperiod. No radio-carbon could be detected outside the pod and seeds of the fruit. Age of the field pea fruits when fed is measured in days after anthesis of their parent flower. Each set of percentage values in the figure represents the average of radioassays of three separate fruits of the same age.
(From Flinn and Pate, 1970)

The speculation was that a proportion of these substrate reserves of the pod might be translocated to the seed at a later stage, and, indeed, a quite substantial loss of dry matter from pods was observed during their later life.

The nutritional significance of the subtending leaf as a major source of carbon for seed development was indicated in the studies of Flinn and Pate (1970), this being in accord with earlier observations by Link and Sudia

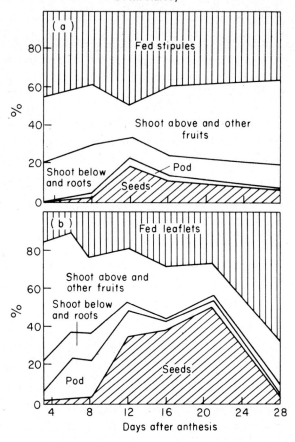

Fig. 12.13. Percentage distribution of radio-carbon remaining in plants of field pea 24 h after the stipules (a) or leaflets (b) of their first flowering node had been fed in the light for 1 h with $^{14}CO_2$ in the middle of their 16-h photoperiod. Proportions of the radioactivity labelled "pod" and "seeds" refer to the fruit at the first flowering node, and the age of the stipules of leaflets when fed is measured in days after anthesis of the flower at their parent node. Three plants with reproductive nodes of the same age were used in each feeding experiment. (From Flinn and Pate, 1970)

(1962), and subsequent reports by Harvey (1973, 1974). Similarly, the finding that the pod did not export significant quantities of ^{14}C to the remainder of the plant was in agreement with the observations from $^{14}CO_2$ feeding studies by Lovell and Lovell (1970) and Harvey (1972b). Flinn (1969) has observed that there is a more general distribution of assimilates to the pods at other nodes from stipules than from leaflets. In contrast with leaflets ^{14}C-export from stipules was not influenced by pod orthostichy.

Szynkier (1974) has examined the short-term competition between pods for foliar ^{14}C-photoassimilate within *P. sativum* cvs 'Syrenka' and 'Alaska'. This was achieved by observing the translocatory effects of removal of either the pod or the subtending leaf at either the first or second fruiting node. For each cultivar tested the translocatory response to removal of the pod at the first fruiting node was to divert ^{14}C assimilates from its subtending leaf to the pod at the second fruiting node. The corresponding response to pod removal at the second fruiting node was a diversion of the ^{14}C-export of that subtending leaf to the pod at the first fruiting node.

Szynkier (1974) suggested that differences in the state of maturation of the respective pods was a major factor in determining the extent to which they received ^{14}C, a further complication being that differences between cultivars might occur in the extent to which ^{14}C assimilates were diverted via the root system prior to entering the pods.

Further work would be required before relating these findings to the situation in the intact plant.

4. The Carbon Balance

The nutritional history of the seed in terms of carbon import can be quantified by taking into account the dry weight gain or loss by the subtending leaf or stipule and measurements of the net flux of carbon from the atmosphere coupled with the known percentage of ^{14}C photosynthate exported by that leaf to the seed. Using this method Flinn and Pate (1970) were able to set out the history for seed carbon at the first fruiting node of *P. arvense* throughout a 22-day period from 5 days after anthesis through to seed maturation. Over the period 12 to 24 days from anthesis the subtending leaf maintained a steady export rate of approximately 11 mg C/24 h to the seed (Fig. 12.14). The corresponding maximum for stipules was 9 mg C/24 h at day 13 in spite of the observations by Flinn (1969) that the stipule and the subtending leaf had a similar photosynthetic efficiency per unit area and a similar surface area. Despite this the stipules were somewhat important contributors during the very early stages of pod growth.

With regard to the pod's own assimilate contribution to the seeds several physiological characteristics must be taken into account before a quantitative comparison can be made with the foliar contribution to seeds. Gas analyser measurements for intact fruits show that during photoperiods early in their life a net uptake of CO_2 can occur (see Chapter 15). However, this will obviously be an underestimation of the pod's rate of photoassimilation since it will also be engaged in reassimilation of CO_2 released to the pod cavity by the respiring seeds (see Flinn and Pate, 1970). In the dark, however, the reverse applies and the carbon loss to the atmosphere from the pod is overestimated, again by an amount equal to the respiratory output of the seed. Thus the carbon contribution from the pod to the seed can be estimated by summation of the pod dry weight change, the net carbon exchange between the pod and the external atmosphere, and the loss of carbon from the respiring seed. This calculation showed that from day 10 to day 25 after anthesis the carbon balance of the pod was consistently positive, the contribution to the enclosed seed ranging from 1·2–19·8 mg C/24 h, with the maximum value at day 18 (Fig. 12.14). The major part of the pod contribution was attributed to the efficient recycling of respired carbon from the developing seed. Indeed a maximum net gain of only 2·0 mg C per photoperiod was obtained in gaseous exchange between fruit and the external atmosphere, but a maximum value of 9·0 mg C/24 h was attained for the rate of mobilization of dry matter accumulated in the earlier stages of pod development. The total contribution from the pod to the seed was thus estimated to be 154 mg C out of a total contribution of 402 mg C by all photosynthetic organs at the fruiting node. These organs together were estimated to contribute approximately two-thirds of the total carbon requirement of seed borne at that node, the remainder was presumed to be made up by the translocation of carbon from elsewhere in the plant to the seed.

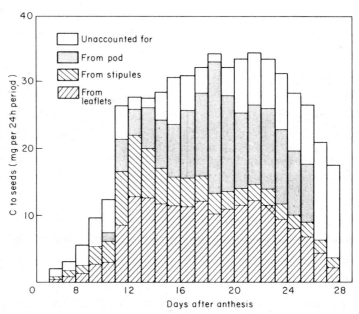

Fig. 12.14. A day-by-day record of the provision of carbon to seeds borne at the first fruiting node of field pea by their pod, and by the stipules and leaflets of their subtending leaf. The estimates include photosynthetically fixed carbon, carbon available through mobilization of dry matter, and in the case of pods, carbon reassimilated from the respiring seeds. These data for transfer are matched against the daily requirement of the seeds for carbon, the latter being calculated by summation of daily losses of carbon by the seeds in respiration and daily gains of carbon by the seed as dry matter. Consequently, the unhatched areas of the histograms (labelled "unaccounted for" represent carbon required by the seeds but not derived from organs at the fruiting node.

(From Flinn and Pate, 1970)

REFERENCES

ACOCK, B., THORNLEY, J. H. M. and WARREN-WILSON, J. (1971). *In* "Potential Crop Production" (P. F. Wareing and J. P. Cooper, eds), pp 43–75. Heinemann, London.
ALI-KHAN, S. T. and YOUNGS, C. G. (1973). *Can. J. Pl. Sci.* **53**, 37–41.
BIERHUIZEN, J. F. (1973). *Act. Hortigothoburg.* **32**, 119–126.
BOOTH, A., MOORBY, J., DAVIES, C. R., JONES, H. and WAREING, P. F. (1962). *Nature, Lond.* **194**, 204–205.
BOUCK, G. B. and CRONSHAW, J. (1965). *J. Cell Biol.* **25**, 79–95.
BRAVDO, B. (1971). *Pl. Physiol.* **48**, 609–612.
CANNY, M. J. (1971). *A. Rev. Pl. Physiol.,* **22**, 237–260.
CANNY, M. J. (1973). "Phloem Translocation". Cambridge University Press.
CARR, D. J. and PATE, J. S. (1967). *Symp. Soc. exp. Biol.* **21**, 559–600.
COOPER, J. P. and WILSON, D. (1970). *Int. Grassld Cong.* **11**, 522–527.
DAVIDSON, D. R. (1971). PhD Thesis, University of Wisconsin, U.S.A.
DAVIES, C. R. and WAREING, P. F. (1965). *Planta* **65**, 139–156.
DEVADAS, C. and BECK, C. B. (1972). *Am. J. Bot.* **59**, 557–567.
DOWDELL, R. J. and DODGE, A. D. (1970). *Planta* **94**, 282–290.
ESCHRICH, W. (1970). *A. Rev. Pl. Physiol.* **21**, 193–214.
FLINN, A. M. (1969). Ph.D. Thesis. The Queens University, Belfast, N. Ireland.
FLINN, A. M. (1974). *Physiologia Pl.* **31**, 275–278.
FLINN, A. M. and PATE, J. S. (1970). *J. exp. Bot.* **21**, 71–82.
GOLDSWORTHY, A. (1970). *Bot. Rev.* **36**, 321–340.
GUNNING, B. E. S., PATE, J. S. and BRIARTY, L. G. (1968). *J. Cell Biol.* **37**, 7–12.
HARDY, R. W. F., BURNS, R. C. and HOLSTEN, R. D. (1973). *Soil Biol. Biochem.* **5**, 47–81.
HARVEY, D. M. (1972a). *Ann. Bot.* **36**, 981–991.
HARVEY, D. M. (1972b). *Rep. John Innes hort. Instn.* 40–43.
HARVEY, D. M. (1973). *Ann. Bot.* **37**, 787–794.
HARVEY, D. M. (1974a). *Ann. Bot.* **38**, 327–336.
HARVEY, D. M. (1974b). *Rep. John Innes hort. Instn.* 28–29.
HARVEY, D. M. and HEDLEY, C. L. (1974). *Lab. Prac.* **23**, 567–568.
HELLMUTH, E. O. (1971). *Photosynthetica* **5**, 190–194.
HIGHKIN, H. R., BOARDMAN, N. K. and GOODCHILD, D. J. (1969). *Pl. Physiol.* **44**, 1310–1320.
JACKSON, W. A. and VOLK, R. J. (1970). *A. Rev. Pl. Physiol.* **21**, 385–432.
KROTKOV, G., RUNEKCLES, V. C., THIMANN, K. V. (1958). *Pl. Physiol.* **33**, 289–292.
LA RUE, T. A. G. and KURZ, W. G. W. (1973). *Can. J. Microbiol.* **19**, 304–305.
LAWRIE, A. C. and WHEELER, C. T. (1973). *New Phytol.* **72**, 1341–1348.
LAWRIE, A. C. and WHEELER, C. T. (1974). *New Phytol.* **73**, 1119–1127.
LEWIS, O. A. M. and PATE, J. S. (1973). *J. exp. Bot.* **24**, 596–606.
LINK, A. J. and SUDIA, T. W. (1962). *Experientia* **18**, 69–70.
LOVELL, P. H. and LOVELL, P. J. (1970). *Physiologia Pl.* **23**, 316–322.
LOVELL, P. H., OO, H. T. and SAGAR, G. R. (1972). *J. exp. Bot.* **23**, 255–266.
MACROBBIE, E. A. C. (1971). *Biol. Rev.* **46**, 429–481.
MANGAT, B. S., LEVIN, W. B. and BIDWELL, R. G. S. (1974). *Can. J. Bot.* **52**, 673–681.
MILLER, J. H. (1960). *Am. J. Bot.* **47**, 532–540.
MILTHORPE, F. L. and MOORBY, J. (1969). *A. Rev. Pl. Physiol.* **20**, 117–138.

MINCHIN, F. R. and PATE, J. S. (1973). *J. exp. Bot.* **24**, 259–271.

MORRIS, D. A. and THOMAS, E. E. (1968). *Planta* **83**, 276–281.

PARK, R. B. and SANE, P. V. (1971). *A. Rev. Pl. Physiol.* **22**, 395–430.

PATE, J. S. (1958a). *Aust. J. biol. Sci.* **11**, 366–381.

PATE, J. S. (1958b). *Aust. J. biol. Sci.* **11**, 516–527.

PATE, J. S. (1966). *Ann. Bot.* **30**, 93–109.

PATE, J. S. (1975). *In* "Crop Physiology" (L. T. Evans, ed.), pp 191–224. Cambridge University Press, London.

PATE, J. S. and FLINN, A. M. (1973). *J. exp. Bot.* **24**, 1090–1099.

PATE, J. S. and GUNNING, B. E. S. (1972). *A. Rev. Pl. Physiol.* **23**, 173.

PATE, J. S. and WALLACE, W. (1964). *Ann. Bot.* **28**, 83–99.

PLAUT, Z. and REINHOLD, L. (1969). *Aust. J. biol. Sci.* **22**, 1105–1111.

SCOTT, T. K. and BRIGGS, W. R. (1960). *Am. J. Bot.* **47**, 492–499.

SETH, A. K. and WAREING, P. F. (1967). *J. exp. Bot.* **18**, 65–77.

SESTAK, Z., CATSKY, J. and JARVIS, P. G. (1971). *In* "Plant Photosynthetic Production. Manual of Methods." (W. Junk, ed.). N.V. Publishers, The Hague.

SKOOG, F. (1973). *Biochem. Soc. Symp.* **38**, 195–215.

SMALL, J. G. C. and LEONARD, O. A. (1969). *Am. J. Bot.* **56**, 187–194.

SMILLIE, R. M. (1962). *Pl. Physiol.* **37**, 716–721.

SMILLIE, R. M. and KROTKOV, G. (1959). *Can. J. Bot.* **37**, 1217–1225.

SMILLIE, R. M. and KROTKOV, G. (1961). *Can. J. Bot.* **39**, 891–900.

SNOAD, B. (1974). *Euphytica* **23**, 257–265.

SZYNKIER, K. (1974). *Acta. Agric. scand.* **24**, 7–11.

THROWER, S. L. (1967). *Symp. Soc. exp. Biol.* **21**, 483–506.

WARK, M. C. (1965). *Aust. J. Bot.* **13**, 185–193.

WARK, M. C. and CHAMBERS, T. C. (1965). *Aust. J. Bot.* **13**, 171–183.

13. Nodulation and Nitrogen Metabolism

Department of Botany, University of Western Australia

I. INTRODUCTION

The nitrogen metabolism of the pea, like that of many other legumes, is dominated by the fact that the plant can enter a nitrogen-fixing relationship with the nodule-forming organism *Rhizobium* and this association can supply virtually all of the nitrogen required to maintain a maximum rate of growth in the host plant. As a result of this, man, in his modern intensive agriculture, is able to rely almost as heavily on nitrogen fixation for nourishment of his pea crops as he must have done, unwittingly, in much earlier times when nitrogen fertilizers were unknown or not readily available. In some cases it is simply a matter of expediency to permit the crop to utilize atmospheric nitrogen, since experience has shown that fully effective strains of *Rhizobium* are present in the soil in which the crop is to be sown and that little or no improvement in crop growth results from adding fertilizer nitrogen (see Reynolds, 1960). However, in other instances, there is clear evidence that the indigenous nodule bacteria are not fully effective in symbiosis and that a measurable benefit to crop growth and nitrogen content can be obtained if the seed is inoculated at sowing with a more compatible strain of *Rhizobium*. According to Mishustin and Shil'nikova (1971) harvest yields of peas in several European countries (e.g. U.S.S.R., Poland, Holland) have been increased by some 20–30% by artificial inoculation with *Rhizobium*; the trials where positive responses were obtained far outweighing those in which no benefit was apparent. In other countries, e.g. U.K. and the U.S.A., inoculation of peas is rarely practised, or deemed necessary. In these cases, the soils may be better endowed with the more effective strains of nodule organism or the plants more able to rely on mineralized soil nitrogen to supplement their symbiosis. Also, it may be general that in those countries where peas are harvested green and immature for freezing, one is less likely to encounter, or be able to detect, symbiotic incompetence than in situations where the crop is not harvested until it is fully mature. In the latter instance a full yield of seed protein is called for and the plants are therefore more likely to be stretched to a limit, as far as their symbiotic productivity is concerned.

From what has been said above it would seem to be entirely appropriate to devote a considerable proportion of this chapter to *Pisum–Rhizobium* relationships, describing, in some detail, the cardinal events of the symbiotic cycle, the functioning and fixation of nodules, and showing how symbiotic performance is affected by host or bacterial properties or by external factors in the nodule: plant environment. These topics are covered in the next two sections of the chapter.

It has long been known that the symbiotic relationship of the pea is by no means an obligatory one and that perfectly satisfactory plant growth can be obtained in the absence of root nodules but in the presence of an appropriate source of combined nitrogen. Section IV therefore deals specifically with the assimilation of forms of inorganic or organic nitrogen known to act as alternative sources for peas and compares their uptake and initial processing with the pattern of assimilation of molecular nitrogen in the root nodule. Special attention is paid to nitrate as a source of nitrogen since this is likely to be the main form of inorganic nitrogen available to peas under normal agricultural conditions. Section V following this considers assimilation of nitrogen under conditions where the nodules are fixing nitrogen at the same time as their parent roots are assimilating a combined form of nitrogen, such as nitrate. This, of course, is the very situation encountered in the reasonably fertile soils in which peas are normally grown.

The intermediary nitrogen metabolism of individual host plant organs and the role in these of specific nitrogenous metabolites form the basis of Section VI of this chapter. This leads to a final section in which the functioning of the whole plant is discussed, dealing in particular with the interrelationships of organs in the overall assimilation and processing of nitrogen. Problems relating specifically to translocation and the relationship between photosynthesis and nitrogen metabolism are mentioned only briefly since these topics are dealt with in Chapter 12. Similarly, there is virtually no discussion on the nitrogen metabolism of germinating seeds nor of fruits, pods and seeds, since these subjects are given space elsewhere in this volume (Chapters 3 and 15).

II. NODULATION AND NITROGEN FIXATION

A. Nodulation of the Seedling

Infection of roots of pea by the nodule-forming organism *Rhizobium* takes place through root hairs (Lie, 1964; Dixon, 1964), so that the sequence of nodulation, like that of other legumes, is inevitably linked to the expansion of various parts of the root system. *Rhizobium* strains effective in nitrogen fixation generally produce large, yet few nodules on a root, and the major fraction of these is to be found on the upper region of the main axis of the root—giving a so-called "crown nodulation". Ineffective strains generally produce smaller nodules, scattered over the entire root system. There is evidence that already-formed nodules have a restrictive influence on later infections since, if the main root is already well covered with large, effective nodules, nodulation of the upper laterals and tertiary roots will tend to be somewhat sparse. Conversely, if primary root nodulation has been restricted, abnormally profuse nodulation of the laterals is likely to occur. This is especially evident if inoculation of the seedling with *Rhizobium* has been delayed. (For further details of rhizobial infection and formation of nodules, see Chapter 5.)

It is quite normal to find over 100 nodules on a single pea root, and to find that the clothing of nodules on the primary root is more intense than that on the lateral roots. The ratio of nodule number: lateral number, and the number of nodules per centimetre of primary root are specific for a host cultivar and are also affected markedly by the *Rhizobium* strain (Pate, 1956). The studies of Gelin and Blixt (1964) and Federova (1966) provide evidence that nodule number in pea is an inherited character, specific to the host genotype, but it is considered unlikely that this trait could be used to advantage in developing lines of pea with unusually high symbiotic potential. Probably both root and nodulation characteristics are controlled by a large number of interacting genes. It seems to be a general rule in legumes for compensatory interactions to exist between nodule size and number, so that nodule mass per root or nodule weight: plant weight tend to be highly reproducible quantities for a particular symbiotic association at a specific time in its development. For instance, in one experiment involving inoculations with progressively more and more dilute suspensions of *Rhizobium*, the subsequent nodulation of 30-day-old pea seedlings (cv. 'Meteor') varied from 25–200 nodules per root but, despite this, no significant differences were observed between inoculation treatments in nodule mass per root, nodule weight: plant weight ratio, and, eventually, in fixation performance (J. S. Pate, unpublished). However, substantial differences can be shown to exist between strains of *Rhizobium* in the mass of nodules which they form on a host, even when such strains prove equally effective on the host in terms of nitrogen fixation. It is clear that these differences must relate to the efficiency with which the bacterial tissues of the different classes of nodule fix nitrogen. Similarly host plant cultivars may exhibit substantial differences in the efficiency of fixation of the nodules they form when inoculated with the same *Rhizobium*.

B. Assaying Nitrogen Fixation

Peas, like many other large seeded legumes, form a considerable fraction of their root system and its nodule complement from the reserves stored in the seed. Indeed, the first-formed nodules have usually commenced fixing nitrogen well before the cotyledons have been drained of nitrogen, so that it is unusual to find any evidence of a "nitrogen hunger" stage in seedling establishment (Pate, 1958).

Pigmentation of the first-formed nodules with haemoglobin is generally taken as a reliable indicator of the start of fixation (Virtanen *et al.* 1947) and, coinciding with this, it becomes possible to detect, for the first time, the fixation product asparagine in the xylem sap exuding from a detopped root system (Wieringa and Bakhuis, 1957; Pate and Wallace, 1964). More recently an even more sensitive technique utilizing gas chromatography has become available for studying fixation in legumes. This utilizes the discovery by Dilworth (1966) and Schollhorn and Burris (1966) that nitrogen-fixing systems will reduce acetylene to ethylene. The activity of the nitrogen-fixing enzyme (nitrogenase) of the nodule can then be assayed by enclosing detached nodules or nodulated roots in an atmosphere containing acetylene and, after a brief interval of time, assaying the atmosphere for traces of ethylene. The sensitivity of the technique is such that nitrogenase activity can be detected in seedling peas approximately 1 day before amino acid analysis of xylem sap would have shown up asparagine, and some 3 days before Kjeldahl analysis for nitrogen would have given reliable proof of a net increment in seedling nitrogen through nodular activity (see Oghoghorie and Pate, 1971). Later in the life of the plant the acetylene assay can prove equally effective for making short-term, quantitative measurements of nitrogen fixation.

The stoichiometry of the acetylene-reduction assay states that the same number of electrons should be involved in reducing one mole of nitrogen gas to ammonia as are required to reduce three moles of acetylene to ethylene. Accordingly the formula given below should hold:

$$\frac{(m \text{ mol } C_2H_2 \text{ reduced to } C_2H_4 \text{ per plant}) \times 28 \text{ (mol. wt. of } N_2)}{3} = m \text{ g N fixed}$$

In practice, comparisons of rates of acetylene reduction with rates of nitrogen fixation measured by Kjeldahl analysis, show that a conversion ratio of $2.4:1$ is appropriate for peas, this conversion factor applying only to assays conducted on freshly harvested, whole nodulated roots (J. Halliday, F. R. Minchin and J. S. Pate, unpublished). A somewhat higher yield of ethylene is obtained if the whole plant is assayed $in\ situ$ (i.e. a conversion ratio closer to $3:1$), but if detached nodules are used, rates of ethylene production are much lower (Oghoghorie and Pate, 1971). These findings suggest that nitrogenase activity is quickly affected if parent root and shoot are not present to provide a continued supply of carbohydrate to the nodules. Marked departure from the $3:1$ ethylene: nitrogen ratio is also observed if nodulated plants have access to combined nitrogen (Oghoghorie and Pate, 1971), or are subject to waterlogging stress (Minchin and Pate, 1975). In both of these instances the acetylene assay appears to underestimate fixation. Of course, the assay becomes technically difficult in a waterlogged environment because of problems in administering the acetylene to roots $in\ situ$ and complications due to acetylene dissolving in the aqueous phase of the culture. Fortunately, the assay product, ethylene, is only sparingly soluble so that it exchanges rapidly with the gaseous phase of the medium from which the gas samples are withdrawn for gas chromatography.

C. Sequence of Events in the Symbiotic Cycle

A detailed study of the nodulation cycle of field pea (*Pisum arvense*) under field conditions has shown that a highly reproducible pattern of synchronization exists between the development of the host and its population of root nodules (Pate, 1958). Cardinal events in the symbiotic cycle such as the time of appearance of nodules, maxima in nodule weight and number on various parts of the root system, the maximum in nodule weight: plant weight ratio, and the time when half the nodule complement has senesced, are found to take place in a very precise sequence, characteristically related to the pattern of leaf production and nitrogen accumulation in the host (see Fig. 13.1).

Growth of the major part of the root and its nodulation is complete by mid-vegetative growth indicating that if these nodules were to become damaged or destroyed, it would be unlikely that a replacement set of nodules could be formed. Disasters of this kind have been recorded for peas when larvae of the weevil *Sitonia lineatus* have eaten the nodules (see Mishustin and Shil'nikova, 1971).

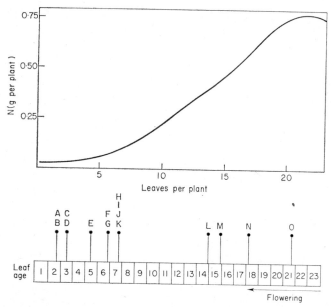

FIG. 13.1. The synchronization of host and symbiotic development in *Pisum arvense* cv. 'New Zealand Maple'. Various features of the nodulation cycle are related to host plant leaf production and nitrogen accumulation. A, maximum primary-root white nodules; B, first nodule turns red; C, 90% of cotyledon nitrogen exhausted; D, maximum secondary-roots white nodules; E, maximum primary-root red nodules; F, maximum haemoglobin concentrations in red nodules; G, maximum nitrogen concentration in red nodules; H, maximum total roots nodules; I, maximum total roots red nodules; J, maximum nodules/plant weight ratio; K, maximum secondary-roots red nodules; L, nodule initiation ceases; M, 50% of red nodules destroyed; N, maximum in red-nodule nitrogen per plant; O, maximum in total plant nitrogen. Redrawn and modified from Pate (1958)

A progressive elimination of the smaller members of the nodule population occurs throughout the latter part of the pre-flowering stage of the growth cycle of field-grown pea plants, possibly an expression of the earlier senescence of nodules formed by less effective strains of *Rhizobium*. However, since this is more than offset by increases in size of the remaining nodules it is not until flowering has started that a maximum mass of nodules is present on the root. At this time also a maximum is recorded in the efficiency of nodular fixation. Degenerative changes in the nodule population coincide with fruit setting, these being expressed as an extensive development of green bile-type pigment in the older portions of the nodules and as a steady decline in nodule fixation efficiency. Eventually, the decay of many nodules on the root occurs (Pate, 1958). At this time also, sections of the root can be seen to be in a state of decay (Salter and Drew, 1965).

D. Nitrogen Fixation over the Life of a Crop

The usual pattern of symbiosis in the tall, late flowering field pea is one in which the accumulation of fixed nitrogen by the host follows a typical sigmoid growth curve (Fig. 13.2, and see Pate and Flinn, 1973) with only 25%

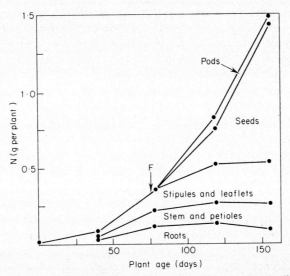

Fig. 13.2. Changes in amount and distribution of nitrogen during the life cycle of the field pea (*P. arvense* L.). Note that at the commencement of flowering (marked F on figure) plants have accumulated only 24% of their final amount of nitrogen and that at maturity 61% of the plant nitrogen is located in the seeds. (Data redrawn from Pate and Flinn, 1973)

of the final amount of nitrogen having accumulated at the commencement of flowering. Since 60–65% of the plant nitrogen is eventually stored in seeds it is clear that nitrogen fixation after flowering must be responsible for providing at least 60% of this seed nitrogen. For this to happen nodules must continue to command adequate supplies of photosynthate even when a larger and larger proportion of the photosynthetic surface has become deployed in nourishing developing fruits. It appears that the stipules of field pea are particularly important in continuing to supply assimilates to the roots, since they are much less deeply committed than the leaflets are in supplying photosynthate to the subtended fruits (Flinn and Pate, 1970). This is discussed further in Chapters 12 and 15. The special role of newly assimilated nitrogen in the later part of the life cycle is brought out in Fig. 13.2, where it can be seen that the exponential rise in seed nitrogen over the period 100–150 days from sowing can not be attributed to any large scale draining of nitrogen from pods, roots, stems or leaves. It must be emphasized that the pattern of nitrogen accumulation expressed by *P. arvense* may not be typical of all peas, and, indeed is unlikely to apply to those dwarf varieties of *P. sativum* which show almost synchronous flowering and much smaller final reserves of nitrogen. Nevertheless it does emphasize the importance of ensuring that the factors within the aerial and especially the rooting environment of the plant remain optimal for nodule activity right until the fruiting stage is complete. It is surely an indictment of our approach to plant physiology that so few studies have been carried out on yield-determining factors in the reproductive stages of a crop plant such as pea, whereas so many overlapping studies have been undertaken at "experimentally easier" times of the life cycle such as germination and early seedling development.

The advent of the acetylene reduction assay for nitrogen fixation has made it possible to assess the symbiotic status of a crop of plants at frequent intervals throughout its development. By comparing these measurements with those for crops in other localities or previous seasons it should therefore be possible to ascertain whether a particular crop's performance is satisfactory or whether fixation is occurring at such a slow rate that it is necessary for the crop to be "rescued" by applying nitrogen fertilizer. Hardy *et al.* (1971) describe three parameters which they regard as being particularly useful when describing and comparing seasonal cycles of fixation in field grown legumes. The first of these is "initiation time", the time after sowing when acetylene reduction assays first show evidence that nitrogen fixation has become active in nodules. Depending on season and variety, initiation time in peas is recorded as varying from 14–30 days. The second quantity is "doubling time", the time taken during the experimental phase of the growth cycle for the rate of fixation in the nodule population to be doubled. Hardy *et al.* (1971) give a value of 2·1 days for doubling time in *P. sativum*, as opposed to values of 8·4–9·0 for soybean (*Glycine max*) or 7·2–10·5 for peanut (*Arachis hypogaea*), but our own experience with *P. arvense* and *P. sativum* in pot or field culture in cool summers at high latitudes is that doubling time can be much longer than this (5–10 days). The third parameter is "termination age", the time at which the exponential phase of increase in nitrogen fixation ceases. Again there is considerable variation, values for dwarf peas grown for freezing lying within the range 30–50 days, whilst in tall, late-flowering varieties fixation continues, albeit more slowly, for 100 days or more (see, for example, Fig. 13.2). If one assumes that the fixation pattern is strictly exponential and that after termination time little or no further nitrogen is fixed, knowledge of the three parameters mentioned above should provide all the information that is

ecessary to compute the seasonal return of fixed nitrogen. It is probably a
eneral feature for varieties with a low (fast) doubling time to have an early
ermination age and varieties of high (slow) doubling time to have a much
ater termination age, but this remains to be tested over a sufficiently wide
ange of genotypes. Effects of onset of flowering and fruiting on the timing of
hese events have also to be evaluated.

II. SYMBIOSIS AND THE ENVIRONMENT

t is now widely accepted that the infection of roots with *Rhizobium* and the
ubsequent initiation of nodules are processes that are considerably more
ensitive to physical factors in the plant's environment than is growth of the
ost root itself, and that, once nodules have formed successfully, their func-
ioning in nitrogen fixation is much more subject to environmental stress than
s the growth of the host plant utilizing a combined source of nitrogen (Pate,
1976). These generalizations certainly extend to the symbiosis of *Pisum*, as
he following paragraphs will endeavour to show.

A. Effects of Temperature

Most cultivars of pea nodulate over a wide range of temperatures (5–30°C),
nodules tending to be more numerous and smaller at the upper than at the
ower or mid-regions of the temperature range (Jones and Tisdale, 1921;
Stalder, 1952; van Schreven, 1959). Studies involving timed transfers of
noculated plants from one temperature to another show that it is the events
of root-hair infection which take place within the first 2 or 3 days after
noculation with *Rhizobium* that have the most exacting temperature
requirements (Lie, 1971). Naturally occurring cultivars and mutants have
been encountered showing total resistance to nodulation by known strains of
Rhizobium (Lie, 1971; Holl, 1973), and in other cultivars such resistance
occurs but is not absolute. For instance, the cultivar 'Iran' is described by
Lie (1971) to be resistant to most bacterial strains at 20°C, but nodulates
normally with them at 26°C, while one particular bacterial strain nodulates
the cultivar at both 20°C and 26°C. Breeding studies suggest that the "resistance to nodulation" trait is genetically based (Holl, 1973; Lie *et al.*, 1976).

Fixation in nodulated pea plants is markedly inhibited by high temperatures and since inordinately low levels of soluble nitrogen are found in plants
grown under such circumstances (J. S. Pate, unpublished) a specific effect on
fixation is suggested. The most likely influence of high temperature is that of
restricting the carbohydrate supply to the nodules either by depressing shoot
photosynthesis, and hence translocation to the nodule, or by increasing
root respiration and thereby decreasing the availability of below ground

carbohydrate to the nodules. The growth of peas, whether nodulated or not is markedly benefited by fluctuating as opposed to constantly maintained temperatures (Highkin, 1960) and this effect is found to be particularly pronounced in respect of nitrogen fixation. This is well brought out in a study conducted by Roponen et al. (1970). They grew plants in a range of temperature regimes incorporating a night temperature of 10°C and a day temperature selected from within the range 15–33°C. These fluctuating environments gave much greater return of fixed nitrogen than did comparable plants grown in any constantly maintained temperature selected from the same temperature range (15–33°C). The extent of benefit from nightly exposure to low temperature varied for the two cultivars. One from Scandinavia (cv. 'Torsdag II') showed greater benefit than the other originating from southern Europe (cv. 'Violetta'), but this applied only if the low night temperature was combined with a day temperature of less than 24°C. With day temperatures above this the response of the cultivars was reversed. The magnitude of the effect of diurnal periodicity was such that fixation per plant was twice as high in 'Torsdag II' with a regime of 20°C (day), 10°C (night), than in a constant temperature of 20°C; whilst in 'Violetta' a regime of 27°C (day), 10°C (night) elicited almost three times the amount of fixation as occurred at a constant temperature of 27°C. The benefit of low night temperatures to nodular fixation has been studied further by Minchin and Pate (1974), using nodule and root respiration measurements and acetylene reduction assays to follow the economy of root and nodule through a daily cycle of activity. Using an 18°C (day), 12°C (night) regime it was demonstrated that fixation was as great, if not greater, at night than during the day, this being possible because pools of sugar and starch accumulated the previous day were consumed during night time (Fig. 13.3). Lawrie and Wheeler (1973) implicated the uninfected cells of the bacterial tissue as sites where temporary reserves of carbohydrate are located, but since nodules can exhibit a five-fold daily change in level of sugar (see Fig. 13.3) it is likely that virtually all tissues of the nodule are involved. It was estimated that pools of carbohydrate are as important as newly arrived translocate in supporting fixation at night (Minchin and Pate, 1974). Extending the study to a constantly maintained temperature regime of 18°C it was discovered that both the daily storage of nodule carbohydrate and the night time depletion of these reserves were as great as in the 18°C:12°C environment, but that the fixation activity at night was much depressed, suggesting that the carbohydrate of the nodule had been used less efficiently (Minchin and Pate, 1974).

Many workers have used the acetylene reduction assay to construct temperature profiles for response of the nitrogenase of various genera of legumes. A broadly based temperature optimum is always displayed and this, in pea spans the range 10–30°C. Clearly then, the nitrogenase is capable of operating with almost equal effectiveness over the normal daily range of temperature

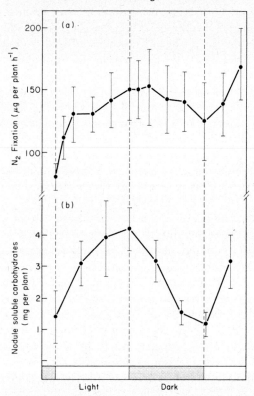

IG. 13.3. Diurnal fluctuations in (a) nitrogen fixation, as measured by acetylene reduc-
on and (b) anthrone-positive soluble carbohydrate of nodules of 28 to 29-day-old
lants of *P. sativum* cv. 'Meteor' grown on a regime of 12 h photoperiod 18°C: 12 h
ark period 12°C. Confidence limits of 95% are indicated for both sets of data. (Data
obtained and redrawn from Minchin and Pate, 1974)

ɔ be experienced during the daily cycle of a temperate environment. It
ɔllows then that availability of photosynthate to nodules is a most likely
ace-setting factor in nitrogen fixation under natural conditions.

. Effects of Light, Carbon Dioxide, Defoliation, Shading and Darkness

'he logic in grouping these factors together is that all are likely to influence
ιe supply of assimilates to root nodules and hence possibly to affect the
urrent rate of nitrogen fixation. The light intensity–carbon dioxide interac-
ons for photosynthesis exhibited by other plants are likely to obtain
ɔr peas, so that carbon dioxide levels will probably limit photosynthesis and
xation at high light intensities, while light intensity itself will be limiting at
ιe lower energy values which are incapable of fully saturating photosynthesis.
Vhen low light intensities are used, spraying of sugar on to leaves can be as

effective as carbon dioxide supplementation in benefiting nodulation ar
nitrogen fixation (van Schreven, 1959). On the other hand, defoliation (
shading severely inhibits nitrogen fixation as assayed by acetylene reductic
(J. Halliday, unpublished). If shoot apices are continuously removed ar
flower buds not allowed to form, fixation can be greatly enhanced and grow
of roots and formation of new nodules can be encouraged to continue muc
longer than in untreated plants reproducing normally (Pate, 1958). Indee
Roponen and Virtanen (1968) reported an eight-fold stimulation of fixatic
if plants are artificially preserved in vegetative condition by removal (
flowers. Presumably this effect is due to diversion of extra assimilates to th
nodules of the non-reproductive plants. If plants are continuously disbudde
at an earlier stage, excessive levels of soluble nitrogen accumulate in th
remaining parts of leaf and stem, and in the roots. Leaves become noticeab
greener and thicker and senesce much more slowly than in untreated plant
Eventually considerable amounts of surplus nitrogen become liberated to th
root environment (J. Halliday and J. S. Pate, unpublished), an effect reminisce
of that reported by Virtanen in his classic studies on nitrogen excretion from pe
(see Virtanen, 1947). Of course, in his experiments, whole plants were involve

Peas have extensive cotyledon reserves, so it is not surprising to find th
nodules form quite effectively on seedlings raised in complete darkne
(McGonagle, 1949). However seedling photosynthesis is subsequently neede
for these nodules to grow properly and to form the haemoglobin pigme
characteristic of actively functioning nodules. A curious non-photosynthet
effect of light on nodulation has been reported by Lie (1969) and in this th
phytochrome system has been implicated. An equally interesting effect is th
of darkness on already nodulated and nitrogen-fixing plants. Even 24 h (
darkness is sufficient to reduce fixation almost to zero (Virtanen et al., 1955
to cause haemoglobin breakdown (Virtanen, 1947) and to disrupt the bacte
oids (Roponen, 1970). Indeed, peas appear to be ultrasensitive in this respec
for other legumes (e.g. annual vetches and lupins) will continue to fix nitroge
throughout a week or more of continuous darkness, although of course the
do this at an ever-diminishing rate (J. S. Pate, unpublished).

C. Effects of pH and Mineral Status of the Rooting Medium

Peas grow best in soils of slightly acidic reaction (pH 6·5) (Small, 1946) an
this is also reflected in their ability to nodulate. According to Lie (1969, 197
the majority of bacterial strains fail to nodulate in media as acid as 5·0, bu
one Rhizobium (No. 313) nodulates abundantly and fixes nitrogen at pH 4·(
As stated above in respect of temperature, it is the early stages in infectio
and nodule development which are especially intolerant (Lie, 1969). Onc
nodules are initiated, plants can be transferred to pH 4·6 or below and sti
continue to fix nitrogen.

Little information exists on the precise mineral requirements for pea symbiosis, but it can be assumed that the balance of nutrients most conducive to host growth is likely to apply also to the symbiotic organs. The special role of molybdenum in nitrogen fixation of the pea has been stressed (Mulder, 1948) and, more recently, a stimulus to nodulation through foliar sprays of molybdenum has been reported (Sharga and Jauhari, 1970). The low tolerance of peas to acidity may, in many soil conditions, reflect also a high requirement for calcium. The relationship of symbiosis to inorganic nitrogen in the rooting medium is discussed separately in a later section (Section V).

D. Effects of Gases and Water in the Rooting Medium

Nodules have a requirement of at least 0·35 ml of water for flushing each milligram of fixed nitrogen to the host plant via the xylem, and estimates made by Minchin and Pate (1973) suggest that this requirement is unlikely to be met fully by water which might be attracted to the nodule through the phloem with a mass flow of assimilates.

It is conceivable that, if the nodule were surrounded by reasonably moist soil, it might fulfil its water budget by absorption through surface layers, but in conditions where the upper layers of soil are dry it is more likely that water must be abstracted laterally from the parent root by proximal parts of the nodule. Osmotic effects on fixation due to salts becoming concentrated on nodule surfaces have been suggested from studies using nutrient solutions (Minchin and Pate, 1975) and these are likely to be a natural part of the drought syndrome of plants growing in a soil well-endowed with nutrients. Sprent (1971) found that in soybean (*Glycine max*) a loss of water from a nodule amounting to more than 20% of its maximum fresh weight caused irreversible structural damage and a total suppression of nitrogen fixation. Water losses less than this cause a reversible depression of nitrogenase activity. It is likely that a similar situation applies to nodules of peas, for these are also very sensitive to drought (see Minchin and Pate, 1975). Indeed, severe drought may lead to nodules being shed from roots, and, even if water is subsequently restored and the plants recover turgidity, it is unlikely that a second set of nodules will form, unless significant new root growth can be accomplished (J. S. Pate, unpublished). Masefield (1968) has suggested from his seasonal studies of nodulation of peas and other annual legumes in England and Nigeria that soil moisture may be the factor of greatest importance in determining the formation and activity of nodules. He maintains that soils are often too dry for maximum nodulation if crops are sown too late in an advancing season.

As a general rule waterlogging has a considerably more dramatic effect on pea symbiosis than does deficiency of water. Waterlogging symptoms involve a restriction of nodulation to all but the surface layers of rooting medium, a low level and efficiency of nitrogenase activity in all but the uppermost nodules, a yellowing of foliage, and an unusually low percentage of nitrogen in plant dry matter (Virtanen and von Hausen, 1936; van Schreven, 1959; Minchin and Pate, 1975). The restricted growth of a waterlogged plant can be markedly improved by supplying nitrate.

Many complex physiological effects are likely to occur within the waterlogged environment, not all of these relating specifically to nodules. However, the dominant effects are likely to be those associated with oxygen deficiency, pea nodules, in particular, showing a marked decline in fixation activity once placed in an environment with oxygen level below 15% (see Bond, 1961). The requirements of pea nodules for oxygen are estimated to be approximately 3 ml of oxygen for each ml of nitrogen gas fixed, the nodules on a root altogether consuming approximately one-quarter of the oxygen required for respiration of the complete root system (see Minchin and Pate, 1973). Effects of oxygen shortage in roots are likely to induce leakage of metabolites or to initiate anaerobic pathways of metabolism, the pea being classed generally as a flooding-sensitive species (e.g. see McManmon and Crawford, 1971). The presence of hydrocarbons (e.g. methane, ethylene) in the waterlogged environment is also likely to cause damage both to roots and to nodules. High levels of carbon dioxide (1% or more) cause inhibition of root growth in peas (Stolwijk and Thimann, 1957).

Relatively minor fluctuations are likely to occur continuously during normal crop growth in the proportions of gas and water in the pore space of the soil and it is possibly of greater agronomic interest to study the influence of these mild effects on symbiotic nitrogen fixation than the more dramatic effects due to unusually severe levels of water excess or deficiency. A pot experiment simulating the effect of varying intensities and frequencies of fluctuation in water availability has shown (Minchin and Pate, 1975) that even brief periods of exposure of plants to an incomplete state of drying out or waterlogging may have significant depressive effects on nitrogen fixation, and that for a maximum rate of fixation to occur it is necessary to apply water at frequent intervals, yet to ensure that the rooting medium drains freely at all times. Plants subjected to water regimes close to but not quite at the optimum do not look in any way unhealthy—they just fix nitrogen at a slower rate. So it would be difficult to judge from visual inspection of a crop whether benefit might be gained from a better adjustment of its water regime. Obviously the most likely time for water deficiencies to occur is at the height of crop growth, in late season. Since fixation is particularly active at this time (see previous section), benefit from irrigation is likely to be especially great (see Salter and Drew, 1965).

V. NITROGEN ASSIMILATION

A. Assimilation of Nitrate

The extraction and assay of the nitrate-assimilating enzyme, nitrate reductase, from tissues of pea has been described (Wallace and Pate, 1965; Jones and Sheard, 1973). The enzyme is soluble and specific to NADH as opposed to NADPH, and can be detected in both leafy shoots and in roots. Studies using sterile cultures of excised roots confirm that the reductase of the root is definitely of plant origin and not due to rhizosphere organisms (Wallace and Pate, 1965; Pate, 1973). Extraction of nitrate reductase from peas does not require the high levels of sulphydryl-protecting agents required for its successful extraction in many other species. This is probably due to the presence of quite large quantities of natural protecting agents, such as glutathione and cysteine, and also to the apparent absence of interfering substances such as phenolic compounds. Indeed, maximum activity of nitrate reductase from etiolated seedling shoots of pea can be obtained by extraction into a cysteine-free solution (Jones and Sheard, 1973), whilst for older seedling tissues only 10^{-5} M cysteine is required (Wallace and Pate, 1965). Jones and Sheard (1973) recorded $2 \cdot 0 \times 10^{-4}$ M as the apparent Michaelis constant of the enzyme for nitrate.

Nitrate reductase offers a classic instance of a plant enzyme system being induced specifically by the presence of its substrate (see Filner et al., 1969). Within 1 or 2 h of applying nitrate for the first time to a pea seedling or older nodulated pea plant, nitrate reductase activity can be detected in root and shoot, and over the next 20 h or so the level of enzyme increases steadily, presumably as more of the inducer becomes available to tissues where the reductase is being synthesized. If bleeding sap is collected from roots during this induction period it can be shown that the appearance and progressive build-up of amides and amino acids in the xylem sap parallels the increase in reductase activity (Fig. 13.4). However, there is a 6–7 h lag between the first detection of nitrate reductase in the roots and the first evidence of elevated levels of amino compounds in the sap, suggesting that there is a considerable period required before reductase action builds up an exportable surplus of reduced nitrogen. Most noticeable is the appearance and eventual dominance of amides, mainly asparagine, in the bleeding sap of nitrate-treated roots (Fig. 13.4). Twenty-four hours after feeding of nitrate, asparagine has become the major constituent, and by then, also, the levels of glutamine, homoserine and allantoic acid have increased, suggesting that they too represent major assimilation products of nitrate reduction (see Wallace and Pate, 1965). This has recently been proved by feeding $^{15}NO_3$ to roots and showing extensive

labelling of all of the above compounds in root tissues and in bleeding sɑ
from roots (J. S. Pate, unpublished).

A constitutive "nitrate reductase" is found in nodules of the pea (Oghoᵣ
horie and Pate, 1971). Probably it is of bacterial origin and similar to th
initially described for soybean nodules by Evans (1954). Its function *in vi*
is not understood—it might even be part of the nodules' fixation system aɾ
not be concerned at all *in vivo* with the reduction of nitrate. A separaɾ
reductase system can be induced in the nodule on exposure of roots to nitraᵣ
(Oghoghorie and Pate, 1971); it is probably a true nitrate reductase. Activiᵣ
of nodules in nitrate reduction can be shown by incubating freshly detacheᵣ
nodules in $^{15}NO_3$ and demonstrating ^{15}N incorporation into soluble aɾ
insoluble organic nitrogen fractions of the nodule (Oghoghorie, 1971).

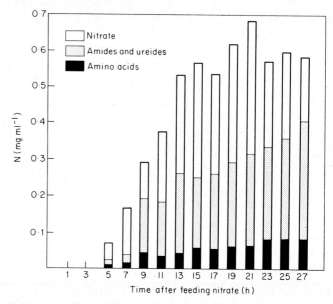

FIG. 13.4. Build-up in nitrate and organic solutes of nitrogen in xylem sap after nitraᵣ
(10 mg NO_3-N per plant) had been fed to un-nodulated pea seedlings raised on a mediuᵣ
devoid of nitrate. Nitrate reductase activity was detected in roots within 2 h and rose t
a maximum at 6 h and was then maintained at or near this level until the end of th
study period. (Data from Wallace and Pate, 1965)

There is good evidence from study of the pattern of changes in nitraᵣ
reductase activity with organ age that the enzyme system reaches maximuɪ
activity early in the life of a root, leaf, or shoot segment and has commenceᵣ
to decline markedly in activity long before levels of protein and solutes staɾ
to fall in the ageing organ (Wallace and Pate, 1965; Carr and Pate, 1967
Similarly, rates of induction are always found to be highest in young tissueᵣ
as the studies of Jones and Sheard (1973) on etiolated shoots of field pea havᵣ

early shown. It can be estimated that the young apical regions of the root system (0–3 cm from the root apices) and the uppermost three leaves and stem segments of the shoot, together contain three-quarters or more of the total nitrate reductase extractable from a plant. Older organs are more concerned with storage of nitrate than with its reduction, mature tissues of roots, internodes and petioles all exhibiting this propensity if high levels of nitrate have been fed to the plant (Oghoghorie, 1971). Leaf laminae, on the other hand, rarely accumulate free nitrate, their nitrate reductase level rising and falling daily, presumably in response to fluctuations in the availability of nitrate from the incoming transpiration stream (Wallace and Pate, 1965; Carr and Pate, 1967). Each day the level of extractable reductase rises from dawn onwards and by noon of a sunny day it is likely to have increased by three to four times its night-time minimum. In the late afternoon reductase activity declines almost as rapidly as it builds up early in the day (Wallace and Pate, 1965).

Changes of this order of magnitude, viewed alongside data on rates of induction after exposure to nitrate, suggest that the pea, like other plants, is capable of an extremely high rate of synthesis and turnover of nitrate reductase (see Huffaker and Peterson, 1974). Jones and Sheard (1974) have shown that light has a marked enhancing effect on nitrate reductase activity in etiolated shoot buds of pea. The mechanism involved is not known but they believe that it operates independently of photosynthesis yet involves phytochrome and a blue-sensitive high energy system.

Nothing appears to be known of the enzyme system handling the nitrite produced by the nitrite reductase enzyme of peas. It can be assumed that a nitrite reductase is involved, similar to that observed for other species. It is generally held (see Miflin, 1974) that the nitrite reductase of higher plant leaves is located in chloroplasts and is stimulated by light. The parallel process in roots, where chloroplasts are absent and light is unlikely to penetrate, remains a mystery.

B. Utilization of Other Forms of Nitrogen

It has been known ever since the classic studies on pea growth in sterile culture by Virtanen and co-workers (see Virtanen, 1947), that peas as well as certain other species can grow quite effectively with organic nitrogen as their sole source of nitrogen. Aspartate and glutamate turn out to be excellent sources of nitrogen (Virtanen and Linkola, 1946), which are equally effective, in fact, as ammonium or nitrate. Later studies (Virtanen and Linkola, 1957; Waris, 1967) have shown that not all amino acids act as acceptable sources of nitrogen for peas and that characteristic morphological disorders, such as shoot dwarfing and abnormal leaf development, can be observed if a particular amino compound is supplied. The compounds α-alanine, β-alanine and phenylethylamine are recorded as eliciting deleterious effects (Virtane and Linkola, 1957) but it has been subsequently shown that the severity o their effect can be ameliorated if sucrose and nitrate are also provided in th medium (Waris, 1967).

Peas will grow, though only slowly, if ammonium or urea is the sole sourc of nitrogen (J. S. Pate, unpublished). Ammonia must be supplied at very lov concentration and preferably to roots whilst urea is best applied to leaves presumably because this is where urease is most active. When supplied t roots urea is only partially degraded and some free urea can be detected i the xylem stream leaving the root (Pate and Wallace, 1964). When ammoniun is fed to roots free ammonium does not spill over to the xylem, but the amid level of root tissue and bleeding sap from the roots becomes much elevate indicating that ammonia has been trapped in these compounds (Pate an Wallace, 1964). When ^{15}N-labelled nitrate (140 mg 1^{-1} NO_3-N) is provide as nitrogen source for isolated roots of pea in sterile culture it can be show that the presence of lower levels of the ammonium ion (40 mg 1^{-1} NH_4-N exerts only a slight inhibitory effect on nitrate uptake, but exerts a muc greater restriction on the reduction of the labelled nitrate (Pate, 1973) Presumably ammonium is the preferred source of nitrogen and compete effectively with nitrate for carbon skeletons.

The pathway for nitrogen in nitrogen fixation has not been properl evaluated for pea nodules but asparagine, glutamine, homoserine and asparti acid appear as major exported products in the xylem sap of nodulated root (Pate and Wallace, 1964) or detached nodules (Minchin, 1973). These fou compounds together contain 80–90% of the fixed nitrogen leaving the root, increasing in amount in the xylem throughout the assimilatory phase of the life cycle of nodulated plants (Pate *et al.*, 1965).

C. Ammonia-assimilating Enzymes

The enzyme system generally held to be responsible for incorporating am monia into the α-amino nitrogen of amino acids of plants is glutamic acid dehydrogenase (GDH):

$$NH_3 + \alpha\text{-oxoglutarate} + \left\{ \begin{array}{c} NADH \\ or \\ NADPH \end{array} \right\} + H^+ \underset{GDH}{\rightleftharpoons}$$

$$\text{L - glutamate} + H_2O + NAD^+$$
$$(NADP^+).$$

Its presence in peas has been demonstrated by several workers, active preparations having been obtained from shoots (Hartmann, 1973), root nodules (Grimes and Turner, 1971) and, especially, from seedling roots (Pahlich and Joy, 1971; Sahulka, 1973, 1974; Joy, 1973). Several isozymes appear to be involved, the enzyme patterns of shoots and cotyledons being similar, but different from that of roots (Hartmann, 1973).

Purification of the enzyme and studies relating to the control of its synthesis and activity have concentrated on the GDH of the seedling root. The purified enzyme exhibits aminating or deaminating activity, in accordance with the reaction stated above (see Pahlich and Joy, 1971). Nevertheless its role *in vivo* is regarded as one of synthesizing glutamate. The bulk of the enzyme is located in mitochondria, is NADH-dependent and is inhibited by its end-product, glutamate. The activity of this fraction is increased ten-fold by added zinc, manganese or calcium, suggesting that some form of ionic regulation might operate *in vivo* (Joy, 1973). The NADPH-dependent GDH of pea roots is probably of cytoplasmic origin, its activity being unaffected by glutamate (Joy, 1973). Regulation of activity and synthesis of GDH by aspartic and glutamic acids, and by sub-lethal levels of hydroxylamine has also been reported (Sahulka, 1973, 1974), so that the picture relating to its functioning *in vivo* is likely to be a complicated one.

Shoots of pea receive nitrogen from roots in the form of nitrate, amides, ureides and amino acids. Free ammonia is not recovered in appreciable amounts in the xylem supply to the shoot, though it is likely to be generated subsequently on reduction of the nitrate or catabolism of amide or ureide. Synthesis of amino acids appears to be especially active in photosynthesizing leaves, these being sites of receipt of the bulk of the nitrogen donated by the root, and, as shown from $^{14}CO_2$ and $^{35}SO_4$ feeding studies, also sites of synthesis of a variety of amino compounds, especially serine, the sulphur-containing amino acids and the aromatic amino acids (see Brennan, 1966; Pate, 1965, 1966, 1968). Analyses of phloem exudates labelled with ^{14}C-translocate from leaves fed with $^{14}CO_2$ suggest that some of the photosynthetically produced amino acids are exported, especially serine and alanine (Lewis and Pate, 1973).

If $^{15}NO_3$, ^{15}N-glutamate, or ^{15}N-(amide labelled) glutamine are fed to pea shoots through the transpiration stream, all substrates are utilized extensively in protein synthesis, donating nitrogen to a wide variety of amino acids in protein and to the main free amino acids of the shoot (Lewis and Pate, 1973). Since the pattern of distribution of label amongst these amino compounds is very similar for all three substrates (Lewis and Pate, 1973), it is likely that common pathways are involved in their assimilation, probably all routed through the same ammonia-assimilating system. Lea and Miflin (1974) provide convincing evidence that it is the chloroplasts of the pea leaf which are likely to be concerned with synthesis of much of the α-amino nitrogen synthesized in a leaf, having succeeded in isolating from chloroplasts a ferredoxin-dependent glutamate synthase capable of forming glutamate from oxoglutarate and glutamine.

L-glutamine $+$ α-oxoglutarate $+$ NADPH $+$ H$^+$ $\rightleftharpoons$

$$\text{glutamate}$$
$$\text{synthase}$$

$$2\text{-L-glutamate} + \text{NADP}^+$$

Lea and Miflin (1974) reason that this enzyme is a more likely candidate for the generation of glutamate in a photosynthesizing tissue than is a conventional GDH system for, unlike GDH, the glutamate synthase system is light-dependent and greatly stimulated by added glutamine or asparagine. The ability of the glutamate synthase to utilize the amide nitrogen of amide as a source of nitrogen fits well with the finding that in most nutritional circumstances, amides transported from the roots, not ammonia, serve as the bulk supply of nitrogen to the shoot. However, as mentioned earlier, there are situations in which nitrate, not amide, is the major nitrogenous solute arriving in the leaf. It is then conceivable that glutamate synthase might operate in conjunction with the leaf-located nitrate reductase and an associated glutamine synthetase. The latter enzyme has been demonstrated in pea chloroplasts (see Lea and Miflin, 1974) and ^{15}N-glutamine is certainly a major primary product of ^{15}NO$_3$ feeding of legume leaves, including those of pea (J. S. Pate, unpublished). Lea and Miflin (1974) point out that glutamine synthetase has a much higher affinity for ammonia than GDH, equipping it to operate in the chloroplast at levels of ammonia well below those known to uncouple chloroplast activity.

The situation regarding ammonia-assimilating enzymes in nodules of pea is not clear, despite the demonstration of GDH and several amino-transferases in the species (Grimes and Turner, 1971). Since glutamine synthetase and glutamate synthase have been demonstrated to occur in nodules of other legumes (see Dilworth, 1974) it is possible that the assimilation pathway for ammonia released from the nitrogen-fixing bacteroids is something akin to that suggested above for chloroplasts.

V. NITROGEN FIXATION AND SOIL NITROGEN

The fact that field-grown peas respond only rarely to dressing with nitrogen fertilizer bears evidence of the general effectiveness of symbiosis in providing adequate supplies of nitrogen to the host plant, although, as mentioned earlier, there is always the strong possibility that mineral nitrogen from the soil might have supplemented symbiosis. Virtanen and Saubert-von Hausen (1952) and Virtanen and Holmberg (1958) devised a rather ingenious comparative method for testing this and assessing quantitatively the relative importance of molecular nitrogen and soil mineral nitrogen to growth of pea plants. They grew seedlings in minus-nitrogen culture in pots, supplying some with a fully effective strain, the others with a strain (H VIII) totally ineffective in fixation. The young seedlings were then planted out into plots of soil.

espite the presence of effective strains in the soil, the nodules already formed the ineffective H VIII strain conferred almost total immunity against rther nodulation, so that it could be assumed that these plants, when finally rvested, contained nitrogen which had all been derived from the soil and t from nodular fixation. By comparing the nitrogen contents of the effecvely and ineffectively nodulated plants they were able to deduce that some % of the nitrogen assimilated by the effectively nodulated plants had come om nodules, the remaining 10% from soil nitrogen. Since the soil in question as a highly fertile one and grew cereal crops quite successfully without added trogen fertilizer they concluded that the peas in question had not been rticularly effective in absorbing inorganic forms of nitrogen from the soil. remains to be seen whether this conclusion applies generally.

Further information on the physiology of nodulated peas exposed to, or veloped continuously in, defined regimes of mineral nitrogen has been stricted to studies carried out in sand and water culture. The nodulation sponse to increasing amounts of nitrate suggests that nodule initiation in ea is somewhat less sensitive to high levels of nitrate supply than is the odulation of many other herbaceous legumes. In one experiment on $P.$ rvense by Oghoghorie and Pate (1971) it was shown that a nitrate level of 00 mg 1^{-1} NO_3-N was needed to suppress nodule numbers to one-half f that on control plants receiving no added nitrogen. Over the range 0–60 mg ml^{-1} NO_3-N nodule numbers were not affected and the nitrate enefited nodule growth relative to that of plants receiving no nitrogen. he physical presence of combined nitrogen in the rooting medium of pea not essential for the inhibition of nodulation, since nitrogen applied to aves (as urea) will also delay and reduce nodulation (Cartwright and Snow, 962).

Once nodules have formed, the continued presence of high levels of comined nitrogen eventually interferes significantly with fixation of the nodules. Jse of $^{15}NO_3$ enables one to resolve the extent to which fixation has been ffected. Supplying the $^{15}NO_3$ to sand cultures at known enrichment throughut the growth period, and studying the dilution of this ^{15}N in total plant itrogen at final harvest, the contribution from fertilizer and nodules can be ssessed. An investigation of this kind employing a range of levels of nitrate n the rooting medium suggests that symbiotic fixation in $P.$ arvense plants is emarkably tolerant of nitrate. A concentration of 120 mg 1^{-1} NO_3-N is equired to reduce fixation to 50% of that in control plants without added itrogen, while a concentration of 315 mg 1^{-1} NO_3-N still allows fixation to roceed at 20% of this control rate (see Oghoghorie and Pate, 1971). Nitrate evels of this magnitude are unlikely to be encountered in normal soils, so it s likely that the pea relies heavily on its nodules under most agricultural and orticultural conditions. A low level of inorganic nitrogen, of the order of 0–40 mg 1^{-1} NO_3-N, turns out to be an optimum economic compromise for nitrogen nutrition since it achieves almost a maximum in yield whilst still permitting fixation to occur at a rate of 70–90 % of that of plants relying solely on symbiotic nitrogen. Levels of nitrate of this order of magnitude might well be available through mineralization of a fertile soil during crop growth, in which case further application of nitrogen as fertilizer would be both undesirable and uneconomical.

Apart from any specific physiological effect which nitrate may have on the infection of the roots by *Rhizobium* and on the early formation of nodules, the plant nitrate reductase enzyme requires the provision of electron donor and a carbon source from the host and, in this respect, nitrate assimilation must be regarded as being potentially competitive with the fixation process. If low levels of nitrate are supplied continuously the bulk of the extractable nitrate reductase is found to be located in the nodulated root, but if higher levels of nitrate are presented this reductase system is apparently saturated and more and more nitrate becomes reduced in the shoot (see earlier section). A specific case described for *P. arvense* by Oghoghorie and Pate (1971) showed that the root reductase was saturated by a level of 70 mg 1^{-1} NO_3-N in the rooting medium, but that the shoot reductase rose linearly with nitrate concentration once this threshold concentration in the rooting medium had been exceeded (Fig. 13.5a). Experimental studies on the translocation of photosynthetically fixed ^{14}C-labelled assimilates to roots and nodules of plants experiencing different levels of nitrate provide evidence to support the hypothesis of competition between root and nodule for translocate. Small and Leonard (1969) working with garden pea and subterranean clover (*Trifolium subterraneum*) found that the labelled photosynthate was fairly evenly distributed between roots and nodules in plants not receiving combined nitrogen but that if nitrogen was added a higher proportion of this photosynthate accumulated in roots than in nodules. Similar results were obtained for field pea (C. G. O. Oghoghorie and J. S. Pate, unpublished), with the additional findings that the proportion of fixed carbon available to the nodulated root as a whole declined markedly as nitrate supply was increased and that, when the specific activities of ^{14}C-labelled assimilate in rootlets and nodules were compared, it was the proximally placed nodules, not the distal root apices, which fared badly when photosynthate was in short supply (Fig. 13.5b). Yet, even with this information, cause and effect relationships are not clear, for one might argue that assimilate starvation is the direct cause of low nodule activity, or, equally convincingly, that the observed effects are nothing more than a reflection of relative sink capacities of root and nodule.

Whatever mechanism may be involved in symbiotic injury by combined nitrogen, acetylene reduction assays provide evidence that the nodule's efficiency in fixation is severely curtailed by high levels of added nitrogen. Studies on field pea seedlings by Oghoghorie and Pate (1971) show that over the

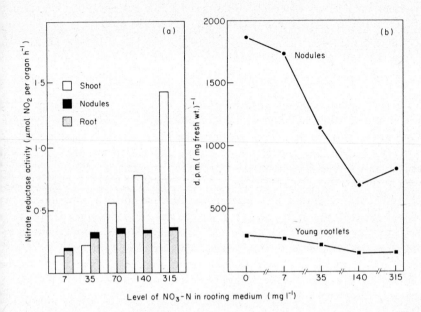

Fig. 13.5. (a) Distribution of nitrate reductase in shoots, roots and nodules of 28-day-old pea plants grown in the presence of a range of levels of nitrate. (Data redrawn from Oghoghorie and Pate, 1971) (b) Specific activity of ^{14}C in nodules and young apical regions of rootlets of *P. arvense* L. plants 6 h after feeding $^{14}CO_2$ to the shoots of plants which had been receiving continuously through their roots different levels of nitrate-nitrogen. Note the non-linear ordinate for nitrate level. (C. G. O. Oghoghorie and J. S. Pate, unpublished)

period 18–30 days after germination, the rate of acetylene reduction in nodules is virtually unaffected by nitrate supply within the range 0–140 mg 1^{-1} NO_3-N, but that above this level fixation efficiency declines sharply. In older plants equally dramatic results can be demonstrated. Thus, Oghoghorie (1971) showed that exposure of plants to 315 mg 1^{-1} NO_3-N caused, within 48 h, a drastic curtailment in nitrogenase activity as measured by acetylene reduction. Conversely, if plants whose fixation had been suppressed by prolonged exposure to this excessively high level of nitrate had their nitrate supply removed from the rooting medium, it took about the same time (50–60 h) before a significant restoration of nitrogenase activity could be confirmed. The time scale of these events suggested that regulation of nodular activity might be achieved by factors operating via the translocatory arrangements to and from the nodule. Useful information might be gained by extending studies of this kind to the field, where variations in moisture supply and microbial activity as well as the additions of fertilizer are likely to give rise to wide fluctuations in a plant's access to combined nitrogen. Tolerance of symbiosis to such extremes may be an especially important factor in determining the degree of autotrophy in respect of nitrogen under crop conditions.

VI. ORGANIC SOLUTES OF NITROGEN

An earlier section dealing with the primary events of nitrogen assimilation has indicated how incoming sources of nitrogen are dealt with by various parts of the plant body and shown how the more active of these centres of primary synthesis are capable of producing exportable surpluses of several key organic nitrogen compounds. Throughout vegetative development, and even for some time after flowering, these soluble forms of nitrogen and derived products continue to accumulate in quite massive amounts in the plant body. For the most part, long-term storage of soluble nitrogen is confined to older tissues of the shoot and principally to the petioles and internodes, and it is possible to show, using ^{15}N, that once fruiting is under way these shoot organs gradually lose nitrogen and make it available to the developing fruits (see Pate and Flinn, 1973). Since only a narrow and highly characteristic range of nitrogenous solutes reach high levels in the soluble phase of the vegetative plant, it can be assumed that each of these has some special role to play as provider of nitrogen and carbon for the eventual synthesis of seed protein. Accordingly this section will deal in turn with the compounds involved and with their significance in plant functioning.

A. The Amides, Asparagine and Glutamine

These usually account for as much as 70–80% of the nitrogen of xylem and phloem sap, and together with homoserine represent the dominant constituents of the ethanol-soluble fraction of leaves, stems and roots of actively growing plants. Indeed, the extent and rate of amide accumulation in the shoot might well be taken as a measure of the degree to which current assimilation of nitrogen in root and shoot has exceeded the demands of the shoot for nitrogen in protein synthesis. In practically all vegetative tissues asparagine is present at considerably higher concentration than glutamine. There are several reasons for this. Firstly, asparagine is the principal amide released from roots in the xylem. Secondly, on arrival in the shoot, it is the less readily metabolized of the two amides and it is the one which contributes less effectively to shoot protein synthesis (Pate et al., 1965; Pate and O'Brien, 1968). And, thirdly, asparagine forms readily in shoots from other amino acids arriving in the transpiration stream (Pate, 1968).

The general conclusion then is that glutamine is the more important source of nitrogen for the current growth of the shoot whereas asparagine remains metabolically or compartmentally inviolate to a large extent, until transported the assimilate stream to the seeds. The latter organs appear to have the special metabolic equipment for utilizing the carbon and nitrogen of asparagine for protein synthesis (see Chapter 15).

Difference in behaviour of the two amides extends to the asynchronous timing of their maxima in the soluble pool of developing organs (Carr and Pate, 1967), and there is a similar lack of phasing in their respective diurnal fluctuations in the soluble pool of the mature leaf. Asparagine levels tend to follow those of most other amino acids in being at a maximum in the leaf at night, whilst glutamine reaches peak quantities in mid-afternoon, at a time when photosynthetically assisted synthesis of protein in the leaf is likely to be fully active (Carr and Pate, 1967).

3. The Non-protein Amino Acids Homoserine and O-acetyl Homoserine

These unusual compounds are quite rare as accumulated products of plants or animals, being known to occur in really large amounts only in certain legumes of the sub-tribe *Viceae* (see Simola, 1968). Virtanen *et al.* (1953) were first to demonstrate the uniquely high levels of homoserine in young pea seedlings, and the identity of the homoserines as L (−)-homoserine and (+)-o-acetyl homoserine was established by Grobbelaar and Steward (1958, 1969). Several studies on these compounds have since added considerably to our knowledge.

A phase of particularly active synthesis of homoserine takes place immediately after germination and, within a week or so, it becomes by far the most abundant free amino compound. At this time it may account for as much as 70% of the total nitrogen, or up to 12% of the dry matter of a seedling (Virtanen *et al.*, 1953; Lawrence and Grant, 1963). Homoserine accumulates in both "wild" and cultivated forms of *Pisum*, maximum intensities of accumulation in root and shoot usually lying within the range 50–75 μmol g f.wt.$^{-1}$ (Lawrence, 1973)—a surprisingly high order of level for a single nitrogenous solute to achieve in healthy plant tissue. Since assimilation of nitrogen from the surrounding medium or from nodules is unlikely to be active within these very young seedlings it must be assumed that the cotyledons, though not themselves containing appreciable amounts of homoserine, must provide the nitrogen and carbon necessary for its synthesis. Feeding studies involving the feeding of ^{14}C-glutamic acid, ^{14}C-aspartic acid, or ^{14}C-pyruvic acid to young pea seedlings show that synthesis of ^{14}C-homoserine from these substrates can take place very actively, especially at phases in seedling development when the overall level of homoserine is increasing (see Larson and Beevers, 1965; Mitchell and Bidwell, 1970b). Roots of the seedlings are held to be the main sites of homoserine synthesis, subsequent translocation to the epicotyl resulting in massive accumulation in the shoot (see Mitchell and Bidwell, 1970b). Homoserine can be recovered in the sap bleeding from decapitated

N

roots of young seedlings (see Bakhuis, 1957; Pate and Wallace, 1964), a finding consistent with the concept of the root as the major centre of homoserine synthesis and the xylem as the avenue of export to the shoot.

Homoserine accumulates in relatively greater amounts in seedlings raised in low as opposed to high temperatures (Grant and Voelkert, 1970). Mitchell and Bidwell (1970a) suggested an indirect pathway for conversion of aspartate to homoserine, this occurring initially via the Krebs cycle, and, thence, in an extra-mitochondrial location in the sequence succinate, fumarate, malate, homoserine (or asparagine). Simola (1968) noted that homoserine synthesis in *Pisum* can also occur from glutamic acid, β-aspartyl phosphate and γ-amino butyric acid so that it is possible that there is a quite complex series of reactions involved in its synthesis from several of the protein amino acids released from cotyledons during germination. ^{14}C-homoserine feeding studies on young seedlings suggest that breakdown of the compound takes place most effectively in shoot tissues in the light (Mitchell and Bidwell, 1970b; Grant and Voelkert, 1972).

Although homoserine may be shown to decrease in relative and absolute amounts in seedlings in the second and third week after germination, there is convincing evidence that its synthesis is resumed once the seedling root and nodules commence to assimilate nitrogen. Thus, it is among the major products to become labelled when nodulated roots receive ^{14}C-labelled sugars from a photosynthesizing shoot and it remains a major constituent of xylem sap throughout the assimilatory phase of the growth cycle (Pate et al., 1965; Pate, 1968). Further evidence supporting its origin in roots comes from the finding that it is usually the most abundant free amino acid of the root, and is concentrated there especially at or near the root tip where nitrate reductase is most active (Carr and Pate, 1967). As well as being exported to shoots via the xylem, homoserine can also be liberated in quantities in root exudates (Boulter et al., 1966; Ayers and Thornton, 1968), being excreted in particularly high quantities from the apical region of the root (Egeraat, 1972). Direct evidence for de novo synthesis in roots comes from its appearance in the tissues and exudates of isolated roots grown in culture (J. S. Pate, unpublished).

There is equally convincing evidence that the shoots of mature plants can synthesize homoserine, for it is a major photosynthetic product in ^{14}CO$_2$ feeding studies of blossom leaflets (Flinn, 1969), and it can be shown to sequester a significant proportion of the ^{15}N supplied to shoots through the transpiration stream as ^{15}N-glutamic acid, ^{15}N-nitrate, or ^{15}N-(amide label)-glutamine (see Lewis and Pate, 1973). As a result of these synthetic activities the amounts of homoserine rise continuously throughout the life cycle, an adult pea plant eventually containing 10 or more times the amount of homoserine as in a young seedling.

Little work has been done on the biochemical or physiological factors affecting the accumulation of homoserine and o-acetyl homoserine in various parts of the plant, or on the mechanisms responsible for the eventual disappearance of these compounds from the fruiting plant. Feeding of ^{14}C-homoserine to vegetative shoots through the xylem results in labelled carbon being transferred to unidentified soluble and insoluble compounds and to amino acids found in protein (Pate et al., 1965). Nevertheless, in comparison with the metabolism of other amino acids, its utilization occurs extremely slowly (Pate and O'Brien, 1968), suggesting that the bulk of it may be compartmentally isolated from the main stream of nitrogen metabolism, at least until flowering occurs. O-acetyl homoserine, by contrast, is to be found only in the early stages of organ development (Carr and Pate, 1967). Only in pods does it remain for a long time as a major constituent. Factors governing its accumulation, metabolism and early disappearance from some organs are even more obscure than in the case of homoserine.

Another equally puzzling feature of the homoserines is the finding that their content tends to be inversely related to that of the other major storage product, asparagine. This is well documented by Grobbelaar and Steward (1969) who found that the content of the two forms of homoserine increased progressively in pea shoots as night temperature was decreased from 26°C down to 4°C whereas the opposite was true of asparagine.

C. Other Non-protein Amino Acids, Amines, Ureides and Peptides

The status of compounds of this nature in the genus *Pisum* has been well documented by Virtanen and Miettinen (1953) and by Simola (1968). The last named author lists pipecolic acid (and its possible precursor α-amino adipic acid), γ- and α-amino butyric acids, amino ethanol and cadaverine as having been detected in peas. The peptides glutathione and γ-glutamylalanine are also recorded, the former especially in roots (Carr and Pate, 1967), the latter accumulating in young seedlings (Virtanen and Berg, 1954). The ureide allantoic acid is also a reasonably common constituent, accounting for some 10% of the total nitrogen of root xylem sap of peas assimilating various sources of nitrogen (Pate and Wallace, 1964). Another interesting amino compound recovered from peas is indol-3-acetyl aspartic acid, a possible storage form of auxin (Simola, 1968). The function and metabolism of all of the above compounds are poorly understood.

D. The Protein Amino Acids

Pool sizes of the protein amino acids in peas tend, on the whole, to be somewhat smaller than those of principal storage products such as the amides and the homoserines. Indeed, analysis of root, stem and leaf tissue (see Pate, 1968) indicates that glycine, alanine, aspartic acid, threonine, serine, glutamic acid and valine are the only protein amino acids which appear to accumulate in amounts significantly in excess of current requirements for protein syn-thesis. In transport, also, protein amino acids seem to have only secondary significance, although the levels of aspartic acid and valine are relatively high in xylem (especially in nitrogen deficient or ineffectively nodulated plants (Wieringa and Bakhuis, 1957), whilst serine, alanine, glycine and aspartic acid levels are high in phloem (Lewis and Pate, 1973). Pool sizes of the aromatic amino acids and sulphur-containing amino acids tend to be low; arginine accumulates in sizeable amounts only in certain stages of seed development (Flinn, 1969), and for a short time after germination (Jones and Boulter, 1968; Shargool and Cossins, 1969).

A variety of interesting nitrogenous solutes exist in peas and the evidence to date suggests that each has a special role to play in plant functioning. Some exist principally as primary products of assimilation, some predominantly as transported forms of nitrogen, some as temporary or long-term storage pro-ducts, and others, such as homoserine and asparagine, have a ubiquity sugges-tive of a major involvement in all aspects of intermediate nitrogen metabolism. Particularly interesting are the events in the life cycle of the plant and its individual organs when metabolism is for a time geared to the rapid synthesis or breakdown of a specific solute. The regulatory controls exercised at these times pose a fascinating challenge for the plant biochemist.

VII. NITROGEN METABOLISM—INTERRELATIONSHIPS OF ORGANS DURING VEGETATIVE DEVELOPMENT

Study of interactions of organs in nitrogen metabolism embraces an under-standing of the overall flow patterns for nitrogen in the plant, information which is available in general terms for *P. arvense* from the ^{15}N tracer studies of Oghoghorie and Pate (1972). A diagram summarizing the principal findings of their study is reproduced here (Fig. 13.6), and when this information is added to earlier studies of nitrogen metabolism of individual organs of the plant, the following picture emerges.

The assimilation of inorganic sources of nitrogen is carried out by three separate classes of organ—nodules which fix atmospheric nitrogen (NF, Fig. 13.6), young parts of the root which reduce nitrate or assimilate other forms of inorganic nitrogen (NR_1, Fig. 13.6), and photosynthetic organs of the shoot which reduce any excess nitrate which spills over to the shoot (NR_2, Fig. 13.6). The changing relationships between these organs induced by changes in the amount and type of nitrogen available in the rooting medium has already been discussed in earlier sections of this chapter.

Distribution of assimilated nitrogen is accomplished by means of two principal avenues of transport. One (labelled 1 in Fig. 13.6) operates through the xylem and its associated transpiration stream and distributes nitrogen

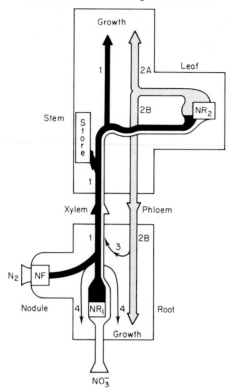

Fig. 13.6. Scheme illustrating the main features of the assimilatory and transport systems for nitrogen in the field pea. The situation represented is a young vegetative plant, effectively nodulated and growing in the presence of a moderately high level of nitrate in the medium.
□ nitrate; ■ aspartate, glutamine, homoserine, asparagine, allantoic acid; □ various amino compounds; NR, nitrate reduction; NF, nitrogen fixation. (For numbering of transport pathways, see text.)

upwards to the shoot principally in the form of amide, ureide and some amino acid. The five major organic compounds involved are asparagine, glutamine, homoserine, aspartic acid and allantoic acid, and, depending on nutritional circumstances, unassimilated forms of inorganic nitrogen may also be present. The second channel of export (labelled 2A, 2B in Fig. 13.6) conveys nitrogen from mature leaves to centres of growth or storage. This is nitrogen which has been acquired in excess by the transpiring leaves and its export takes place in the phloem alongside the translocation of other assimilates, such as sugars, formed in photosynthesis. It is possible to provide an inventory of the nitrogenous solutes present in this phloem stream by analysing the sap bleeding from cut phloem, and it is then found that not only are the nitrogenous compounds typical of xylem transport present, but also other amino acids (e.g.

glutamic acid, serine, alanine, glycine, valine, γ-amino butyric acid, phenyl-alanine, tyrosine and histidine) which are not present in sizeable amounts in xylem. The carbon of these is likely to have been fixed photosynthetically and the nitrogen they contain is likely to have formed either from the photo-synthetic assimilation of nitrate, or from the breakdown of amides in the leaf (see Section IV). Note that the phloem stream of nitrogen is depicted in Fig. 13.6 as being distributed in a bidirectional fashion, upwards (2A) to the apical region of the shoot, and downwards (2B) to the roots. The extent to which each individual leaf is occupied in upward or downward transport of nitrogen is likely to vary with its age and position on the stem. Judging from informa-tion on the transport of ^{14}C-photosynthate from leaves (see Chapter 12), it is likely that upper leaves will specialize in upward transport to the adjacent shoot meristem, while lower leaves will be prominent in nourishing the roots.

A dominant feature of the scheme depicted in Fig. 13.6 is the cycling of newly assimilated nitrogen from nodulated root to shoot and back again to growing parts of the root (Pathway 1→2B in Fig. 13.6). Indeed the studies of Oghoghorie and Pate (1972) suggest that this pathway is many times more important in nourishing outlying parts of the root than is the direct downward transfer of nitrogen from nodule or centre of root assimilation to the distally located root apices (Pathway 4 in Fig. 13.6). An indication of the magnitude of this cycling of nitrogen through the shoot comes from a long-term ^{15}N-feeding study of Pate and Flinn (1973). Feeding a young plant for 1 week ^{15}NO$_3$ and studying the subsequent fate of this ^{15}N in plant organs over a subsequent period of 37 days, it was shown that older parts of the shoot lost back to the root 0·61 mg of the ^{15}N acquired during the week of feeding, this loss amounting to some 8·6% of the 7·1 mg ^{15}N which each plant had assimil-ated. Assuming that a similar rate of recycling were to have taken place throughout the whole period of vegetative development, Pate and Flinn (1973) calculated that the shoot would have made available to the roots approxi-mately one-quarter of their total requirement for nitrogen, notwithstanding the fact that the roots were continuously surrounded by nitrate and could therefore have been engaged in nitrogen assimilation for their own private use. In a plant having access only to atmospheric nitrogen, the nodule→shoot →root recycling of nitrogen is likely to be of even greater significance in contributing to root nutrition. Furthermore, since the phloem stream may well contain nitrogenous solutes formed only sparingly in roots, it is possible that this cycled nitrogen may have special nutritional qualities in addition to its general significance quantitatively as a bulk source of nitrogen.

The root appears to be relatively inactive in transferring nitrogen from the downward translocation stream across to the stream of nitrogen leaving the root in the xylem (i.e. Pathway 2B→3→1 in Fig. 13.6). The key experiment in this connection is the one conducted by Oghoghorie and Pate (1972) in which they fed a lower leaf of a plant for 1 week with ^{15}NO$_3$ and then decapi-ated the plants and collected bleeding (xylem) sap from the root stumps. The enrichment with ^{15}N of this sap and the various parts of the plant were then examined. The bleeding sap from the roots contained ^{15}N at only 0·027 atm % excess whereas the soluble and insoluble nitrogen of the root were labelled much more intensely, respectively at 2·23 and 4·93 atm % excess. Over the 12 h during which the bleeding sap was collected only 0·03 μg ^{15}N was recovered in the exported sap, and if it is assumed that an equivalent rate of export of ^{15}N were to have taken place throughout the week during which the ^{15}N had been fed it can be estimated that the total amount of ^{15}N returned to the shoot would represent a mere 0·1 % of the nitrogen received by the root from the shoot during the same period of time. Oghoghorie and Pate (1972) regard this lack of communication between downward and upward streams of nitrogen in the root to constitute a major obstacle to the free circulation and mixing of nitrogen in the plant. It should be mentioned that this apparent lack of cycling of nitrogen through roots is in striking contrast with the translocatory profile for carbon. As mentioned in Chapter 12, photosynthetically fixed carbon passes quite freely through the roots, downwards as translocate largely in the form of sucrose and then back to the shoot again attached to assimilatory products of nitrogen. Some 20% of the carbon received by the root from a shoot may cycle in this manner (see Pate, 1975).

The relationships of the individual segments of the shoot with the rest of the plant has been the subject of several investigations of nitrogen metabolism in peas (see Brennan, 1966; Carr and Pate, 1967; Pate, 1968; Oghoghorie and Pate, 1972). In early life, a leaf or internode segment grows largely at the expense of carbon and nitrogen donated from older regions of the shoot, this principle being readily demonstrated using $^{14}CO_2$ or $^{15}NO_3$ as tracer fed to the older leaves of a plant. Leaves increase steadily in soluble and insoluble nitrogen until they have reached about three-quarters of their final maximum in fresh weight, and have become fully active in export of photosynthate. Thereafter they commence to lose nitrogen, at first relatively slowly, but as the leaf loses its photosynthetic activity and begins to yellow, the losses become much more rapid. Storage tissues of stem and petiole also show net losses of nitrogen during ageing but these losses tend to be delayed until much later in organ life than in the case of sister leaves (Pate, 1968). Storage tissues of the stem also tend to have high ratios of soluble: insoluble nitrogen, so that they act principally as reservoirs of asparagine and glutamine, their two main solutes of nitrogen. Leaflets and stipules, on the other hand, have much more protein nitrogen than soluble nitrogen and are therefore likely to provide a variety of amino compounds to other parts of the plant as these reserves of protein break down during senescence. It is difficult to devise labelling studies which might indicate what specific amino acids are mobilized from leaves during senescence, although the phloem bleeding technique might be of value in this direction.

Taken as a whole, the losses of nitrogen which take place during the progressive ageing and senescence of segments of the shoot can be regarded as effecting quite a substantial contribution of nitrogen to growing regions of the vegetative plant, including, as indicated earlier, the transport of a sizeable amount of nitrogen to the root system. But the more important contribution which these organs make is that of receiving, processing and re-exporting the nitrogen released from the root. This activity cannot be detected, of course, from sequential analysis for nitrogen during the life of shoot organs, since a quite rapid throughput of nitrogen is concerned, involving probably the daily turnover of certain classes of protein and, in part, the cycling of soluble nitrogen between incoming xylem stream and the outgoing phloem stream. The former activity is likely to be the prerogative of the photosynthetic tissues of leaf mesophyll and stem cortex, since in these photosynthesis is likely to assist in synthesis and turnover of short-life proteins, whereas the cycling of nitrogenous solutes from xylem to phloem is likely to proceed most effectively in the minor veins of the leaf and in the nodes of the stem. In peas, transfer cells are present in these situations and are likely to mediate such an exchange (see Gunning and Pate, 1974).

An approach to the problem of measuring how active stem and leaf tissues are in short-term cycling of incoming nitrogen can be made by the feeding of $^{15}NO_3$. Thus, Pate and Flinn (1973) fed *P. arvense* plants in mid-vegetative growth (12-leaf stage) with $^{15}NO_3$ and studied the distribution of this ^{15}N immediately after feeding and then again 35 days later, by which time a further nine leaves had expanded and the first flower primordia had become visible. The changes in ^{15}N and total N content were recorded for the 12 lowest sets of shoot segments, storage-type organs (stem segment and petiole) being recorded separately from photosynthetic organs (stipules and leaflets). By the end of the 35-day period all ages and classes of organ had lost considerable percentages of the ^{15}N they had acquired during the week of feeding, and these losses occurred regardless of whether the relevant shoot segment had shown a net loss or gain of nitrogen during the period of study. Nevertheless there was a tendency for percentage losses of ^{15}N to be highest from the older shoot segments, these being quite mature at the time of feeding and either losing or, at most, showing only a modest gain of, total nitrogen during the course of the experiment. The turnover of ^{15}N from the older parts turned out to be of the order of 2–3 % per day, whereas in the youngest segments of the shoot this turnover rate was less than 1 % per day.

The net result of this cyclic turnover of nitrogen within the maturing segments of the shoot is that the nitrogen atoms entering the plant early in its life are likely to be passed in sequence from one age class of organ to progressively younger age groups of organ and thereby be incorporated into many generations of cellular structures, proteins and soluble reserves before being ultimately liberated, say, for incorporation into the protein reserves of a developing seed. This subject receives further comment in Chapter 15.

Finally, consideration must be given to the place of the growing apices of the plant in the nutritional scheme which has been presented in this chapter. It should be apparent that the sustenance of these growing regions involves the participation of a variety of metabolic events leading from inorganic sources of nitrogen right through to the synthesis of a complex array of nitrogenous solutes, and that the consummation of these events and the eventual transport of nitrogenous nutrients to the growing region requires the involvement, at one stage or other, of virtually every organ of the plant. For example, as the ^{14}C-labelling studies of Pate (1966) have indicated, roots, leaves and stems all contribute amino compounds to the shoot apex, each class of organ contributing its own characteristic set of amino compounds, yet all sources of nitrogen arriving together in a common phloem stream to the apex. Then, with the additional amino compounds which form in the apical region of the shoot itself, the full range and balance of monomers for the synthesis of protein is obtained. The mechanisms whereby these specific rations of nitrogen are made up, and the reasons for the somewhat different synthetic propensities of the contributing organs requires a much more detailed knowledge of enzymatic regulation in these organs than is available at present. The forces of integration, the mechanisms for regulating supply and demand, and the means of achieving such a great degree of nutritional homeostasis in donor and receptor organs remains a most challenging area of study, not only in peas, but in all multicellular plants.

REFERENCES

AYERS, W. A. and THORNTON, R. H. (1968). *Pl. Soil* **28**, 193–207.
BAKHUIS, J. A. (1957). *Nature, Lond.* **180**, 713.
BOND, G. (1961). *Z. allg. Mikrobiol.* **1** (2), 93–99.
BOULTER, D., JEREMY, J. J. and WILDING, M. (1966). *Pl. Soil* **24**, 121–127.
BRENNAN, H. (1966). M.Sc. Thesis, Queens University, Belfast, N. Ireland.
CARR, D. J. and PATE, J. S. (1967). *Symp. Soc. exp. Biol.* **21**, 559–600.
CARTWRIGHT, P. M. and SNOW, D. (1962). *Ann. Bot.* **26**, 257–259.
DILWORTH, M. J. (1966). *Biochim. biophys. Acta* **127**, 285–294.
DILWORTH, M. J. (1974). *A. Rev. Pl. Physiol.* **25**, 81–114.
DIXON, R. O. D. (1964). *Arch. Mikrobiol.* **48**, 166–178.
EGERAAT, A. W. S. M. VAN (1972). Ph.D. Thesis, University of Wageningen, Netherlands.
EVANS, H. J. (1954). *Pl. Physiol.* **29**, 298–303.
FEDEROVA, L. N. (1966). *Agrochimya Poshvovedenie* **4**, 31–39.
FILNER, P., WRAY, J. L. and VARNER, J. E. (1969). *Science, N.Y.* **165**, 358–367.
FLINN, A. M. (1969). Ph.D. Thesis, Queens University, Belfast, N. Ireland.
FLINN, A. M. and PATE, J. S. (1970). *J. exp. Bot.* **21**, 71–82.
GELIN, O. and BLIXT, S. (1964). *Agri. Hort. Genet.* **22**, 149–159.
GRANT, D. R. and VOELKERT, E. (1970). *Phytochemistry.* **9**, 985–990.

N*

GRANT, D. R. and VOELKERT, E. (1972). *Phytochemistry* **11**, 911–916.
GRIMES, H. and TURNER, S. (1971). *Pl. Soil* **35**, 269–273.
GROBBELAAR, N. and STEWARD, F. C. (1958). *Nature, Lond.* **182**, 1358–1359.
GROBBELAAR, N. and STEWARD, F. C. (1969). *Phytochemistry* **8**, 553–559.
GUNNING, B. E. S. and PATE, J. S. (1974). *In* "Dynamic Aspects of Plant Ultrastructure" (A. W. Robards, ed.). McGraw-Hill, London, U.K.
HARDY, R. W. F., BURNS, R. C., HEBERT, R. R., HOLSTEN, R. O. and JACKSON, E. K. (1971). *Pl. Soil* (Special Volume), 561–590.
HARTMANN, T. (1973). *Planta* **111**, 129–136.
HIGHKIN, H. R. (1960). *Cold Spring Harb. Symp. quant. Biol.* **25**, 231–238.
HOLL, F. B. (1973). *Pl. Physiol.* **51**, Supplement 35.
HUFFAKER, R. C. and PETERSON, L. W. (1974). *A. Rev. Pl. Physiol.* **25**, 363–392.
JONES, F. R. and TISDALE, W. B. (1921). *J. agric. Res.* **22**, 17–31.
JONES, R. W. and SHEARD, R. W. (1973). *Can. J. Bot.* **51**, 27–35.
JONES, R. W. and SHEARD, R. W. (1974). *Can. J. Bot.* **52**, 1433–1435.
JONES, V. M. and BOULTER, D. (1968). *New Phytol.* **67**, 925–934.
JOY, K. W. (1973). *Phytochemistry* **12**, 1031–1040.
LARSON, L. A. and BEEVERS, H. (1965). *Pl. Physiol.* **40**, 424–432.
LAWRENCE, J. M. (1973). *Photochemistry* **12**, 2207–2209.
LAWRENCE, J. M. and GRANT, D. R. (1963). *Pl. Physiol.* **38**, 561–566.
LAWRIE, A. C. and WHEELER, C. T. (1973). *New Phytol.* **72**, 1341–1348.
LEA, P. J. and MIFLIN, B. J. (1974). *Nature, Lond.* **251**, 614–616.
LEWIS, O. A. M. and PATE, J. S. (1973). *J. exp. Bot.* **24**, 596–606.
LIE, T. A. (1964). Ph.D. Thesis, University of Wageningen, Netherlands.
LIE, T. A. (1969). *Pl. Soil* **31**, 391–406.
LIE, T. A. (1971). *Pl. Soil* (Special Volume), 117–127.
LIE, T. A., LAMBERS, D. H. R. and HOUWERS, A. (1976). *In* "Symbiotic Nitrogen Fixation in Plants" (P. S. Nutman, ed.), pp 319–334. I.B.P. International Synthesis Meeting, Edinburgh, 1973. Cambridge University Press, London.
MCGONAGLE, M. P. (1949). *Proc. R. Soc. Edinb.* B. **63**, 219–229.
MCMANMON, M. and CRAWFORD, R. M. M. (1971). *New Phytol.* **70**, 299–306.
MASEFIELD, G. B. (1968). *Expl Agric.* **4**, 335–338.
MIFLIN, B. J. (1974). *Planta* **116**, 187–196.
MINCHIN, F. R. (1973). Ph.D. Thesis, Queens University, Belfast, N. Ireland.
MINCHIN, F. R. and PATE, J. S. (1973). *J. exp. Bot.* **24**, 259–271.
MINCHIN, F. R. and PATE, J. S. (1974). *J. exp. Bot.* **25**, 295–308.
MINCHIN, F. R. and PATE, J. S. (1975). *J. exp. Bot.* **26**, 60–69.
MISHUSTIN, E. N. and SHIL'NIKOVA, V. K. (1971). "Biological Fixation of Atmospheric Nitrogen". Macmillan, London.
MITCHELL, D. J. and BIDWELL, R. G. S. (1970a). *Can. J. Bot.* **48**, 2001–2007.
MITCHELL, D. J. and BIDWELL, R. G. S. (1970b). *Can. J. Bot.* **48**, 2037–2042.
MULDER, E. G. (1948). *Pl. Soil* **1**, 94–119.
OGHOGHORIE, C. G. O. (1971). Ph.D. Thesis, Queens University, Belfast, N. Ireland.
OGHOGHORIE, C. G. O. and PATE, J. S. (1971). *Pl. Soil.* (Special Volume), 185–202.
OGHOGHORIE, C. G. O. and PATE, J. S. (1972). *Planta* **104**, 35–49.
PAHLICH, E. and JOY, K. W. (1971). *Can. J. Biochem.* **49**, 127–138.
PATE, J. S. (1956). Ph.D. Thesis. Queens University, Belfast, N. Ireland.
PATE, J. S. (1958). *Aust. J. biol. Sci.* **11**, 516–527.
PATE, J. S. (1965). *Science, N.Y.* **149**, 547–548.

PATE, J. S. (1966). *Ann. Bot.* **30**, 93–109.
PATE, J. S. (1968). *In* "Recent Aspects of Nitrogen Metabolism in Plants" (E. J. Hewitt and C. V. Cutting, eds). Academic Press, London and New York.
PATE, J. S. (1973). *Soil Biol. Biochem.* **5**, 109–119.
PATE, J. S. (1975). *In* "Crop Physiology—Some Case Histories" (L. T. Evans, ed.). Cambridge University Press.
PATE, J. S. (1976). *In* "Symbiotic Nitrogen Fixation in Plants" (P. S. Nutman, ed.), pp 335–360. I.B.P. International Synthesis Meeting, Edinburgh, 1973. Cambridge University Press, London.
PATE, J. S. and FLINN, A. M. (1973). *J. exp. Bot.* **24**, 1090–1099.
PATE, J. S. and O'BRIEN, T. P. (1968). *Planta* **78**, 60–71.
PATE, J. S. and WALLACE, W. (1964). *Ann. Bot.* **28**, 83–99.
PATE, J. S., WALKER, J. and WALLACE, W. (1965). *Ann. Bot.* **29**, 475–493.
REYNOLDS, J. D. (1960). *Agriculture, Lond.* **66**, 509–513.
ROPONEN, I. E. (1970). *Physiologia Pl.* **23**, 452–460.
ROPONEN, I. E. and VIRTANEN, A. I. (1968). *Physiologia Pl.* **21**, 655–667.
ROPONEN, I. E., VALLE, E. and ETTALA, T. (1970). *Physiologia Pl.* **23**, 1198–1205.
SAHULKA, J. (1973). *Biologia Pl.* **15**, 137–139.
SAHULKA, J. (1974). *Biologia Pl.* **16**, 230–233.
SALTER, P. J. and DREW, D. H. (1965). *Nature, Lond.* **206**, 1063–1064.
SCHÖLLHORN, R. and BURRIS, R. H. (1966). *Fedn Proc. Fedn Am. Socs exp. Biol.* **58**, 213.
SCHREVEN, D. A. VAN (1959). *Pl. Soil* **11**, 93–112.
SHARGA, A. N. and JAUHARI, O. S. (1970). *Madras Agric. J.* **57**, 216–221.
SHARGOOL, P. D. and COSSINS, E. A. (1969). *Can. J. Biochem.* **47**, 467–475.
SIMOLA, L. K. (1968). *Acta bot. fenn.* **81**, 4–60.
SMALL, J. (1946). "pH and Plants". Balliere, Tindall and Cox, London.
SMALL, J. G. C. and LEONARD, D. A. (1969). *Am. J. Bot.* **56**, 187–194.
SPRENT, J. I. (1971). *New Phytol.* **70**, 9–17.
STALDER, L. (1952). *Phytopath. Z.* **18**, 376–403.
STOLWIJK, J. A. J. and THIMANN, K. V. (1957). *Pl. Physiol.* **32**, 513–520.
VIRTANEN, A. I. (1947). *Biol. Rev.* **22**, 239–269.
VIRTANEN, A. I. and BERG, A. M. (1954). *Acta. chem. scand.* **8**, 1089–1090.
VIRTANEN, A. I. and HAUSEN VON S. (1936). *J. agric. Sci., Camb.* **26**, 281–287.
VIRTANEN, A. I. and HOLMBERG, A. (1958). *Suomen. Kemist.* (B). **31**, 98–102.
VIRTANEN, A. I. and LINKOLA, H. (1946). *Nature, Lond.* **158**, 515.
VIRTANEN, A. I. and LINKOLA, H. (1957). *Suomen Kemist.* (B). **30**, 220–221.
VIRTANEN, A. I. and MIETTINEN, J. K. (1953). *Biochim. biophys. Acta.* **12**, 181–187.
VIRTANEN, A. I. and SAUBERT-VON HAUSEN, S. (1952). *Pl. Soil* **4**, 171–177.
VIRTANEN, A. I., JORMA, J., ERKAMA, J. and LINNASALMI, A. (1947). *Acta chem. scand.* **1**, 90–111.
VIRTANEN, A. I., BERG, A. and KARI, S. (1953). *Acta chem. scand.* **7**, 1423–1424.
VIRTANEN, A. I., MOSIO T., and BURRIS, R. H. (1955). *Acta chem. Scand* **9** 184–186
WALLACE, W. and PATE, J. S. (1965). *Ann. Bot.* **29**, 655–671.
WARIS, H. (1967). *Ann. Acad. Sci. fenn.* A. **106**, 1–66.
WIERINGA, K. T. and BAKHUIS, J. A. (1957). *Pl. Soil* **8**, 254–261.

14. The Physiological Genetics of Flowering

I. C. MURFET

Botany Department, University of Tasmania

I. INTRODUCTION

The overall behaviour of a plant and the range of responses which it may show to various environmental factors are features basically determined by the underlying genotype. An understanding of the genetic control of the flowering process is therefore fundamental to a better understanding of the physiology. However, it is equally clear that a consideration of the developmental physiology can make and has made an important contribution towards elucidation of the genetic control.

The genetics of flowering in *Pisum* has attracted interest for a long time
and Mendel (1866) in his classic paper referred to the flowering time of
hybrids as standing almost exactly between the time of the two parents. This
blending type of inheritance was not particularly suited to Mendel's analytical
technique which relied on a high degree of dominance and the recognition of
distinct differences. In fact, flowering time and flowering node often show
continuous variation in crosses grown under field conditions. Biometrical
analyses of the data have indicated control of flowering by simple additive
polygenic systems with little or no dominance, interaction or transgression
(Clay, 1935; Rowlands, 1964; Watts *et al.*, 1970; Snoad and Arthur, 1973a,b).
These conclusions do not translate readily into functional terms and from a
physiological standpoint it is obviously desirable to identify individual genes
controlling various aspects of the flowering process. However, such analyses
do not exclude the possibility of a contribution by major genes which may
simply have escaped detection under these circumstances. For example, by
choosing environmental conditions or treatments which favour the expression
of a particular gene or gene combination a difference can be magnified perhaps
to the extent of obtaining a distinct segregation. This approach was successfully employed by Barber (1959) who used short days and absence of vernalization to maximize the difference in flowering node between his early and late
varieties. By so doing he demonstrated a monohybrid difference with dominance of lateness, the dominant allele simultaneously conferring an ability to
respond to both photoperiod and vernalization. This approach has recently
been taken one step further by Marx (1968, 1969) and Murfet (1971a) both
of whom have used controlled environmental conditions and the measurement of two variables in order to distinguish several discrete differences in
flowering behaviour.

The early sections of this chapter are devoted to providing a genetic foundation from which to view the physiological responses and mechanisms covered
in the later sections, to establishing a perspective on the flowering criteria and
to detailing some of the problems encountered by a worker in this field.

II. VARIETIES USED

The taxonomy of the genus *Pisum* is covered in Chapter 2. Flowering studies
have mostly used material from the groups referred to by Lamprecht (1956a)
as ecotypes *sativum* and *arvense* including garden, canning, freezing, field and
fodder peas but some weed peas and primitive cultivars from ecotypes *humile*
and *abyssinicum* have also been used (Rosen, 1944; Snoad and Arthur, 1973b).
Some of the cultivars used in physiological studies are listed in Table 14.I.
It can be seen that one cultivar may have two names. On the other hand one
name may be used for what are essentially two cultivars, e.g. wilt resistance

was bred into cv. 'Greenfeast' over a decade ago and the complement of flowering genes may have been altered in the process. Genetic impurity within a cultivar is not uncommon (Rowlands, 1964). For example cv. 'Alaska' generally flowers at the ninth or tenth node and shows little response to photoperiod or vernalization (Haupt, 1952; Reath and Wittwer, 1952; Barber, 1959; Marx, 1969). However, strains flowering as low as node 7 (Leopold and Guernsey, 1953, 1954) and as high as node 17 (Nakamura, 1965) have been reported, with response to both photoperiod and vernalization in the latter case. One batch of 'Alaska' received at Hobart was separable into three distinct phenotypes under short days. These variations suggest heterogeneity at one or more major loci. Heterogeneity is not always readily apparent. For example, cv. 'Massey' is phenotypically homogeneous under a range of environments and physiological tests but was found by crossing to be heterogeneous at the major flowering locus E (Murfet, 1971a). Phenotypic similarity therefore does not prove genetic homogeneity. On the other hand, phenotypic diversity, although a strong indication, is not absolute proof of genetic heterogeneity. For example, the discontinuous distribution of the flowering node in line 61a (Table 14.II) results from incomplete penetrance of gene Sn not genetic impurity. These remarks illustrate some of the difficulties in obtaining repeatable data and in comparing the results of different workers. Maintaining your own stocks of genetically known pure lines is of some help in this regard.

III. MEASUREMENT OF FLOWERING BEHAVIOUR

The two criteria most commonly used as measures of flowering behaviour are flowering time (FT—days from sowing to opening of the first fully developed flower) and flowering node (FN—number of the stem node at which the first flower is initiated). Flowering time is a compound variable since it is dependent firstly on the apex switching to the production of flower primordia in the leaf axils and secondly on the subsequent development of the primordia into mature open flowers. Step 1 may proceed in some cases without step 2 occurring. Therefore in recording FN it is necessary to check each leaf axil for the presence of an abortive flower bud. This introduces a third measure of flowering: the time at which the first flower primordium is initiated. The relative merits and usefulness of these criteria depend on the circumstances. Both FN and FT are easily measured and often show a high correlation (Tedin, 1897; Wellensiek, 1925a; Rowlands, 1964) but this may be broken by genetic or environmental factors causing abortion of flower buds or differentially influencing the general rate of growth (Hänsel, 1954a,b; Murfet, 1971a). The time of flower initiation is difficult to measure since it requires destructive dissection of the apex. However, there are circumstances where it can be

TABLE 14.I. Pea varieties used in physiological studies of flowering. A list of some pea cultivars and lines used in physiological research on flowering and their phenotypes and genotypes (confirmed or supposed) after Murfet (1971a,b, 1973a—see Fig. 14.2). Phenotypes of Geneva lines after Marx (1968—see Fig. 14.1). Line numbers refer to Hobart lines.

Cultivar or Line	Synonym	Phenotype	Genotype	Most Frequent Users
Alaska	Geneva 13	ED (El/L)	lf e sn hr*†	Barber, Leopold, Marx, Nakamura, Reath
Alderman	Telephone	L	Lf Sn hr*	Haupt, Köhler
Greenfeast	Lincoln?	L	Lf e Sn hr*†	Amos, Barber, Crowden, Paton, Sprent
Rheinländerin Kleine	Meteor	ED	lf e sn hr*	Haupt, Köhler, Reid
Massey		ED	lf E/e sn hr†	Amos, Barber, Crowden, Moore, Murfet, Paton, Sprent
Supreme		L	Sn hr*	Barber, Bonde, Marx, Moore
Telephone	Alderman	L	Lf Sn hr	Barber, Hänsel, Highkin, Went
Unica		L	Sn hr*	Hänsel Murfet
Vinco	Line 51	ED	lf E sn Hr†	Barber, Highkin
Zelka		L	Sn*	Murfet
Line 2	Graue niedrige	L	Lf E Sn hr	Murfet
Line 7	Vilmorin's "Acacia"	VEI	lf^a E Sn hr	Murfet
Line 8	Lamm line 8a	EI	lf E Sn hr	Murfet, Wellensiek‡
Line 16	Wellensiek's "Dominant"	LHR	$Lf^d E$ Sn Hr	Murfet, Reid
Line 22	Massey	ED	lf E/e sn hr	Murfet, Reid
Line 24	Greenfeast	L	Lf e Sn hr	Murfet, Reid
Line 53		L	lf e Sn hr	Murfet, Reid
Line 58		ED	lf e sn hr	
Line 59	Massey	ED	lf E Sn hr	Murfet

Line		Genotype	
Lines 65, 65E	(FN 11–22)	*Lf e sn hr*	Murfet, Reid
Geneva 1326	G	*Lf E Sn Hr**	Marx
Geneva B1912	I2	*lf E Sn hr**	Marx
Geneva C66–105	I1	*lf e sn Hr**	Marx
Geneva G11	K	*Lf E Sn hr**	Marx
Geneva C66–104	G2	*lf E Sn Hr*	Marx

* Probable but unconfirmed genotype.
† Cultivar may be heterogeneous.
‡ Wellensiek has also used mutant forms of "Dominant" flowering at medium and low node numbers.

TABLE 14.II. Effect of cotyledon removal. Distribution of node of first flower in the early initiating line 60 ($lf\ E\ Sn\ hr$) and the low penetrance late line 61a ($lf\ e\ Sn\ hr$) following cotyledon removal at days 4–6. (After Murfet, 1973b)

Line	Photo-period	Cotyledon Status	Node of First Flower																			
			10	11	12	13	14	15	16	17	18	19	20	21	22	23	24	25	26	27	28	29
61a	8	2*			4	7	2	1	2									5	8	4	0	1
61a	8	1*				2	1										12	4	3	3	3	
61a	8	0	1	9	8																	
60	8	2									1	2	9	5	0	0	1	2				
60	8	0										1	13	9	3	4	4					
60	18	2		9	6	3																
60	18	0		4	8	8																

* Penetrance: control 0·54 and one cotyledon removed 0·91; difference significant at the 0·001 level.

helpful in deciding if a treatment has a direct effect on the flowering process (Collins and Wilson, 1974a), e.g. with treatments which concomitantly influence FN and the growth rate. Nevertheless, it can be just as misleading as FN or FT if used inadvisedly, and comparison between treatments with different growth rates must be made with caution (Haupt, 1969).

Peas have a single main axis with the first two nodes above the cotyledons bearing scale leaves. Secondary axes may develop from the lower nodes and lateral branches from the higher nodes depending on the genotype (Blixt, 1972) and the environment (Nakamura, 1965). In flowering studies under controlled environment conditions, laterals are generally excised to prevent competition with the main axis. When scoring it is obviously important to detect every node, including the lowest concealed in the soil, and to avoid including secondary axes since they mostly flower at a lower node than the main axis (Tedin and Tedin, 1923). Most workers record FN by counting the first scale leaf as node 1. However, others commence counting at the second scale leaf (e.g. Marx, 1968) or the first true leaf (Lamprecht, 1956b; Gottschalk, 1960; Monti and Scarascia Mugnozza, 1967) and addition of one and two nodes respectively is necessary to bring their data into line with the first system. Once the pea apex enters the flowering state it usually continues to lay down flower primordia until growth ceases. However, it may temporarily revert to a vegetative condition either spontaneously (this is particularly prevalent in certain genotypes—see Murfet, 1971b) or where a factor promoting flowering acts for only a limited period (see Sections VI A5 and VII B).

Although FN or FT can be a sufficient measure of flowering behaviour there are many circumstances where data on time of flower initiation, rate of node formation, abortion of floral development, reversion to a vegetative condition, total number of reproductive nodes and yield are helpful or even essential in understanding the effect of a particular gene or treatment.

IV. PHENOTYPIC CLASSIFICATION

In the field in temperate regions the earliest cultivars flower as low as node 5 and the latest somewhere around node 26, with a continuous spectrum of varieties in between (Tedin, 1897; G. A. Marx, unpublished). Subdivision into classes under field conditions is therefore rather arbitrary but generally speaking those cultivars with a mean FN less than 13 are considered early and the rest late. Various more complex, but still arbitrary, systems have been proposed on the basis of FN (see Wellensiek, 1925b) or FT (Kopetz, 1938) in the field. The commercial grower classifies the cultivars into first early, early, mid-season, late, etc. categories which are defined in relation to reference cultivars and based as much on relative date of harvest as the variables FN or FT. Such categories are of more practical than physiological significance.

The lowest node at which flower initiation occurs (FN 5) is not altered by environment although the subsequent development of the bud may be greatly affected. The early/late cut-off point (FN 13) suggested in the field classification above roughly coincides with the point at which the FN starts to show a marked response to environment. Under a controlled environment of short days (8 h) and warm nights (18°C) we have obtained individual plants with a FN as high as 176. (FT values in the phytotron have spanned the range 23–409 days). Marx (1968, 1969) has used controlled photoperiod and two variables, the number of vegetative and reproductive nodes, to distinguish and define four distinct response classes which he calls I, G2, K and G (Fig. 14.1).

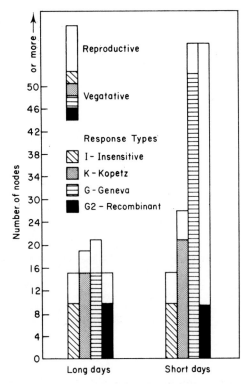

FIG. 14.1. Number of vegetative and reproductive nodes typical of the four response classes I, K, G and G2 under long-day and under short-day conditions. (Schematic, after Marx, 1968)

Plants of the I-type have a low FN, are day-neutral and form only a small number of reproductive nodes under either long or short photoperiods. G2 plants are similar to I plants in long days but in contrast they show a very prolonged reproductive phase and delayed senescence in short days. K-type plants flower a few nodes later than I and G2 plants in long days; their FN is delayed in a limited quantitative manner by short days but their reproductive phase remains short. G-type plants are similar to K-type plants in long days but their vegetative phase and life span are greatly prolonged by short days.

In a somewhat similar scheme Murfet (1971a) used photoperiod and the variables FN and FT to distinguish six phenotypic classes—ED, EI, L, LHR, VEI and VL (Fig. 14.2). A standard variety was assigned to each class in order to provide a floating reference point to accommodate environmental fluctuation. Class VL has now been merged with class LHR (Murfet, 1975e).

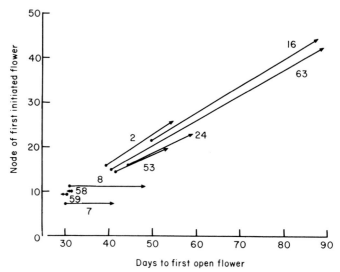

FIG. 14.2. Node of first flower and flowering time for lines 7 (lf^a E Sn hr), 59 (lf E sn hr), 58 (lf e sn hr), 8 (lf E Sn hr), 53 (lf e Sn hr), 24 (Lf e Sn hr), 2 (Lf E Sn hr), 63 (lf e Sn Hr) and 16 (Lf^d E Sn Hr) under short (8 h) and long (18 h) photoperiods. The arrows point to the 8 h co-ordinates. Four distinct response classes are visible; early developing (ED—59, 58), early initiating (EI—8), late (L—53, 24, 2) and late high response (LHR—63, 16). A fifth class, very early initiating (VEI—7), is not obviously distinct but has proved useful in crosses. The increased time to first open flower in lines 7 and 8 in short days results from retarded development or abortion of the first formed flower buds.
(After Murfet, 1971a)

The FN of classes ED (early developing), EI (early initiating) and VEI (very early initiating) is low and unaffected by photoperiod but short photoperiods considerably increase FT in EI and VEI plants as a result of retarded development or abortion of the lower flower buds. VEI plants flower at nodes 5–8. Pure ED and EI lines usually have a FN of 9–12 but the minimum frequency region in the flowering node distribution of crosses (e.g. 59 × 53 in Fig. 14.3) sometimes requires extension of the EI class to include a FN of 16 (Murfet, 1971a,b). L (late) type plants flower around node 15 in long days and are delayed some 5–15 nodes in short days with a commensurate increase in FT.

LHR (late high response) plants flower with the L class in long days but show an extensive 25–50 node delay in short days although this delay may not be fully manifest at temperatures below 15°C. A reduction of the photoperiod response at low temperatures is also evident in L-class varieties (Fig. 14.4).

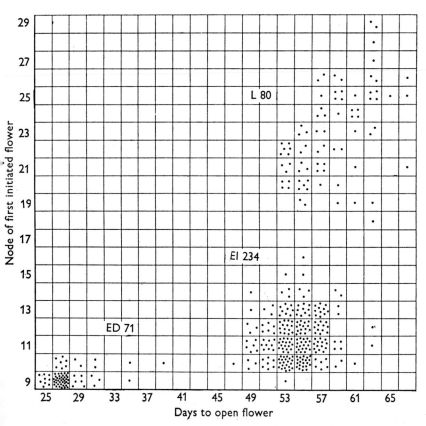

FIG. 14.3. Three distinct phenotypic classes ED (early developing), EI (early initiating) and L (late) resulting from segregation of the gene pairs E/e and Sn/sn on an $lf\ hr$ background. Each dot represents the flowering node and flowering time of a single plant. The expected ratio is 4 ED ($E/e\ sn$): 9 EI ($E\ Sn$): 3 L ($e\ Sn$). Photoperiod 8 h. Data from cross 59 × 53 (Murfet, 1971a)

The classes L and LHR correspond phenotypically with classes K and G respectively but the relationship of classes ED, EI, I and G2 is not yet clear (G. A. Marx and I. C. Murfet, unpublished). These class systems work fairly well and have proved very useful as a basis for genetical and physiological studies. However, resolution of the classes is not always perfect. This is due in part to the need for careful adjustment of the photoperiod, temperature, light intensity, plant spacing, nutrient supply and growth medium to provide

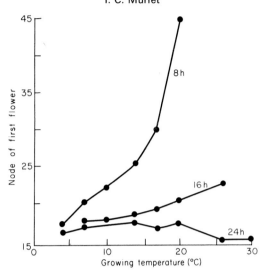

Fig. 14.4. Node of first initiated flower for the late cv. 'Greenfeast' (*Lf e Sn hr* ?) grown under photoperiods of 24 h, 16 h or 8 h and at various temperatures. (After Paton, 1968)

optimum conditions for resolving say a supposed ED/EI segregation, and in part to the very wide range of genetic variation in *Pisum* so that genotypes occur which do not fit neatly into a particular class. For example, we have selected a line (L 65) which has a high FN typical of the L class but its response to photoperiod is much smaller than is characteristic of this class.

V. GENETIC CONTROL

Many papers have been published on the genetics of flowering in peas and many genes proposed; some well founded and some not so well founded. One of the problems is the continuous distribution of such variables as FN and FT in many circumstances, and the consequent need for arbitrary classification; but there is also a failure in some cases to verify the proposals through the use of higher generations and confirmatory crosses. In addition, the majority of proposals have been put forward without the connection between previous results being established. Therefore one of the tasks facing the geneticist today is the resolution of this confusion, not only by establishing the status and relationship of the various Mendelian genes which have been proposed, but also by bringing together the results of the Mendelian and biometrical analyses. The results of these analyses may be stated in different terms but they refer to the very same genes in most cases. One way to do this is to conduct under field conditions a biometrical analysis of diallel crosses between lines already genotyped under controlled environment conditions

Establishing the relationship of some of the major genes is hampered by the fact that the material originally used appears no longer to be available. But these are problems for the geneticist. Since this book is principally concerned with physiology, rather than reviewing the genetics in detail, I will bias the account towards a background for the physiology with an occasional aside on possible gene relationships. (Some more detailed remarks on gene relationships are included as an appendix.)

Evidence of a gene dominant for late flowering showing linkage with the basic gene for anthocyanin (A) emerges consistently from the earlier papers (Lock, 1905; Tschermak, 1910; Hoshino, 1915; Tedin and Tedin, 1923; Rasmusson, 1935; Pellew, 1940). The evidence was particularly clear in some F_3 and F_4 progenies raised by Hoshino where a distinct segregation in FT occurred. Mostly, however, the distributions in these field grown crosses were continuous, although the Tedins reported one F_2 in which the FN distribution was so markedly bimodal as to leave no doubt regarding its monohybrid origin. Pellew reported tentative evidence of multiple alleles determining flowering at the 18th, 15th and 9th node respectively (presumably under English field conditions). The major locus principally concerned in these studies was probably the one now known as Lf (White, 1917; Murfet, 1971b) and recent evidence indicates that there are at least four alleles at this locus for which the symbols lf^a, lf, Lf and Lf^d have been suggested (Murfet, 1975e). The dominance hierarchy ascends in the order shown. This series seems to be particularly concerned with determining the minimum length of the vegetative period, with minimum FN values of 5, 8 and 11 respectively for the alleles lf^a, lf and Lf.

Photoperiodism in peas began to attract the attention of geneticists in the 1950s. Day-neutral and long-day strains were observed within the cv. 'Kvithamar brytsukkerert' and the difference between the two strains proved to be monogenic, with dominance of the long-day trait (Bremer, 1953; Bremer and Weiseth, 1961). Barber (1959) used short-day conditions to distinguish a dominant gene with several effects: it increased the flowering node, conferred the ability to respond to photoperiod and vernalization, delayed the appearance of the first leaf with four leaflets and decreased the internode length. Rowlands (1964) and Murfet (1971a,b) both found evidence of a major gene having definite similarities with that described by Barber. It seems likely, though not in every case certain, that these reports refer to the gene defined by Murfet (1971b) as Sn (see the appendix).

Genetic analysis of the photodependent response classes I, G2, K and G is not yet concluded. So far it has been shown that a cross between an I-type and a G-type line can produce all four response classes in the F_2 and a cross between two different I-type lines can produce a G2-type F_1, with segregation into a 9 G2:7 I ratio in the F_2 (Marx, 1968, 1969). These results show that the G2-type of response is controlled by two major loci and that a dominant allele must be present at each locus before the characteristic prolonged reproductive phase can be exhibited in short days. The data also indicate that a third locus is involved in the step G2 to G.

Crosses between the classes ED, EI, L and LHR have revealed two further major genes, *E* and *Hr*, in addition to *Lf* and *Sn* (Murfet, 1971a,b, 1973a). Genotype *lf e sn hr* is ED. Genotype *lf e Sn hr* is an L-type with the characteristic responses to photoperiod and vernalization, and genotype *lf e Sn Hr* is an LHR type in which these responses are further magnified. The strong interaction between genes *Sn* and *Hr* is illustrated in Fig. 14.5. Gene *E* is epistatic to *Sn* in terms of FN and genotypes *lf E Sn hr* and *lf E Sn Hr* are EI.

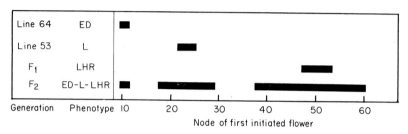

FIG. 14.5. The range of flowering node values for a cross between lines 64 (*lf e sn Hr*) and 53 (*lf e Sn hr*) illustrating the strong interaction between genes *Sn* and *Hr* in the F$_1$ and segregation of the F$_2$ into three distinct phenotypic classes ED, L and LHR. The observed numbers were in accord with the expected ratio of 4 ED : 3 L : 9 LHR. Photoperiod 8 h. (After Murfet, 1973a)

Lf is epistatic to *E* and genotypes *Lf E Sn hr* and *Lf E Sn Hr* are L and LHR respectively. Genes *E*, *Hr* and *Lf* do not cause any large effects by themselves; genotype *lf E sn hr* is ED, *lf e sn Hr* is ED with some EI tendencies, and *Lf e sn hr* shows characteristics of both the ED and L classes. The phenotypes and genotypes of several lines are given in Table 14.I. A distinct three-class segregation produced by selfing genotype *lflf, Ee, Snsn, hrhr* is shown in Fig. 14.3. Gene *Sn* has not been located but *Lf* is in linkage group 1 about 10 recombination units from *A*, *Hr* is in group 3 about 4–6 units from *M*, and *E* is in group 6 about 20 units from *P*. (A linkage map for *Pisum* appears in Fig. 2.1).

The concept that flowering is controlled by both major and minor genes has developed steadily over the years (Tedin and Tedin, 1923; Rasmusson, 1935; Barber, 1959; Rowlands, 1964; Marx, 1968, 1975a; Murfet, 1971b, 1975e). The basic response classes described above appear to be determined primarily by major genes but minor genes also play a role in this respect. For example, by selecting for a system of polygenes modifying the penetrance of *Sn* it is possible to develop a line with the complement of major genes *lf e Sn hr* in which a proportion of the plants regularly classify as EI instead of L (see line 61a, Table 14.II). Again by selecting for quantitative genes it is possible to raise substantially the FN of genotype *Lf e sn hr* so that the phenotype changes from what is essentially ED to something akin to an L-type with an atypically small response to photoperiod. Quantitative systems also appear to contribute to within-class variation, firstly by modifying the expression of the major flowering genes, and secondly by some other mechanism. For example, it is possible to select two L-type lines with the same complement of major genes (*lf e Sn hr*) which flower on average some 10 nodes apart. Of course the effects of the major genes may also become quantitative in the field but their contribution in that case should not be discounted since significant effects for all four genes have been observed under certain long-day conditions (*Sn*—Murfet, 1973a; *E*—Murfet, 1975b; *Lf* and *Hr*—Murfet, 1975e). Finally, at least some of the quantitative variation in FN and FT is caused by Mendelian genes whose major effect is on some quite different trait (e.g. internode length—see below).

Genes *Lf*, *E*, *Sn* and *Hr* mostly influence FT indirectly by altering FN, but *Sn* and *Hr* can directly influence FT in short days by retarding or aborting floral development, e.g. in genotypes *lf E Sn hr*, *lf e sn Hr*, *lf^a E Sn hr* and *lf^a E Sn Hr* (see Section VIB). Genotype *lf^a E Sn hr* (line 7) also has a later FT in long days than its very low FN would suggest (Fig. 14.2). Development of the first-formed flower buds seems to be held back with the result that the first two or three inflorescences may open almost simultaneously as described by Marx (1972). It is not known whether this effect is directly related to the presence of gene *lf^a* or whether flower initiation simply has occurred so early in the life cycle that the plant has had insufficient opportunity to build up the resources necessary to ensure a normal rate of floral development. There also appear to be other genes and gene systems which disturb the usual relationship between FN and FT but in most cases they probably act through changes in the rate of growth of the plant as a whole rather than a specific effect on floral development (Hänsel, 1954a,b, 1963; Snoad and Arthur, 1973b). Rasmusson (1935) and Rowlands (1964) found that *Le* (tall) segregates tended to have an earlier FT than *le* (dwarf) segregates. In contrast Barber (1959) found that the *Le* allele tended to increase FN. These apparently contradictory findings seem to be explicable in terms of a more rapid rate of growth in *Le* plants (Marx, 1975b). A somewhat analogous situation seems to occur with the combination of length genes conferring the semi-tall phenotype known as crypto-dwarf (*le la cry^c*). Crypto-dwarf segregates tend to have a higher rate of node formation than dwarf segregates (Dalton and Murfet, 1975) and this can lead to an earlier than average FT relative to their FN (see the crypto-dwarf line 8 in Fig. 14.2—long-day data). However, a potential reduction in the FT of crypto-dwarf segregates tends to be offset in the ED and L classes by an associated increase in FN. The loci *La* and *Cry* may also have a direct influence on the flowering process since the frequency of vegetative reversion in EI plants and the degree of dominance shown by *Lf* over *lf^a* are both influenced by segregation of the dwarf (*le*, *La* and/or *Cry*) *vs* slender (*le la*

cry^s) length form (Murfet, 1971b, 1975e). Further relevant information may be found in the latter part of Section XD.

Several flowering variants have been obtained by the use of mutagenic agents (Gottschalk, 1960; Knavel, 1967; Monti and Scarascia Mugnozza 1967, 1970; Sidorova *et al.*, 1974; Wellensiek, 1961, 1964, 1969, 1972a,b 1973a). Wellensiek reported a series of three multiple alleles No^h, No^m and no determining flowering at a high, medium and low node respectively, and he subsequently utilized these in physiological studies. These mutant lines have the advantage that they probably differ from the ancestral line only with respect to a single gene and any physiological differences can be attributed to a particular gene with considerable confidence. By the use of mutation techniques and a wider range of naturally occurring material, we should be able to gain a more complete coverage of the genes controlling flowering in peas particularly if more sensitive and specific scanning techniques are used.

VI. EFFECT OF LIGHT AND DARK

A. Flower Initiation

1. Intact Plants

A photoperiod effect on flowering in peas was first reported by Kopetz (1938, 1941, 1943) who observed that early cultivars were essentially day-neutral whereas the flowering of late cultivars was significantly delayed by short days. Kopetz recorded FT but his data undoubtedly reflect the direct effect of photoperiod on flower initiation which was subsequently reported by Hänsel (1954a). Early varieties have been reported to initiate flower primordia in continuous darkness (Borgström, 1939; Leopold, 1949; Haupt, 1952). Late varieties certainly require some light before they will flower but the question of whether an obligate long-day requirement exists in *Pisum* has not yet been completely settled. Wellensiek's (1969, 1973b) multiple allelic series no, No^m and No^h confers an increasing response to short days but he is of the opinion that even No^h plants will eventually flower in short days if they can be kept alive for sufficient time, and the same is probably true of the latest lines studied at Hobart. Our L-class plants are definitely quantitative long-day plants but plants of the LHR-class seem to approach an obligate (qualitative) long-day requirement at high temperatures. For example, genotype Lf^a E Sn Hr eventually flowered after producing over 100 nodes when grown in an 8 h photoperiod with 18°C nights, but unfortunately the behaviour at even higher night temperatures has not been tested. Gene Sn by itself confers a quantitative response to photoperiod. Gene E largely overrides the effect of Sn on flower initiation and the flowering node of the EI genotype lf E Sn hr is unaffected by photoperiod. Gene Lf overrides E and restores the photoperiod response, which is further magnified by the addition of *Hr*. Neither *E*, *Hr* nor *Lf* by themselves render flower initiation sensitive to photoperiod in intact plants, but if the flowering node of genotype *Lf e sn hr* is raised to about node 18 in long days, as for instance in line 65, by selecting for a later background of polygenic modifiers a small response to photoperiod is detectable. Line 65 was found to flower about 2·5 nodes earlier in a 24-h photoperiod than in an 8-h photoperiod when the long-day photoperiod included 16 h of high intensity light (Murfet and Reid, 1974). However, the bulk of this response seems to be a growth-related phenomenon since if the same total quantity of light energy is supplied in the two photoperiods, the difference in FN is only 0·6 of a node (G. M. Rowberry, unpublished). Line 65 is therefore remarkable in having a higher FN in long days than many late varieties carrying *Sn*, while exhibiting a true photoperiod response only one-tenth to one-twentieth as large as the *Sn* lines.

Both Hänsel (1954a) and Haupt (1957, 1969) have suggested that the absence of a photoperiod response in early varieties is not so much a consequence of a particular genetic situation but rather follows automatically because flower initiation takes place so rapidly after germination that there is no opportunity for the seedling to respond to photoperiod. This is certainly true of very early genotypes flowering at the sixth to seventh node (e.g. *lf*^{a} *E Sn hr*) since 6–7 nodes are already present in pea embryos (Haupt, 1952; Blixt and Gelin, 1965). Similarly, in the early cvs 'Kleine Rheinländerin' and 'Massey', flowering is physiologically determined at about day 5 or 6, which is prior to emergence, although flower primordia may not be visible until the 7–9th day (Haupt, 1952; Johnston and Crowden, 1967). Unfortunately, the situation is not so simple in other genotypes. Line 65E (*Lf e sn hr*), for example, flowers about three nodes later than 'Massey' and is probably not determined until the 12th day but it remains photoperiodically insensitive even if exposed to light from the start of germination. Similar treatment can stimulate the late line 53 (*lf e Sn hr*) to flower some three nodes earlier than line 65E (Murfet and Reid, 1974). Paton (1971) had previously found that the cotyledons of cv. 'Greenfeast' developed the competence to perceive a light stimulus by day 4, so time is not the only factor. In the succeeding paragraph we will examine evidence that the cotyledons of early varieties actively promote flowering, and this may be an important factor contributing to the photoperiodic insensitivity of early varieties, particularly if it is conceded that the cotyledons exert a predominant influence over the early growth and development of the young seedling.

2. Decotylized Plants

Pea cotyledons are large compared with the rest of the embryo. They actively contribute to the growth of the plant until some 15–17 nodes have been initiated, although their contribution relative to that of the shoot itself must diminish rapidly towards the end of their functional life. It is obviously cogent for the physiologist to study the role of the cotyledons in the flowering process by such techniques as cotyledon removal and the grafting of cotyledon-bearing stocks to scions of different genotypes. These techniques are particularly relevant in determining whether the cotyledons act as a source of stimulatory and/or inhibitory substances, and the majority of results are deferred to the section on flowering hormones (Section VII). For the moment we will confine our attention to the photoperiod response which cotyledon removal engenders in some early flowering genotypes.

Pea seedlings can usually survive cotyledon removal from day 4 onwards without special conditions but removal before this time requires culture of the embryos on sterile agar slopes containing carbohydrates and other nutrients (Haupt, 1952; Paton and Barber, 1955). Cotyledon removal in early flowering genotypes generally results either in no change in the FN or an increase in the FN which may be quite substantial depending on the genotype, light and nutrient conditions. The FN of cv. 'Massey' (*lf E/e sn hr*) was not altered when the embryos were excised immediately after imbibition but removal of cotyledons at day 4 sometimes resulted in a delay of one or two nodes. This delay appears to be favoured by high light intensities but was reported not to be influenced by photoperiod (Barber and Paton, 1952; Paton and Barber, 1955; Moore, 1964, 1965; Johnston and Crowden, 1967). The effect of cotyledon removal on the FN of genotype *lf e sn hr* (Hobart lines 58 and 68) varies from no significant effect to a small delay which was significantly greater in short days for line 68 (Murfet, 1973b; Murfet and Reid, 1973; Wall *et al.*, 1974; Reid and Murfet, 1975a). There are reports of a two node delay and half node promotion in cv. 'Alaska' as a result of cotyledon removal (Moore, 1964; Collins and Wilson, 1974b). The FN of decotylized plants of cv. 'Kleine Rheinländerin', rose from 11 in continuous light to 17 in a 4-h photoperiod; this delay could be mimicked in continuous light by lowering the light intensity to 200 lx but in total darkness the flowering node fell to less than 13 (Haupt, 1957). Cotyledon removal in Hobart lines 60 (*lf E Sn hr*) and 64 (*lf e sn Hr*) under long days caused at most a two node increase in FN values but under short days the decotylized plants flowered on average 5·5 to 9 nodes later than the intact plants (Table 14.II; Murfet, 1973b; Wall *et al.*, 1974; Reid and Murfet, 1975a).

In brief, cotyledon removal leads at most to small increases in the FN of cvs 'Massey' and 'Alaska' and lines 58 and 68. There is no certainty that these minor changes are anything more than a by-product of the drastic alteration in the growth habit as a result of cotyledon removal. With cv. 'Kleine Rheinländerin', substantial increases in FN occur in decotylized plants but only at very short photoperiods or low light intensities, and even here there is an element of doubt as to whether we are dealing with a true photoperiod response or what could be called photo-dependent or growth-dependent ffects. However, with line 60 FN values as high as node 25 were obtained in ecotylized plants grown under an 8-h photoperiod of high intensity sunlight. These vigorous plants flowered some 12–14 nodes later than their long day counterparts and a true photoperiod response is indicated.

These results suggest that the cotyledons of line 60 actively stimulate flowering since the intact plants flower at a low node which is unaffected by photoperiod. The photoperiod response in decotylized plants of line 60 can be explained by the assumptions that gene *E* operates only in the cotyledons but that gene *Sn* operates in both the cotyledons and shoot (Murfet, 1971c, 1973b). The decotylized plants would then show the quantitative response to photoperiod characteristic of the late flowering genotype *lf e Sn hr*. However, gene *Sn* is not responsible for the photoperiod response in decotylized plants of line 64 and 'Kleine Rheinländerin' since they carry the recessive *sn* allele. If these two varieties have their cotyledons removed under identical short-day conditions, line 64 shows by far the larger increase in FN (Reid and Murfet, 1975a). This evidence, together with the fact that intact plants of genotype *Lf e sn Hr* occasionally initiate flowers as late as node 20 (Murfet, 1973a), suggests that the gene *Hr* may exert a flower-delaying action independently of *Sn*. On the other hand, it has been suggested (Murfet, 1971b,c) that recessive *sn* may retain some physiological activity and this could explain the behaviour of line 64 and the lesser responses of line 68 and 'Kleine Rheinländerin'. The weak flower-delaying activity of *sn* is assumed to be overcome in intact plants by the strong promotory influence of the cotyledons, but in line 65 the combined effect of *Lf* and its modifiers defers flower initiation to the point where cotyledonary influence has diminished and the small potential for a photoperiod response conferred by *sn* can then be realized (Murfet and Reid, 1974).

3. Light Intensity

It is important to distinguish here between treatments involving a low light intensity for the entire light cycle and cases where the plants are given a sufficient period of high intensity light (e.g. 8 h) to sustain essentially normal growth with different intensities of light supplied as photoperiod extensions or as flashes in the long night. Treatments of the first kind, as described above, can markedly alter the FN of decotylized plants. However, Marx (1969) found no difference in the FN of intact K- or G-type lines between the relatively high light intensities of 7 and 30 klx. Recent studies by J. B. Reid (unpublished) using the second procedure have shown that photoperiodic induction in intact plants of line 63 (*lf e Sn Hr*) is almost independent of light intensity (Table 14.III).

4. Critical Photoperiod

As Wellensiek (1969, 1973b) has pointed out, the term "critical photoperiod" is not strictly applicable to peas. He uses the term "critical duration of

TABLE 14.III. Effect of light intensity during photoperiod extension or night break. Node of first initiated flower in line 63 ($lf\ e\ Sn\ Hr$) plants given a daily cycle of 8 h of daylight followed by either 16 h of incandescent light at various intensities or 16 h of darkness interrupted midway by 1 h of red light at various intensities. (Data of J. B. Reid, unpublished)

Intensity mW cm^{-2}	16-h Incandescent			Intensity μW cm^{-2}	1-h Red Flash		
	$\bar{x}$	s.e.	n		$\bar{x}$	s.e.	n
8·2	14·08	0·25	24	56	25.55	0·41	20
1·8	14·87	0·17	23	20	25·28	0·32	18
0·2	15·36	0·36	22	5	24·47	0·24	15
0·06	15·25	0·36	24	1	27·67	0·54	18
0	50+			0	50+		

exposure" to describe a relationship which seems to hold good for peas over a range of photoperiods from 14 to 17·5 h. However, it may be useful to retain use of the term "critical photoperiod" by defining it in a manner appropriate to the situation.

The FN of late peas shows little change as the photoperiod is decreased from 24 to approximately 20 h (J. B. Reid and I. C. Murfet, unpublished) but as the photoperiod is further decreased the FN begins to rise slowly at first and then more steeply somewhere between a photoperiod of 16 and 12 h (Barber, 1959; Bremer and Weiseth, 1961; Marx, 1969; Wellensiek, 1969, 1973b). The photoperiod at which FN starts to show a marked increase corresponds, in a sense, to the "critical daylength" of species possessing an obligate requirement. For a more precise measure of the critical photoperiod in peas we could take the photoperiod at which the rate of change in the response curve is at a maximum (not the slope itself). Preliminary studies suggest that the critical photoperiod may lie between 13 and 15 h for line 6. (LHR) at a temperature of 17·5°C (J. B. Reid and I. C. Murfet, unpublished). There is clear evidence that the photoperiod response is reduced by low temperatures (Fig. 14.4; Barber, 1959; Paton, 1968; Wellensiek, 1969) and may even be nullified if vernalization is followed by continued cold nights (Murfet and Reid, 1974), but there is no evidence of whether the critical photoperiod, as defined above, is altered by temperature. Likewise little is known concerning the effects of plant age or genotype on critical photoperiod. The minimum effective dark period also needs further study in view of the proposal that Sn activity is suppressed by light.

5. Number of Inductive Cycles

The data in Fig. 14.6 show that older plants respond to a smaller number of inductive cycles than young plants. Gene Lf confers a requirement for an increased number of inductive cycles in the young plant but in the absence of Hr these Lf plants eventually initiate flowers about the same time as lf

plants. Gene *Hr* does not increase the number of cycles required but it prolongs the vegetative phase in line 63 so that there is a period of some 2 months
in which one inductive cycle will cause flowering in 50–100% of the plants.
This enables line 63 to be used in studies akin to those conducted on plants
of other species exhibiting an obligate photoperiod requirement. Plants of
line 63 caused to flower by a small number of inductive cycles may subsequently revert to a vegetative state if returned to short days, and Barber
(1959) observed a similar response by cv. 'Zelka'.

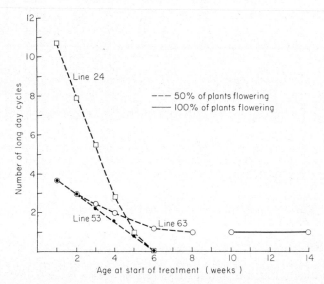

FIG. 14.6. The effect of age on the number of long day cycles (first cycle 32 h then
multiples of 24 h) required to induce flowering in lines 53 (*lf e Sn hr*), 24 (*Lf e Sn hr*)
and 63 (*lf e Sn Hr*). The plants were grown in 8-h short days before and after treatment.
The night temperature was 17°C. The 50% data were obtained in part by extrapolation.
(Data of J. B. Reid and I. C. Murfet, unpublished)

6. Light Quality; Light Breaks

Marx (1969) found that extension of a 9 h short day (daylight) by 6 h of light
from fluorescent tubes (peak output 640–660 nm) significantly reduced FN
in several K-type lines. However, the same treatment failed to induce flowering in a G-type line. This may indicate that genetic differences exist in the
critical photoperiod or it could be interpreted as evidence of differences in the
quality of light required by some genotypes. Nakamura (1965) reported that
interrupting the long night with a 1-h light break at 500 lx from a mixed
incandescent/fluorescent source was equally as effective as long days in promoting flowering of the cv. 'GW', either treatment reducing FN by three
nodes in relation to the short-day controls. Some recent data of J. B. Reid
(unpublished) show the effect on line 63 of a regime of 8 h of daylight followed

TABLE 14.IV. Spectral dependence and night breaks. Percentage of plants induced to flower and flowering node for line 63 (*lf e Sn Hr*) plants given 8 h of daylight and various light treatments during the remaining 16 h of a 24-h cycle; experiment 1 started on day 6 stopped on day 56; experiment 3 started on day 20 stopped on 52; and experiment 6 started on day 20 stopped on 63.
(Data of J. B. Reid, unpublished)

Experiment	Treatment	% Flowering	Flowering Node x̄	s.e.	n
1	16L	100	14·5	0·3	21
1	16R	100	28·2	0·5	19
1	16FR	100	18·2	0·5	17
1	16B	0	42·9	0·4	19
3	16L	100	18·8	0·2	12
3	7D, 1R, 8D	100	28·0	0·5	19
3	8D, 2FR, 6D	25	49·9	2·2	24
3	7D, 1R, 2FR, 6D	100	31·6	0·3	20
3	16D	0	57·1	2·8	8
6	16L	100	19·1	0·3	7
6	7·75D, 0·25RH, 8D	100	32·8	0·6	20
6	16D	0	63·2	1·1	11

L   mixed incandescent/fluorescent light.
R, RH red light at an intensity of 20 μW cm^{-2}, 50 μW cm^{-2}.
FR   far-red light.
B   blue light.
D   darkness.

by 16 h of light of various wavelengths (Table 14.IV). Eight hours of daylight is of course a non-inductive photoperiod and it seems that extending this photoperiod with an additional 16 h of blue light is likewise ineffective in inducing flowering. Extension of the photoperiod with 16 h of light from a mixed incandescent/fluorescent source was found to be the most effective of the treatments tested in inducing flowering, but 16 h of far-red light was nearly as effective, and 16 h of red light also resulted in flowering although a larger number of cycles was required in the case of red light before induction was completed. However, red light was much more effective than far-red light when given as a light break about midway through a 16-h night and caused 100% induction even at low intensities (Table 14.III) and in breaks as short as 15 min (Table 14.IV). In contrast, only 25% of plants flowered in response to a 2-h break of far-red light. The inductive effect of the red light flash was partially reversed by a subsequent period of far-red light. These results suggest the existence of two light-controlled processes; one a phytochrome-mediated switch mechanism, the other a reaction proceeding under the continued stimulus of far-red light. (Further information relating to phytochrome in peas is to be found in Chapter 11.)

B. Floral Development

Genotypes *lf E sn hr*, *lf e sn Hr*, *lf E Sn hr* and *lf E Sn Hr* all initiate flower primordia at an early node but while development in genotype *lf E sn hr* is little influenced by photoperiod [this genotype is typical of the ED (early developing) class], the other three genotypes display an increasing tendency towards retarded development or abortion of their lower flower buds when grown in short days. Gene *Sn*, for example, is responsible for a substantial increase in the time to first open flower (FT) in genotype *lf E Sn hr* under short days and this effect is the basis of the distinct segregation into the ED and EI (early initiating) classes illustrated in Fig. 14.3. Gene *Hr* also tends to delay floral development in short days. The effect is not as pronounced as that conferred by *Sn*, and genotype *lf e sn Hr* is closer to an ED than an EI phenotype, but a significant effect of *Hr* on FT is detectable in short days on both an *sn* (Murfet, 1973a) and an *Sn* (Murfet, 1975c) background. Wellensiek (1969) found that floral development in a medium-late line was suppressed either by short days or by a combination of low light intensity and high temperature in long days. In fact a combination of short days and either low light intensity or high night temperature can lead to abortion of flower buds even in ED lines. This suggests that development of the flower bud is influenced by the general level of photosynthetic products and reserves of metabolites as well as by some more specific circumstance determined through genes *Sn* and *Hr*.

VII. EVIDENCE FOR FLOWERING HORMONES

Much time and effort has been devoted over the years to the search for the endogenous substances which regulate flowering in higher plants. Despite these efforts, the flowering hormones have remained elusive and we must rely heavily on circumstantial evidence. Attention at first centred around the concept of a universal flowering stimulus but as von Denffer (1950) pointed out the disappearance of a flower inhibitor may have the same consequences as the formation of a flower promoter and indeed it is often difficult to distinguish between the two situations. It is now generally accepted that different plants may have evolved different methods of regulating flowering; that a substance which promotes flowering in one plant may inhibit flowering in another; that the phytochrome control system may be utilized in different ways by different plants, and so on. Certainly there is now strong evidence that in some plants flowering is largely regulated by a flower inhibitor (e.g. *Fragaria*, Guttridge, 1969). In other plants there is evidence of both promotory and inhibitory hormones, e.g. the leaves of *Lolium* appear to produce a flower inhibitor in short days and a flower promoter in long days (Evans,

o

1969a). A detailed discussion of the evidence in general for flowering hormones is given by Lang (1965) and Evans (1969b).

The techniques of cotyledon removal, grafting and leaf masking or removal have generally proved the most useful in studying the hormonal control of flowering in *Pisum*. These techniques enabled Barber (1959) to propose that the *Sn* gene was responsible for the photoperiod response in peas through the formation of a graft-transmissible flower inhibitor in the cotyledons of late varieties (Barber and Paton, 1952; Paton and Barber, 1955) which was destroyed by long days. It now seems that the *Sn* gene also produces inhibitor in the shoot as well as the cotyledons (Murfet, 1971c) and that long days reduce the level of inhibitor by suppressing the activity of gene *Sn* rather than through destruction of inhibitor (Murfet and Reid, 1974). Haupt (1957, 1969) and Köhler (1965) originally put forward evidence of a flower promoter in the cotyledons of early peas. While leaving open the possibility of a flower inhibitor, they proposed that flowering in peas is determined by one of two ways; either "by induction", as for example in early varieties through the action of their own cotyledons, or "autonomously", as for example in late varieties through the effluxion of time. Again recent evidence supports their suggestion of a flower promoter in the cotyledons of early peas (Murfet 1971c, 1973b) and I have suggested a compromise hypothesis in which flowering is determined by the "balance" at the shoot apex between a flower inhibitor and a flower promoter. The gene *E* is considered to reduce the level of inhibitor in the cotyledons and the series of alleles lf^a, lf, Lf, and Lf^a are pictured as conferring an increasing sensitivity of the apex to the flower inhibitor (e.g. the ratio of promoter to inhibitor necessary to trigger flowering is greatest in an apex carrying Lf^a). The "balance" hypothesis retains the concept of autonomous determination in the sense that *Sn* activity appears to diminish with age. It is postulated that gene *Hr* tends to block this ageing effect thus delaying the achievement of a correct "balance" for flower initiation (Murfet, 1973a).

Efforts to isolate and characterize the flowering hormones in *Pisum* by means of extracts and diffusates have met with little success. Small promotions of flowering have been reported for diffusates of both vernalized and unvernalized seeds (Highkin, 1955) and small delays and promotions of the FN have been achieved by applying aqueous extracts of the cv. 'Dwarf Telephone' (Moore and Bonde, 1962). Extracts of yeast and pea cotyledons were found to significantly increase the FN of excised embryos of cv. 'Kleine Rheinländerin' but these flowering delays appear to be related to the nitrogen content in general of the extract rather than the level of any particular substance (Haupt, 1952, 1954b). The next few sections are devoted to a more detailed analysis of the evidence for flowering hormones in *Pisum*.

A. Cotyledon Removal: Cuttings

The technique of cotyledon removal has already been introduced in Section VI A2 and we saw there that in some early varieties cotyledon removal leads to a substantial increase in FN under certain circumstances (e.g. short days). Since the FN of the intact controls was not increased by these same circumstances we inferred that the cotyledons, for example of cv. 'Kleine Rheinländerin' or line 60, normally evoke flowering in intact plants by acting as a source of flower promoter.

Turning now to the late flowering genotypes we find that cotyledon removal generally results in a reduction of the FN (e.g. Moore, 1964, 1965) and this effect has been put forward as evidence that their cotyledons act as a source of flower inhibitor (Barber and Paton, 1952; Paton and Barber, 1955; Johnston and Crowden, 1967; Amos and Crowden, 1969). Stem cuttings of late varieties also flower at a lower node and again this has been interpreted as showing leaching of a flower inhibitor (Sprent and Barber, 1957). However, this interpretation has been fairly widely challenged on the grounds that cotyledon removal and cutting do not necessarily result in a direct promotion of the flowering process. For example, both treatments set the growth back and although the FN may be reduced relative to intact plants, the FT and time of flower initiation may actually be later in decotylized plants (Moore, 1964, 1965; Collins and Wilson, 1974b) or cuttings (Haupt, 1955b). Haupt (1969) has suggested that the lower position of the first flower is an indirect effect resulting from a general reduction in the rate of growth and my own view is similar (Murfet, 1973b). This interpretation is supported by the general observation (Paton, 1969) that factors which speed up the growth of late plants tend to raise the FN (see length factors, gibberellins, Section IX) and factors which reduce the growth rate tend to lower the FN, for example the growth retardant B995, severe defoliation or small seed size (Sprent, 1967; Paton, 1967; Reid and Murfet, 1974a). These remarks hold only for late varieties; with early varieties, factors which reduce the rate of development may in fact raise the FN (Moore and Anderson, 1966; J. B. Reid, unpublished). More recently Amos (1974) has proposed that the effect of cotyledon removal in late varieties is achieved through alteration in the growth pattern so that fewer nodes are present in the apical bud at the time a leaf requirement for flowering is satisfied.

Line 61a provides an unusual opportunity to study the effect of cotyledon removal on a single genotype with the peculiar property of displaying two distinct phenotypes, early initiating (EI) and late (L). The results provide a striking example of the interplay of factors in the cotyledons and in the shoot and a chance to show how the "balance" hypothesis and theory of gene action may be applied to a particular situation. The major gene content of line 61a (*lf e Sn hr*) would normally specify a late phenotype but line 61a was selected for a polygenic background leading to incomplete penetrance of gene *Sn*. As a result, about 50% of the plants normally classify as EI and about 50% as L (Table 14.II) but I again emphasize that the EI and L plants are genetically identical (and probably very alike in a physiological sense as well). Removal of a single cotyledon is sufficient to cause a significant increase in the proportion of late plants, and with both cotyledons removed, all plants are late (Table 14.II). These results indicate that the penetrance modifiers operate in the cotyledons where they tend to create a hormonal balance in favour of flowering. (Their effect is therefore similar to, but weaker than, that of gene *E*.) The distinct variation in the phenotype of line 61a is believed to result from the threshold nature of the flowering process and not from any discontinuity in the underlying hormonal situation. A theoretical model is presented in Fig. 14.7. The hormonal "balance" is pictured as lying closely around the critical threshold at the time nodes 12–14 are physiologically determined. If the stimulatory influence of the cotyledons is sufficient to produce a "balance" at the apex favourable to flowering while the shoot is still small the plant will classify as EI; if not, then production of inhibitor by *Sn* in the shoot (under short days) will maintain the vegetative state until *Sn* activity diminishes through the ageing process with the result that the plant will classify as L. On this model the immediate effect of cotyledon removal on line 61a is to create a situation less favourable to flowering, even in those plants which would have been late if left intact, a conclusion exactly opposite to that of Paton and Barber (1955) and others. Nevertheless, cotyledon removal leads to a reduction of the flowering node in the late region (nodes 18–29 in Table 14.II) and the following alternative explanation is proposed.

The age of a plant can be measured in chronological terms, in developmental terms, such as number of nodes formed, or in genetic terms such as which notes are playing on the genetic keyboard. My own view is that cotyledon removal has slowed down the rate of node formation relative to the passage of time recorded chronologically with the result that ageing leads to phasing out of *Sn* activity in late varieties after fewer nodes have been formed than would otherwise occur in intact controls. This type of reasoning is also applicable to certain other situations, for example, when interpreting the effect on flowering of vernalization or seed weight.

B. Grafting

Grafting is a particularly useful procedure for establishing one of the diagnostic properties of a phytohormone, namely its transmissibility. The grafting techniques described for peas by Haupt (1954a), Paton and Barber (1955) and Köhler (1965) are not particularly difficult and usually give a high rate of

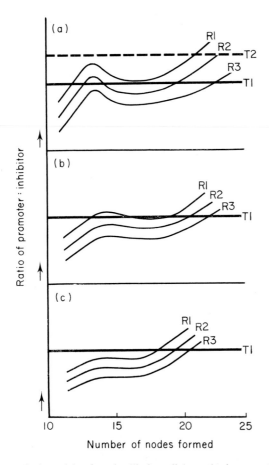

Fig. 14.7. A theoretical model using the "balance" hypothesis to explain the variable penetrance and response to cotyledon removal in line 61a (*lf e Sn hr*, see Table 14.II). The curves R1–3 represent the ratio of flower promoter to flower inhibitor at the shoot apex as various nodes are being laid down. Flowering is evoked in *lf* apices by a ratio in excess of the threshold Tl. The situation with intact plants is represented in (a). The first increase in the ratio is produced by a stimulatory pulse from the cotyledons and the second increase results from diminution of inhibitor production in the shoot as the plant ages. Ratio R1 results in stable flowering at low node (phenotype EI); R2 results in early flower initiation (EI) followed by a limited return to a vegetative condition; R3 results in the first flower being initiated at a high node (phenotype L). With intact plants about 50% of plants are EI; with one cotyledon removed (b) only about 10% of plants are EI and with both cotyledons removed (c) all plants are late. The differences between curves R1, 2 and 3 result from environmental variation and internal noise, not genetic variation. This model can be applied to other cases. For example an *E Sn* stock induces early flowering in *lf Sn* scions but not *Lf Sn* scions (Fig. 14.8). The ratio curve is probably like R2 in (a) in both cases but the first peak fails to reach the higher threshold T2 in *Lf* apices.

success, but some problems may be encountered with vernalized scions, partners of widely different ages, scions from seedlings less than 4 days old (plumules less than 8 mm) or with what appears to be incompatibility between some
genetically different partners. Grafts made between young seedlings (when the
plumules are around 15 mm) often show noticeable elongation within 1 day
of grafting, obvious signs of success by the 5th day and active growth by the
7th to 9th day. Grafts showing these characteristics have been referred to as
"vigorous" grafts as opposed to "slow" grafts which take several weeks to
get under way (Murfet, 1971c). Slow scions behave more like decotylized
plants and they may give distinctly different results from vigorous grafts
between the same two partners. Vigorous and slow grafts should therefore be
distinguished when recording and analysing the results. We have found it
useful to grow decotylized controls with graft experiments. Apart from providing useful information on flowering, measurement of their rate of growth
is an additional aid in distinguishing between vigorous and slow grafts. Even
vigorous grafts lose about 1 week of growth relative to ungrafted controls,
and self-grafts therefore represent a vital control on the experimental grafts.
Nevertheless, vigorous self-grafts usually show little or no difference in FN
from the ungrafted controls.

Experimental grafts have produced two quite different types of response
referred to by Köhler (1965) as quantitative and qualitative responses. The
term quantitative is used to describe cases in which the FN values of the
experimental graft and the self-graft differ significantly but the scion is not
caused to flower in a different phenotypic class. Responses of this type were
first reported by Barber and Paton (1952) who found that stocks of the late
cv. 'Telephone' delayed flowering in scions of the early cv. 'Massey' by about
two nodes. In contrast 'Massey' stocks promoted flowering in 'Telephone'
scions by two nodes. A quantitative response is illustrated in Fig. 14.8 by the
graft of line 59 (early) scions to line 24 (late) stocks (59/24). A qualitative
response was first reported by Köhler (1965) who found that scions of the
late cv. 'Alderman' (AL) could be induced to flower as early as node 11 by
grafting to stocks of the early cv. 'Kleine Rheinländerin' (KR). In comparison
the AL/AL self-grafts flowered at about node 22. Not all AL/KR grafts
flowered at a low node and the distribution of FN values was in fact bimodal
and discontinuous with a zero frequency region at node 17. This situation
corresponds closely to the distinct segregation into the early initiating and
late classes observed in cross 20 (Fig. 14.3) and line 61a (top line of Table
14.II). The term "qualitative" response is therefore justified by the discontinuity between the early (FN less than 17) and late classes. Examples of qualitative responses from late to early and early to late are illustrated in Fig. 14.8
by the grafts 24/59 and 60/24 respectively. In fact the graft results obtained
with lines 59 and 24 (Murfet, 1971c) closely parallel Köhler's results for
cvs KR and AL. For example AL/KR grafts showing the qualitative response

flowered a few nodes higher than the KR/KR self-grafts and the same sort of difference occurred between the grafts 24/59 and 59/59. However, grafts 53/59 and 59/59 flowered at a similar node (Fig. 14.8). Since the late lines 24 and 53 have genotypes *Lf e Sn hr* and *lf e Sn hr* respectively we may infer that gene *Lf* is responsible for the difference and that AL possibly carries both *Lf* and *Sn*. Again with grafts AL/KR and 24/59 many of the scions induced to flower at a low node, flowered only for a short interval before again reverting to a vegetative condition. They subsequently commenced to flower for a second time at about the same node as those scions which had not shown a qualitative response to grafting. The proportion of AL scions showing vegetative reversion was much higher in short as opposed to long days, and in young as opposed to old scions. However, at this point it is appropriate to turn our attention more directly towards the evidence for flowering hormones in peas.

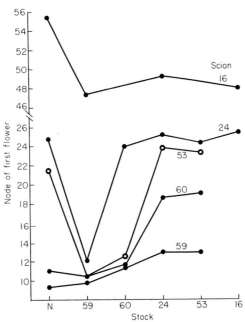

FIG. 14.8. Node of first initiated flower for lines 59 (*lf E sn hr*), 60 (*lf E Sn hr*), 53 (*lf e Sn hr*), 24 (*Lf e Sn hr*) and 16 (*Lfᵈ E Sn Hr*) either grafted in various combinations (at age 4 days, plumules 8–16 mm) or ungrafted (N). Photoperiod—short days. The flowering node values fall within three distinct regions: low (FN 9–13), medium (FN 18–26) and high (FN 46–56). Changes in the flowering node of a scion from medium to low (e.g. 24/59) or low to medium (e.g. 60/53) are referred to as qualitative responses. Significant changes within a region are referred to as quantitative responses (e.g. the flowering nodes of 59/60 and 59/53 are significantly increased in relation to 59/59). The stock sequence *E sn, E Sn, e Sn* shows increasing antagonism to flowering. The scion sequence *lf Sn, Lf Sn, Lfᵈ Sn* shows increasing resistance to early induction. (After Murfet, 1971c, 1975e)

The cotyledon removal studies with cv. KR and line 60 provide fairly convincing evidence that the cotyledons of these early varieties act as a source of flower promoter. The graft studies reinforce that conclusion. For example, stocks of line 60 will induce early flowering in scions of the late line 53. Again stocks of the early line 59 will induce early flowering in scions of line 24. Self-grafted, intact and decotylized plants of line 24 all flower at a high node. The cotyledons themselves are further implicated as the source of the floral stimulus since KR stocks lacking one or both cotyledons are unable to induce early flowering in AL scions under short-day conditions. In contrast, KR stocks with both cotyledons present not only induced early flowering in AL scions under both long- and short-day conditions but the actual node at which flowering occurred was not influenced by the photoperiod. These results therefore provide additional support for the proposition that it is the overwhelming stimulatory influence of the cotyledons which is the important factor in causing insensitivity to photoperiod in intact early varieties. Further evidence for a flower promoter in peas may be found in papers by Haupt (1954a, 1957, 1958) and Köhler (1965). Perhaps the most convincing evidence for a flower inhibitor in peas comes from the fact that scions of the early line 58 (*lf e sn hr*) grafted to rootstocks (with cotyledons) of the late line 53 (*lf e Sn hr*) flowered under short days at least three nodes later than either intact, decotylized or self-grafted plants of line 58 (Murfet and Reid, 1973). These results suggest that the line 53 cotyledons actively delayed flowering by supplying a graft-transmissible flower inhibitor formed through the agency of gene *Sn*. The fact that gene *Sn* is dominant for late flowering is not evidence in itself that *Sn* is directly forming a flower inhibitor. As Haupt (1969) has pointed out, *Sn* could achieve an inhibitory function by suppressing the formation of a flower promoter. The remaining evidence is purely circumstantial.

Graft studies provide further information on the possible physiological action and sites of operation of the genes *Lf*, *E*, *Sn* and *Hr* (Murfet, 1971c, 1975e). Gene *Sn* appears to act in the shoot as well as the cotyledons since scions of genotype *lf E Sn hr* (line 60, EI) flowered in the late class (FN 19) when grafted to stocks of lines 24 or 53 but these same stocks delayed flowering only to node 13 in scions of genotype *lf E sn hr* (line 59, ED). In contrast gene *E* appears to function only in the cotyledons where it reduces the effect of *Sn*. For instance, stocks of genotype *E Sn* were intermediate between stocks of genotype *e sn* and *e Sn* in their ability to cause a quantitative delay in flowering of *sn* scions or to induce a qualitative response in scions of late lines (see Fig. 14.8). These results suggest that *E* lowers the level of inhibitor in the cotyledons or at least renders the hormonal "balance" emerging from the cotyledons less antagonistic to flowering. Stocks of genotype *E Sn* belonging to lines 7 (*lf^a E Sn hr*), 60 (*lf E Sn hr*), 2 (*Lf E Sn hr*) and 16 (*Lf^a E Sn Hr* all seem to have about the same effect, even though the ungrafted lines range from the earliest to the latest flowering in our Hobart collection (see Fig. 14.2). These results indicate that genes *Lf* and *Hr* have no effect in the stock and by implication must act in the shoot.

The effect of the *Lf* series is in fact strikingly obvious in the shoot (Fig. 14.8). Line 7 (*lf^a*) scions do not appear to be influenced by any stock, which is not surprising since they are believed to initiate flower primordia prior to grafting stage. The late line 53 (*lf*) is induced to flower early by either an EI stock (line 60, *E Sn*) or an ED stock (*E/e sn*) but the late line 24 (*Lf*) is only induced by the ED stocks and even these do not induce early flowering in line 16 (*Lf^a*). In addition vegetative reversion is both more common and more extensive in *Lf* scions than it is in *lf* scions. These results suggest that sensitivity to inhibitor increases in the order *lf^a*, *lf*, *Lf*, *Lf^a* or in other words these alleles specify a requirement for an increasingly higher ratio of promoter to inhibitor at the apex before flower initiation can occur. The *Lf* alleles also appear to determine the minimum period of vegetative development. Plants bearing dominants *E*, *Sn* and *Hr* may have anything from five to over 100 vegetative nodes. However while *lf^a* may flower as low as node 5, *lf* has not been observed to flower below node 8 nor *Lf* below node 11 (a single exception at 10 is known to Murfet). These minima hold for such diverse treatments as vernalization, exposure to continuous light, induction by grafting, etc. The lowest FN so far observed for *Lf^a* is 18 but this genotype has not been studied sufficiently to determine the true minimum. If a true juvenile phase exists in which flowering is precluded even if the hormonal "balance" is above the critical threshold, then the inability of *Lf^a* shoots to respond when grafted to ED stocks is understandable since cotyledonary influence probably would have waned by the time the scion was capable of responding.

Finally we will see how the "balance" hypothesis may be applied to some of the graft results. The ratio of promoter to inhibitor in ungrafted plants of line 53 (*lf e Sn hr*) probably follows a pattern something like the lowest curve in Fig. 14.7(c), resulting in flowering at a high node. With ungrafted plants of line 60 (*lf E Sn hr*) the ratio probably follows a curve like R1 in Fig. 14.7(a), resulting in stable flowering at a low node. In the graft 53/60 the curve may be like R2 in Fig. 14.7(a). A pulse of promoter from the line 60 cotyledons results in early flowering, but reversion to a vegetative condition follows an influx of inhibitor produced by the shoot itself. However, line 60 stocks fail to induce early flowering in scions of genotype *Lf e Sn hr* (line 24), not because the ratio of promoter to inhibitor is different from the graft 53/60, but because the early peak in the curve R2 fails to reach the higher critical threshold (T2 in Fig. 14.7(a)) in a scion of genotype *Lf*. With the graft 59/53 the early scion of genotype *lf E sn hr* is presumed to have a very limited capacity to produce inhibitor and flowering follows as the inhibitory influence of the

o*

line 53 cotyledons (*e Sn*) diminishes. This appears to be about node 14 or 15 in *lf* shoots.

C. The Role of the Leaf

Leaves function as a manufacturing centre and as a receptive organ for the light stimulus. As such we might expect them to play a prime role in the formation of the flowering hormones and as the organ translating the light/dark cycle into chemical terms. On the "balance" hypothesis, at a given photoperiod each leaf may be exporting a certain level of both promoter and inhibitor. Removal of a leaf will reduce the absolute level of these substances in the plant, but not necessarily their relative proportions. This may explain why leaf removal appears to have little effect on flowering in peas in most circumstances. Removal of the lower leaflets from cv. 'Greenfeast' in long days was reported to cause little change in the flowering node (Sprent, 1966a) but removal of both the lower leaflets and stipules did result in a small (1·6 node) delay which was reduced if defoliation was severe enough to lower the rate of node formation (Paton, 1967). The leaf area can also be altered genetically (see Chapters 2 and 12). For example the recessive mutant *tl* causes all tendrils to be replaced by leaflets but segregation for this gene appears to have no significant effect on flowering (Murfet, 1975e). The gene combination *af st Tl* gives a phenotype almost devoid of leaves but again there seems to be no marked effect on flowering although there are indications of some minor effects (G. A. Marx, unpublished).

The relative proportions of the hormones contributed by a leaf might be expected to vary with the genotype, the physical environment in which the leaf is functioning and possibly the age of the leaf. Certainly there is clear evidence in support of the first two premises. Paton (1971) showed that masking the cotyledons or the first foliage leaf of cv. 'Greenfeast' delayed both the completion of photoperiodic induction and the FN without having any effect on the rate of node formation. The cotyledons achieved the competence to respond to light before the epicotyl hook opened and thereafter appeared to function in much the same way as foliage leaves. Paton has suggested that exposed leaves and cotyledons play a quantitative role in the induction of flowering. In long-day conditions the late line 53 (*lf e Sn hr*) and the early line 58 (*lf e sn hr*) usually flower about five nodes apart, that is, when the seeds are germinated underground (in effect with the cotyledons masked). However, when the cotyledons and plumule were exposed to continuous light from the start of germination, the two lines were found to flower at the same node (Murfet and Reid, 1974). These results suggest that light suppresses the formation of flower inhibitor by the *Sn* gene. Applying this concept to Paton's results for cv. 'Greenfeast' (*Lf e Sn hr*), we would expect *Sn* activity to con-

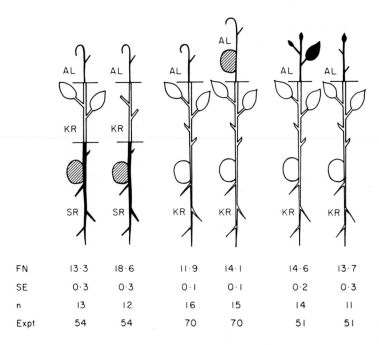

FN	13·3	18·6	11·9	14·1	14·6	13·7
SE	0·3	0·3	0·1	0·1	0·2	0·3
n	13	12	16	15	14	11
Expt	54	54	70	70	51	51

FIG. 14.9. Pairs of grafts (left and right) showing leaves in both a stimulatory and inhibitory role. FN is the mean node of first flower, SE the standard error of the mean and n the number of plants. SR indicates a vegetative stock of the late cv. 'Schweizer Riesen' (*Sn*), KR flowering stocks or interstocks of the early cv. 'Kleine Rheinländerin' (*sn*), and AL young vegetative scions of the late cv. 'Alderman' (*Sn*). The centre pair shows the inhibitory effect of 'Alderman' cotyledons. Photoperiod 9 h. (After Haupt, 1958)

tinue in a masked leaf and therefore contribute to the observed flowering delay. Finally the graft studies of Haupt (1958) elegantly illustrate that a leaf may play both a promotory and an inhibitory role in *Pisum* (Fig. 14.9). Under short-day conditions a leafy interstock of the early cv. 'Kleine Rhein-länderin' (KR) promoted flowering in a scion of the late cv. 'Alderman' (AL) much more effectively than a defoliated interstock. On the other hand, the presence of either cotyledons or leaves on the AL scion tended to counteract the promotive influence of KR stocks. These results give further support to the concept that flowering in peas is determined by the balance between promotory and inhibitory components.

VIII. TEMPERATURE

A. Effects on Flowering

The temperature at which the plants are grown has little effect on the FN of early varieties (Haupt, 1952; Barber, 1959) but with late varieties the FN can be influenced substantially by temperature, in particular under short-day conditions although a small effect may still occur even in continuous light (Stanfield et al., 1966; Paton, 1968; Fig. 14.4). Conversely, low temperatures (2–4°C) may greatly reduce or even eliminate the response to photoperiod normally shown by late peas (Fig. 14.4, Murfet and Reid, 1974). We have suggested that the loss of the photoperiod response results from a severe reduction in the activity of the Sn gene at low temperatures. Paton (1968) has recognized in cv. 'Greenfeast' three types or patterns of response to temperature using the variables FN and TFI (time of flower initiation) as criteria for classification. In Type 1 there is a high temperature promotion of both FN and TFI in continuous light above 20°C. In 8-h or 16-h photoperiods raising the temperature always increases FN but TFI is at first reduced (Type 2) then increased (Type 3). He explained these responses largely in terms of a leaf requirement for flowering. However, they may also be interpreted under the "balance" hypothesis if we assume firstly, that the reaction producing inhibitor has a higher temperature coefficient than the reaction-controlling the formation of promoter, and secondly, that inhibitor production is suppressed by continuous light but 8 h of darkness are sufficient to permit some activity of gene Sn (see Sections VII C, VI A4). Other consequences of growth at low temperatures are introduced in the next section on vernalization.

Paton (1957) also observed a thermoperiodic response in cv. 'Greenfeast'. Breaking a 16-h night into two 8-h periods at different temperatures accelerated the time of flower initiation relative to a constant night temperature. The larger the temperature difference the more the TFI was accelerated. The effect was independent of the temperature during the 8-h light period or the order in which the high and low temperatures were applied during the dark period. Unfortunately the effect on FN was not reported.

B. Vernalization

Germinating the seed at low temperatures (2–7°C) lowers the flowering node in late varieties but causes no change nor even a slight increase in the FN of early varieties (Highkin, 1956; Went, 1957; Moore and Bonde, 1958; Barber, 1959; Highkin and Lang, 1966). The later the variety the larger is the response to vernalization (Wellensiek, 1969). Vernalization treatments as short as 5 days may produce a significant response and the maximum response is usually achieved by cold treatments lasting about 4–6 weeks. Highkin (1956) found that the flower-promoting effect of vernalization was completely counteracted in the cv. 'Unica' if vernalization was followed immediately by 10 days of high temperature (30°C) but this treatment only partially counteracted the effect of vernalization on cv. 'Zelka'. Germination of 'Zelka' at 20 or 26°C resulted in loss of the ability to respond to vernalization after 3 days at 20°C and 2 days at 26°C. In contrast, vernalization of cv. 'Greenfeast' was reported still to be effective after the seedlings had grown for 10 days at about 20°C (Amos and Crowden, 1969).

Three important questions arise:

1. What is the genetic control of the vernalization response?
2. By which mechanism(s) is the response achieved?
3. At what site(s) does it occur?

Barber (1959) concluded that the vernalization and photoperiod responses were conferred by the same gene, Sn. This effect of Sn has been confirmed but the dominant gene Lf has also been found to confer a potential to respond to vernalization, the magnitude of the response being influenced by the background of Lf modifiers, that is to say, those Lf e sn hr lines having the highest FN also show the greatest response to vernalization (Murfet and Reid, 1974). Vernalization has also been reported to promote flowering in decotylized plants of early varieties; by up to three nodes in the case of 'Kleine Rheinländerin' (Haupt and Nakamura, 1970) and by five nodes in line 60 (Wall et al., 1974). The response in decotylized plants of line 60 (lf E Sn hr) can be understood in terms of gene Sn which is released from the epistatic effect of gene E when the cotyledons are removed. On the other hand, the vernalization response in 'Kleine Rheinländerin' does not seem to be attributable to either Lf or Sn since both these genes appear to be absent. However, recent studies have revealed that a small positive response to vernalization is also manifest in scions of genotype lf e sn hr if their flowering is delayed by grafting to stocks of genotype lf e Sn hr (Reid and Murfet, 1975b). It seems, therefore, that while certain genes confer a marked response to vernalization, most genotypes probably have the potential to show at least a small response, provided appropriate circumstances can be found to reveal it. Conversely, the substantial vernalization response normally displayed by the late line 53 (lf e Sn hr) can be eliminated by exposing the plants to continuous light from the start of germination, and in the same line vernalization followed by continued cold (2–4°C) nights can eliminate the photoperiod response (Murfet and Reid, 1974). These results support Barber's (1959) observation that vernalization and long days act in a competitive manner. However, while he suggested that both factors may destroy a flower inhibitor produced by gene Sn we favour the view that both factors tend to repress Sn activity.

The delaying action which unvernalized stocks of cv. 'Greenfeast' exert on scions of cv. 'Massey' is largely eliminated by vernalization of the stock and this has been interpreted as a low temperature repression of inhibitor synthesis in 'Greenfeast' cotyledons (Paton, 1969). Repression of *Sn* activity could also explain the small response shown by 'Greenfeast' embryos decotylized prior to vernalization (Amos and Crowden, 1969) since *Sn* operates in the shoot as well as in the cotyledons. However, vernalization also seems to condition the apex so that it responds more readily to a floral stimulus, since flower initiation in vernalized 'Greenfeast' follows immediately photoperiodic induction has been completed, whereas in unvernalized plants there is a 3-day interval between the completion of induction and the evocation of flowering (Paton, 1969). The increased readiness to flower in vernalized apices of 'Greenfeast' bears a similarity to the proposed effect of substituting allele *lf* for *Lf*, namely, a reduction in the threshold ratio of promoter to inhibitor required to evoke flowering. However, if vernalization does also act through the same mechanism as *lf* it does not fully cancel the effect of *Lf* since the phenotype of genotype *Lf E Sn hr* is not changed from L to EI (Murfet and Reid, 1974). A vernalization response also appears to occur at the apex of *lf* shoots (Fig. 14.10) but the existence of an even earlier allele, *lf^a*, leaves open the possibility that environment could modify the action of *lf* towards *lf^a*.

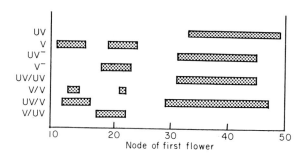

Fig. 14.10. Distribution of node of first initiated flower for line 63 (*lf e Sn Hr*) plants unvernalized (UV), vernalized for 35 days at 2–4°C (V), decotylized (UV⁻ and V⁻) or grafted in various ways, e.g. vernalized scion on vernalized stock V/V. Photoperiod 8 h; grafting and cotyledon removal performed at day 5 for UV plants and day 35 for V plants. (After Reid and Murfet, 1975b)

Some recent data (Reid and Murfet, 1975b) show very clearly that vernalization has an effect in both the shoot and the cotyledons of line 63. The FN values for line 63 presented in Fig. 14.10 fall into three distinct regions— low (FN 10–16), intermediate (FN 17–24) and high (FN 29–49). Unvernalized plants flower in the high region whether intact (UV), decotylized (UV⁻) or grafted UV/UV. Promotion of flowering to the low region appears to result from an effect of vernalization in the cotyledons since two-thirds of the unvernalized scions grafted to vernalized stocks (UV/V) flowered in the low region but none of the V/UV or V⁻ plants did. On the other hand, flowering in the intermediate region appears to result from some effect of vernalization on the shoot (embryonic leaves or apex) since all of the V⁻ and V/UV plants flowered in this region but none of the UV/V ones did. The vernalization effects in the stock and the shoot resulted in a promotion of flowering by 26 and 20 nodes respectively. We suggest the effect in the stock results from low temperature repression of Sn activity, and the effect in the shoot from a reduction in the threshold ratio of promoter to inhibitor necessary to evoke flowering. The bimodal distribution of the FN values in treatments V and UV/V may be understood in terms of the theory already given in relation to line 61a (Fig. 14.7). The magnitude of the response to vernalization in line 63 ($lf\ e\ Sn\ Hr$) is much larger than the 2–6 node reduction in the FN normally observed in line 53 ($lf\ e\ Sn\ hr$) or 'Greenfeast' ($Lf\ e\ Sn\ hr$) but the underlying changes are not necessarily larger in line 63. The larger visible effect is probably a consequence of the prolongation of Sn activity through the influence of gene Hr. For example, if the curves for lines 53 and 63 in Fig. 14.6 are taken to represent the decrease in inhibitor level with age, a change in the critical threshold at the apex which leads to a small reduction in the FN of line 53 (steeply declining curve) may result in a relatively large promotion of flowering in line 63 (gently sloping curve).

In summary, the results to date indicate at least two different effects of vernalization. The first arises from low temperature repression of Sn activity. It is most obvious in the cotyledons but is assumed to occur at any site of Sn activity and is probably a continuation of the same temperature response curve which at high temperatures leads to a high rate of inhibitor production and a magnification of the photoperiod response. The second effect occurs in the shoot, possibly at the apex where vernalization may reduce the threshold ratio of promoter to inhibitor necessary to evoke flowering. However, a second mechanism may also be involved in the shoot. If the ageing process is not slowed to the same extent as the rate of node formation during the vernalization treatment, some reduction in the FN may result from phasing-out of Sn activity after fewer nodes have been developed.

IX. NUTRITION

A reduced level of nutrients, in particular N, Mg or possibly P, has been reported to promote flowering by one or two nodes in intact and decotylized late varieties (Nakamura, 1965; Sprent, 1966b, 1967). Nutrient factors of course influence various aspects of plant growth but the effect on FN was not associated with any change in the rate of node formation. Flowering in intact early varieties is generally little influenced by nutrient conditions but sucrose and arginine have been reported to delay flower initiation in cv. 'Alaska' (Leopold and Guernsey, 1953). Cotyledon removal after imbibition renders an early variety more susceptible to nutrient conditions. For example, decotylized plants of 'Kleine Rheinländerin' flowered at a higher node when cultured on an agar medium rich in sucrose or nitrogen (Haupt, 1952, 1954b, 1955a). This occurred whether the nitrogen was supplied as inorganic nitrate, as a mixture of amino acids or in the form of yeast extract or extract of pea cotyledons. Increased N also stimulated growth, in particular when supplied in organic form, but again there was no evidence that the effect on FN was associated with a change in the rate of node formation. However, low sucrose levels tended to reduce both the flowering node and the rate of node formation. Nitrogen and sugar appear to influence flowering independently.

X. GROWTH REGULATORS

A. Interactions between Growth Regulators

Attempts to directly isolate and identify the endogenous substances regulating flowering in peas have met with little success. The technique of gel electrophoresis appears particularly useful with regard to identifying protein or enzyme differences between genotypes or within the same genotype at different ages or under different environmental conditions. Although we have obtained at Hobart some very interesting results using this technique, a lack of repeatability in the results so far has prevented us from reaching any firm conclusions. However, the question of gene action can also be approached indirectly. By studying the patterns of response to exogenous application of various growth regulators we may be able to find effects which mimic the action of a particular gene or gain some clues as to the identity of the proposed endogenous regulators of flowering. The interaction between growth regulators is also relevant with respect to the "balance" hypothesis since the relative level of various growth substances has been shown to determine certain morphogenetic steps. For example, the ratio of auxin to cytokinin appears to determine whether a culture of tobacco callus tissue will differentiate root or shoot organs (Skoog and Miller, 1957). Again the relative levels of endogenous auxin, ethylene and gibberellin play a decisive role in determining whether male or female flowers are developed in certain species of cucurbits although recent evidence suggests that the critical factor is probably the ethylene: gibberellin balance, with the endogenous concentration of auxin functioning indirectly by influencing the endogenous concentration of ethylene (Laibach and Kribben, 1950; Peterson and Anhder, 1960; Galun *et al.*, 1965; Robinson *et al.*, 1969; Byers *et al.*, 1972). Exogenous application of auxin increases the production of ethylene by cucumber plants (Shannon and De La Guardia, 1969) and this same mechanism may explain why high concentrations of auxin can delay flowering in peas.

3. Auxins

Auxin was first reported to promote flower initiation in peas (Borgström, 1939). However, Leopold and Guernsey (1953, 1954) later reported that auxins could either promote or delay flowering in cv. 'Alaska' depending on the temperature. Flowering was promoted by several nodes if the seed was first soaked in auxin solution (naphthalene acetic acid [NAA], β-naphthoxy-acetic acid, indol-3yl acetic acid [IAA], indol-3yl butyric acid or p-dichlorophenoxyacetic acid) and then given 3–15 days of cold treatment (3°C), but if the seeds were germinated at 18°C, 1 mg l^{-1} of NAA caused a three node delay. (The observed low FN and sensitivity to season suggest that Leopold and Guernsey were not using the common strain of 'Alaska'.) Indol-3-yl acetic acid at 25 mg l^{-1} had no effect on the flowering of decotyl-ized 'Kleine Rheinländerin' (Haupt, 1952), and L. I. Slade and D. M. Paton (unpublished) observed no consistent effect of IAA on cvs 'Massey' or 'Tele-phone'. Reid and Murfet (1974b) found a dose of 1 mg of IAA applied to the seed of line 58 caused a significant increase in the FN (Fig. 14.11) but doses of 5–500 μg were ineffective. The delay was possibly mediated by the well-known effect of auxin stimulating the production of ethylene.

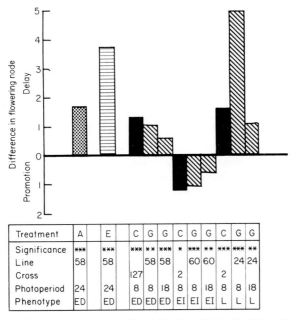

Treatment	A	E	C	G	G	C	G	G	C	G	G
Significance	***	***	***	**	***	*	***	**	***	***	**
Line	58	58		58	58		60	60		24	24
Cross			127			2			2		
Photoperiod	24	24	8	8	18	8	8	18	8	8	18
Phenotype	ED	ED	ED	ED	ED	EI	EI	EI	L	L	L

Fig. 14.11. Effect on flowering node of indol-3yl acetic acid (A) at 1 mg per seed, Ethrel (E) at 40 mg l^{-1}, gibberellic acid (G) at 5 μg per seed and segregation for the crypto-dwarf (C) vs dwarf length form in crosses 8 × 24 (2) and 8 × 58 (127). *, **, *** difference significant at the 5%, 1% and 0·1% levels respectively. (Data from Murfet, 1971b; Reid and Murfet, 1974; Dalton and Murfet, 1975)

C. Ethylene

Concentrations as low as 1 mg 1^{-1} of Ethrel (active ingredient the ethylene-releasing compound 2-chloroethylphosphonic acid) significantly raised the FN of line 58 (*lf e sn hr*) and 40 mg 1^{-1} caused a 3·7 node delay (Fig. 14.11; Reid and Murfet, 1974b). Certain parallelisms were noted between the effects of Ethrel and gene *Sn*; both delay flower initiation, retard floral development, reduce internode length and raise the node of the first leaf with four leaflets. Ethylene production in peas (Goeschl *et al.*, 1967) and *Sn* activity (Murfet and Reid, 1974; Table 14.IV) both appear to be suppressed by light through the action of a phytochrome switch. However, the parallelisms could be fortuitous and there is certainly no conclusive evidence linking *Sn* with ethylene metabolism.

D. Gibberellins

Gibberellic acid (GA_3) is the most widely studied of the gibberellins as far as flowering in peas is concerned. It inhibits development of the flower bud into an open flower and may cause complete abortion of a bud if it is present in sufficient concentration while the young primordium is developing (Brian *et al.*, 1958; Barber *et al.*, 1958; Sprent, 1966b; Wellensiek, 1973b). If applied after flower initiation, GA_3 speeds up the appearance of the first flower bud in dwarf varieties since it increases not only the size of the internodes but also the number of internodes expanded in a given time (Brian *et al.*, 1958; Sprent, 1966b). Because GA_3 alters the rate of development, small changes in the flowering node are not necessarily indicative of a direct effect on the flowering process. For example, in long days 5 μg of GA_3 applied to the dry seed of line 24 caused an increase of one node in the FN but a decrease of half a day in the time to open flower (Dalton and Murfet, 1975). However, in cv. 'Massey', GA_3 was found to increase all three parameters FN, FT and TFI (time to initiation of the first flower primordium), and a true delay in flowering is indicated (Barber *et al.*, 1958). [Collins and Wilson (1974a) apparently were unaware that we had checked all three parameters in cv. 'Massey' when they claimed that Murfet and Barber (1961) were not justified in referring to the effect of GA_3 as a "delay in flowering".] The effect of GA_3 on the rate of development in cv. 'Massey' is in fact quite complex. Although application of GA_3 to the seed led to more unfolded leaves being present at any given time, when the nodes present in the apical bud were also included in the count, the untreated plants actually had a higher total number of nodes than the treated plants at all stages prior to initiation of the twelfth node (I. C. Murfet, unpublished).

The response to GA_3 clearly varies with both the genotype and the environment. GA_3 generally has a minimal effect on the FN of tall (*Le, La* and/or

Cry) peas although some small increases and a couple of marked decreases in FN values have been reported (Bonde and Moore, 1958; Barber *et al.*, 1958; Wellensiek, 1973b). On the other hand, the flowering node of dwarf (*le, La* and/or *Cry*) peas is generally increased by GA_3 but again there are exceptions. For example, GA_3 consistently reduces the FN of the EI line 60 (*lf E Sn hr*) by one-half to one node (Dalton and Murfet, 1975) and in one instance GA_3 was found to reduce the FN of an *No^m na* line by four nodes (Wellensiek, 1973b). GA_3 generally increases the FN of *sn* lines (e.g. *lf e sn hr, lf E sn hr, Lf e sn hr*) by one or two nodes. The effect of GA_3 on late varieties was at first thought to be independent of photoperiod (Barber *et al.*, 1958) but more recently we have consistently found GA_3 to cause substantial increases in both the FN (3–5 nodes) and the FT of genotypes *lf e Sn hr, Lf e Sn hr* and *Lf E Sn hr* under short days. In long days, however, the same treatment seems to cause only small changes in the FN including in one instance a small reduction in the FN of genotype *Lf E Sn hr* (Dalton and Murfet, 1975). Some evidence of this interaction between GA_3 and photoperiod is also apparent in the data of Wellensiek (1973b). No consistent pattern of interaction between vernalization and GA_3 emerges from the literature (Barber *et al.*, 1958; Wellensiek, 1973b) but GA_3 does seem to increase the FN of the late cv. 'Dwarf Telephone' more effectively when applied after, rather than before, vernalization (Moore and Bonde, 1958).

Gibberellins A_1, A_4, A_5, A_7 and A_8 all significantly increased the flowering node of cv. 'Massey' when applied as a 2-μg dose to the dry seed; GA_9 was ineffective at all dose rates tested (0–10 μg). The least active were GA_1 and GA_4, but GA_8 was as effective as GA_3 in delaying flowering whilst being only 30% as active as GA_3 in respect of its stimulatory effect on stem length (P. J. Dalton, unpublished). Indeed gibberellins sometimes significantly increase the flowering node without having any detectable effect on stem length as exemplified by the effect of allo-gibberic acid on cv. 'Massey' (Murfet and Barber, 1961) or GA_3 on decotylized plants of cv. 'Dwarf Telephone' (Moore, 1965).

In summary, although GA_3 sometimes promotes flowering in peas, it does not appear to evoke flowering under circumstances not conducive to flowering as it does in some other long-day species and in many circumstances it is definitely inhibitory to flowering. [Application of GA_3 did enable survival and eventual flowering of the very compact *na No^m* line under an 8-h photoperiod in which untreated plants sometimes died before flowering, but this should not be interpreted as a promotion of flowering since untreated plants surviving these conditions were found to flower at an earlier node and time than the GA_3 treated plants (Wellensiek, 1969, 1973b). The marked general improvement in the growth of *na* plants in response to exogenous gibberellin suggests that this genotype is severely deficient in endogenous gibberellin.]

We may now return to a more detailed exploration of the relationship between gibberellin and the genes governing internode length and flowering. As we have already seen, the response to GA_3 depends on the status of the genes Le/le and Sn/sn. The loci Le, La and Cry determine four different length forms—normal tall (Le, La and/or Cry), dwarf (le, La and/or Cry), crypto dwarf ($le\ la\ cry^c$, semi-tall) and slender (Le or le, $la\ cry^s$, very long internodes). Continued application of GA_3 to a dwarf produces a phenocopy of slender and Brian (1957) has suggested that the Le locus may control gibberellin synthesis, with gibberellin reversing the inhibition of growth by La and Cry (Further relevant information may be found in Chapters 2 and 9.) The Le gene has been reported to promote flowering time by speeding growth (Rasmusson, 1935; Marx, 1975a) but to have little effect on flowering node or possibly even to increase the FN in Sn plants (Barber, 1959). This pattern shows some homology with the effect of applied gibberellin on dwarf peas. In addition the flower-delaying gene Sn was found to reduce internode length (Barber, 1959) and more recently the presence of Sn was reported to obscure the segregation of an unidentified length gene (Murfet, 1975d). The expression of this length gene was found to be influenced at internodes laid down well in advance of the first flower primordium suggesting interlocking of the relevant metabolic pathways at a fairly basic level. A relationship between the length factors, flowering and gibberellin is strikingly illustrated by the data in Fig. 14.11. It can be seen that crypto-dwarfs (solid black) in a segregating progeny tend to flower at a higher node than dwarfs in the ED and L classes but in the EI class crypto-dwarfs tend to flower at a lower node. The response of dwarfs to GA_3 (diagonal stripe) follows an identical pattern. If Brian's theory is correct, the effect of gibberellin on flowering can be viewed as the result of a hormonal state which arises from the blocking action of GA_3 on genes La and Cry. The smaller flowering response to gibberellin in tall plants would then be expected if the La and Cry genes are already partially blocked. A reduced flowering response to gibberellin would also be predicted for slenders and crypto-dwarfs since the dominant genes La and Cry are absent, and preliminary data for the EI class (P. J. Dalton, unpublished) support this prediction. However, gibberellin effects on flowering in peas are not always associated with an effect on internode length and this raises the possibility of a second type of effect perhaps achieved through a different mechanism and/or a different gibberellin.

XI. CONCLUDING REMARKS

In writing this chapter I have been struck by the multiplicity of cause and effect relationships and by the limitations of too singular an approach either to the problem itself or to the interpretation of the wealth of data available. Too narrow a genetic base at the start of an investigation imposes limits on the interpretation which even the most thorough physiological study is powerless to resolve. The brief remarks which follow serve to illustrate these themes and to highlight some of the more important findings.

The *Sn* gene straight away provides an example of a multiple effect since it is a major gene for flowering and a minor gene for internode length; conversely, genes *Le*, *La* and *Cry* are major genes for length and minor genes for flowering. Gene *Sn* also influences the number of leaflets per leaf, floral development and the length of the reproductive phase. Dominant *Sn* confers a marked response to photoperiod but a smaller response can be elicited in *sn* plants. This may be through an alternative mechanism but we must consider the possibility that the recessive alleles of important major genes such as *Sn* and *Le* are not necessarily inactive. Indeed these loci may well govern the formation of substances so vital to the plant that complete loss of activity at the locus would prove lethal. On the other hand there may be alternative, but less efficient, pathways supplying the substance. This line of reasoning leads us to the concept that the flowering hormones themselves may be present to some extent throughout the life of the plant and that the sudden transition to the flowering state is a product of the threshold nature of the flowering process rather than the sudden appearance of a highly active stimulatory substance.

Our understanding and interpretation of flowering in peas has been considerably altered by the realization that the cotyledons are capable of perceiving a light stimulus in the same way as foliage leaves and that the *Sn* gene appears to produce inhibitor in the shoot (presumably the leaves) as well as in the cotyledons. Light appears to have several different effects, possibly controlling *Sn* activity through a phytochrome switch, destroying inhibitor through a far-red mechanism and influencing flowering through the general level of metabolites and growth phenomena. A vernalization response is conferred in a major way by *Sn* and to a lesser extent through *Lf* and its modifiers, but small responses can also be elicited in *lf sn* plants. Vernalization appears to promote flowering by at least two different mechanisms, namely by regulating *Sn* activity and by increasing the readiness of the apex to flower. There has been in the past some uncertainty over the relationship between the proposed flower inhibitor in peas and the photoperiod and vernalization responses. The explanation probably lies in the multiplicity of cause and effect relationships outlined above. We see that a (not the) response to both photoperiod and vernalization is conferred by gene *Sn* and the response to both factors appears to be manifest through the flower inhibitor as a result of repression of *Sn* activity by light and low temperatures. However, photoperiod and vernalization responses can also be separated genetically. For instance line 65 (*Lf e sn hr*) has a high FN in long days characteristic of a late variety (FN 17–18) but the true photoperiod response in this line seems to be only about 0·6 of a node while the response to vernalization can be 5–7 nodes. In late lines carrying *Sn*, the response to photoperiod (5–15 nodes) is usually much larger than the response to vernalization.

There is evidence of both a flower promoter and a flower inhibitor in peas and it is proposed that the level of these hormones may vary in a quantitative manner with flowering being evoked by the gradual achievement at the apex, of a balance (or ratio) of promoter to inhibitor in excess of a critical ratio. So far most of the genes which have been identified seem to be concerned with regulating flowering through the inhibitor and it is a little disconcerting that we have not yet identified genes for certain concerned with the flower promoter. Indeed, it must be emphasized little is really known about the true nature of the flowering hormones, the gene products, the light control mechanisms or the effects of plant age. The *Lf* series of multiple alleles appear to have a rôle in determining the minimum age at which flowering can occur; genes *Sn* and *Hr* help to determine the maximum life of the plant through their interrelated effects on flowering and senescence. Conversely age itself seems to play a part in the flowering process. The *Lf* alleles are particularly interesting since they appear to draw a blind across the window of events until a minimum number of nodes have been laid down, something analogous to a radio alarm clock—until the switch is sprung you don't hear the programme! The proposed models and schemes of gene action discussed in this chapter enable some understanding of the problem and they have led to some successful predictions of experimental results but in many ways the flowering process remains as enigmatic and as challenging as ever.

ACKNOWLEDGMENTS

I am particularly indebted to Professors W. Haupt and G. A. Marx for helpful comments and advice concerning the manuscript and to my colleagues P. J. Dalton, J. B. Reid and G. M. Rowberry for permission to use unpublished results.

APPENDIX

The true relationship between genes can only be resolved by crossing experiments, but certain circumstantial evidence enables some comment to be made on possible relationships. The *Lf* locus equals, by definition: S_1 of Murfet (1971b); by inference from properties and linkage data, *A* of Hoshino (1915) [resymbolized *Lf* by White (1917)]; *Xa* of Rasmusson (1935); *L* of Pellew (1940) [resymbolized *Li* by Blixt (1969)]; "Gengruppe B" of Hänsel (1954b) and *Pra* of Monti and Scarascia Mugnozza (1967, 1970, 1972). Both *pra* and the very early flowering mutant of Gottschalk (1960) have similar FN values o lf^a. The allele lf^a was isolated from an old line of peas tracing back to Vilmorin's 'Acacia' variety and the very early forms ib and iba of Lamprecht 1956b) also have this variety in their ancestry. Finally the cross 'Rapid $\times$ Stensart' (Tedin and Tedin, 1923) is almost identical in appearance to the cross 7 $\times$ 2 in long days (Murfet, 1975e) which is of the type lf^a E Sn hr $\times$ Lf E Sn hr. Ironically therefore the symbol Sn was first introduced by the Tedins for what was probably an Lf/lf^a segregation but unfortunately no linkage data are available as 'Rapid' and 'Stensart' are both white flowered.

Gene E is not equivalent to Hoshino's B [resymbolized Ef by White (1917)], a possibility raised by Blixt (1972). Ef is incorrectly referred to in the literature as a dominant early gene; Hoshino used his B gene as a latening factor (see Murfet, 1975a). Blixt (1972) suggested resymbolizing E as Efd but the symbol E is now agreed (Murfet and Blixt, 1974).

Gene Sn is by definition equal to gene S_2 of Murfet (1971b) and is equivalent to the symbol Sn as used by Barber (1959). However Lf was also probably segregating in Barber's 'Massey' $\times$ 'Greenfeast' cross which appears to be of the form Lf e Sn hr $\times$ lf e sn hr. Although dihybrid, this cross segregates 3 late: 1 early [see the cross between lines 58 and 24 ('Greenfeast') by Murfet (1971b)] with genotypes lf e sn hr and Lf e sn hr classifying as ED and lf e Sn hr and Lf e Sn hr classifying as L. The visible segregation is due to Sn and the properties typical of an L-type plant can therefore be conferred by Sn alone. A cross of the type lf E Sn hr (EI) $\times$ Lf E Sn hr (L) also gives a 3 late:1 early (EI) ratio in F_2 but in this case the visible segregation is determined by Lf. The Ph gene for photoperiod response mentioned briefly by Barton et al. (1964) shows a tight linkage of 4–8 % (D. W. Barton and G. A. Marx, unpublished) with the seed marbling gene M. A linkage of 3–6 % has been reported for Hr and M (Murfet 1973a, 1975e). Genes Ph and Hr are therefore probably identical and evidence is accumulating (G. A. Marx, unpublished) that the large response to photoperiod in classes G2 and G is conferred by the combination Sn Hr. The Dn gene for the long-day habit described by Bremer and Weiseth (1961) may be the same as Sn or Hr but without crossing data the relationship of Dn and also No^h (Wellensiek, 1964) remains hidden.

REFERENCES

Amos, J. J. (1974). Ph.D. Thesis, University of Tasmania, Hobart, Tasmania.
Amos, J. J. and Crowden, R. K. (1969). *Aust. J. biol. Sci.* **22**, 1091–1103.
Barber, H. N. (1959). *Heredity, Lond.* **13**, 33–60.
Barber, H. N. and Paton, D. M. (1952). *Nature, Lond.* **169**, 592.
Barber, H. N., Jackson, W. D., Murfet, I. C. and Sprent, J. I. (1958). *Nature, Lond.* **182**, 1321–1322.
Barton, D. W., Schroeder, W. T., Provvidenti, R. and Mishanec, W. (1964). *Pl. Dis. Reptr* **48**, 353–355.
Blixt, S. (1969). *Pisum Newsl.* **1**, 23–60.

BLIXT, S. (1972). *Agri Hort. Genet.* **30**, 1–293.

BLIXT, S. and GELIN, O. (1965). *Radiat. Bot.* **5**, Suppl. 251–262.

BONDE, E. K. and MOORE, T. C. (1958). *Physiologia Pl.* **11**, 451–456.

BORGSTRÖM, G. (1939). *Bot. Notiser* 830–838.

BREMER, A. H. (1953). *In* "Report of the 13th International Horticultural Congress, 1952" (P. A. Synge, ed.) Vol. 2, pp 643–647. Royal Horticultural Society, London.

BREMER, A. H. and WEISETH, G. (1961). *In* "Proceedings of the 15th International Horticultural Congress, 1958" (J. Garnaud, ed.) Vol. 1, pp 426–435. Pergamon Press, Oxford.

BRIAN, P. W. (1957). *Symp. Soc. exp. Biol.* **11**, 166–181.

BRIAN, P. W., HEMMING, H. G. and LOWE, D. (1958). *Ann. Bot.* **22**, 539–542.

BYERS, R. E., BAKER, L. R., SELL, H. M., HERNER, R. C. and DILLEY, D. R. (1972). *Proc. natn. Acad. Sci. U.S.A.* **69**, 717–720.

CLAY, S. (1935). *J. Pomol.* **13**, 149–189.

COLLINS, W. J. and WILSON, J. H. (1974a). *Ann. Bot.* **38**, 175–180.

COLLINS, W. J. and WILSON, J. H. (1974b). *Ann. Bot.* **38**, 181–188.

DALTON, P. J. and MURFET, I. C. (1975). *Pisum Newsl.* **7**, 5–7.

DENFFER, D. VON (1950). *Naturwissenschaften* **37**, 296–301 and 317–321.

EVANS, L. T. (1969a). *In* "The Induction of Flowering: Some Case Histories" (L. T. Evans, ed.) pp 328–349. Macmillan, Melbourne, Australia.

EVANS, L. T. (1969b). *In* "The Induction of Flowering: Some Case Histories" (L. T. Evans, ed.), pp 457–480. Macmillan, Melbourne, Australia.

GALUN, E., IZHAR, S. and ATSMON, D. (1965). *Pl. Physiol.* **40**, 321–326.

GOESCHL, J. D., PRATT, H. K. and BONNER, B. A. (1967). *Pl. Physiol.* **42**, 1077–1080.

GOTTSCHALK, W. (1960). *Züchter* **30**, 32–42.

GUTTRIDGE, C. G. (1969). *In* "The Induction of Flowering: Some Case Histories" (L. T. Evans, ed.), pp 247–267. Macmillan, Melbourne, Australia.

HÄNSEL, H. (1954a). *Züchter* **24**, 77–92.

HÄNSEL, H. (1954b). *Züchter* **24**, 97–115.

HÄNSEL, H. (1963). *Züchter* Sonderheft 6, 15–24.

HAUPT, W. (1952). *Z. Bot.* **40**, 1–32.

HAUPT, W. (1954a). *Z. Bot.* **42**, 125–134.

HAUPT, W. (1954b). *Ber. dt. bot. Ges.* **67**, 75–83.

HAUPT, W. (1955a). *Ber. dt. bot. Ges.* **68**, 107–120.

HAUPT, W. (1955b). *Planta* **46**, 403–407.

HAUPT, W. (1957). *Ber. dt. bot. Ges.* **70**, 191–198.

HAUPT, W. (1958). *Z. Bot.* **46**, 242–256.

HAUPT, W. (1969). *In* "The Induction of Flowering: Some Case Histories" (L. T. Evans, ed.), pp 393–408. Macmillan, Melbourne, Australia.

HAUPT, W. and NAKAMURA, E. (1970). *Z. PflPhysiol.* **62**, 270–275.

HIGHKIN, H. R. (1955). *P. Physiol.* **30**, 390.

HIGHKIN, H. R. (1956). *Pl. Physiol.* **31**, 399–403.

HIGHKIN, H. R. and LANG, A. (1966). *Planta* **68**, 94–98.

HOSHINO, Y. (1915). *J. Coll. Agric. Tohoku imp. Univ. Sapporo* **6**, 229–288.

JOHNSTON, M. J. and CROWDEN, R. K. (1967). *Aust. J. biol. Sci.* **20**, 461–463.

KNAVEL, D. E. (1967). *J. Hered.* **58**, 78–80.

KÖHLER, G. D. (1965). *Z. PflPhysiol.* **53**, 429–451.

KOPETZ, L. M. (1938). *Gartenbauwissenschaft* **12**, 329–334.

KOPETZ, L. M. (1941). *Gartenbauwissenschaft* **16**, 178–187.

KOPETZ, L. M. (1943). *Gartenbauwissenschaft* **17**, 255–262.
LAIBACH, F. and KRIBBEN, F. J. (1950). *Ber. dt. bot. Ges.* **62**, 53–55.
LAMPRECHT, H. (1956a). *Agri Hort. Genet.* **14**, 1–4.
LAMPRECHT, H. (1956b). *Agri Hort. Genet.* **14**, 195–202.
LANG, A. (1965). *Handb. PflPhysiol.* **15**, 1380–1536.
LEOPOLD, A. C. (1949). *Pl. Physiol.* **24**, 530–533.
LEOPOLD, A. C. and GUERNSEY, F. S. (1953). *Am. J. Bot.* **40**, 46–50.
LEOPOLD, A. C. and GUERNSEY, F. S. (1954). *Am. J. Bot.* **41**, 181–185.
LOCK, R. H. (1905). *Ann. R. bot. Gdns Peradeniya* **2**, 357–414.
MARX, G. A. (1968). *BioScience* **18**, 505–506.
MARX, G. A. (1969). *Crop Sci.* **9**, 273–276.
MARX, G. A. (1972). *Pisum Newsl.* **4**, 28–29.
MARX, G. A. (1975a). *Pisum Newsl.* **7**, 26–27.
MARX, G. A. (1975b). *Pisum Newsl.* **7**, 30–31.
MENDEL, G. (1866). *Verh. naturf. Ver. Brunn* **4**, 3–47. Reprinted (1951) in *J. Hered.* **42**, 3–47.
MONTI, L. M. and SCARASCIA MUGNOZZA, G. T. (1967). *Genet. agr.* **21**, 301–312.
MONTI, L. M. and SCARASCIA MUGNOZZA, G. T. (1970). *Genet. agr.* **24**, 195–206.
MONTI, L. M. and SCARASCIA MUGNOZZA, G. T. (1972). *Genet. agr.* **26**, 119–127.
MOORE, T. C. (1964). *Pl. Physiol.* **39**, 924–927.
MOORE, T. C. (1965). *Nature, Lond.* **206**, 1065–1066.
MOORE, T. C. and ANDERSON, J. D. (1966). *Pl. Physiol.* **41**, 238–243.
MOORE, T. C. and BONDE, E. K. (1958). *Physiologia Pl.* **11**, 752–759.
MOORE, T. C. and BONDE, E. K. (1962). *Pl. Physiol.* **37**, 149–153.
MURFET, I. C. (1971a). *Heredity, Lond.* **26**, 243–257.
MURFET, I. C. (1971b). *Heredity, Lond.* **27**, 93–110.
MURFET, I. C. (1971c). *Aust. J. biol. Sci.* **24**, 1089–1101.
MURFET, I. C. (1973a). *Heredity, Lond.* **31**, 157–164.
MURFET, I. C. (1973b). *Aust. J. biol. Sci.* **26**, 669–673.
MURFET, I. C. (1975a). *Pisum Newsl.* **7**, 41.
MURFET, I. C. (1975b). *Pisum Newsl.* **7**, 41–42.
MURFET, I. C. (1975c). *Pisum Newsl.* **7**, 42–43.
MURFET, I. C. (1975d). *Pisum Newsl.* **7**, 44–45.
MURFET, I. C. (1975e). *Heredity, Lond.* **35**, 85–98.
MURFET, I. C. and BARBER, H. N. (1961). *Nature, Lond.* **191**, 514–515.
MURFET, I. C. and BLIXT, S. (1974). *Pisum Newsl.* **6**, 43.
MURFET, I. C. and REID, J. B. (1973). *Aust. J. biol. Sci.* **26**, 675–677.
MURFET, I. C. and REID, J. B. (1974). *Z. PflPhysiol.* **71**, 323–331.
NAKAMURA, E. (1965). Special Report of the Laboratory of Horticulture, Shiga Agricultural College, Kusatsu.
PATON, D. M. (1957). *Pl. Physiol.* **32**, Suppl. 9–10.
PATON, D. M. (1967). *Nature, Lond.* **215**, 319–320.
PATON, D. M. (1968). *Aust. J. biol. Sci.* **21**, 609–617.
PATON, D. M. (1969). *Aust. J. biol. Sci.* **22**, 303–310.
PATON, D. M. (1971). *Aust. J. biol. Sci.* **24**, 609–618.
PATON, D. M. and BARBER, H. N. (1955). *Aust. J. biol. Sci.* **8**, 231–240.
PELLEW, C. (1940). *J. Genet.* **39**, 363–390.
PETERSON, C. E. and ANHDER, L. D. (1960). *Science, N.Y.* **31**, 1673–1674.
RASMUSSON, J. (1935). *Hereditas* **20**, 161–180.
REATH, A. N. and WITTWER, S. H. (1952). *Proc. Am. Soc. hort. Sci.* **60**, 301–310.
REID, J. B. and MURFET, I. C. (1974a). *Pisum Newsl.* **6**, 44–45.

430   I. C. Murfet

REID, J. B. and MURFET, I. C. (1974b). *Aust. J. Pl. Physiol.* **1**, 591–594.
REID, J. B. and MURFET, I. C. (1975a). *Pisum Newsl.* **7**, 47–48.
REID, J. B. and MURFET, I. C. (1975b). *J. exp. Bot.* **26**, 860–867.
ROBINSON, R. W., SHANNON, S. and DE LA GUARDIA, M. D. (1969). *BioSciencᵉ* **19**, 141–142.
ROWLANDS, D. G. (1964). *Genetica* **35**, 75–94.
ROSEN, G. VON (1944). *Hereditas* **30**, 261–400.
SHANNON, S. and DE LA GUARDIA, M. (1969). *Nature, Lond.* **223**, 186.
SIDOROVA, K. K., KHVOSTOVA, V. V. and UZHINTZEVA, L. P. (1974). *Pisum Newsl* **6**, 47.
SKOOG, F. and MILLER, C. O. (1957). *Symp. Soc. exp. Biol.* **11**, 118–131.
SNOAD, B. and ARTHUR, A. E. (1973a). *Euphytica* **22**, 327–337.
SNOAD, B. and ARTHUR, A. E. (1973b). *Euphytica* **22**, 510–519.
SPRENT, J. I. (1966a). *Nature, Lond.* **209**, 1043–1044.
SPRENT, J. I. (1966b). *Ann. Bot.* **30**, 779–790.
SPRENT, J. I. (1967). *Ann. Bot.* **31**, 607–618.
SPRENT, J. I. and BARBER, H. N. (1957). *Nature, Lond.* **180**, 200–201.
STANFIELD, B., ORMROD, D. P. and FLETCHER, H. F. (1966). *Can. J. Pl. Sci.* **46**, 195–203.
TEDIN, H. (1897). *Sver. Utsädesför. Tidskr.* **7**, 111–129.
TEDIN, H. and TEDIN, O. (1923). *Hereditas* **4**, 351–362.
TSCHERMAK, E. VON. (1910). *Verh. naturf. Ver. Brünn* **49**, 169–191.
WALL, B., REID, J. B. and MURFET, I. C. (1974). *Pisum Newsl.* **6**, 50–51.
WATTS, L. E., STEVENSON, E. and CRAMPTON, M. J. (1970). *Euphytica* **19**, 405–410.
WELLENSIEK, S. J. (1925a). *Genetica* **7**, 1–64.
WELLENSIEK, S. J. (1925b). *Biblphia genet.* **2**, 343–476.
WELLENSIEK, S. J. (1961). *In* "Effects of Ionising Radiations on Seeds", pp 321–326. International Atomic Energy Agency, Vienna.
WELLENSIEK, S. J. (1964). *Radiat. Bot.* **5**, Suppl. 393–397.
WELLENSIEK, S. J. (1969). *Z. PflPhysiol.* **60**, 388–402.
WELLENSIEK, S. J. (1972a). *Pisum Newsl.* **4**, 59.
WELLENSIEK, S. J. (1972b). *Pisum Newsl.* **4**, 60.
WELLENSIEK, S. J. (1973a). *Scientia Hort.* **1**, 77–83.
WELLENSIEK, S. J. (1973b). *Scientia Hort.* **1**, 177–192.
WENT, F. W. (1957). *Chronica bot.* **17**, 115–123.
WHITE, O. E. (1917). *Proc. Am. phil. Soc.* **56**, 487–588.

15. Fruit and Seed Development

J. S. PATE

Department of Botany, University of Western Australia

A. M. FLINN

School of Biological and Environmental Studies, New University of Ulster, N. Ireland

I. INTRODUCTION

Our understanding of the biology of the pea plant expands rapidly; even a cursory glance through the literature reveals that the number of papers published annually in this area has increased many fold over the past few decades, and that by far the larger proportion of these have the highly

practical object of improving the yield and quality of produce from th
reproductive plant.

It is somewhat disappointing to discover how little plant physiologist
have become involved in this work. Instead, their efforts on peas have beei
concentrated in other directions, notably in the compiling of the most detailec
accounts of chemical and other changes which accompany fruit and see
maturation; in the tracing of assimilates from vegetative to reproductiv
units of the plant; in studies of biosynthesis of reserves in seeds; and ii
investigating the roles of various plant parts in providing different classes o
nutrients to the seeds. Although much of this material is highly relevant tc
our basic understanding of the processes of fruit maturation, it has failed tc
unravel the dynamic aspects of fruit·and seed growth and to specify whicl
regulatory forces within the whole plant shape events during the closin
stages of the life cycle.

This chapter attempts a synthesis of present knowledge of the fruiting
processes of peas. Section II describes the morphology and structure of th
constituent parts of fruit and seed; Section III deals with events concernec
with the laying down of the all important reserve materials in cotyledons o
the seed; and Section IV discusses the nutritional interrelationships of specifi
organs of the fruit and the role of vegetative organs in providing much o
the fruit's nourishment.

II. STRUCTURE, GROWTH AND COMPOSITION OF THE FRUIT AND ITS PARTS

A. Flowering, Pollination, Fertilization and Seed-Set

Despite the extremely wide range of variation in growth habit and flowering
behaviour that exists in *Pisum*, all known forms of the genus show an axillary
and sequential pattern of reproduction. Flowering occurs obsessively once
the obligatory period of vegetative growth has been completed.

The pea is self-fertile and its flowers are usually self-pollinated (Cooper,
1938; Attia, 1958). Pollination takes place in the late bud stage, 24–36 h
before the flower is fully open, and, by the time of full blossom, fertilization
has taken place and the zygote has started to divide (Cooper, 1938). The
embryo sac has a complement of three antipodals, two synergids, an egg cell
and a large, binucleate primary endosperm cell. Pollen tubes grow along the
upper suture of the ovary, several penetrating a micropyle but only one
finally entering an embryo sac (Cooper, 1938).

It is usual for all ovules of a pea pod to be fertilized, but a considerable
proportion of them may fail to develop into mature seeds. Linck (1961),
studying cv. 'Alaska', recorded abortion rates of 30–50% and noted that
embryo failure occurred with highest frequency in the proximal and distal

nds of the carpel. Blunt-podded varieties are regarded to be less prone to
mbryo failure than are ones with pods of tapered outline, so abortion may
esult from space restrictions in the ends of the pod. There is no evidence
hat varieties selected for high ovule number per pod exhibit abnormally
iigh frequencies of ovule abortion so it is unlikely that shortage of germinated
iollen or insufficient space for growth of pollen tubes are normal causes of
ailure in seed-set. Nevertheless, high frequencies of abortion of ovules at the
ire-fertilization stage can be observed in peas grown under adverse environmental conditions (unpublished observations).

3. The General Pattern of Growth

A typical set of growth measurements for the main parts of a pea fruit is
;hown in Fig. 15.1. A major feature is the precocious growth of the pod with
respect to that of its enclosed seeds, the pod having reached more than half
its final size (fresh weight) before the seeds enter their exponential phase of
growth. The growth of the seed coat is completed considerably in advance
of that of the embryo, and the relatively small size and short life of the
endosperm relative to other parts of the fruit is illustrated (Fig. 15.1). The
drying-out of the pod in its later life causes major losses in fresh weight, but
total nitrogen, dry matter and minerals also decrease so that more than a loss

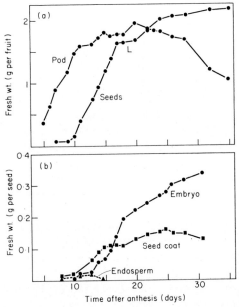

FIG. 15.1. Growth of pod and contained seeds (a) and parts of seed (b) from fruit
borne at the first blossom node of the field pea, *Pisum arvense* L. The "lag" in the fresh
weight curve for the seeds is marked L. (Data redrawn from Flinn and Pate, 1968)

of water is involved (see Flinn, 1969). The seeds may also show a levelling-off if not a decline, in fresh weight on approaching maturity, this being correlated with a loss of moisture content in the seed coat and, possibly also, in the embryo. But water losses are partly compensated for by weight gains due to deposition of reserve materials in the cotyledons (e.g. see data of Matthews, 1973a,b; Smith, 1973). The final stages of filling cotyledon cells with starch and protein take place under conditions where there must be considerable dehydration of some cells.

The seeds of certain varieties of pea exhibit a biphasic or diauxic pattern of growth; that is, two distinct periods of growth increase are separated by a lag of a few days during which growth is temporarily arrested (e.g. see Fig. 15.1). Carr and Skene (1961) discuss this phenomenon at length, pointing out that it may be much more widespread among seeds than has hitherto been realized. Certainly, if sampling of the seed population for growth measurements is not frequent and if sampling errors are large, the existence of a "lag" phase in a growth curve may easily be missed. But there are several sets of growth data for peas (e.g. those of Bain and Mercer, 1966; Scharpe and Parijs, 1973; Frydman *et al.*, 1974) in which these criticisms hardly apply yet in which an uninterrupted sigmoidal curve of growth is recorded. Moreover, even if a lag phase is not obvious, the growth curve need not necessarily be sigmoidal. For instance, the dry weight curve for cotyledon growth of *Pisum sativum* illustrated by Millerd and Spencer (1974) shows an abrupt change in growth rate in mid-growth of the seed, an earlier exponential phase of increase being replaced by one showing a slower, and essentially linear increase in growth. This change in rate is not observed, however, in the fresh weight curve these authors present (see Fig. 15.6). Burrows and Carr (1970) point out that in its early life the embryo of *P. arvense* grows relatively faster than the seed, but that after the lag in growth, allometric relationships change, seed and embryo then continuing to grow at the same rate.

It is interesting that all of the more pronounced examples of diauxic growth come from plants raised in cool, outdoor conditions where the cycle of seed development may be almost twice as long as under optimum growing conditions. Under these slow-growing conditions factors outside the seed (e.g. rate of translocation) may impose unusual restriction on the pace and pattern of embryo growth and development.

There are several possible interpretations of what causes the "lag" in seed growth in legumes. Carr and Skene (1961) support a purely physical interpretation, noting that the lag coincides with the time when the embryo first completely fills the embryo sac and when there are likely to be quite definite space restrictions on further expansion of the enlarging cotyledons. On the other hand, it may be that the temporary lull in growth is nutritionally based and relates primarily to the disappearance of the endosperm with its high content of cytokinins and other nutrients (see Burrows and Carr, 1970).

Yet another interpretation would be that the lag marks basic changes in cellular activity of the embryo. Certainly the lag coincides with the end of cell divisions in the cotyledons (Smith, 1973) and with a transition from a phase of expansion dominated by solute accumulation to a non-expansive one in which insoluble reserves start to accumulate in earnest. Accumulation of reserves occurs, however, at the expense of vacuolar and cytoplasmic space in the cotyledon cells.

C. The Pod

Rapid increases in pod length and width occur during early growth and these are accompanied by a thickening of the pod wall. The pod then inflates to form a hollow envelope, presumably through differential growth of its inner and outer layers. This expansion is normally accompanied by considerable sclerification of the cells underlying the inner epidermis, thus forming a fibrous endocarp (Flinn, 1974). The mesocarp consists of large parenchymatous cells with thin walls, the exocarp a single layer of thick-walled epidermal cells (Esau, 1965). The orientation of these thickened cells in the pod wall is held to be related to the dehiscent properties of legume fruits as they dry out (see Fahn and Zohary, 1955); but splitting and twisting of the walls is not very noticeable in the garden pea, Man having selected forms with essentially non-shattering fruits.

Chloroplasts are present in cells underlying the outer epidermis and also in the cells of the inner epidermis lining the pod cavity. The positioning of these suggests, respectively, the photosynthetic capture of carbon dioxide from the external environment and the reassimilation of respired carbon dioxide in the pod's interior. Gas exchange on the pod's outer surface is likely to be facilitated by the presence of stomata, although their density is much lower than on the surfaces of stipules or leaflets (Flinn, 1969).

The vasculature of the fruit is typical of that of legumes. Two adjacent longitudinal veins, equivalent to marginal veins of the carpel, supply traces alternately to the seeds. A vein, equivalent to the mid-rib vein of the carpel, runs along the pod's lower surface. All three of these veins converge at the distal and proximal ends of the pod. A network of minor veins traverse the lateral walls of the pod, interconnecting freely with the longitudinal veins. All vascular tissue is composed of phloem and xylem elements, although xylem elements are noticeably sparse in the vein branches supplying the individual seeds.

By the end of the inflation phase of growth the pod has attained its maximum fresh weight, and the levels of starch, sucrose, protein and soluble nitrogen are at their highest (Bisson and Jones, 1932; McKee et al., 1955a; Raacke, 1957; Flinn and Pate, 1968). The soluble amino acid pool of the pod is dominated by three compounds, homoserine, asparagine and o-acetyl

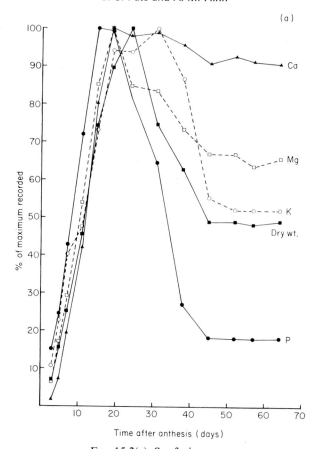

FIG. 15.2(a). See facing page.

homoserine, which together account for some three-quarters of the soluble
N in the pod (Flinn and Pate, 1968). In the early life of the pod the homo
serines are the commonest components; later, as the pod approaches sen
escence, asparagine supplants them.

Continually changing patterns of free amino acids and soluble proteins
as revealed by acrylamide gel-electrophoretograms, indicate major metabolic
changes during pod development (Flinn and Pate, 1968). A detailed under
standing of this developmental programme is not yet possible but severa
attempts have been made to correlate changes in the activity of specific
enzymes with particular phases in pod growth. For example, Bowling and
Crowden (1973) have observed changes in peroxidase activity which appear
to be related to the lignification of cells in the endocarp of the pod.

A recent study of changes in the mineral composition of ageing pea pods
(P. Hocking, unpublished) suggests that nutrient elements build up at relatively

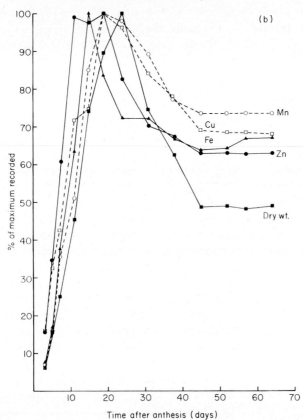

Fig. 15.2(a) and (b). Changes with age in dry matter and mineral content of pods of *P. sativum*, cv. 'Greenfeast'. Data for each commodity are expressed as percentages of the maximum recorded during the life of the pot. (a) The macronutrients phosphorus, calcium, magnesium and potassium; (b) the micronutrients copper, zinc, manganese and iron. Note how certain elements increase relatively in advance of dry weight, others relatively behind dry weight gain. Big differences are to be seen in the efficiency of mobilization of minerals during pod senescence. The significance of pod mobilization in nutrition of the seeds is shown in Table 15.1. (Unpublished data kindly provided by P. Hocking, University of Western Australia)

different rates during early growth of the pod, and are then mobilized from the pod with differing degrees of efficiency during its senescence. The situation for eight essential elements is summarized in Fig. 15.2, changes in content being expressed on a common percentage basis and in comparison with changes in dry weight of the pod. Some elements (Zn, P, Fe, Cu) appear to increase somewhat in advance of dry weight, others (Mn, Mg, K) parallel with the curve for dry weight gain, while Ca falls slightly behind. Then, in later pod life, when pods lose approximately 50% of their dry matter, comparable losses of nutrients are 10% for Ca, 27% for Mn, 32% for Cu and Fe,

34% for Mg, 37% for Zn, 48% for K and 82% for P. If these mobilized minerals are captured by the seeds, the pod must be viewed as a substantial contributor of minerals to the developing seed. The quantitative implications of this are dealt with later.

D. The Endosperm

The endosperm of pea, like that of most legumes, is short-lived and remains in a free nuclear condition and liquid state throughout its life (Kapoor, 1966). During early post-fertilization development the volume of the endosperm increases rapidly, this being achieved by rapid outgrowth of the embryo-sac wall into regions of the seed opposite the micropyle (Fig. 15.3). The maximum volume of the liquid endosperm contents is of the order of 15–50 μl per seed, depending on variety and seed size (unpublished data). The final size and shape of the enlarged region of the endosperm conforms to the space eventually occupied by the cotyledons (Fig. 15.3). Increase in volume of endosperm cytoplasm fails to keep pace with the overall volume increase of the endosperm, so that the spherical region of the endosperm becomes virtually all filled with liquid and has only a thin layer of cytoplasm lining its walls (Fig. 15.3). However, the tubular, micropylar region, containing embryo and suspensor, shows little evidence of vacuolation and possesses the densest cytoplasm to be found in the endosperm (Kapoor, 1966). Nuclear divisions in the endosperm at first keep pace with increases in volume of its cytoplasm, but later the nuclei cease to divide, and enlarge, exhibiting extreme polymorphism. 6C and 12C nuclei are observed in advanced stages of endosperm development (Kapoor, 1966). The ultrastructure of endosperm is well described in a study by Marinos (1970). Characteristic are the high electron density of cytoplasm, numerous ribosomes and mitochondria of great size and internal complexity. These features are most noticeable adjacent to the embryo and

---

FIG. 15.3. Diagrammatic representations of the spatial relationships of endosperm, embryo and maternal tissues in the developing seed of pea (not drawn to scale). 1—embryo; 1a—cotyledon; 1b—plumule; 1c—radicle; 2—basal cell; 3—suspensor cell; 4—endosperm; 4a—liquid region of endosperm; 4b—cytoplasmic region of endosperm; 4c—cellulosic strands traversing the cytoplasmic space between embryo and outer wall of embryo sac; 4d—extra embryonic sheath surrounding suspensor and embryo; 5—maternal tissue of seed; 6—micropyle.
(a) Early embryo stage at time of maximum development of endosperm.
(b) Enlarged region of endosperm of (a), showing extra embryonic sheath, anchoring strands and shorter wall ingrowths in the endosperm. Details of embryo are omitted.
(c) Stage when embryo first fully fills the endospermic cavity. Note remains of suspensor.
(d) Cellular details of inset of (c) to show layer of epidermal transfer cells (T) on cotyledon (1a), remains of endosperm (4), and nucellus tissue (5).
[Structural details derived from the studies of Marinos (1970) and Gunning and Pate (1974)]

suspensor. The endosperm wall is, in essence, a greatly enlarged megaspore wall, and, when mature, it possesses a whorled arrangement of cellulose microfibrils. Fields of holes in the wall are discernible and, although not carrying plasmodesmatal connections, they may be regarded as equivalent to the pit fields in the boundary walls of other plant cells. Their presence might expedite apoplastic transfer from nucellus to endosperm.

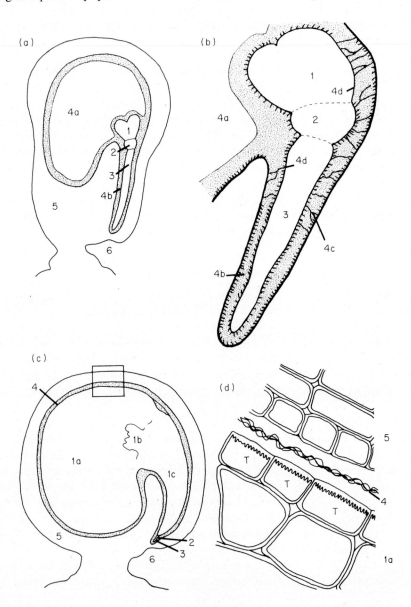

The endosperm persists for only a relatively short period of seed development, yet the cycle of build-up and loss within it of quite sizeable pools of solutes suggests a nutritional significance of some magnitude during early endosperm development. Sucrose is the main sugar (Flinn, 1969) and ammonia (40–100 mM) is the main nitrogenous component (unpublished data). Homoserine, alanine and glutamine are its major amino acids and amides (Flinn and Pate, 1968); malate its principal organic anion. The low pH (4·2–4·4) of endoplasmic fluid is undoubtedly related to its high ammonia content. Endospermic fluid is reported to be of high cytokinin content (Burrows and Carr, 1970).

It is not clear what parts of the endosperm are actively engaged in throughput of solutes to the young embryo. It may be that only the micropylar region in cytoplasmic contact with the embryo is so engaged and that the liquid-filled portion distant from the young embryo represents a relatively static reservoir in which waste products, or solutes arriving in excess of embryo demands, are temporarily accumulated. However, the complete disappearance of the endosperm contents in later embryo development suggests that embryo and seed coat are together capable of absorbing and utilizing all of the previously accumulated solutes. A better understanding of nutritional interrelationships at these stages of development might further greatly our understanding of the regulatory mechanisms of early embryo growth.

E. Embryology and the Cytological Surroundings of the Young Embryo

The embryology of *Pisum* is well described by Cooper (1938) and Reeve (1948) and the account below is derived from their findings. Early divisions of the zygote generate a two-celled suspensor attached to a basal cell, which is in turn attached to the embryo proper. The orientation of these structures in the embryo sac is illustrated in Fig. 15.3. The most noticeable feature of early development is the great elongation of the suspensor cell, the effect of this being to push the embryo away from the micropylar end of the embryo sac. Finally, the radicle end of the embryo becomes lodged within the bend of the endosperm sac, and from this position the cotyledons can readily expand and encroach upon the enlarged spherical region of the endosperm (Fig. 15.3). Each suspensor cell contains 64 nuclei at maturity, degeneration of the suspensor being in evidence once the embryo starts to enlarge, and little remaining of it by the time of differentiation of cotyledons, hypocotyl and epicotyl. (*Pisum* differs in this respect from other legume genera, e.g. *Phaseolus*, in which the suspensor apparatus continues to enlarge and remains in an apparently healthy state until much later in embryo development—e.g. see Clutter and Sussex, 1968.)

Reeve (1948), describing the events of histogenesis in *Pisum*, found that the epicotyl apex is initiated before the cotyledons expand, and that procambial differentiation occurs in the cotyledonary and epicotyl traces before the seed is fully mature. However, the pea is one of those plant species in which differentiation of recognizable vascular elements is not accomplished before the onset of dormancy, although this is one of the first structural events to take place after imbibition of the germinating seed (Flinn and Smith, 1967).

A striking feature of embryogenesis in legumes generally is the wall thickenings of a cellulosic nature which develop within the endosperm, particularly in the region occupied by the suspensor and young embryo (Gunning and Pate, 1974). The situation in *Pisum* is well described by Marinos (1970) who noted that the lower region of the embryo, the outer surfaces of the basal cell and the suspensor cells become clothed in an extra-embryonic sheath, presumed to be laid down by the cytoplasm of the endosperm (Fig. 15.3). Eventually cellulosic strands are formed from this sheath and extend across the endosperm to the opposing outer wall (Fig. 15.3). Marinos (1970) interpreted these cross-linking strands as structures for anchoring the lower region of the young embryo and for positioning it correctly in the endospermic cavity. Shorter, irregular ingrowths, also of cellulosic construction and lined with plasma membrane, are formed from the endosperm wall into its cytoplasm, these being particularly frequent in the micropylar region of the endosperm (see Fig. 15.3). Equivalent structural specializations are found widely in the reproductive apparatus of many families and taxa of plants (see Gunning and Pate, 1974), their general significance being interpreted as providing a large and effective surface for membrane-mediated flux of solutes across the cytoplasmic discontinuities between different nuclear generations of the plant. In the situation here in peas these discontinuities are between diploid mother tissue (nucellus) and triploid endosperm, and between the endosperm and the developing (diploid) tissue of the embryo.

During the time when the endosperm is expanding, and its "vacuole" filling with solutes, very little embryo growth takes place, although its cells are dividing rapidly (Bain and Mercer, 1966; Smith, 1973). Most cell divisions in the embryo have ceased by the time the endosperm has attained its maximum volume. Then, over a space of only a few days, the cotyledon cells enter a phase of rapid cell enlargement and this results in the cotyledons quickly swelling to fill virtually all of the space originally filled by liquid endosperm (see Smith, 1973). By the end of this phase the storage parenchyma cells of the cotyledons have become highly vacuolate, so that it is likely that resorption of the liquid endosperm involves a direct transfer of some of its solutes to a second site of temporary storage in the vacuoles of the embryo. The mature cotyledons show a gradient of increasing cell size in their storage parenchyma from outer (abaxial) to inner (adaxial) regions. This is an expression of the earlier cessation of cell division on the inner faces of the cotyledons (Smith, 1973).

Just before the cotyledons fill the endosperm cavity, the cells of their abaxial (outer) epidermis commence to develop wall ingrowths. The epidermal cells are then classified as "transfer cells" (Flinn, 1969; Gunning and Pate, 1969, 1974). The zones of wall ingrowths are entirely restricted to the outer walls of the epidermal cells so that when the epidermis eventually contacts the boundary wall of the endosperm (Fig. 15.3), the greatly enlarged surface of membrane which it exposes is likely to increase the efficiency with which the cotyledons absorb solutes secreted by the surrounding nucellus (Fig. 15.3). In certain legumes (e.g. *Vicia*), but not, apparently, in *Pisum*, the endothelium (inner layer of nucellus) also develops zones of ingrowths on the walls of its cells and these "transfer cells" may also facilitate exchange with the embryo (see Pate and Gunning, 1972). As the cotyledons dry out and mature, and after most of the cotyledonary reserves have been accumulated, the abaxial epidermis of pea cotyledons develops further wall accretions, and these have the effect of completely obliterating the original investment of wall ingrowths. These later thickenings may afford the cotyledon a measure of protection against dehydration, but, since they are essentially of a hydrophilic character, the cotyledons are still capable of imbibing water rapidly during germination.

III. ACCUMULATION OF RESERVES IN DEVELOPING SEEDS

A. Ultrastructural Features of Cotyledon Storage Cells

Several interesting structural changes accompany maturation of the storage cells of the cotyledon. Bain and Mercer (1966) have shown that the phase of cell expansion and vacuolation is the time when endoplasmic reticulum and ribosomes become prominent; when mitochondria and plastids appear; and when the nuclei of the cotyledon cells become greatly lobed and swollen. It has subsequently been shown by Feulgen histochemistry that nuclear enlargement is associated with massive endo-reduplication of the DNA of the cotyledon nuclei (Scharpé and Parijs, 1973; Smith, 1973). Storage parenchyma of the cotyledon have DNA levels of from 4C up to the 64C level, highest levels being encountered in the larger cells of the inner adaxial region, while the smaller cells of the abaxial epidermis of the cotyledon rarely exceed the 2 or 4C level. As Smith (1973) points out, the overall effect of endo-reduplication is to achieve a fairly constant level of DNA per unit cell volume over a large range of cell size. The possible significance of these giant nuclei in synthesis of storage protein is discussed in a later section.

Bain and Mercer (1966) have also provided information on the ultra-structural events leading to synthesis of storage reserves in the cotyledons. Small starch grains are visible in the mid-phase of cell expansion, as are minute lipid deposits and traces of protein, but the main phase of reserve accumulation does not commence until after the cotyledons have reached their maximum size. Starch grains then become prominent, one per plastid, and each grain eventually fills its plastid. In doing so it compresses lamellae and stroma against the outer, limiting membrane of the plastid. At maturity the starch grains are about 10 μm in length (Fig. 15.4), and the lamellae of their plastids have disappeared.

Storage protein is laid down in pea embryos in discrete, membrane-bound bodies, 1–2 μm in diameter (see Fig. 15.4 and Varner and Schidlovsky, 1963; Bain and Mercer 1966; Swift and Buttrose, 1973). Protein bodies are far more numerous than starch grains and both forms of reserve eventually fill the cotyledon cell so completely that cytoplasm becomes restricted to a thin peripheral layer and minute crevices between the packages of reserve materials (see Bain and Mercer, 1966; Flinn, 1969; Swift and Buttrose, 1973). Protein and starch are also laid down in other organs of the embryo, but nowhere in so great a density as in the cotyledons. There is a noticeable gradient of increase in size of storage bodies from the outer abaxial to the inner regions of the cotyledons (Flinn, 1969; Smith, 1973). This gradient parallels those of cell size and nuclear size.

The final stage of seed development involves the dehydration and shrinkage of the cotyledons. Once this happens studies of fine structure become especially difficult because the seed is by then very hard and brittle, and rehydration artefacts inevitably occur if water-based fixatives are employed before sectioning. For instance, starch grains have the appearance of being highly disintegrated ("exploded") when sectioned following normal aqueous fixation (e.g. see Bain and Mercer, 1966), but if 100% glycerol is employed as mounting medium and fresh sections are used (Swift and Buttrose, 1973), the starch grain of the dry seed is revealed as a considerably more intact body. Freeze-etching techniques show that both it and the protein body are enclosed with complex boundary structures, which possess sufficient structural integrity to enable them to become quickly reorganized on hydration of the air-dry seed. Swift and Buttrose (1973) have observed that following imbibition of a seed the convoluted boundary structure of the protein body expands into a complex system of tubules (see also Chapter 3).

The problems encountered using aqueous fixatives lead one to question the validity of ultrastructural studies on mature cotyledons, particularly those depicting a massive fragmentation of membranes and a disorganization of endoplasmic reticulum and mitochondria (e.g. see Bain and Mercer, 1966). Nevertheless, since cotyledon cells lose solutes for some hours after imbibition of a dormant seed, and since it takes a day or so, after germination, before aerobic metabolism is restored and hydrolases become active (see Pate, 1975; also Chapter 3), it does appear that some quite catastrophic form of structural damage occurs when the seed dries out, and that it takes some time after rehydration before cellular integrity can be restored.

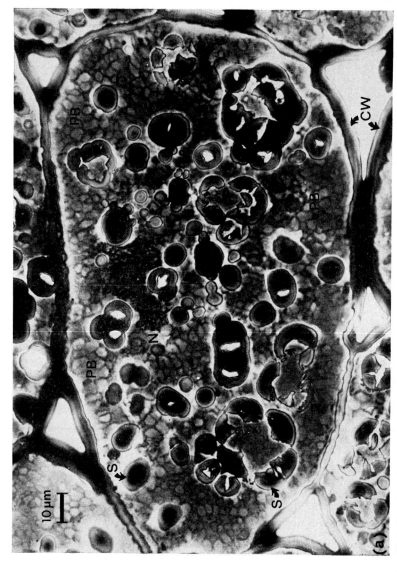

FIG. 15.4. Starch grains, protein and lipid bodies in the storage cells of the cotyledons of *P. sativum* (cv. 'Greenfeast').

(a) Light micrograph of an air-dry glutaraldehyde fixed cotyledon parenchyma cell. Section stained with Periodic-acid-Schiff's Reagent, fast green and toluidine blue; phase contrast. A 10 μm marker indicates the level of magnification.

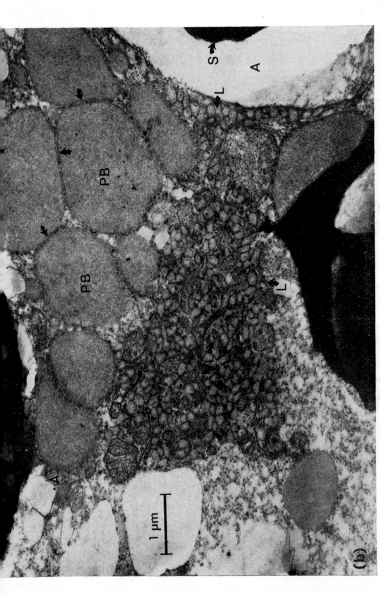

(b) Electron micrograph of part of an air-dry parenchyma cell. The unlettered arrows indicate material lying between, and possibly linking protein bodies. A 1 μm marker indicates level of magnification.

Symbols: S—starch grains; PB—protein body; CW—cell wall; A—amyloplast; L—lipid body; N—nucleus.

(Figures kindly supplied by Drs J. G. Swift and M. S. Buttrose)

B. Carbohydrate and Lipid Reserves of the Seed

The major storage carbohydrate of the cotyledons of the garden pea (*P. sativum*) is starch and its synthesis begins on a large scale at the end of the endospermic stage and continues almost up until seed desiccation and the onset of dormancy (McKee *et al.*, 1955b; Turner *et al.*, 1957; Rowan and Turner, 1957; Bain and Mercer, 1966). Round-seeded varieties of pea have about 45% of their seed dry weight deposited as starch, wrinkle-seeded varieties only 35%. Wrinkle-seeded forms have a higher ratio of amylose to amylopectin than do the round-seeded forms (see Yarnell, 1962).

In the field pea (*P. arvense*) the period of active starch synthesis is shorter and is eventually supplanted by a phase when hemicelluloses are laid down in great quantity in the walls of the cotyledon cells (Flinn and Pate, 1968). Hemicelluloses eventually comprise about 40% of the dry matter of the embryo, starch only some 25% (Fig. 15.5). Smith (1973) has shown that field pea cotyledons have acquired a full complement of starch grains by the time cell division has ceased, so that the main phase of starch accumulation is one of grain filling and not of proliferation of new grains. The larger adaxial storage cells of the cotyledons contain greater numbers of grains than do the smaller abaxial cells, while starch is absent from the epidermis and from most of the procambial cells of the cotyledon.

The level of free sugar remaining in mature pea seeds usually lies within the range 3–7% of dry matter. As might be expected those cultivars high in sugar tend to be of low protein content. Sucrose is recorded by all workers as the main free sugar of the developing seed but other sugars, notably fructose, glucose, galactose, raffinose, stachyose and verbascose can be detected in chromatographic separations of the ethanolic extracts of young seeds.

Sugar changes in the immature pea seed have been a favourite subject for study, in view of the importance of sweetness as a criterion of palatability in fresh or frozen green peas. It has been observed in many studies that free sugars achieve an early maximum in seed growth but then fall sharply once starch synthesis gets under way (Bisson and Jones, 1932; McKee *et al.*, 1955b; Danielsson, 1956, 1959; Turner *et al.*, 1957; Rowan and Turner, 1957; Bain and Mercer, 1966). The quantitative importance of this pool of sugars as precursor material for starch formation has been stressed, as has its possible role as a controlling agent in switching on or regulating the synthesis of starch. But the build-up of a large pool of free sugars is by no means a universal feature of pea seed development (e.g. see data of Flinn and Pate, 1968; Fig. 15.5), and even within the one cultivar the pattern of sugar and starch accumulation is very markedly influenced by factors within the plant's environment (Robertson *et al.*, 1962). It may well be that the changes in the sugar: starch ratio which occur during development are nothing more than manifestations of the ever changing balance between the rate of arrival of sugars as translocate and the rate of their disappearance in polysaccharide production. Of course, it is to be expected that sugar levels will be high in early development, for a significant proportion of the volume of the seed is then occupied by liquid endosperm or highly vacuolated cells, which are ideal depositories for solutes. The phenomenally large build-up of sugars and soluble nitrogen in seeds grown at low night temperatures, and the delayed appearance of starch and protein under such circumstances (Robertson et al., 1962), bears witness to the extent to which precursor solutes can continue to accumulate in a developing seed if synthesis of the relevant macromolecules is inhibited.

Pectins—linear polymers consisting primarily of residues of D-galacturonic acid—are synthesized in the later phases of seed growth and may account for over 10% of the final dry matter of the seed (McKee et al., 1955b). As the amount of pectin increases in the cotyledon more rapidly than does the area of its cell walls, it seems plausible that pectin, like hemicellulose, may have a storage function. It is significant that the galactose which is present in the maturing cotyledons declines in amount during the active period of pectin deposition (Turner et al., 1957).

A number of workers have investigated the biochemical pathway of starch synthesis in the pea (Danielsson, 1956; Turner and Turner, 1957; Turner et al., 1957; Rowan and Turner, 1957). Early findings have suggested that starch phosphorylase activity might be implicated, if only because of its linear relationship with the rate of starch synthesis (Turner and Turner, 1957). It now appears that this enzyme is more likely to be involved in the breakdown of starch than in its synthesis, and that starch synthesis probably proceeds by the transfer of glucosyl residues from nucleoside diphosphate sugars (see Preiss et al., 1967; Hassid, 1967). The provision of these sugars for starch synthesis probably proceeds through the freely reversible sucrose synthetase reaction (Hassid, 1967).

Compared with other grain legumes such as peanut, soybean and lupin, the pea does not store significant amounts of fat in its seeds; a fat content of 1% of dry matter is recorded by Farrington (1974) for peas. Nevertheless, fat globules are discernible in cotyledon cells, sometimes appearing to be enclosed by a quite definite boundary membrane (Mollenhauer and Totten, 1971). The lipid droplets forming in the late stages of maturation of the seed as it enters dormancy are said to be associated with the disappearance of lipoprotein membranes (Bain and Mercer, 1966).

C. Nitrogen-containing Compounds of the Seed

By no means all of the nitrogen of the dry pea seed is in protein, some 2–10% of it being in an ethanol-soluble form, largely as amino acids and small peptides, and a further 5–10% as non-proteinaceous, insoluble nitrogen,

presumably as fibre and large peptides (unpublished results). The seed coat
is particularly rich in non-protein nitrogen, probably because of its lignified
nature. It is general for the concentration of nitrogen in soluble form to fall
steadily as the seed matures, but the total amount per seed usually remains
fairly constant throughout development (see Fig. 15.5 and McKee *et al.*,
1955a; Flinn and Pate, 1968).

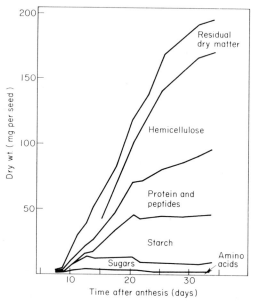

Fig. 15.5. Accumulation of insoluble reserves and changes in major organic solutes in
the developing seed of *P. arvense* L. Note the extensive accumulation of hemicellulose
and the early cessation of starch build-up in this species. All data expressed on a dry
weight basis.
   (Data obtained from Flinn and Pate, 1968; Flinn, 1969; and unpublished results)

   Substantial changes in the levels and relative proportions of individual amino
acids have been recorded during seed development, some reaching maxima
early and being associated with the temporary reserves of endosperm and seed
coat, others not reaching a maximum until the phase of intensive protein
synthesis in the cotyledons. The amino acid sequences described for *P.
sativum* (McKee *et al.*, 1955a) are sufficiently different from those described
for *P. arvense* by Flinn and Pate (1968) to suggest that there might be substantial differences in amino acid metabolism between the "species". Yet
sufficient similarities are demonstrated in the sequences of maxima of amino
acids in two seasons of growth of the one cultivar of *P. arvense* to lead one to
conclude that the programme of build-up and disappearance of different
species of amino acid are reproducible features of development and hence of
metabolic significance.

Osborne and Harris (1907), Osborne (1926) and later, Danielsson (1949, 1952), provided the pioneer studies of the storage proteins of peas, finding that they behaved as classic globulins in being soluble in dilute salt but not in water. Two major storage proteins—vicilin and legumin—were described by Osborne, and were later designated by Danielsson as a 7 s fraction "vicilin" (mol. wt. approximately 186 000), and a larger 12 s fraction "legumin" (mol. wt. 331 000). A water-soluble "albumin" fraction completed the protein complement described for the seed, it tending to appear in seeds somewhat in advance of the globulins (Danielsson, 1952). It was observed that the ratio of vicilin:legumin declined during ripening, suggesting that the two globulins did not form at equal rates (Danielsson, 1952). The "distinctness" of the two globulin fractions was substantiated on the basis of amino acid composition and sulphur content (Danielsson and Lis, 1952).

Since these early studies new advances in techniques have allowed significant further progress to be made towards a proper purification and characterization of the seed proteins. Varner and Schidlovsky (1963) describe procedures for recovering a pellet of protein bodies from cotyledons extracted with 0·1 M Tris buffer (pH 7·1). Albumin-type proteins proved to be associated with the supernatant and are therefore likely to be of cytoplasmic origin. DEAE-cellulose chromatography, polacrylamide gel-electrophoresis and ultra-centrifuge analysis have been used by Grant and Lawrence (1964) in an attempt to resolve further the globulins of pea. By using urea, formamide and sodium dodecyl sulphate (SDS) as dissociating agents, the globulins can be split into sub-units. SDS treatment, for example, leads to the appearance of 12 defined bands on electrophoresis, four of these from vicilin, six from legumin and the remaining two from an unidentified source. This suggests some degree of heterogeneity in sub-unit structure.

Amino acid analyses of hydrolysed pea globulins show that they have a somewhat similar composition, but that legumin is richer in arginine and the sulphur amino acids (Boulter and Derbyshire, 1971). The globulins have higher proportions of glutamyl, arginine and amide residues but less of the sulphur amino acids than companion albumins.

The great diversity of albumin-type proteins of pea is obvious from the studies of Fox et al. (1964) and the changes in electrophoretically distinguishable proteins during ageing of the seed are detailed by Flinn and Pate (1968).

The advent of immuno-chemical techniques such as immuno-diffusion and immuno-electrophoresis has greatly facilitated study of seed storage proteins, typing of peas having been accomplished by Kloz and Turkova (1963) and Dudman and Millerd (1975). Dudman and Millerd (1975) report a large measure of success in separating, identifying and comparing the legumin and vicilin of pea and other legumes using immuno-chemical techniques. Antiserum prepared against purified legumin from pea or *Vicia faba* reacts so as to suggest that proteins closely related to it are distributed widely throughout the Fabaceae, implying that considerable conservatism must have been expressed during evolution of the legumin proteins of this plant family (Dudman and Millerd, 1975). Vicilin-type proteins, immuno-chemically indistinguishable from one another, are present in *P. sativum* and *V. faba*, and both species share a "third protein", whose mobility in immuno-electrophoresis is intermediate between that of vicilin and legumin. This third protein is regarded as a possible second form of vicilin (see Millerd, 1975). Immuno-chemical means of detection have enabled Millerd and her co-workers to determine the sequence of appearance of the three storage proteins in pea cotyledons, the timetable reported being 9 days after flowering for the appearance of vicilin; 1 day later for legumin (see Fig. 15.6); and some days later still for the third protein. Globulins can be detected in all organs of the embryo, but not in tissues of fruit or vegetative plant (Millerd, 1975). Basha (1974) (see Millerd, 1975) and Beevers and Poulson (1972) have also followed changes in protein levels during cotyledon development of pea. They agree with other workers that albumins are synthesized some time in advance of globulins, Basha (1974) estimating a ratio of albumins:globulins of $1:1\cdot4$. According to Basha (1974), carbohydrate moieties (glucose, mannose and glucosamine) are attached to the reserve proteins of pea. Using SDS and dithiothreitol as dissociating agents, Basha (1974) demonstrated that the sub-unit structure of each protein changes during the maturation of the cotyledon. This finding poses interesting problems when attempting to unravel the sequence of biosynthetic events in the laying down of specific proteins in the developing seed. A further complication is suggested from studies of Graham and Gunning (1970). Using a fluorescent antibody technique to localize legumin and vicilin in the protein bodies of the developing cotyledons they show that, while most bodies react for both proteins, certain of them fail to give a fluorescent reaction for either type of globulin. These non-reactive bodies are suggested to be stores of albumins. Although the study of Graham and Gunning relates to *V. faba*, it is possible that a similar situation exists in the closely related genus *Pisum*.

D. Biosynthesis of Reserve Proteins in Cotyledons

It is probably correct to assume that the biosynthetic pathways leading to the accumulation of storage proteins in a seed such as pea proceed along the same general lines suggested for the synthesis of other specific proteins in living organisms. If this be so it should be possible to obtain information on the transcription of genes coded specifically for the storage proteins, and the RNA polymerases and messenger RNAs associated with such transcription; to study the means by which information for the specific sequences of amino acids for the proteins becomes translated; and to prescribe the cellular structures and routes which mediate the synthesis and eventual accumulation of

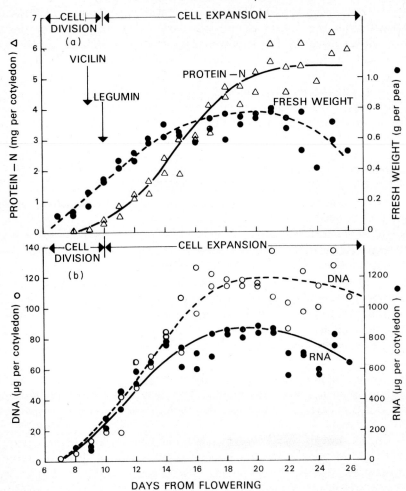

FIG. 15.6. Timetable of changes in nucleic acids and protein in developing cotyledons of *P. sativum* L. (cv. 'Greenfeast').
(a) Timing of cell division, cell expansion and the appearance of vicilin and legumin, relative to growth and total protein accumulation.
(b) Changes in content of nucleic acid (RNA and DNA) during development of cotyledons.
(Figure redrawn with permission from Millerd and Spencer, 1974)

protein within the storage body. Millerd (1975) discusses each of these areas of investigation in a recent review of the biochemistry of legume seed proteins, and it is largely from her account that the information relating to *Pisum* is presented here.

A study of RNA polymerase activity during development of pea cotyledons has been made by Millerd and Spencer (1974), paying special attention to the possible relationship between the endo-reduplication of the nuclear DNA in cotyledon cells (see earlier section) and their activity in protein synthesis. When the endogenous RNA polymerase activity of isolated nuclei was measured it was found that RNA polymerase activity per unit DNA declines throughout the time when endo-reduplication of the cotyledon DNA is occurring. But the template activity of isolated chromatin, assayed using non-limiting quantities of bacterial RNA polymerase, remains reasonably constant over the same period of development. They concluded from their results that the level of RNA polymerases is probably the factor limiting the rate of transcriptional activity *in vivo*, that the extra DNA formed during endo-reduplication is not repressed to any greater extent than the DNA of the 2C nucleus, and that the extra DNA reduplicated does not promulgate a commensurate increase in RNA synthesis and transcriptional activity in the developing cotyledon. Millerd and Spencer (1974) prefer to view the extra DNA of the mature cotyledon cell as having a function as a storage reserve of deoxynucleosides, and do not support the conclusion of Scharpé and Parijs (1973) that DNA beyond the 2C level might exercise a direct gene dose effect by providing multiple extra copies of the cistrons coded for synthesis of storage proteins. The issue is somewhat clouded by the differences obtained by various groups of workers in the time courses of increase in DNA, RNA and protein in the cotyledon cells (see Fig. 15.6 and data of Scharpé and Parijs, 1973; Smith, 1973). It may be that different experimental conditions enforce quite different patterns of accumulation of nucleic acids and protein and, in so doing, bring into play quite different types of regulatory control at the transcriptional and/or translational level. This may well be one case in which effects of environment on seed development are well worthy of detailed study.

Papers by Beevers and Poulson (1972) and Poulson and Beevers (1973) provide information on the RNA metabolism and associated amino-acid-incorporating activity of developing pea cotyledons. Their studies are especially significant in reporting success in the development of a cell-free system active in amino acid incorporation into protein. Judging from the operational characteristics of this system, it would appear to be essentially the same as that in other plants. They established a positive relationship between translational activity and the polysome content of their ribosomal preparations. As the cotyledon matured and its ability to synthesize protein declined the polysome content fell and, at the same time, evidence was obtained that levels of mRNAs might be limiting rates of protein synthesis. Beevers and Poulson observed qualitative and quantitative changes during development in the messenger-like species of RNA obtained from cotyledon ribosomes, but failed to detect any major RNA constituents which might be assigned specifically to synthesis of globulins. This is not surprising since the globulin molecule is composed of several different sub-units, each with in individual coding and each likely to have a special structural rôle to play in the configuration of the complete macroprotein. Since synthesis of globulins cannot be detected outside the embryo the question must be raised about the nature of the processes which regulate and perform synthesis of this group of proteins. It has been suggested that histones serve this function in the pea (Bonner et al., 1963) but considerable differences of opinion still exist as to the precise function of the histones in gene regulation.

The conclusion from the labelling studies of Wheeler and Boulter (1966) on V. faba is that protein bodies of legumes are unlikely to engage in the synthesis of significant amounts of protein. If this be so it becomes necessary to determine the precise ultrastructural events outside the protein bodies which lead to synthesis and accumulation of globulins. An electron auto-radiographic study on cotyledon slices of V. faba by Bailey et al. (1970) suggest that the likely route for amino acid incorporation is by means of the endoplasmic reticulum and thence, probably as already formed protein—sub-units or macroprotein—to the site of final deposition in the protein body. According to these authors the Golgi apparatus is not implicated in any major fashion in protein synthesis, and the through-put time from reticulum to protein body is likely to be of the order of 25 min.

Obviously, the biochemistry of storage protein synthesis in peas is still extremely sketchy, and can, as yet, provide no proper framework for building biochemical or genetical studies aimed at manipulating protein quantity or quality in the seed. Study has been severely limited by an inability to develop the correct procedures for isolating, purifying and characterizing the proteins and their sub-units, and until this situation improves progress will inevitably be slow.

E. Growth Substances of the Seed

The developing seed shows a changing content of substances showing cyto-kinin activity (Burrows and Carr, 1970). Three maxima in cytokinin con-centration are recorded by these workers, one timed with the attainment of a maximum volume of liquid endosperm in the seed; the other two larger maxima being synchronized with maximum rates of growth in seed and embryo respectively. Significantly perhaps, there is a fall in cytokinin content during the lag phase of seed growth. The chemical identities of the cytokinins extracted by these workers have not been determined, nor has it been assessed whether the changes observed are of significance in the establishment ot reserves of cytokinins in the dormant seed.

The situation with respect to gibberellins has been described by Reinhard and Konopka (1967) and, more recently, by Frydman and MacMillan (1973) and Frydman et al. (1974). Reinhard and Konopka (1967) isolated at least three different gibberellins from developing seeds. One, identified as GA_5,

was designated as the major gibberellin. Frydman and her colleagues identi
fied GAs 20, 29, 9, 17, 38 and 44, abscisic acid (ABA) and dihydrophasei
acid in pea seeds and made measurements of the major gibberellins (GA_9
GA_{17}, GA_{20}, GA_{29}). It is clear from the sequence of attainment of maxima in
the various gibberellin species that complex biosynthetic relationships are
likely to be involved. The precise rôle of these gibberellins in the develop
mental process remains to be established as does their significance in the
establishment of the bound reserves of gibberellin in the seed. (N.B. the
chemical structures of some of the growth substances found in pea plants are
illustrated in Chapter 9.)

A range of auxin-like compounds including 4-chloroindol-3yl acetate
(Marumo et al., 1968) and D-4-chlorotryptophane derivatives (Marumo and
Hattori, 1970) have been isolated from immature pea seeds, but little is known
of their rôle in seed growth, or as reserves of indol compounds. Seth and
Wareing (1967) have demonstrated the potent effect of indol-3yl acetate in
directing the accumulation of ^{32}P in defruited peduncles of French bean
(Phaseolus vulgaris) and the enhancement of this effect by added gibberellin
and cytokinin. A similar situation might be obtained in vivo in respect of the
growth substances present in pea-seed pods.

The synthesis of growth substances in fruits or cotyledons grown in
isolated culture is dealt with in the section on nutrition (Section IV D).

F. Minerals of the Seed

The time courses for build-up of specific mineral elements in the developing
pea seed have been studied by P. Hocking (unpublished data). As in the case
of the pod (see Fig. 15.2), Zn is accumulated precociously relative to dry
weight gains, while other elements (e.g. Na, Mg and Ca) accumulate more
slowly. There is little sign of loss of nutrient elements as the seeds dry out and
enter dormancy. Comparisons of the data obtained for pods and seeds show
that the period when the pea pod is losing its mineral reserves most rapidly
coincides exactly with the time of fastest gain of minerals by the seed. It is
highly probable, therefore, that minerals mobilized from pods are translocated
to the seeds, this source of supply being probably of greatest importance for the
more mobile elements such as P, K, Mg and Zn. The extent to which the
seeds in a pod are likely to benefit, and the proportion of their total requirements that they might obtain from the senescent pod are set out in the
accompanying table (Table 15.I).

Quarterman (1973), in making a comparison between animal and plant
products in respect of their availability of trace elements to humans, points
out that peas, like other pulses, are high in Fe, Cu and Zn. Their P content
as phytate is also high (phytate-P at 88 mg per 100 kcal). The large and well-
balanced set of minerals provided by the cotyledons allow young pea

TABLE 15.I. Senescence losses of minerals from the pod of *Pisum sativum* cv. 'Greenfeast') and the possible significance of these losses in nutrition of the seed. (Unpublished data kindly provided by P. Hocking)

	Final amount of element in seeds of the fruit (wt. per fruit)	Net loss of element during senescence of pod of fruit (wt. per pod)	Loss from pod as a proportion of requirement of seeds (%)
Macronutrients (mg)			
Potassium	26·7	3·7	13·9
Phosphorus	16·7	2·1	12·6
Magnesium	2·67	0·36	13·5
Calcium	2·35	0·28	11·9
Micronutrients (μg)			
Iron	134·5	12·4	9·2
Zinc	67·9	4·5	6·6
Manganese	32·2	3·0	9·3
Copper	16·3	0·84	5·2

seedlings to grow quite healthily for a considerable time without any added nutrients (see Chapter 3) and indeed, certain micronutrient reserves (e.g. those of Mo and B) are sufficiently large to permit two or more successive generations of peas to grow quite normally without any further addition of the micronutrient (Woodbridge, 1969). Another effect of the large seed reserves is the possibility of "carry over" effects from one generation to the next, due to the provision within the seed of an abnormally high or low amount of a specific macronutrient. This effect is described for P in the experiments of Austin (1966).

IV. NUTRITION OF THE FRUIT AND SEED

A. Patterns and Pathways of Assimilate Transfer

This topic is dealt with in detail elsewhere in this volume (see Chapter 12), so only the major features will be summarized. Studies on pea using $^{14}CO_2$ feeding (e.g. Linck and Sudia, 1962; Flinn and Pate, 1970; Harvey, 1972, 1974) have shown that although each blossom leaf is deeply committed to supplying photosynthate to its subtended fruit, during its early life it supplies quite sizeable amounts of photosynthate to other parts of the plant. Stipules make a larger contribution to the subtended fruit than do the companion leaflets, not necessarily because the stipules are less active photosynthetically, but because they participate more than leaflets do in transporting assimilate up and down the stem. This is because stipules possess less direct vascular connections with the fruit stalk than do leaflets. Some cross-feeding can occur between a blossom leaf and fruits subtended above or below it, the evidence being that a particular leaf is more likely to feed fruits inserted above it than those located on the opposite sides of the stem (Brennan, 1966). Again, vascular architecture seems to dictate patterns of nutrient flow in the pea.

Unlike the blossom leaf, the pod is entirely committed to transport to its seeds, the extent of this involvement increasing in proportion to the mass of seeds present (Lovell and Lovell, 1970). Seeds do not seem to be capable of significant photosynthesis whilst in their pod despite their intense green colour, but if $^{14}CO_2$ is injected into the internal cavity of an illuminated fruit both the chloroplast-containing epidermis lining this cavity and the outer tissues of the pod assimilate $^{14}CO_2$ and return labelled photosynthate back to the seeds (Flinn, 1969). There is no doubt that the pod has a highly significant rôle in conserving and recycling carbon lost in seed respiration (see Chapter 12), its gas cavity being rich in CO_2 ($0.15–1.5\%$ v/v).

The transfer of leaf-applied isotopes other than ^{14}C has been studied using $^{32}PO_4$ (Linck and Swanson, 1960) and ^{65}Zn (Sudia and Linck, 1963). As with photosynthetically fixed carbon, the dominant sink for a label from a blossom leaf is the subtended fruit, although some transfer of isotope to fruits at other nodes can occur. The translocation of K, Ca and P into pea fruits has been studied using the electron probe X-ray microanalyser (Läuchli, 1968). Long-distance transport of these elements through the fruit stalks appears to take place mainly in the phloem.

The flow of photosynthate from leaves is complemented by an upward transport from roots and nodules of minerals, solutes of nitrogen and growth substances. Tracer studies, involving the introduction of ^{15}N- or ^{14}C-labelled substrates to shoots through the transpiration stream, have established that nutrients move first mainly to centres of transpiration, especially the mature leaves, and are then subsequently mobilized to young organs of the shoot, and to fruits and seeds (Pate et al., 1965; Lewis and Pate, 1973). This export from leaves appears to occur through the phloem, and accompanies the passage of photosynthetically produced materials from the leaf and the mobilization of solutes from senescent tissues.

Thus, the fruit receives through its vascular connections with the rest of the plant solutes derived from virtually every mature region of the pea plant. Because of low stomatal densities and the low dilution of xylem fluids, phloem translocation is clearly more important than transpiration in supplying the necessary nutrients to a fruit.

B. The Specific Nutrients Carried to Fruits and Seeds

The pea is one of those few species from which it is possible to collect the liquid contents of both xylem and phloem elements. Xylem fluids can be recovered from the bleeding of roots, nodules or petioles (Pate and Wallace, 1964; Brennan et al., 1964; Minchin, 1973), while small amounts of phloem sap can be obtained from cut pedicels or peduncles of young fruits (Lewis and Pate, 1973). The xylem contains certain growth substances (Carr and Burrows, 1966) and ions, but its principal solutes are nitrogenous (Pate, 1962). Asparagine, glutamine, aspartic acid, homoserine and allantoic acid together comprise 90% of the xylem nitrogen when plants are active in assimilating nitrogen (see Chapter 13). The phloem carries large quantities of sugars, virtually all sucrose. This last constituent is present in phloem sap at up to 19% on a weight:volume basis. Indeed, sugar carries up to 88% of the carbon moving in the phloem stream to the fruits; most of the remainder is present in amino acids (Lewis and Pate, 1973). The main amino acids and amides in phloem are (in order of abundance): asparagine, serine, glutamine, homoserine, alanine, aspartic acid, glycine and o-acetyl homoserine. Of these, serine, alanine and glycine can all be labelled intensely in phloem sap after $^{14}CO_2$ has been fed to the adjacent blossom leaf, and are thus likely to be contributed to phloem from sites of photosynthesis. Other amino acids of the phloem, and especially the amides asparagine and glutamine, are likely to originate from the root and to transfer across from xylem to phloem in their passage through stems and leaves (see Chapters 12 and 13).

Considerable rearrangement of the carbon and nitrogen delivered by the incoming phloem and xylem must occur in the seeds before reserves of protein and carbohydrate can be laid down. Sucrose is the obvious starting material for starch, hemicellulose and pectin syntheses, and the presence of quite large pools of monosaccharides in embryos (Flinn, 1969) but not in the entering phloem stream (Lewis and Pate, 1973), suggests that breakdown of sucrose occurs after delivery and close to the sites of synthesis in the storage cotyledons. The amino acid balance of phloem is considerably different from that of storage protein, so that certain amino acids which are scarce in the phloem, but abundant in the seed, must be synthesized in the seed itself. Arginine, lysine, histidine, tyrosine, phenylalanine and the leucine(s) fall within this category, and, as might be expected, these achieve high specific activity in seeds after $^{14}CO_2$ is fed to the blossom leaf or ^{15}N-labelled amide is fed to fruiting shoots through the transpiration stream (Lewis and Pate, 1973). The inference from these labelling studies is that sucrose and amides, uncommitted sources of carbon- and amino-groupings respectively, are the compounds arriving in sufficient quantity to be utilized as raw materials for synthesis of these "extra" amino acids required by the seeds. A complication which must not be overlooked is the establishment of pools of free amino acids within pod, seed coat and endosperm. The importance of these as bulk sources of nitrogen to the developing seeds is looked at in the next section

C. The Significance of Different Organs and Assimilate Sources in the Feeding of Seeds

When fully mature, approximately half of the dry matter (Brouwer, 1962 and two-thirds of the total nitrogen (Pate and Flinn, 1973) of a pea plant are located in seeds, and since mobilization of materials from vegetative parts does not assume importance until the final third of the plant's growth cycle, it becomes clear that solutes must be transported to seeds at this time with great rapidity and efficiency. A study of the fate of radio-carbon from $^{14}CO_2$, and the labelled nitrogen of $^{15}NO_3$ fed either early or late in the plant's life cycle (Pate and Flinn, 1973) gives some indication of the relative importance of new and old assimilates in the feeding of the fruit. The two elements behave somewhat differently. Carbon assimilated in early vegetative growth is subsequently transferred to seeds with a very low efficiency (2%), low recovery in the seed being mainly due to the heavy respiratory losses of labelled carbon before flowering commences, and to the fact that most of the ^{14}C which does survive in the plant until maturity has become incorporated into insoluble structural material from which it cannot be readily retrieved during organ senescence (see Chapter 12). So, it may be general that photosynthesis before flowering has little direct relevance to the carbon nutrition of the seeds. However, as mentioned earlier, photosynthates formed during the fruiting period are donated to seeds with high efficiency suggesting a bulk, direct transfer of sucrose with little consumption of this photosynthate by vegetative parts.

The element nitrogen contrasts with carbon in being mobilized to seeds at all times with high efficiency (Schilling and Schalldach, 1966). Thus, Pate and Flinn (1973) recorded a 51% transfer to seeds of early fed nitrogen (^{15}N) and a 74% transfer of ^{15}N fed after flowering, the implication being that the bulk of the nitrogen resources of a plant are readily available to its seeds, regardless of where they are located or when they were elaborated during the life cycle.

A tentative budget for nitrogen (Fig. 15.7) can be set out for fruits of the field pea by combining data for nitrogen distribution between plant organs at different times of the life cycle with data for efficiency of transfer of nitrogen to seeds as determined using ^{15}N. The study of Pate and Flinn (1973) shows that by the time of anthesis of the first flower, plants have accumulated only 24% of their final amount of nitrogen. Since the efficiency of transfer to seeds of this element is 51%, and because seeds eventually contain 61% of the total nitrogen present at plant maturity, it can be estimated that the nitrogen assimilated before flowering provides, at best, only one-fifth of the seed's

(a) Carbon balance of seeds at first blossom node

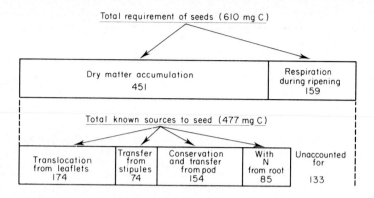

(b) Nitrogen balance of seeds at first blossom node

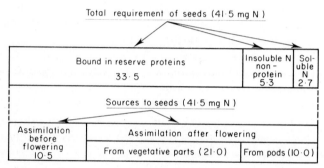

FIG. 15.7. (a) Carbon and (b) nitrogen requirements of the seeds of the fruit of the first blossom node of field pea (*P. arvense* L.) and estimations of how these requirements are met from sources outside the seed. (Estimations made from data of Flinn and Pate, 1970; Pate and Flinn, 1973)

N.B. A more detailed account of day by day transfer of carbon to seeds is provided in Chapter 12.

requirement for nitrogen. The remainder represents nitrogen assimilated during reproductive growth (see Fig. 15.7). Of course the situation might be somewhat different in dwarf varieties of garden pea, especially those whose fruiting period is curtailed by genetic factors inducing premature shoot senescence. The significance of pods in providing nitrogen to their seeds can also be estimated. McKee *et al.* (1955a) and Flinn and Pate (1968) arrived at the conclusion that the net loss of nitrogen from the pod during seed filling could be held responsible for providing the nitrogen to one seed in a pod of four or five seeds (see Fig. 15.7).

As mentioned in the chapter on translocation (Chapter 12) the budget for carbon is a complicated one largely because respiratory losses and photosynthetic gains of carbon occur during fruit and seed development. Leaflets and stipules generate an exportable surplus of photosynthate from the time of anthesis of the subtended flower right through until the resulting seeds are fully mature. As mentioned before stipules are likely to yield less carbon to the seeds than are the leaflets.

Pod photosynthesis appears to be particularly concerned with the re-assimilation of CO_2 lost to the pod interior by the respiring seeds. A young fruit, with only small seeds, can regularly maintain a net intake of CO_2 from the atmosphere during the photoperiod, but when larger seeds are present, the fruit often fails to reach compensation point when illuminated (see Flinn and Pate, 1970; Flinn, 1974). It would be interesting to know whether the fleshy walls of the pod can engage in dark fixation of CO_2, akin to the Crassu-lacean acid metabolism displayed by certain genera of plants with succulent stems or leaves. A mechanism of this sort might increase greatly the efficiency of conservation of carbon by the fruit.

A generalized budget for carbon for a developing pea fruit is also shown in Fig. 15.7. It can be seen that photosynthesis at the blossom node is responsible for providing approximately two-thirds of the carbon required by the seeds during their period of ripening. A further, small fraction arrives attached to nitrogen supplied from the root system, and the remainder can be assumed to be made up of contributions from photosynthesis at other blossom nodes. The amounts transferred to seeds on a day by day basis are outlined in Fig. 12.14 in Chapter 12, and it can be seen there that the blossom node is most effective in meeting the requirements of its fruit during the middle period of fruit development. Earlier or later in the growth cycle the fruit must draw heavily on other undefined sources of carbon.

D. Nutritional Requirements of Fruit and Seed when Cultured in Isolation

The foregoing account has made clear that when fruits develop normally on a pea plant they receive a highly sophisticated diet of organic and inorganic nutrients derived from virtually all mature regions of the plant. It therefore becomes of considerable interest to know to what extent these nutritional requirements are obligatory, and to what extent the fruit and its contained seeds can function autotrophically if forced to do so in isolated culture.

Attempts to culture isolated fruits in sterile conditions and to specify their nutrient requirements have been made by several groups of workers. Fruits excised at the "flat" stage (10 days after anthesis) grow successfully in a simple medium consisting of mineral nutrients and 5% sucrose in agar. After a period of a week or two in sterile culture the seeds in the fruit reach full size and if subsequently harvested can be shown to be capable of germination. This system has been used to demonstrate that seeds of isolated fruits exhibit a net synthesis of gibberellins (Baldev et al., 1965) and of cytokinins (Hahn et al.,

974). Also, during a period of isolated culture of young cotyledons, the onset of globulin synthesis has been detected using immuno-chemical techniques (Millerd *et al.*, 1975). In the studies of this last group of workers, an amino acid and vitamin mixture was added as supplement to the basal medium of sucrose and minerals; agar was dispensed with. Under these conditions the fruit and seeds increased in fresh weight almost as fast as they would have done if left attached to the plant.

Excised fruits can provide only equivocal information on the synthetic qualities of the contained seeds since the pod itself might generate a quite sophisticated array of organic nutrients and supply these to the seed. An attempt to make more definitive studies on the nutritional performance of the seed *per se* has been made by Millerd *et al.* (1975) in their studies of *in vitro* culture of isolated cotyledons of pea. Cotyledons were removed from seeds at a time when the embryo had three-quarters filled the seed coat, and the cotyledons were cultured in the same sucrose–minerals–vitamins–amino acid medium as had been used for culture of excised fruits. In periods of culture of up to 4 days, considerable quantities of chlorophyll, starch, DNA, RNA and protein were synthesized, and the two storage proteins, vicilin and legumin, became detectable in the cotyledon. If a nitrogen source was omitted from the medium protein synthesis occurred at only half the rate as when the full supplement of amino acids was present. However, since some protein synthesis did occur in the minus-nitrogen medium it appeared that endogenous pools of amino acids could be utilized, at least, that is, when starvation prevailed.

The experiments of Millerd and her colleagues provide interesting information on the sources of nitrogen acceptable to the cotyledon for protein synthesis. The regular medium contains a mixture of 18 amino acids, but an equal rate of protein accumulation can be achieved if this is replaced by an equivalent amount of nitrogen entirely in the form of asparagine. Glutamine is less suitable, maintaining protein synthesis at a level only marginally greater than in a minus-nitrogen medium, whilst ammonia and urea are totally unsuitable as sole sources of nitrogen. The preference of seeds for asparagine over glutamine contrasts with the situation in leaves and stems in which glutamine is more readily utilized (see Chapter 13). These findings fit well with the fact that asparagine is the principal nitrogenous constituent supplied to the developing fruit through both xylem and phloem. It would be interesting to see whether other major constituents of the phloem—serine, homoserine and alanine—are utilized by pea cotyledons as effectively as is asparagine or whether, like glutamine, they support only low rates of protein synthesis.

The phloem of legumes rarely carries unreduced nitrate, and since seeds are likely to be nourished almost entirely by the phloem, it is most unlikely that this form of nitrogen would represent a normal substrate for the developing embryo (see Pate *et al.*, 1974). Consistent with this are the findings that seeds of legumes rarely contain detectable levels of free nitrate and that cotyledons taken directly from seeds developing on intact plants show little or no nitrate reductase activity (Pate, unpublished). Nevertheless, the ability to synthesize the inducible enzyme system for reducing nitrate is present in the cotyledon, for if cultured for 2 days in a medium containing nitrate, activity is detected (Millerd *et al.*, 1975). A similar inducible nitrate reductase has been demonstrated for developing bean (*Phaseolus*) cotyledons by Lips *et al.* (1973). It may be general that isolated embryos of legumes can be induced to accept nitrate and grow at a normal rate using it as a sole source of nitrogen. If this proves to be so it is clear that cotyledons are potentially capable of manufacturing all of the amino acids they need for protein synthesis.

The general impression from these few attempts at *in vitro* culture, is that the fruit and embryo of the pea have a capacity for primary and secondary syntheses which is considerably greater than would ever be suspected from a study of the manner in which they are nourished when attached to the intact plant. What remains to be determined is whether enforcing a fruit or seed to utilize relatively simple nutrients condemns it to a slower rate of growth and engenders radical differences in seed composition compared with the normal product maturing on an intact plant. The natural system employed by the plant has the inherent advantages of accomplishing biochemical events related to seed nourishment at sites outside the seed and, to an extent, at times outside the normal span of seed development. The precise nature of these advantages remains to be appreciated.

E. Regulatory Processes in Fruit Maturation: Sink–Source Interactions

Major questions still remain unanswered about the processes involved in the co-ordination of the nutritional activities of plant organs. In particular the regulatory aspects of sink–source relationships are poorly understood and are still, in fact, a matter of great controversy (see Neales and Incoll, 1968). However, evidence from a variety of plants, including the garden pea, has suggested that assimilation rates in the source organ may be altered by changes in the size of a major sink. These studies have depended on either the removal of a sink organ (tomato: Moss, 1962; potato: Burt, 1964; Nosberger and Humphries, 1965; wheat: King *et al.*, 1967) or on the removal of part of the assimilatory area (garden pea: Kursanov, 1934; apple: Maggs, 1964; dwarf bean, maize and willow: Wareing *et al.*, 1968). Less drastic treatments have involved cooling of the root (sugar beet: Habeshaw, 1973) and shading part of the plant's assimilatory area (soybean: Thorne and Koller, 1974). In general the evidence supports the view that assimilatory areas, in particular leaves, do not operate continuously at the peak photosynthetic rates of which they are capable in a given environment, but are constrained by factors which have in turn influenced the size and activity of the sink regions which they supply. Thus an increase in demand of a sink may result in a rise in the assimilation rate of the source organ; a decrease in demand may lead to a fall in assimilation rate.

In the pea the onset of flowering and subsequent fruit growth leads to a rapid doubling in the photosynthesis of the whole plant (Lawrie and Wheeler, 1974). The complex nutritional relationship existing, in the reproductive plant, between the leaves, as primary source organs, and the photosynthetically less-effective parts of the plant, as sinks, was recognized by Kursanov (1934). His experiments showed that the removal or shading of the leaves led to an increased rate of assimilation by the pod, suggesting that under normal conditions pod photosynthesis is repressed by the plentiful supply of assimilates from the leaves. However, these experiments, in common with most other work on sink–source relationships, involve considerable manipulation of parts of the plant. It therefore seems pertinent to ask about the regulatory nature of the relationship between leaf and adjacent fruit in the intact and unmanipulated plant. Recent work (Flinn, 1974) suggests that leaflet photosynthesis in pea is regulated by the pattern of assimilate demand from the subtended fruit during its development. Fruits of cv. 'Onward' show three well-defined peaks in assimilate demand, the first being due to the rapid elongation of the pod; the second, probably, to inflation growth of the pod; and the final peak to rapid growth of the seeds (see Fig. 15.8). Leaflet photosynthesis rises and falls in response to the swings in demand for assimilates by the developing pod, but responds to a lesser extent to the demand for carbon by the maturing seeds. This last observation is explicable in the light of the demonstration that late growth of seeds may be substantially dependent on sources of carbon from outside the node. Leaflets at vegetative nodes, and nodes at which the flower has been removed at anthesis, do not show the sharp peaks in photosynthetic activity characteristic of leaflets at the intact fruiting node. Thus, the nutritionally coupled leaflet–fruit unit presents itself as a system which may be exploited to reveal the mechanisms involved in sink–source relationships.

An important feature of source–sink interactions which must be borne in mind is the fact that one source rarely feeds only the one sink, and that if the major sink is removed, or its demands for assimilates toned down, other sinks, lower in the hierarchy, may command increasing shares of assimilate. Effects of this kind are evident in the vegetative stages of the life cycle (see Chapter 12; and Pate, 1975) but a particularly dramatic example is provided for pea fruits in the studies of Lovell and Lovell (1970). Their study of the pod as a provider of photosynthate showed it to be committed exclusively to its seeds under normal circumstances, but if the seeds were removed and $^{14}CO_2$ fed to the pod, ^{14}C-labelled assimilates were then exported from the pod to other parts of the plant. Combined with this effect they found that the proportion of ^{14}C exported from a pod without seeds was only one-quarter of that evident when seeds were present. The results suggested that the presence of seeds in a pod may have exercised a stimulatory effect on pod activity in translocation and possibly, as mentioned above, a stimulus also to its photosynthetic performance.

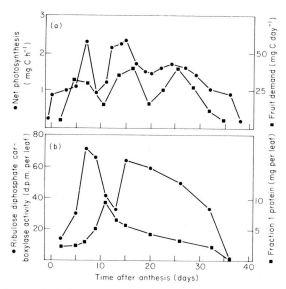

FIG. 15.8. Modulations in net photosynthesis (a) and ribulose diphosphate carboxylase activity (b) in blossom leaflets of *P. sativum* (cv. 'Onward'). Modulations in photosynthesis are in relation to the demands for carbon by the fruit borne at that node. Changes during development in the level of Fraction I protein of the leaflets are recorded in (b), and are seen not to match the fluctuations in the carboxylase very well.
(Data from Flinn, 1974; and Flinn, unpublished)

While the importance of sink regulation of assimilation rate of a source organ is now firmly established, there is rather less agreement as to how this form of control is excised. Gaastra (1959, 1963) has suggested that at saturating light intensities leaf photosynthetic rate may be determined by the physical resistance of the mesophyll. Similarly, Barrs (1968) proposed that stomatal resistance may be the dominant factor under conditions of CO_2 limitation. Other workers have suggested that the prime factor in sink control may be product inhibition, in which accumulating assimilates in the leaf depress its current rate of photosynthesis (Hartt, 1963; Humphries, 1963; Warren Wilson, 1966; Habeshaw, 1973). However, it appears unlikely that changes in sugar and starch levels in the source leaf are major mediating factors in many sink–source relationships. If this produce-inhibition hypothesis holds for peas one might expect to find raised levels of soluble hexoses in

ource leaflets correlated with phases of low fruit growth and depressed source
hotosynthesis. In fact there is no evidence for this.

Changes in the hormonal balance between sink and source, acting through
he initial carboxylation step of photosynthesis, seem to be a plausible enough
andidate for regulation of source activity (Woolhouse, 1968; Wareing et al.,
968; Thorne and Koller, 1974), and some initial experiments carried out on
he fruiting node of pea tend to support this view (unpublished). Fluctuations
n the activity of unpurified leaflet ribulose, 1,5-diphosphate carboxylase were
ound to match closely both the patterns for net photosynthesis and assimilate
demand for pod growth (see Fig. 15.8). However, measurements of Fraction I
protein in the adjacent leaflets did not show this relationship, presenting the
possibility that control is exercised through the carboxylase by factors
operating at the substrate level. Further exploration and testing of these
suggestions is called for.

From what is said above it should be obvious that plant physiologists
studying peas have still very little to offer towards solving basic problems
concerned with the regulation of fruit growth and seed filling by forces within
or outside the reproductive organs. It is of paramount importance to know
which processes are really responsible for limiting the rate of reproductive
development, so as to provide the plant breeder with information which would
enable the improvement of yield within the species. Until this information
becomes available, direction and purpose in pea breeding will be severely
restricted.

REFERENCES

ATTIA, M. S. (1958). *Alex. J. agric. Res.* **6**, 93–98.
AUSTIN, R. B. (1966). *Pl. Soil* **24**, 359–368.
BAILEY, C. J., COBB, A. and BOULTER, D. (1970). *Planta* **95**, 103–118.
BAIN, J. M. and MERCER, F. V. (1966). *Aust. J. biol. Sci.* **19**, 49–67.
BALDEV, B., LANG, A. and AGATEP, A. O. (1965). *Science, N.Y.* **147**, 155–157.
BARRS, H. D. (1968). *Physiologia Pl.* **21**, 918–929.
BASHA, S. M. M. (1974). Ph.D. Thesis, Graduate College, University of Okla-
    homa, U.S.A.
BEEVERS, L. and POULSON, R. (1972). *Pl. Physiol.* **49**, 476–481.
BISSON, C. S. and JONES, H. A. (1932). *Pl. Physiol.* **7**, 91–106.
BONNER, J., HUANG, R. C. and GILDEN, R. V. (1963). *Proc. natn. Acad. Sci.
    U.S.A.* **50**, 893–900.
BOULTER, D. and DERBYSHIRE, E. (1971). *In* "Chemotaxonomy of the Legum-
    inosae" (J. Hasborne, ed.), pp 285–308. Academic Press, London and New
    York.
BOWLING, A. C. and CROWDEN, R. K. (1973). *Aust. J. biol. Sci.* **26**, 679–684.
BRENNAN, H. (1966). M.Sc. Thesis, Queens University, Belfast, N. Ireland.
BRENNAN, H., PATE, J. S. and WALLACE, W. (1964). *Ann. Bot.* **28**, 527–540.
BROUWER, R. (1962). *Neth. J. agric. Sci.* **10**, 361–376.

BURROWS, W. J. and CARR, D. J. (1970). *Physiologia Pl.* **23**, 1064–1070.

BURT, R. L. (1964). *Aust. J. biol. Sci.* **17**, 867–877.

CARR, D. J. and BURROWS, W. J. (1966). *Life Sciences* **5**, 2061–2077.

CARR, D. J. and SKENE, K. G. M. (1961). *Aust. J. biol. Sci.* **14**, 1–12.

CLUTTER, M. E. and SUSSEX, I. M. (1968). *J. Cell. Biol.* **39**, 26a.

COOPER, D. C. (1938). *Bot. Gaz.* **100**, 123–132.

DANIELSSON, C. E. (1949). *Biochem. J.* **44**, 387–410.

DANIELSSON, C. E. (1952). *Acta chem. scand.* **6**, 149–159.

DANIELSSON, C. E. (1956). *Physiologia Pl.* **9**, 212–219.

DANIELSSON, C. E. (1959). *Acta. Agric. scand.* **9**, 181–188.

DANIELSSON, C. E. and LIS, H. (1952). *Acta chem. scand.* **6**, 139–148.

DUDMAN, W. F. and MILLERD, A. (1975). *Biochem. Syst. Ecol.* **2**.

ESAU, K. (1965). "Plant Anatomy", pp 589–590. Wiley, New York.

FAHN, A. and ZOHARY, M. (1955). *Phytomorphology* **5**, 99–111.

FARRINGTON, P. (1974). *J. Aust. Inst. agric. Sci.* **40**, 99–108.

FLINN, A. M. (1969). Ph.D. Thesis, Queen's University, Belfast, N. Ireland.

FLINN, A. M. (1974). *Physiologia Pl.* **31**, 275–278.

FLINN, A. M. and PATE, J. S. (1968). *Ann. Bot.* **32**, 479–495.

FLINN, A. M. and PATE, J. S. (1970). *J. exp. Bot.* **21**, 71–82.

FLINN, A. M. and SMITH, D. L. (1967). *Planta* **75**, 10–22.

FOX, D. J., THURMAN, D. A. and BOULTER, D. (1964). *Phytochemistry* **3**, 417–419.

FRYDMAN, V. M. and MACMILLAN, J. (1973). *Planta* **115**, 11–15.

FRYDMAN, V. M., GASKIN, P. and MACMILLAN, J. (1974). *Planta* **118**, 123–132

GAASTRA, P. (1959). *Meded. LandbHoogesch. Wageningen* **59**, 1–68.

GAASTRA, P. (1963). *In* "Environmental Control of Plant Growth" (L. T. Evans, ed.), p. 113. Academic Press, New York and London.

GRAHAM, T. A. and GUNNING, B. E. S. (1970). *Nature, Lond.* **228**, 81–82.

GRANT, D. R. and LAWRENCE, J. M. (1964). *Archs Biochem. Biophys.* **108**, 552–561.

GUNNING, B. E. S. and PATE, J. S. (1969). *Protoplasma* **68**, 107–133.

GUNNING, B. E. S. and PATE, J. S. (1974). *In* "Dynamic Aspects of Plant Ultrastructure" (A. W. Robards, ed.), pp 441–480. McGraw Hill, London.

HABESHAW, D. (1973). *Planta* **110**, 213–226.

HAHN, H., DE ZACKS, R. and KENDE, H. (1974). *Naturwissenschaften* **61**, 170.

HARTT, C. E. (1963). *Naturwissenschaften* **50**, 666.

HARVEY, D. M. (1972). *Ann. Bot.* **36**, 981–991.

HARVEY, D. M. (1974). *Ann. Bot.* **38**, 327–335.

HASSID, W. Z. (1967). *A. Rev. Pl. Physiol.* **18**, 253–280.

HUMPHRIES, E. C. (1963). *Ann. Bot.* **105**, 175–183.

KAPOOR, B. M. (1966). *Genetica* **37**, 557–568.

KING, R. W., WARDLAW, I. F. and EVANS, L. T. (1967). *Planta* **77**, 261–276.

KLOZ, J. and TURKOVA, V. (1963). *Biologia Pl.* **5**, 29–40.

KURSANOV, A. L. (1934). *Planta* **22**, 240–250.

LÄUCHLI, A. (1968). *Planta* **83**, 137–149.

LAWRIE, C. and WHEELER, C. T. (1974). *New Phytol.* **73**, 1119–1128.

LEWIS, O. A. M. and PATE, J. S. (1973) *J. exp. Bot.* **24**, 596–606.

LINCK, A. J. (1961). *Phytomorphology* **11**, 79–84.

LINCK, A. J. and SUDIA, T. W. (1962). *Experientia* **18**, 69–70.

LINCK, A. J. and SWANSON, C. A. (1960). *Pl. Soil* **12**, 57–68.

LIPS, S. H., KAPLAN, D. and ROTH-BEJERANO, N. (1973). *Eur. J. Biochem.* **37**, 589–592.

LOVELL, P. H. and LOVELL, P. J. (1970). *Physiologia Pl.* **23**, 316–322.
MCKEE, H. S., NESTEL, L. and ROBERTSON, R. N. (1955a). *Aust. J. biol. Sci.* **8**, 467–475.
MCKEE, H. S., ROBERTSON, R. N. and LEE, J. B. (1955b). *Aust. J. biol. Sci.* **8**, 137–163.
MAGGS, D. H. (1964). *J. exp. Bot.* **15**, 574–583.
MARINOS, N. G. (1970). *Protoplasma* **70**, 261–279.
MARUMO, S., HATTORI, H., ABE, H. and MUNAKATA, K. (1968). *Nature, Lond.* **219**, 959–960.
MARUMO, S. and HATTORI, H. (1970). *Planta* **90**, 208–211.
MATTHEWS, S. (1973a). *Ann. appl. Biol.* **73**, 211–219.
MATTHEWS, S. (1973b). *Ann. appl. Biol.* **75**, 93–105.
MILLERD, A. (1975). *A. Rev. Pl. Physiol.* **26**, 53–72.
MILLERD, A. and SPENCER, D. (1974). *Aust. J. Pl. Physiol.* **1**, 331–341.
MILLERD, A., SPENCER, D., DUDMAN, W. F. and STILLER, M. (1974). *Aust. J. Pl. Physiol.* **2**, 51–59.
MINCHIN, F. R. (1973). Ph.D. Thesis, The Queen's University, Belfast, N. Ireland.
MOLLENHAUER, H. H. and TOTTEN, C. (1971). *J. Cell Biol.* **48**, 395–405.
MOSS, D. N. (1962). *Crop. Sci.* **2**, 366–367.
NEALES, T. F. and INCOLL, L. D. (1968). *Bot. Rev.* **34**, 107–125.
NOSBERGER, J. and HUMPHRIES, E. C. (1965). *Ann. Bot.* **29**, 579–588.
OSBORNE, T. B. (1926). "The Vegetable Proteins." Longmans Green, London.
OSBORNE, T. B. and HARRIS, I. F. (1907). *J. biol. Chem.* **3**, 213–217.
PATE, J. S. (1962). *Pl. Soil* **17**, 333–356.
PATE, J. S. (1975). *In* "Crop Physiology—Some Case Histories" (L. T. Evans, ed.), pp 191–224. Cambridge University Press.
PATE, J. S. and FLINN, A. M. (1973). *J. exp. Bot.* **24**, 1090–1099.
PATE, J. S. and GUNNING, B. E. S. (1972). *A. Rev. Pl. Physiol.* **23**, 173–196.
PATE, J. S. and WALLACE, W. (1964). *Ann. Bot.* **28**, 83–99.
PATE, J. S., SHARKEY, P. J. and LEWIS, O. A. M. (1974). *Planta* **120**, 229–243.
PATE, J. S., WALKER, J. and WALLACE, W. (1965). *Ann. Bot.* **29**, 475–493.
POULSON, R. and BEEVERS, L. (1973). *Biochim. biophys. Acta.* **308**, 381–389.
PREISS, J., GHOSH, H. P. and WITTKOP, J. (1967). *In* "The Biochemistry of Chloroplasts" (T. W. Goodwin, ed.), Vol. II, pp 131–153. Academic Press, New York and London.
QUARTERMAN, J. (1973). *Qualitas Plantarum Pl. Fds hum. Nutr.* **23**, 171–190.
RAACKE, I. D. (1957). *Biochem. J.* **66**, 113–116.
REEVE, R. M. (1948). *Am. J. Bot.* **35**, 591–602.
REINHARD, E. and KONOPKA, W. (1967). *Planta* **77**, 58–76.
ROBERTSON, R. N., HIGHKIN, H. R., SMYDZUK, J. and WENT, F. W. (1962). *Aust. J. biol. Sci.* **15**, 1–15.
ROWAN, K. S. and TURNER, D. H. (1957). *Aust. J. biol. Sci.* **10**, 414–425.
SCHARPÉ, A. and VAN PARIJS, R. (1973). *J. exp. Bot.* **24**, 216–222.
SCHILLING, G. and SCHALLDACH, J. (1966). *Albrecht-Thaer-Arch.* **10**, 895–907.
SETH, A. K. and WAREING, P. F. (1967). *J. exp. Bot.* **18**, 65–77.
SMITH, D. L. (1973). *Ann. Bot.* **37**, 795–804.
SUDIA, T. W. and LINCK, A. J. (1963). *Pl. Soil* **19**, 249–254.
SWIFT, J. G., BUTTROSE, M. S. (1973). *Planta* **109**, 61–72.
THORNE, J. H. and KOLLER, H. R. (1974). *Pl. Physiol.* **54**, 201–207.
TURNER, D. H. and TURNER, J. F. (1957). *Aust. J. biol. Sci.* **10**, 302–309.
TURNER, J. F., TURNER, D. H. and LEE, J. B. (1957). *Aust. J. biol. Sci.* **10**, 407–413.

VARNER, J. E. and SCHIDLOVSKY, G. (1963). *Pl. Physiol.* **38**, 139–144.

WAREING, P. F., KHALIFA, M. M. and TREHARNE, K. J. (1968). *Nature, Lond* **220**, 453–457.

WARREN-WILSON, J. (1966). *Ann. Bot.* **30**, 383–402.

WHEELER, C. T. and BOULTER, D. (1966). *Biochem. J.* **100**, 53.

WOODBRIDGE, C. G. (1969). *Proc. Am. Soc. hort. Sci.* **94**, 542–544.

WOOLHOUSE, H. W. (1968). *Hilger J.* **11**, 7–12.

YARNELL, S. H. (1962). *Bot. Rev.* **28**, 465–537.

16. The Pea as a Crop Plant

J. S. PATE

Department of Botany, University of Western Australia

I. USAGE, PRODUCTION AND IMPORTANCE

Archaeological discoveries of seeds of peas in the early dwellings of man (e.g. see Cole, 1961; Hawkes and Wooley, 1963; Janick *et al.*, 1969) tell us that his history of association with the genus *Pisum* dates at least from Stone Age times and probably centres on the land areas impinging on the eastern end of the Mediterranean (see also Chapter 2). This general area is quite rich in other fairly large-seeded members of the legume family as well as in certain primitive forms of *Pisum* (Ben-Ze'ev and Zohary, 1973). It is likely that the pea commended itself to cultivation because it lacked bitter or poisonous principles, it showed a higher than normal level of digestibility than most legume seeds, and its overall edaphic requirements matched well the conditions employed by man in his first attempts at cultivation, especially of cereals. The character of soft seededness (i.e. the absence of dormancy factors due to an impermeable seed coat) would have also been a factor in favour of ready acceptance as a crop plant, although it is debatable whether this character was present in the wild forms of pea or was selected for subsequently once the species had entered cultivation (see Chapter 2). In any event, the introduction of peas into primitive agriculture must have taken place rapidly, for, alongside the cereals, they had become in widespread and common usage by Greek or Roman times (Cole, 1961).

Peas are now grown in temperate regions the world over, including the higher elevations of the tropics. Cultivars are known which can succeed as winter annuals in relatively hot countries such as India. Others exhibit considerable frost hardiness. The short growth cycle (usually 80–100 days) is an attractive feature, say in the short growing season of a high latitude summer.

Most cultivated peas are used in the form of dry seed and as supplements to human diet, mainly in the form of split peas for soup or pottage, or as canned products. About nine million hectares are devoted to the production of dry peas world-wide (FAO Production Yearbook, 1974), the U.S.S.R. and China together contributing more than three-quarters of the total. It ranks third or fourth in worldwide production amongst the grain legumes (Dawson, 1974; Farrington, 1974). As a general rule, pea cultivation is favoured in dry land areas, where the weather is cool and moisture is abundant in early growth (autumn or spring), but where rainfall is minimal or absent during the later stages of crop development.

On a global basis, the area devoted to dry peas is many times greater than that devoted to the production of green peas. Green pea production is also more limited from the standpoint of geographical distribution. Only quite recently, and only in the developed countries, has the pea become important as a green vegetable, being consumed as a fresh or, more often, as a processed product, either canned or frozen. Commercial production is contract-based and built upon a highly sophisticated technology that ensures that sowing and harvesting times are staggered so that a succession of crops from different farmers are all processed at the peak of edible quality. Prime quality—characterized mainly by size, sweetness, succulence, tenderness and uniformity—is ephemeral, lasting literally only a matter of hours.

Because the pea is a nitrogen-fixing legume, its value as a green manure crop has long been recognized. However, its use in that capacity diminished drastically as relatively inexpensive commercial fertilizers became plentiful and as systems of land management placed greater emphasis on cash returns from each cultivation. Nevertheless, interest in the pea as a soil-building crop may once again increase as mineral fertilizers become less available and more expensive. Even now, where the pea is grown commercially as a crop for processing, an effort is usually made to ensure that most of the plant is returned to the soil after the green shelled peas have been harvested. The return in these circumstances is of course less than when the whole plant is dug in since some 60–70% of the plant's nitrogen is removed with the seed (see Chapter 15).

One has only to compare some of the primitive forms of pea with the modern cultivars used for vining and processing to appreciate what successful strides have been made by plant breeders in removing many undesirable traits. Some of the ways in which these aims have been accomplished are described elsewhere in this volume (Chapter 2), and it is not the purpose of this chapter to recapitulate this material. Instead it is hoped to present an overview of the cultivated species at its present level of sophistication, with the purpose of suggesting or speculating how further improvement of yield and quality might be obtained. Plant physiologists have not participated to any major degree in the transformation of the species into its modern forms; so we are largely ignorant of the factors most likely to act as barriers to the further increase of its productivity. Nevertheless, certain areas of inadequacy are all too obvious: the sections which follow will consider these in turn, and discuss possible means of overcoming them.

II. PROBLEMS IN HUSBANDRY

A. Germination and Early Establishment

Like many other legumes and vegetable crops, seeds of pea are subject to quite rapid loss of viability when stored; seedlings formed from aged seed stocks of certain cultivars are of low vigour, and exhibit a significantly higher proportion of abnormal plants than do younger stocks, presumably as a result of physiological or genetic damage in storage. High seed moisture during storage has an adverse effect on viability, this being compounded quantitatively with temperature as illustrated in the viability nomograph presented by Roberts and Roberts (1972) (Fig. 16.1). It is not known how widely their statement holds for peas. The material presented in Chapter 4 suggests that there may be large variations between varieties.

The potential grower is usually protected by seed certification which ensures that seed for sale meets a certain minimum germination requirement (usually set for peas as 70 or 80% germination in a laboratory test), but, as pointed out elsewhere in this volume (Chapter 4), the results of such tests offer an extremely poor indication of the ability of a sample of pea seed to germinate and become successfully established under field conditions. A far better form of assessment would be one based on the extent of release of solutes from the dry seed when steeped in water, for, the greater the tendency to exude organic solutes and electrolytes, the greater the likelihood of infection during pre-emergence stages, particularly if the peas are to be sown early in spring in a cold soil of heavy texture (Bradnock and Matthews, 1970). If a "steep water" test were to become universally accepted and be rigorously standardized on an international basis, there would certainly be much less chance of misunderstanding when seed stocks are exported from one region or country to another. It seems quite clear that the tendency towards pre-emergence mortality is high in certain cultivars, particularly in the wrinkled varieties (Marshall, 1966; Perry and Harrison, 1970), and that provenances of the one cultivar may differ widely in seed "leakiness" (and hence in mortality under field conditions). Conditions during ripening on or off the vine seem to determine whether the resulting seed will be prone to leakiness or not (Matthews, 1973). Dryness of haulms or freshly harvested seed at high

J. S. Pate

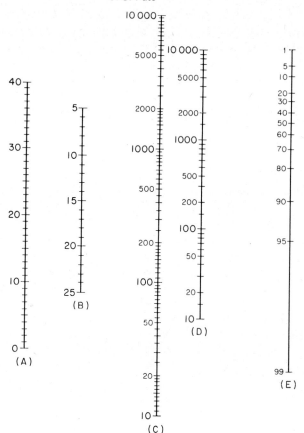

(A) Temperature °C
(B) Moisture content, % wet weight
(C) Moisture viability period, days (pivot line)
(D) Period for viability to fall to percentage indicated, days
(E) Percentage viability

FIG. 16.1. Viability nomograph for peas (*Pisum sativum*). Redrawn from Roberts and Roberts (1972)

temperatures is to be avoided, particularly if the seeds are harvested in a relatively unripe state; the development of "hollow heart" condition under these circumstances (see Harrison and Perry, 1973; and references in Chapter 4) might be viewed as a case of mismanagement. Proneness to leak solutes is held to be associated with lack of structural integrity in the dry seed, possibly combined with an inability of the cotyledon cells, when imbibed, to establish membrane integrity at a sufficiently fast rate to prevent a massive efflux of leached solutes. But there is still a sufficient measure of disagreement in this area to suggest the need for further investigation. It is not established, for example, whether the propensity for leakiness is determined at the macroscopic or ultramicroscopic level; nor is it known how imporant micropylar features are in conditioning the extent of leakiness, possibly by governing the rate of inrush of water during imbibition (see Manohar, 1970; Buttrose, 1973). One wonders whether the problem might be attacked by devising a coating for the seed combining a substance decreasing the rate of imbibition (e.g. see Perry and Harrison, 1970) with others discouraging the establishment of microflora in the neighbourhood of the developing seed. Fungicidal dressings have proved to be of limited use in damp soils and with relatively leaky stocks of seed. The two common pathogens which cause problems at the pre-emergence stage are *Fusarium* spp and *Pythium* spp and the evidence is that these usually infect the seedling first through the cotyledons (Matthews, 1971; Nyvall and Haglund, 1973). It is therefore essential to establish conditions in which the health of the cotyledons is ensured, at least until the seedling shoot has broken the surface.

3. Sowing Time and Density, Row Spacing, Canopy Characteristics

Several studies have been carried out to determine how these factors need be arranged to obtain optimum yield of seed (see Bundy, 1971). The larger haulm size and longer time to maturity of peas grown for dry seed usually demands a somewhat wider spacing and lower sowing density than is normal for peas harvested unripe for freezing (King, 1966). Gritton and Eastin (1968) found that the highest yield of peas for freezing came from a 9-cm row spacing and 1.66×10^6 seeds ha^{-1}, which was the densest spacing they employed; King (1966) recorded a 20-cm row width and 2.5×10^6 seeds ha^{-1} as being optimal. Meadley and Milbourn (1970) found that sowing densities spanning the range $0.43–1.72 \times 10^6$ seeds ha^{-1} did not result in significant differences in yield of harvestable peas, although the highest densities gave rise to great wastage of pods by abscission. Weed problems can become acute at low sowing densities (Marx and Hagedorn, 1961), whilst damage by disease and cost of seed may become important factors if too high densities are selected.

Sowing time is particularly important when crop contracts are to be staggered amongst or within farms, as they normally are in the pea vining industry (see Dobben, 1963). Under English conditions, for example, Bland (1971) states that 18th March to 29th April is the normal sowing period. In a study of the fate of differently timed sowings, Milbourn and Hardwick (1968) found that late sowings (mid-April onwards) tended to have lower yields because the haulms bore fewer podding nodes, fewer pods per podding node, and the pods which did form matured more slowly than those of earlier sowings. Cultivars adapted for early-, mid-, and late-season sowing, are, of course, included in the overall programme of a processing factory.

The untidy growth habit of pea makes it especially difficult to study effects of canopy structure on yield. Kornilov and Kostina (1965) concluded that under Russian conditions, a leaf area index (LAI) value of 7·6–9·9 must be attained if maximum yields are to result; this is a much higher range of values than has been recorded for pea crops in the U.S.A. (Eastin and Gritton, 1969) or England (Meadley and Milbourn, 1970). The general lack of any dramatic response to density of sowing may well be due to the fact that the crop rarely achieves, or is too slow to achieve, its optimum canopy density. Indeed Eastin and Gritton (1969) maintained that unless an LAI value greater than 3·5 is attained, there is little benefit to be gained from changing canopy structure. They also found that a given unit of LAI appears to be more efficient in producing dry matter after flowering than earlier in crop growth, the type of result to be expected if LAI values fail to reach the optimum value. Snoad (1974), after experimenting with leafless peas (see Chapter 12), declared them to be at no disadvantage as regards yield over conventional, fully leafed forms, a rather surprising result in view of the much lower LAI values to be expected from varieties with tendrils in place of leaflets. He maintained that any disadvantage that the leafless cultivar might have in terms of lower leaf area seemed to be offset by better standing ability and harvestability, better light penetration and hence more uniform ripening of pods, and better resistance to attack by pests.

C. Effects of Environment

Peas possess what appear to be quite exacting requirements in respect of seasonal changes in temperature during their growth cycle. Studies in controlled environments have shown the species to be strongly thermoperiodic, such that growth at constant temperatures is always much inferior to that in comparable environments in which the day temperatures exceed night temperatures by 6–10°C (Highkin, 1960). The deleterious effects of growth at constant temperature appear to be transmissible from one generation to another, resulting, after five or six successive generations, in plants of abnormally low vigour; it then requires several generations of growth at fluctuating temperatures for vigour to be restored to its natural level (Highkin, 1960). Stanfield et al. (1966), working on cv. 'Dark Skinned Perfection' found that a day/night temperature range of from 21°C/10°C to 23°C/16°C was optimal for dry matter accumulation at growth stages up to the sixth node, but that after this the optimum shifted to a cooler range (16°C/10°C–18°C/13°C (day/night). Plants were recorded as growing especially tall after being exposed at germination to high temperatures (Highkin and Lang, 1966), whereas low temperatures in early growth tended to promote tillering and a dwarf habit. Brouwer (1959) noted that pod yield and leaf area per plant were higher at 10°C than at 17°C in cv. 'Virtus', but that the growth cycle was much longer at the lower temperature. In a further study (Brouwer and van Vliet, 1960) the optimum root temperature for this cultivar was defined as 22°C for early growth, the optimum falling somewhat in the later stages of growth. Growth temperatures under field conditions have been shown to affect quality, Mack (1973) having noted that warm soils were prone to give seeds of poor quality, though often they produced a high yield.

Crops of peas are extremely sensitive to exposure to periods of abnormally hot dry weather and appear to be most easily affected once they have reached full bloom and are commencing to fill their lower pods. Periods of three or more consecutive days with day temperatures exceeding 30°C are defined as the conditions leading to serious reductions in yield (Lambert and Linck, 1958; Karr, 1959), although it is often difficult to distinguish effects of high temperature from accompanying effects of water stress. Certainly, peas respond well to irrigation in warm dry weather, especially after flowering when the crop's water requirement is at a maximum (see Salter and Drew, 1965; Maurer et al., 1968; Lenka and Gautam, 1972). Such problems may well limit the cultivation of peas in sub-tropical climates unless it is possible to develop varieties more resistant to heat and more provident of water (see Bhardwaj et al., 1971).

Despite the general vulnerability of germinating peas to cold wet conditions and the sensitivity of seedlings to waterlogging (see Chapter 13) cultivars are known (e.g. cv. 'Austrian Winter') which show a reasonable measure of winter hardiness. The inheritance of this character has been studied (Markarian and Andersen, 1966; Dressler, 1973), the hardy varieties and hardy individuals within a variety tending to exhibit weak apical dominance and to branch freely at ground level. According to Dressler (1973) winter hardy varieties sown at the end of September in Germany have the capacity to mature much earlier than do any of the spring-sown varieties. They require a wide row spacing because of their branched habit, and the sowing time in autumn has to be carefully gauged to ensure that seedling size is optimal for a maximum survival rate through the ensuing winter (see Dowker, 1969). Overwintering of pea crops is at present practised on only a limited scale in those countries experiencing severe frost and snow in winter, and the evidence is that it is a much less predictable enterprise than that involving the considerably more hardy winter cereals. On the other hand, overwintering in sub-tropical countries (e.g. Australia, India) is an easy proposition. The lack of sensitivity of most cultivars to daylength allows the plant to be sown in the autumn and harvested in early spring or summer before the hot dry weather sets in.

D. Agencies of Disease

Several species of parasitic fungi attack peas and cause significant damage under certain circumstances. Pre-emergence mortality due to *Fusarium* or

Pythium has already been mentioned, both of these organisms causing damage to peas in virtually every country in which they are now grown (e.g. see Flentjé and Saksena, 1964; Kaiser *et al.*, 1971; Teranashi and Namekata, 1972). Evidence of heritable resistance to these pathogens has been described (Matthews and Dow, 1971; Muehlbauer and Kraft, 1973), although particularly virulent races of both organisms are known, to which many or all of a tested group of cultivars may be found to succumb. Differing degrees of susceptibility of pea cultivars to the root-rot organism, *Aphanomyces*, suggests genetic linkage with undesirable horticultural characters, thus complicating the development of resistant cultivars (Marx *et al.*, 1972). A novel but scarcely practicable method suggested by Wallen (1955) for eliminating the seed-borne pathogen *Ascochyta pisi* is to store seed stocks for 7 years; the host then outlives the disease organism. Several species of downy mildew affect leaves of pea; a common one in certain areas, *Peronospora viciae*, is particularly damaging if high humidities are experienced during crop growth (Wilson, 1971). Powdery mildew caused by *Erysiphe* is another fungus causing damage in certain areas, especially late in the growth cycle.

Of the many forms of animal life causing damage to pea crops, the effects due to aphids are probably the most widespread and severe, especially if instrumental in the spread of viruses such as "leaf mosaic" and "fizzle top" (pea seed-borne mosaic virus, Psbmv). According to Gorshevikova and Popov (1973), varieties whose foliage is low in amino acid but high in sugars are the least susceptible, but it is unlikely that any of the present-day cultivars are really resistant to attack. The weevil *Sitonia lineatus* can ravage pea crops, especially in dry weather (Mulder, 1948). Adults feed on foliage, the larvae on root nodules. The pea weevil *Bruchus* is important locally in certain seed-producing areas (e.g. Western U.S.A.).

A disease known as "marsh spot" and originally incorrectly described as of fungal origin, is quite common in peas, and readily discernible as a characteristic browning of the seedling plumule. It is associated with a deficiency of manganese in the seed (Reynolds, 1955; Sharp and Blunt, 1955). For a more general account of diseases in peas, and efforts made in breeding plants for resistance, the reader is referred to Hagedorn (1973).

III. THE INADEQUACY OF PEAS AS PRODUCERS OF PROTEIN

It seems quite right to argue that the value of the pea to a rapidly expanding world population should remain as a dry seed crop, offering a bulk source of seed protein for man and his animals. According to figures quoted by Farrington (1974), the world average yield of peas is 237 kg protein ha^{-1}, whilst soils of equivalent fertility and geographical location can, on average, support production of 506 kg protein ha^{-1} from soybean (*Glycine max*), 412 kg ha^{-1} from vetches (*Vicia* spp), 294 kg ha^{-1} from annual lupins (*Lupinus* spp), and 280 kg ha^{-1} from broad beans (*V. faba*). It should be pointed out that these comparisons are to an extent unfair since a pea crop occupies the land for only three months or so, whilst the other crops listed have longer life cycles. Nevertheless, there is obviously considerable scope for improvement in this respect as far as pea is concerned, especially where it is the only crop that can be grown within a given season.

In comparison with other cultivated grain legumes the pea has a relatively low content of seed protein, 20–25% by weight in seed dry matter being regarded as average for the species, and the range 15–35% probably coverin all cultivars and mutants whose "crude" protein content has been measured (Pate, 1975). Moreover, unlike soybeans and peanuts, its content of oil is too low to warrant extraction. It is likely that low protein values within the species as a whole reflect basic genetic and physiological tendencies which weight the synthesis of carbohydrates such as sugar, hemicellulose and starch in favour of nitrogen-containing reserves such as free amino acids, peptides and proteins. But it is equally likely that the situation in modern cultivars results from continued selection by man for highly palatable forms, in which tenderness and sweetness in the fresh or cooked state has carried a heavier premium than any specific nutritional value of the seed. However, within the past decade or so concerted efforts have been made to reverse this trend, and to develop varieties of pea showing high and consistent yields of good quality protein (see Chapter 2). Most of this work has been fostered at the genetic level, prompted, no doubt, by the recognition that the Leguminosae as a whole are likely to offer the best potential of any family of higher plants as concentrated sources of seed protein.

Dealing first with the total content of protein in the seed, and not with its quality, there is now abundant evidence from many laboratories that protein level in the seed is genetically controlled, making it possible to select lines by breeding or mutation which exhibit consistently higher percentages of protein in dry matter than their parent genotypes (see Pesola, 1955; Yarnell, 1962; Conev and Ivanova, 1968; Berdysev, 1969; Verbickij, 1970; Müller and Gottschalk, 1973; and many articles published in "The Pisum Newsletter"). Of course, percentage protein in the seed need not necessarily be the best character to select for when aiming to increase protein yields per unit area of crop, for several of the lines showing highest protein levels have performed poorly on a seed size or seed yield per plant basis when grown under field conditions (see publications in "The *Pisum* Newsletter"). Also, since protein content of the seed varies quite widely with environmental conditions of growth (e.g. see Robertson *et al.*, 1962), lines having a high percentage of protein in one environment need not necessarily behave in this fashion when grown in other environments. In fact, a major difficulty in assessing the results of breeding programmes is to disentangle genetic influences on protein level from those caused by the environment (see Johnson and Lay, 1974).

It is important to realize that difficulties can arise when attempts are made to compare published results for protein contents from different laboratories. Estimates are usually made by Kjeldahl estimation of total seed nitrogen and multiplication of the nitrogen value by a correction factor (usually 6·25) to calculate a so-called "crude" protein level. Even if applied consistently to a range of seed samples within the same laboratory the information provided can be misleading, for the use of a common correction factor fails to take into account variations between cultivars in the relative amounts of amide and basic amino acid residues in their protein, and totally ignores variations in the level of non-protein nitrogen present in the seed as fibre, free amino acids and other substances. Differences of this nature can amount to an error of up to 10% of the estimated protein value.

As well as improving total protein content by breeding there is the possibility of altering the nutritional status of the parent plant so as to encourage synthesis of extra protein in seeds. The most obvious suggestion would be to supply the fruiting plant with extra fertilizer nitrogen, but studies on pea and other legumes (J. S. Pate and P. J. Sharkey, unpublished) have shown that while this can result in very large increases in soluble nitrogen in the vegetative parts of the plant, protein level in seeds is, at best, only slightly increased. Indeed, extra nitrogen may even have a depressing effect on total seed yield by encouraging extra vegetative growth and less flowers and fruits. Our studies with another legume, *Lupinus albus* (see Pate *et al.*, 1975), have provided evidence that the composition of phloem sap supplying fruits is scarcely altered at all in respect of its ratio of sucrose:amino acids when effectively nodulated plants are supplied with varying levels of fertilizer nitrogen, and it may be that this innate conservatism in regulating the loading of solutes on to the translocation stream engenders near uniformity in final seed composition despite extreme variations in the overall nutrition of parent plants. The same conservatism is observed within the individual plant, for if seeds are collected and analysed from each of the fruiting nodes of a stem, little difference is observed in percentage nitrogen in dry matter amongst the seed samples. Seeds from early formed pods may be significantly larger than those formed at the end of the reproductive cycle, but composition is surprisingly constant (J. S. Pate, unpublished). A rather promising approach to increasing protein levels in the seed is that of spraying suitable chemicals on to the plant during the time when it is commencing to mobilize materials to its developing seeds. The group of chemicals called *S*-triazines, to which atrazine and simazine belong, would appear to be potentially useful in this connection. Although used primarily as weed killers, their effect when applied at a very low concentration is apparently to alter the enzymatic balance of a plant in favour of the production of nitrogenous compounds as opposed to carbohydrates. The ultimate effect of this on seed composition in legumes is one of elevating protein and decreasing starch, this effect being quite dramatic in *Phaseolus* (Singh *et al.*, 1972) and also noticeable to a lesser extent in peas (Wu, 1971).

Protein quality in peas and other legumes is usually judged by measuring the balance between various amino acid residues in the seed protein(s) and comparing the results with values for some recognized "standard" or "reference" protein (e.g. whole egg protein, the FAO reference protein, or the "standard" protein of Smith, 1966). Comparisons of this sort for peas are provided by Boulter and Derbyshire (1971), Boulter *et al.* (1973), and Johnson and Lay (1974), the values quoted by the first of these groups of workers being reproduced in Table 16.I. It is generally agreed (see Dawson, 1970) that legume seed proteins are nutritionally poor in sulphur-containing amino acids

TABLE 16.I. The amino acid composition of seed proteins of *Pisum sativum* in comparison with that of a "Standard" Animal Protein. (All data from Boulter and Derbyshire, 1971)

Amino Acid	Albumin Fraction	Globulin Fraction		Standard Animal Protein
		Vicilin	Legumin	
Arginine	5·0	7·3	10·5	5·79
Histidine	2·6	2·1	2·8	6·06
Lysine	7·7	7·9	4·9	6·21
Tyrosine	2·37	3·0	3·3	4·79
Tryptophane	2·4	–	1·1	2·1
Phenylalanine	6·3	6·2	4·9	5·4
Cystine	5·4	0·35	0·71	1·1
Methionine	1·01	0·22	0·65	1·9
Serine	5·4	5·8	4·5	5·45
Threonine	4·6	3·4	2·9	4·89
Leucine	5·1	9·2	8·1	8·3
Isoleucine	4·8	5·1	4·0	4·6
Valine	5·4	4·6	4·6	5·8
Glutamyl[a]	7·4	19·3	19·4	10·9
Aspartyl[a]	11·7	12·0	12·5	9·7
Alanine	5·3	3·0	3·7	5·2
Proline	6·1	3·5	4·3	5·6
Glycine	4·8	3·1	3·4	3·57

[a] amide + parent amino acid.

(methionine and cystine) and since this deficiency applies particularly to the globulins (see Table 16.I), there is the possibility of improving nutritional value in respect of sulphur amino acids by increasing the seed's content of albumins relative to globulins. However, only a small proportion of the protein is in the form of albumin, so a quite drastic change in composition would have to be effected to make a worthwhile change in amino acid balance. It is doubtful whether a change of this order could be achieved when it is remembered that the albumin fraction is predominantly cytoplasmic, whereas the globulin is apparently restricted to protein bodies (see Chapter 15). Nevertheless, as expected, existing varieties of high albumin content do show up best in bioassays of nutritional quality using rats as acceptor animals (Bajaj *et al.*, 1971). A significant improvement in quality might also be made by increasing the ratio of the globulin legumin to its partner globulin vicilin, since legumin is the richer in sulphur amino acids (Table 16.I). An even more exciting possibility is to search for mutants whose globulins are composed of subunits unusually rich in those essential amino acids which are normally deficient in the seed. This line of research has not been explored in any depth, although the recent demonstration by Davies (1973) of considerable variation in the types and proportions of sub-units in globulins of peas, and his success in altering them by selection and breeding, provides great encouragement for biochemical investigations.

Finally, there remains, as in the case of protein quantity, the possibility of improving protein quality by manipulating the nutrition of the parent plant. An obvious project, for example, would be to test whether a radical improvement of the availability of sulphur to plants would increase significantly the content of sulphur amino acids in the seed. Possibly improved sulphur status of vegetative parts of the plant would not be reflected at all in seed composition at the protein level. However, it is possible that seeds might fill with reduced forms of sulphur and that this might be of nutritional benefit to animals, even though not incorporated in the seed protein. The hoped-for possibility is, of course, that the effect of improved sulphur supply would be relayed to the seed and result in the formation of sub-units abnormally rich in sulphur amino acids.

IV. TOWARDS A RADICAL IMPROVEMENT IN OVERALL YIELD OF SEED

It has already been mentioned that, in general, the foliar canopy of the pea crop seems to be insufficiently dense to achieve maximum photosynthetic returns, particularly over the early- and mid-stages of the crop's growth cycle. It is in this area, therefore, where the deficiency of the species seems greatest, that further research is likely to pay greatest dividends. Nevertheless, other rate-limiting processes may well be operative under certain growth conditions, and the most important of these are likely to be concerned with the formation and filling of fruits, and with the production by the time of harvest of a maximum number of seeds in the correct state of ripeness for harvesting. Problems in this area are not likely to be particularly great with crops harvested for dry seed, since practically all of the late-formed pods will have time to mature as the vine dries out. On the other hand, in crops for freezing and processing, seeds only in a rigidly defined condition of development are acceptable, and any improvement in the uniformity in achieving this condition will obviously increase the profitability of the crop.

The basic problem in breeding peas for processing is one of having to convert the natural characteristic of sequential flowering and pod-filling to a more synchronous reproductive habit. (A pea crop in flower is shown in Fig. 16.2.) The line of approach adopted by plant breeders has already been mentioned in this volume (Chapter 2), the methods used being to select for

FIG. 16.2. A vining pea crop in eastern England at the time of flowering. (Photograph kindly supplied by the Processors and Growers Research Organization, Peterborough, England)

more synchronous flowering (Scheibe, 1954, 1968; Marx, 1966); for increased pod number per node (see Drufhout, 1972); for more seeds per pod (Marx and Mishanec, 1962); and for varieties with twin-branched shoots (see "*Pisum* Newsletter"). Compounding of several of these features within the one cultivar has already been shown to improve greatly the plant's capacity to yield a maximum number of seeds in optimal condition for harvest, but results to date suggest that this beneficial effect is often offset by a reduction in seed size. It is likely therefore that in the cultivars tested some other factor, such as rate of mobilization of nutrients to the seeds, sets limits to the absolute weight of harvestable seeds, and that until this barrier is raised, yields will continue to be unsatisfactory. Quantitative techniques are now available for analysing the spread in maturity of field-grown peas used for processing (see Schippers, 1969; Drufhout, 1972), so that, if new high-yielding cultivars are developed with promising tendencies towards simultaneous fruit maturation, the means for assessing their performance quantitatively under field conditions will be readily available. One might argue that the "ideotype" which one should strive for should not only exhibit an almost simultaneous maturation of its early fruits, but that once the set of these fruits has taken place further apical growth of the shoot should cease in order to prevent the formation of further vegetative structures and flowers. These younger structures are likely to have no useful purpose and might even be inhibitory to the filling of the already-formed pods. Studies by Marx (1966, 1968) show that an early induction of apical senescence is controlled by the genetic constitution of the plant as well as being influenced by its environment, so it is likely that improvements in this connection could be effected by judicious selection. Indeed, the fasciated varieties which have been developed (see Marx and Hagedorn, 1962) already combine apical senescence with uniformity of maturity, and may well possess an additional advantage in having a fleshy stem from which an especially large reservoir of nutrients can be mobilized during seed ripening.

The present situation is clearly one which calls for confidence and hope, since the genetic plasticity of the species is seemingly adequate for manipulation, and the desired directions for improvement are well defined. All that appears to be required is a mixing by the plant breeder of the correct balance of physiological and morphological attributes, and, once this has been achieved, one might expect a range of cultivars to be produced far superior to the better-yielding of the present-day varieties.

REFERENCES

Bajaj, S., Michelsen, O., Lillevik, H. A., Baker, L. R., Bergen, W. and Gill, J. L. (1971). *Crop Sci.* **11**, 813–816.

Ben-Ze'ev, N. and Zohary, D. (1973). *Israel J. Bot.* **22**, 73–91.

Berdysev, A. P. (1969). *Selekts Semenov.* **2**, 36–37.

BHARDWAJ, S. N., SHARMA, P. N. and NATH, V. (1971). *Indian J. agric. Sci.* **41**, 894–900.

BLAND, B. F. (1971). "Crop Production: Cereals and Legumes". Academic Press, London and New York.

BOULTER, D. and DERBYSHIRE, E. (1971). *In* "Chemotaxonomy of the Leguminosae". (J. Harborne, D. Boulter and B. Turner, eds), pp 285–308. Academic Press, London and New York.

BOULTER, D., EVANS, I. M. and DERBYSHIRE, E. (1973). *Qual. Plantarum Pl. Fds hum. Nutr.* **23**, 239–250.

BRADNOCK, W. T. and MATTHEWS, S. (1970). *Hort. Res.* **10**, 50–58.

BROUWER, R. (1959). *Jaarb. Inst. biol. scheik. Onderz. LandbGewass.* 17–26.

BROUWER, R. and VAN VLIET, G. (1960). *Meded. Inst. biol. scheik. Onderz. LandbGewass.* **108**, 23–26.

BUNDY, J. W. (1971). *In* "Potential Crop Production" (P. F. Wareing and J. C. Cooper, eds). Heinemann, London.

BUTTROSE, M. S. (1973). *Protoplasma* **77**, 111–122.

COLE, S. (1961). "Neolithic Revolution." 2nd edn. British Museum of Natural History, London.

CONEV, D. and IVANOVA, M. (1968). *Grandin. Lozar. Nauk. Hort. Viticult.* (*Sofija*) **5**, 37–48.

DAVIES, D. R. (1973). *Nature New Biology* **245**, 30–32.

DAWSON, R. C. (1970). *Pl. Soil.* **32**, 655–673.

DOBBEN, W. H. VAN (1963). *Jaarb. Inst. biol. scheik. Onderz. LandbGewass* 41–49.

DOWKER, B. D. (1969). *Euphytica* **18**, 398–402.

DRESSLER, O. (1973). *Z. PflZücht.* **69**, 221–230.

DRUFHOUT, E. (1972). *Euphytica* **21**, 460–467.

EASTIN, J. A. and GRITTON, E. T. (1969). *Agron. J.* **61**, 612–615.

FARRINGTON, P. (1974). *J. Aust. Inst. agric. Sci.* **40**, 99–108.

FLENTJÉ, N. T. and SAKSENA, H. K. (1964). *Aust. J. biol. Sci.* **17**, 665–675.

GORSHEVIKOVA, O. L. and POPOV, K. I. (1973). *Biol. Nauki.* **16**, 89–95.

GRITTON, E. T. and EASTIN, J. A. (1968). *Phytochemistry* **8**, 557–559.

HAGEDORN, D. J. (1973). *In* "Breeding Plants for Disease Resistance" (Nelson, ed.). Penn. State Univ. Press, U.S.A.

HARRISON, J. G. and PERRY, D. A. (1973). *Ann. appl. Biol.* **73**, 103–109.

HAWKES, J. and WOOLEY, L. (1963). "History of Mankind. Vol. 1. Prehistory and the Beginnings of Civilization". Allen and Unwin, London.

HIGHKIN, H. R. (1960). *Cold Spring Harb. Symp. quant. Biol.* **25**, 231–238.

HIGHKIN, H. R. and LANG, A. (1966). *Planta* **68**, 94–98.

JANICK, J., SCHERY, R. W., WOODS, F. W. and RUTTAN, V. W. (1969). "Plant Science". Freeman, San Francisco.

JOHNSON, V. A. and LAY, C. L. (1974). *J. agric. Fd Chem.* **22**, 558–566.

KAISER, W. J., OKHOVAT, M. and MOSSAHEBI, G. H. (1971). *Iran J. Pl. Path.* **7**, 9–12.

KARR, E. J. (1959). *Am. J. Bot.* **46**, 91–93.

KING, J. M. (1966). *Misc. Publs* **8**, 4.

KORNILOV, A. A. and KOSTINA, V. S. (1965). *Fiziologiya Rast.* **12**, 551–553.

LAMBERT, R. G. and LINCK, A. J. (1958). *Pl. Physiol.* **33**, 347–350.

LENKA, D. and GAUTAM, O. P. (1972). *Indian J. agric. Sci.* **42**, 476–480.

MACK, A. R. (1973). *Can. J. Soil Sci.* **53**, 59–72.

MANOHAR, M. S. (1970). *Sci. Cult.* **36**, 57–58.

484   J. S. Pate

MARKARIAN, D. and ANDERSEN, R. L. (1966). *Euphytica* **15**, 102–110.
MARSHALL, H. H. (1966). *Can. J. Pl. Sci.* **46**, 545–555.
MARX, G. A. (1966). *Crop Sci.* **9**, 494–496.
MARX, G. A. (1968). *BioScience* **18**, 505–506.
MARX, G. A. and HAGEDORN, D. J. (1961). *Weeds* **9**, 494–496.
MARX, G. A. and HAGEDORN, D. J. (1962). *J. Hered.* **53**, 31–43.
MARX, G. A. and MISHANEC, W. (1962). *Proc. Am. Soc. hort. Sci.* **80**, 462–467.
MARX, G. A., SCHRODER, W. T., PROVVIDENTI, R. and MISHANEC, W. (1972). *J. Am. Soc. hort. Sci.* **97**, 619–621.
MATTHEWS, S. (1971). *Ann. appl. Biol.* **68**, 177–183.
MATTHEWS, S. (1973). *Ann. appl. Biol.* **73**, 211–219.
MATTHEWS, P. and DOW, P. (1971). *Rep. John Innes hort. Instn* **62**, 31–32.
MAURER, A. R., ORMROD, D. P. and FLETCHER, H. F. (1968). *Can. J. Pl. Sci.* **48**, 129–137.
MEADLEY, J. T. and MILBOURN, G. M. (1970). *J. agric. Sci.* **74**, 273–278.
MILBOURN, G. M. and HARDWICK, R. C. (1968). *J. agric. Sci.* **70**, 393–402.
MUEHLBAUER, F. J. and KRAFT, J. M. (1973). *Crop Sci.* **13**, 34–46.
MULDER, E. G. (1948). *Pl. Soil* **1**, 179–212.
MÜLLER, H. and GOTTSCHALK, W. (1973). *In* "Nuclear Techniques for Seed Protein Improvement". Organized by joint FAO/IAEA Division of Atomic Energy in Food and Agriculture and the Gesellschaft fur Strahlen-und Umwelt-forschung, pp 235–253. International Atomic Energy Agency, Vienna.
NYVALL, R. F. and HAGLUND, W. A. (1973). *Phytopathology* **62**, 1419–1424.
PATE, J. S. (1975). *In* "Crop Physiology. Some Case Histories" (L. T. Evans, ed.). Cambridge University Press.
PATE, J. S., SHARKEY, P. J. and LEWIS, O. A. M. (1975). *Planta* **122**, 11–26.
PERRY, D. A. and HARRISON, J. G. (1970). *J. exp. Bot.* **21**, 504–512.
PESOLA, V. A. (1955). *Acta agr. Fenn.* **83**, 125–132.
REYNOLDS, J. D. (1955). *J. Sci. Fd Agric.* **6**, 725–734.
ROBERTS, E. H. and ROBERTS, D. L. (1972). *In* "Viability of Seeds" (E. H. Roberts, ed.). Chapman and Hall, London.
ROBERTSON, R. N., HIGHKIN, H. R., SMYDZUK, J. and WENT, F. W. (1962). *Aust. J. biol. Sci.* **15**, 1–15.
SALTER, P. J. and DREW, D. H. (1965). *Nature, Lond.* **206**, 1063–1064.
SCHEIBE, A. (1954). *Pfl. Zücht.* **33**, 23–30.
SCHEIBE, A. (1968). *Fld Crop Abstr.* **20**, 1089.
SCHIPPERS, P. A. (1969). *Neth. J. agric. Sci.* **17**, 272–278.
SHARP, W. O. and BLUNT, C. F. (1955). *Fmr Stk Breed.* **69**, 52–55.
SINGH, B., CAMPBELL, W. F. and SALUNKHE, D. K. (1972). *Am. J. Bot.* **59**, 568–572.
SMITH, M. H. (1966). *J. theoret. Biol.* **13**, 261–282.
SNOAD, B. (1974). *Euphytica* **23**, 257–265.
STANFIELD, B., ORMOROD, D. P. and FLETCHER, H. F. (1966). *Can. J. Pl. Sci.* **46**, 195–203.
TERANISHI, J. and NAMEKATA, T. (1972). *Biologico* **38**, 91–92.
VERBICKIJ, N. M. (1970). *Sborn. Trud. Aspirant molod. nauc. Sotrud, vses, nauc-issled, Inst. Rasten.* **15**, 237–241.
WALLEN, V. R. (1955). *Pl. Dis. Reptr.* **9**, 674–677.
WILSON, V. E. (1971). *Pl. Dis. Reptr.* **55**, 730.
WU, M. T. (1971). Ph.D. Thesis. Utah State University, Logan, Utah, U.S.A.
YARNELL, S. H. (1962). *Bot. Rev.* **28**, 465–537.

Subject Index

A

Abortion of flowers and fruits, 387, 390, 392, 405, 422, 432–433

Abscisic acid (ABA), 38, 68, 136, 143, 147, 247–248, 251, 252, 276–277, 281–286, 454

Abscisins, 247–248

Acacia varieties, 30, 239, 427

"Accelerator α", 244

Accumulation, of reserves in seeds, 442–450

Acer pseudo-platanus (sycamore), 165

Acetate
  excretion by tendrils, 311

Acetyl aspartic acid, 245

O-acetyl homoserine, 373–375, 435–436, 457

Acetylene reduction
  assay for nitrogen fixation, 337, 352–353, 356, 370–371

Acid phosphatase, 59, 68, 69

Acid ribonuclease, 59, 60, 62, 63, 64, 69, 163, 172

Aconitase [citrate (isocitrate) hydro-lyase], 320

Action spectrum
  of de-etiolation, 297
  of IAA oxidation, 303

Adaptation, of stem sections to light, 308–309

Adenine, 143

Adenosine 3′,4′-cyclic monophosphate (cyclic AMP), 258

Adenosine triphosphate (ATP), 310, 320, 323, 329

Adenosine triphosphatase (ATPase), 59, 68, 159, 160, 167, 310

Adventitious roots, 139, 143–144

Ageing, 98, 402–403, 406, 408, 410–411, 414, 419, 426

Ageotropic mutant, 37, 240, 246–247

Alanine, 329, 367, 376, 378, 440, 457, 479

Alanyl glycine dipeptidase, 175

'Alaska' cultivar, 55–58, 61, 66–67, 69–70, 73–75, 89, 121–123, 126–127, 207–208, 236, 265–273, 275, 279–280, 282–283, 293–298, 305, 310, 323, 335, 344, 387–388, 400, 420–421, 432

"Albumin", in seeds, 449, 479–480

Alcohol dehydrogenase (ADH), 49

'Alderman' cultivar, 388, 410, 415

Allantoic acid, 363, 377, 457

Allium cepa (onion), 166

Allometry, of internode growth, 307–309

'Alsweet' cultivar, 89

Amino acids
  synthesis of, 367

Amino-acyl transfer RNA synthetases, 9

α-Amino adipic acid, 375

Amino butyric acid, 375, 378

Amino ethanol, 375

Amino-transferases, 368

Ammonia
  in endosperm, 440
  as a nitrogen source, 366

AMO1618, 242, 252

Amoeba proteus, 1

Amylase, 59, 62, 68–69

Amylopectin, in seeds, 40, 446

Amylopectin-1, 6-glucosidase, 59, 64

Amylose, in seeds, 40, 446

Anaerobic respiration, of germinating pea seeds, 6, 49

Anatomy
  of cotyledons, 441
  of nodules, 146
  of transition zone, 215
  of shoot apex, 185–186

Anthesis, 338

Anthocyanin, 32–34, 395

Antimetabolite, effects on root growth, 171